CMOS Plasma and Process Damage

Kirk Prall

CMOS Plasma and Process Damage

Springer

Kirk Prall
KP Semiconductor LLC
McCall, ID, USA

ISBN 978-3-031-89028-4 ISBN 978-3-031-89029-1 (eBook)
https://doi.org/10.1007/978-3-031-89029-1

This Springer imprint is published by the registered company Springer Nature Switzerland AG
The registered company address is: Gewerbestrasse 11, 6330 Cham, Switzerland

Preface

I recently retired from a 40-year career in the semiconductor industry, working at Micron and Philips-Signetics (now NXP). This book is intended to address a particularly troublesome subject that tended to pop up a few times a year during the development of CMOS technology. Plasma and process damage is an often overlooked subject that can wreak havoc with development schedules, timelines, yield, reliability, etc.

This subject has been studied since the early 1980s when high-energy processing started to dominate chip manufacturing. High-energy processes have the propensity to damage CMOS chips. This book covers this topic and several related topics.

Plasma and process damage affects many engineering disciplines during CMOS development. This book is intended to be a practical working guide for engineers of all disciplines. Process integration engineers must provide design and electrical rules that minimize the risk of plasma and process damage. Chip designers and layout engineers must be aware of the subject and provide practical mitigation with layout and circuit optimization. Parametric test structures need to be designed to detect and characterize process and plasma damage. Parametric and functional test engineers need to understand damage mechanisms and optimize testing procedures for detection. Reliability engineers must have product and parametric test procedures in place to detect plasma damage as part of reliability testing. Equipment manufacturers need to understand the problem and deliver problem-free fab equipment to chip makers. Process engineers must develop and optimize recipes that minimize damage. Manufacturing engineers must maintain and monitor the chip fabrication line to avoid excursions.

Chapter 1 introduces the subject and provides some background information. The common defects in CMOS manufacturing are reviewed. The topics covered in this book are selected from a defect list, focusing on the stranger and complex defects.

Chapter 2 introduces the antenna effect that commonly occurs in plasma and process damage. The antenna effect amplifies the damage and exacerbates plasma damage problems. Several types of antennas are reviewed, including direct and indirect antennas, source–drain antennas, wéll antennas, and damage across power domains.

Chapter 3 covers practical damage protection circuits, including diodes and transistors. Protection of specific chip components, including gate dielectrics, metal–insulator–metal capacitors, wells, triple wells, and memory wordlines using protection circuits can help mitigate damage. The use of protection circuits to optimize a CMOS design for damage mitigation is discussed.

Chapter 4 discusses the statistical nature of plasma and process damage electrical results, showing multiple types of distributions. The damage observed during product qualification is covered. The cumulative damage caused by repeated damage throughout the process flow is reviewed. The damage caused by processing and tool excursions is covered.

Chapter 5 covers the large spectrum of time-zero electrical problems that occur due to plasma and process damage, ranging from catastrophic structural damage to subtle parametric shifts. The similarity of process and plasma damage to radiation damage and electrical stress is discussed. Catastrophic damage, including electrical arcing and electrostatic damage during processing, is covered, along with gate dielectric breakdown mechanisms. Transistor degradation mechanisms caused by damage are described, including threshold voltage shift, interface state generation, trapping, and multiple leakage mechanisms. Subtle electrical effects such as transistor mismatch, ring oscillator speed degradation, and noise are described. The failure of the CMOS isolation due to trapping-related threshold voltage shifts is reviewed. The degradation of various CMOS parameters, such as contact resistance, high-k capacitance, and radiation hardness, is discussed.

Chapter 6 discusses the latent damage after process and plasma damage, which can create reliability risks for CMOS chips operating in the field. Most CMOS reliability mechanisms can be degraded by damage, including hot electron effects, bias instabilities, and dielectric lifetimes. The reliability of CMOS metallization can be degraded by process damage, including damage to intermetal dielectrics, corrosion, electrical migration, and stress migration. The physics of latent damage recovery is covered.

Chapter 7 covers the electrical behavior of single-atom defects and their effects on CMOS, including stress-induced leakage currents and giant random telegraph signal (RTS). These defects can be aggravated by process damage. The types of defects in this category are reviewed.

Chapter 8 covers process damage in specific types of CMOS technology that are very susceptible to process and plasma damage. These sensitive technologies include nonvolatile memories, CMOS imagers, DRAM memories, metal–insulator–metal (MIM) capacitors, low-k dielectrics, finfets, and silicon on insulator (SOI) technology.

Chapter 9 covers the specific process steps that can cause plasma and process damage. There are a large number of processing techniques that can induce damage, ranging from simple steps, such as deionized water rinse to complex high-density plasma systems. Many different processing tools are covered, including ion implantation, plasma deposition, plasma etching, through-silicon vias, and bond pad processing. The general concepts of plasma process relevant to damage are covered. Non-plasma processes known to cause damage, such as wet processes, spin processes, and resist processing, are covered.

Chapter 10 describes techniques to measure process and plasma damage inline on uncompleted product wafers, short process flows, and simple insulator on silicon test wafers. These techniques can greatly accelerate learning and problem solving.

Chapter 11 provides a comprehensive survey of relevant parametric test structures necessary to analyze plasma damage problems thoroughly. Methods and uses of the test structures are reviewed.

Chapter 12 provides recommendations for design rules to minimize damage through practical design and layout methodologies.

Chapter 13 describes parametric testing methodologies for the detection and analysis of damage, including time-zero damage detection and quick techniques to detect latent reliability damage.

Chapter 14 covers the complex subject of hydrogen in CMOS process technology. Hydrogen can hurt or help, depending on the situation. Hydrogen interacts with plasma and process damage, complicating the analysis of damage. The physics and mechanisms of hydrogen in a CMOS wafer consisting of 16 different phenomena are described. These hydrogen mechanisms include drift and diffusion of hydrogen, mobile hydrogen during device operation, passivation and depassivation of interface states, acceptor and donor depassivation, interaction with silicon and oxygen vacancies, interaction with thermal donors, interaction with contaminants, bias temperature instability, hydrogen donors and acceptors, hydrogen-related physical defects, hydrogen emission from films, hydrogen interaction with metals, deuterium behavior, and enhanced low-dose rate radiation damage.

Chapter 15 provides a review of metal contaminants in CMOS technology. The common sources of metal contaminants are examined, along with likely culprits among fabrication equipment and consumable supplies. The relevant important properties of metals that impact electrical behavior are reviewed. The worst-case contaminants, copper and iron, are examined in detail, along with their specific behavior in silicon. Techniques for distinguishing between iron and copper are described. Case studies of metal contamination in DRAMs and CMOS imagers are presented.

Chapter 16 covers mobile ion contamination in CMOS technology. The physics of mobile behavior in dielectrics is described, covering drift and diffusion. The electric effects of mobile ion contamination in CMOS are surveyed, including threshold voltage shifts, MOS capacitor degradation, diode degradation, gate dielectric degradation, and intermetal dielectric degradation. Product reliability failures are reviewed. Electrical techniques for measuring mobile ion contamination are taught, including transistor measurements, isolation transistor measurements, triangular voltage sweep (TVS), thermally stimulated ionic current (TSIC), and capacitance–voltage (C-V) methods. A review of the effects of mobile ions on very sensitive nonvolatile memories is presented. Die seal structures inside the product CMOS that can block mobile ions are covered. The analytical chemical techniques used to detect and quantify mobile contamination are described. Modern and historical techniques to mitigate mobile ion contamination, along with mobile ion barriers, are reviewed.

Chapter 17 provides a survey of fixed charge in CMOS devices. A fixed charge is a mechanism by which a charge is permanently trapped in dielectrics in a wafer. The charge shifts the active and isolation threshold voltages, degrading the circuit operation. The electrical symptoms of fixed charge are described with common causes in silicon dioxide. The dependence of fixed charge on the silicon crystalline orientation is reviewed. Carbon as a source of fixed charge is described. The fixed charge interaction with hydrogen and diffusion barriers is examined. Fixed charge in spin-on dielectrics, silane-based dielectrics, and nitrided dielectrics is described. The fixed charge polarity as a function of the film properties is described, along with the use of aluminum oxide as an engineered film to target the polarity and quantity of charge. Fixed charge in through-silicon vias is described as a case study. Finally, techniques to mitigate fixed charge are reviewed.

I hope you like this book.

McCall, ID, USA Kirk Prall
June 2025

Acknowledgments

I took advantage of my friends by asking them to review this book. I am very grateful to Roberta Bottini, Roy Colclaser, Akira Goda, Wayne Kinney, Ken Marr, Chandra Mouli, and Krishna Parat. I would like to thank Roy Colclaser, who guided me through my PhD and is still suffering PTSD from reading my dissertation. I would like to thank Akira Goda for his friendship and constant tutoring on NAND memories. I am grateful to Wayne Kinney for his 40-year friendship and for tutoring me in physics. I am grateful to Ken Marr for his unrelenting search for the truth and root causes. I would like to thank Chandra Mouli for his friendship and device-related expertise. I am grateful to Krishna Parat for his comprehensive review of this book. His review greatly enhanced the quality of this book. This book would have been better if Krishna had written it.

Special thanks to Dan Fleetwood, who helped expand my understanding of hydrogen in silicon.

Special thanks to Olivia Chitranjan, Brian Halm, Charles Glaser, and Hemalatha Velarasu from Springer Nature for guiding me through the labyrinth of publishing.

Contents

1 **Background** ... 1
1.1 Introduction ... 1
1.2 Spectrum of Process Damage ... 2
1.3 Commonly Damaged CMOS Structures ... 3
1.4 Survey of Process Damage Defects ... 5
References ... 7

2 **The Antenna Effect** ... 9
2.1 Introduction ... 9
2.2 Direct Antenna Effect ... 10
2.3 Indirect Antenna Effect ... 11
2.4 Source-Drain Antenna Effects ... 12
2.4.1 Differential Antennas ... 14
2.5 Capacitive-Coupled Antenna Effect ... 17
2.6 Virtual Antenna Effects ... 18
2.7 Well Coupling-Related Damage ... 19
2.7.1 Bulk CMOS Well Coupling-Related Damage ... 19
2.7.2 Triple Wells ... 21
2.8 Across Power Domains' Damage ... 23
2.9 The Reverse Antenna Effect ... 25
2.10 Remote Antennas ... 26
References ... 27

3 **Diode and Transistor Protection** ... 29
3.1 Introduction ... 29
3.2 Diodes for Protection from Well Damage ... 33
3.3 Transistor Clamp Protection ... 33
3.4 Diode Operating Under Processing Conditions ... 34
3.5 Diode Protection for Silicon-on-Insulator Technology ... 36
3.6 Diode Protection of MIM Capacitors ... 38
3.7 Protection of Memory Wordlines ... 39
References ... 41

4 Signatures of Process Damage 43
4.1 Introduction 43
4.2 Signatures of Process Damage 43
4.3 Statistical Signature of Process Damage 45
4.4 Cumulative Damage 45
4.5 Process Damage by Excursions 49
4.6 Burn-In Yield Correlation 52
References 53

5 Electrical Signatures of Process Damage 55
5.1 Introduction 55
5.2 Similarity of Process Damage and Electrical Stress 55
5.3 Similarity of Process Damage and Radiation Damage 56
5.4 Time-Zero Process Damage Electrical Effects 59
5.4.1 Electrical Static Discharge Process Damage (ESD) 59
5.4.2 Dielectric Damage Caused by Process Damage 60
5.4.3 Dielectric Breakdown 61
5.4.4 Enhanced Breakdown Due to Radius of Curvature 63
5.5 Interface States 64
5.5.1 Gain (*gm*) Degradation 64
5.6 Trapping 67
5.6.1 Threshold Voltage Shifts 68
5.7 Transistor Leakage Mechanisms 71
5.7.1 Introduction 71
5.7.2 Subthreshold Conduction Degradation 74
5.7.3 Gate-Induced Diode Leakage Degradation 75
5.7.4 Lightly Doped Drain (LDD) Region Damage 77
5.7.5 Punch-Through Degradation 81
5.8 Transistor Degradation Mechanisms 82
5.8.1 Process Damage-Induced Transistor Mismatch 82
5.8.2 Process Damage Noise Degradation 83
5.8.3 Ring Oscillator Degradation 88
5.9 Other Signatures 89
5.9.1 Contact Resistance Degradation 89
5.9.2 Degradation in MOS Capacitance-Voltage Characteristics 89
5.9.3 Impact of the Gate Dielectric on Process Damage 90
5.9.4 Isolation Edge Leakage 95
5.9.5 High-k Dielectric Dependence 98
5.9.6 Degradation of Radiation Hardness 101
5.9.7 Increased Low-Radiation Dose Rate Sensitivity 101
5.9.8 Degradation of Electrostatic Discharge Circuits 102
5.9.9 Diode Leakage 103
References 104

6 Latent Damage and Reliability Degradation ... 109
6.1 Introduction ... 109
6.2 Damage Recovery ... 111
6.3 Reliability Degradation by Process Damage ... 114
6.3.1 Introduction ... 114
6.3.2 CMOS Transistor Failure Modes ... 116
6.3.3 Degradation in Dielectric Lifetime ... 118
6.3.4 Hot Carrier Lifetime Degradation ... 120
6.3.5 Negative Bias Temperature Instability (NBTI) Degradation ... 121
6.4 Metallization Reliability Degradation ... 122
6.4.1 Introduction ... 122
6.4.2 Intermetal Dielectric Degradation ... 123
6.4.3 Corrosion ... 125
6.4.4 Electromigration and Stress Migration ... 126
References ... 128

7 Atomic-Level Defects and Electrical Effects ... 131
7.1 Introduction ... 131
7.2 SiO_2-Silicon Atomic-Level Defects ... 132
7.3 High-k Dielectric Single-Atom Defects ... 134
7.4 Stress-Induced Leakage Current ... 136
7.5 Giant RTS ... 139
References ... 141

8 Technology-Specific Process Damage ... 143
8.1 Introduction ... 143
8.2 Process Damage to Nonvolatile Memory Cells ... 144
8.2.1 Stress-Induced Leakage in Nonvolatile Memories ... 150
8.2.2 Data Retention ... 151
8.2.3 Charge Gain ... 154
8.2.4 Over-Erase and Over-Programming ... 154
8.2.5 Endurance, Cycling Window Shift, or Closure ... 157
8.2.6 Erratic Programming and Erasing ... 158
8.2.7 Electrical Disturbance Mechanisms (Read Disturb, Erase Disturb, Program Disturb) ... 159
8.2.8 NOR Read Disturbance ... 161
8.2.9 NAND Read Disturb ... 162
8.2.10 Giant Random Telegraph Signal ... 163
8.2.11 Parasitic Charge Trapping ... 164
8.3 Process Damage in CMOS Imagers and CCDS ... 164
8.3.1 Introduction ... 164
8.3.2 Atomic-Level Defects in CMOS Imagers ... 166
8.3.3 Process Damage in CMOS Imagers ... 167
8.3.4 Dark Current ... 168

8.3.5 High-Energy Light Damage in CMOS Imagers 171
8.3.6 White Pixel Process Damage 173
8.4 Process Damage in DRAM Memories 174
8.4.1 DRAM Refresh 175
8.4.2 Variable Retention Time in DRAMs 176
8.4.3 DRAM Row Hammer Problem 181
8.5 Metal-Insulator-Metal Capacitor Damage 183
8.5.1 Introduction 183
8.5.2 Process Damage to MIM Capacitors 184
8.5.3 Process Damage Induced by MIM Capacitors 185
8.6 Process Damage in Low-k Dielectric Films 189
8.6.1 Introduction 189
8.6.2 Low-k Dielectric Damage Examples 190
8.7 Damage in Finfet Technology 194
8.8 Silicon-on-Insulator Technology 196
8.8.1 Introduction 196
8.8.2 Effects of Buried Oxide 197
8.8.3 Propagation Between Floating Bodies 199
8.8.4 SOI Antenna Ratios 200
References 201

9 Common Sources of Process Damage 205
9.1 Introduction 205
9.2 Accelerated Degradation Under Processing Conditions 205
9.3 Common Sources of Process Damage 208
9.4 Ion Implant Damage 208
9.4.1 Ion Implantation Charging Damage 208
9.4.2 Flood Gun 211
9.4.3 Implant Damage Excluding Charging 212
9.4.4 Implant Structural Damage 212
9.4.5 Ion Implantation Through the Gate 215
9.5 Triboelectric Charging in Wet Processes 216
9.6 Damage Due to Spin and Deionized Water Processes 217
9.7 Resist and Masking-Related Damage 218
9.7.1 Resist and Masking Overlap-Related Damage 220
9.7.2 Plasma Ashing 221
9.8 Electron and Ion Beam Damage 223
9.9 Bond Pad Damage 225
9.10 General Considerations for Plasma Processing 225
9.10.1 Equivalent Circuit for a Plasma System 226
9.10.2 Alternating Current Bias 227
9.10.3 Across-Wafer Uniformity 227
9.10.4 Electrostatic Chuck 230
9.10.5 Plasma Ignition and Extinction 233
9.10.6 Recipe Transients 233

9.10.7 Magnetic Fields 235
9.10.8 Plasma Damage in Exposed Silicon Regions 236
9.10.9 Wafer Bevel and Backside Substrate Effects 236
9.10.10 Chamber Cleaning 239
9.10.11 Topology Effects 240
9.11 Plasma Film Deposition 241
9.11.1 Plasma Dielectric Deposition 241
9.11.2 Plasma Conductor Deposition 244
9.11.3 High-Density Plasma Deposition 244
9.12 Sputter Deposition 244
9.13 Plasma Etch-Induced Process Damage 247
9.13.1 Introduction 247
9.13.2 Overetch 248
9.13.3 Layout and Topology Dependent Process Damage 249
9.13.4 Etch Rate Process Damage 254
9.13.5 Spacer Etch Damage 254
9.13.6 Surface Etch Damage 257
9.13.7 Contact and Via Etches 258
9.13.8 Surface Charge Creation 261
9.14 Through-Silicon Vias, Backside Power, and Advanced Packaging Damage 261
9.15 Exposure to High-Energy Light 263
9.16 Damage Recovery by Exposure to High-Energy Light 267
References 268

10 Inline Process Damage Measurements 273
10.1 Introduction 273
10.2 Surface Potential Measurements 273
10.3 Surface Parametric Measurements 274
10.4 Short Loop Flows 279
References 280

11 Process Damage Test Structures 283
11.1 Introduction 283
11.2 Bond and Probe Pads 283
11.3 Inline Probe Pads 284
11.4 Gate Area and Edge Antennas 286
11.5 Well Antenna Structures 287
11.6 Source-Drain Antennas 287
11.7 Matching Transistor Structures 288
11.8 Light Damage Structures 288
11.9 Ring Oscillator Structures 288
11.10 Multiple Metal Layer Detection 288
11.11 Contact and Via Structures 289
11.12 Resist Antenna Structures 290
11.13 Diode Protection 291

11.14 Comb Structures. 292
11.15 Alternative Transistor Layout 292
11.15.1 Introduction 292
11.15.2 Dog Bone Field Edge Layout. 293
11.15.3 Donut Devices to Suppress Field Edge Leakage 293
11.16 Through-Silicon Via Test Structures 294
11.17 Structures for Metal Insulator Metal Capacitors 294
11.18 Memory Array Wordline Structures. 295
11.19 Structures for Metallization Damage. 295
References. 296

12 Design Rules Related to Process Damage. 299
12.1 Introduction 299
12.2 Process Module-Specific Design Rules. 299
12.2.1 Well Design Rules 302
12.2.2 Gate-Related Design Rules. 302
12.2.3 Source-Drain-Related Design Rules. 302
12.2.4 Metal-Related Design Rules. 303
12.2.5 Contact and Via-Related Design Rules. 304
12.2.6 Metal-Insulator-Metal Capacitor-Related Design Rules. 305
12.2.7 Bond Pad-Related Design Rules. 305
12.2.8 Through-Silicon Via-Related Design Rules 306
12.2.9 Technology-Specific Design Rules. 306
References. 306

13 Parametric Damage Testing Strategy and Procedures 309
13.1 Introduction 309
13.2 Plasma Damage Detection Strategy. 309
13.3 Parametric Testing 310
13.4 Diagnostic Revealing Stress. 312
13.5 Reliability Testing 315
13.6 Diagnostic Analysis 315
References. 316

14 Role of Hydrogen 319
14.1 Introduction 319
14.2 Hydrogen Diffusion and Drift 324
14.3 Passivation and Depassivation of Interface States 326
14.4 Mobile Hydrogen. 329
14.5 Dopant Passivation and Depassivation. 332
14.5.1 Acceptor Passivation and Depassivation by Hydrogen 332
14.5.2 Donor Passivation and Depassivation by Hydrogen. 334
14.6 Hydrogen Interaction with Silicon and Oxygen Vacancies 335
14.7 Hydrogen Interaction with Thermal Donors 335

14.8 Hydrogen Interaction with Contaminants ... 336
14.8.1 Metals, Chalcogenides, Etc ... 336
14.8.2 Carbon ... 337
14.9 Bias Temperature Instability ... 338
14.10 Hydrogen Degrades Noise in CMOS ... 339
14.11 Hydrogen-Based Donors and Acceptors ... 340
14.12 Physical Defects Caused by Hydrogen ... 340
14.13 Hydrogen Barriers ... 341
14.13.1 Introduction ... 341
14.13.2 Alumina ... 342
14.13.3 Silicon Nitride ... 343
14.13.4 Hydrogen and Water Trapped Beneath Barriers ... 345
14.14 Hydrogen Emission from Silicon Nitride Films ... 345
14.15 Hydrogen Interaction with Metals ... 349
14.16 Deuterium ... 350
14.17 Enhanced Low-Dose Rate Radiation Damage ... 351
References ... 352

15 Metallic Defects ... 355
15.1 Introduction ... 355
15.2 Common CMOS Contamination Sources ... 357
15.3 Common Effects of Metal Contamination ... 357
15.4 Electrical Activity of Metal Ion Contaminants ... 358
15.5 Relevant Important Properties of Metals ... 363
15.5.1 Introduction ... 363
15.5.2 Trap Energy Location in Bandgap ... 364
15.5.3 Diffusivity ... 367
15.5.4 Carrier Capture Cross Section ... 367
15.5.5 Solid Solubility ... 368
15.5.6 Minority Carrier Lifetime and Diffusion Length ... 369
15.6 The Worst Contaminants ... 372
15.6.1 Copper ... 372
15.6.2 Iron ... 378
15.6.3 Distinguishing Between Copper and Iron Contamination ... 379
15.7 Case Studies for Metal Contamination ... 381
15.7.1 DRAM Refresh ... 381
15.7.2 Tungsten Contamination in CMOS Imagers ... 384
References ... 386

16 Mobile Ion Contamination ... 389
16.1 Introduction ... 389
16.2 Physics of Mobile Ion Contamination ... 391
16.2.1 Common Mobile Ions ... 392
16.2.2 Drift and Diffusion of Mobile Ions ... 393
16.2.3 Dielectric Properties Affecting Mobile Ion Migration ... 395

16.3 Electrical Symptoms of Mobile Ion Contamination 396
16.4 Threshold Voltage Shifts 397
16.4.1 Introduction 397
16.4.2 Isolation Failure 399
16.4.3 CMOS Transistor Degradation 400
16.5 MOS Capacitor Degradation 402
16.6 Diode Degradation 403
16.7 Gate Dielectric Degradation with Mobile Ions 403
16.8 Intermetal Dielectrics Degraded by Mobile Ions 404
16.9 Product-Level Reliability Failures 405
16.10 Electrical Detection of Mobile Ions 406
16.10.1 Electric Measurement Introduction 406
16.10.2 Measurement Sensitivity and Capability 407
16.10.3 Detection of Mobile Ion Contamination Using a CMOS Transistor 409
16.10.4 Triangular Voltage Sweep 410
16.10.5 Detection of Mobile Ions in Intermetal Dielectrics 415
16.10.6 Thermally Stimulated Ionic Current Measurements 417
16.10.7 Mobile Ion Detection Using Capacitance-Voltage Techniques 419
16.11 Mobile Ion Degradation of Nonvolatile Memories 421
16.12 Die Seal Techniques to Prevent Mobile Ion Contamination 424
16.13 Analytical Chemical Detection of Mobile Ions 425
16.13.1 Secondary Ion Mass Spectrometry 426
16.13.2 Vapor Phase Decomposition 428
16.14 Mobile Ion Mitigation 428
16.14.1 Historical Contamination 428
16.14.2 Modern Mitigation 430
16.14.3 Mobile Ion Barriers 430
References 432

17 Fixed Charge 435
17.1 Introduction 435
17.1.1 Electrical Symptoms of Fixed Oxide Charge 436
17.2 Wafer Crystal Orientation 438
17.3 Thermal Silicon Dioxide Fixed Charge 439
17.4 Fixed Charge in Deposited Dielectrics 441
17.4.1 Introduction 441
17.4.2 Fixed Charge Caused by Carbon Incorporation 443
17.4.3 Interaction with Hydrogen and Diffusion Barriers 444
17.4.4 Fixed Charge in Spin-On Dielectrics 445
17.4.5 Fixed Charge in Silane-Based Dielectrics 445
17.5 Fixed Charge in Nitrided Dielectrics 446

17.6 Fixed Charge Polarity 447
17.7 Negative Fixed Charge in Alumina 448
17.8 Fixed Oxide Charge in Through-Silicon Vias 449
17.9 Fixed Charge Mitigation Techniques 451
References 452

Index 455

Chapter 1
Background

1.1 Introduction

Modern CMOS chip fabrication requires processing the wafer through dozens of distinct types of processing equipment (tools) and hundreds or thousands of potentially unique process steps that can potentially damage the wafer. This damage can take many forms, ranging from catastrophic physical damage to the wafer to latent reliability problems potentially causing chip failure in the field. Process damage has different mechanisms that are specific to equipment and processing conditions. This damage is known by several names, including process-induced damage (PID), plasma damage (PD), plasma processing-induced damage (PPID), etc. Plasma damage is sometimes used as a generic term for process damage, even if the damage is caused by a process that uses no plasma, such as deionized water processing. This book defines a plasma process as any processing step using ionized species. Therefore, this book attempts to use only the terms "process damage" and "plasma damage," where "process damage" is used as a generic term, and "plasma damage" is used for processes using plasmas (Cheung, 1999).

Plasma damage became an increasingly important topic in the 1970s and 1980s as the industry transitioned from mostly thermal and wet chemical processes to high-energy plasma processes. An excellent book on plasma damage was published by Cheung in 1999 (Cheung, 1999). This is the only book in existence known to the author. A dedicated conference on this subject called the International Symposium on Plasma and Process-Induced Damage was held from 1996 to 2003. A few good dissertations on this subject have been published by Zhong (2001) and Park (1998). This book attempts to provide an updated and comprehensive view of the subject.

Process damage can cause havoc with yield, reliability, performance, etc. Process damage can be subtle and easily mistaken for poor chip design, lack of process margin, variability, testability problems, etc. Process damage can remain hidden in a mature process, degrading a semiconductor company's economic success. The

K. Prall, *CMOS Plasma and Process Damage*,
https://doi.org/10.1007/978-3-031-89029-1_1

process damage can be dependent on multiple factors, including electrical charging of the wafer, high-energy light (UV or X-ray), temperature, etc.

Process damage results in device degradation, which is remarkably similar to the effects of electrical overstress caused by the voltage or current stress of the devices. Therefore, electrical overstress is commonly used to simulate the effects of process damage on device characteristics. Process damage characteristics are also similar to those of nuclear radiation damage.

Plasma damage is caused by a combination of the plasma processing system, the process recipe, the chip design, and the CMOS wafer. The complexity of this phenomenon is partly due to the interaction of the wafer controlled by the CMOS manufacturing company and the processing tool controlled by the tool vendor. Combining all these factors in modeling or simulation is difficult and not often achieved (Lee, 2023). Most process and plasma damage research tends to be empirical due to the complexity of attempting to model the processing tool, plasma conditions, plasma interaction with the wafer, the structures on the wafer surface, damage-related device physics, and reliability physics. As a result, this book uses only a limited amount of mathematics.

There is typically a trade-off between avoiding process damage and manufacturing efficiency. High-power and high-energy processes have high throughput rates and lower costs but are more likely to cause process damage by providing rapid energy transfer to the wafer. Equipment manufacturers optimize their tools and processes to minimize process damage, but there is a strong interaction with the chip maker's technology requiring coordination with the equipment manufacturer. Process damage problems are often discovered late in the technology development cycle, potentially causing significant delays in the time-to-market for a CMOS technology. A CMOS manufacturing fab must carefully qualify manufacturing equipment to control process damage.

1.2 Spectrum of Process Damage

Process and plasma damage can create a spectrum of different damage mechanisms. Figure 1.1 shows multiple categories of process damage roughly sorted by the energy required to create the category from the highest to lowest energy necessary to create the defect. High-energy damage events can structurally damage the wafer surface through electrostatic discharge, arcing, and structural melting. These high-energy events usually result in chip failure yield loss and chip failure. These types of damage are easily observed as physical damage in failure analysis. Slightly less energetic events can cause dielectric degradation or breakdown. Parametric shifts in the device parameters require even less energy. Low-energy events can cause transistor mismatch or noise. Nevertheless, lower energy events can cause latent, challenging-to-detect reliability hazards.

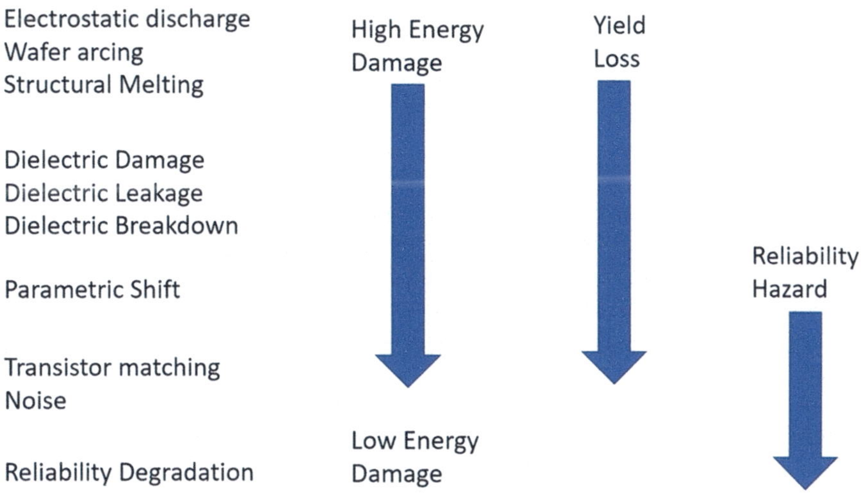

Fig. 1.1 Spectrum of possible damage due to plasma and process damage

Table 1.1 Common CMOS structures damaged by processing

MOS transistor and capacitor damage is caused mainly by damage to the gate dielectrics (high-k dielectrics may be more sensitive)
Diode
Metal-insulator-metal (MIM) capacitors
Isolation
Low-k dielectrics
Parametric test structures (worse with no protection)
Nonvolatile memory transistors
DRAM
Memory array wordlines
CMOS imagers
SOI and Finfet silicon channels
Contacts and vias
Metallization and related dielectrics

1.3 Commonly Damaged CMOS Structures

The structures on the CMOS wafer that are commonly damaged by processing are typically the structures that can withstand the least amount of voltage or current stress. Additionally, the sensitivity of structures to damage plays a significant role in the severity and extent of the damage. Table 1.1 shows a partial list of CMOS structures commonly degraded or destroyed by process damage.

The gate dielectric is extremely sensitive to process damage. Gate dielectric damage can be observed in dozens of different parametric measurements. Dielectric damage can result in a spectrum of different effects on the CMOS transistor, ranging from slight parametric shifts to catastrophic time zero failures with a variety of problems in between. Damage can create hidden, latent reliability problems that are not observed with routine test procedures. The gate dielectric is typically optimized to achieve the minimum thickness possible to increase transistor performance, making it one of the most susceptible elements in a CMOS process.

A metal-insulator-metal (MIM, also called metal oxide metal, MOM) capacitor is commonly added to a CMOS process to provide a chip design with a high-performance capacitor. The MIM capacitor is also optimized for minimum thickness to achieve a high capacitance, making it susceptible to process damage.

The dielectrics used in CMOS isolation to separate transistors, such as shallow trench isolation (STI), can trap charge, shifting the electrical behavior of the isolation and leading to parametric shifts or electrical failure of the isolation. The isolation tends to fail first as a function of the damage due to its greater sensitivity.

Low-k dielectrics are commonly used in the metallization layers of CMOS processes. This dielectric is a porous, low-density film necessary to achieve a low dielectric constant, enabling increased performance and reduced power consumption. However, exposure to diverse types of processing can degrade the dielectric constant and change the material properties.

Parametric test structures tend to be very susceptible to damage because they are generally unprotected by electrostatic discharge protection circuits on the bond pads. As a result, process damage to the parametric structures can create misleading data that may not match the actual product performance. In addition, the bond pads can collect a large amount of charge, damaging the devices connected to the pads. Therefore, careful experimentation and analysis are necessary to segregate the parametric structure process damage from the undamaged product-level electrical performance.

Nonvolatile memory transistors such as floating gate transistors, silicon oxide nitride oxide silicon (SONOS), and other related structures are sensitive to process damage. These devices are designed to trap charge to achieve nonvolatility and retain data without power. A process-induced current passing through these transistors adds or removes charge from these structures' charge-trapping regions. This trapping creates threshold shifts identical to storing the memory state, resulting in an abnormal state. The abnormal state may remain during further wafer processing and may be observable electrically after the wafer has completed processing. The abnormal state may be neutralized or reduced by thermal or light exposure during subsequent wafer processing following the damaging process step. Process damage can also damage sensitive dielectrics in the nonvolatile cell, causing yield or reliability problems.

Dynamic RAM (DRAM) operates by storing minimal amounts of charge on extremely small capacitors. As a result, DRAMs are sensitive to atomic-level defects that can be caused by process damage. These defects can cause leakage and charge loss, destroying the data stored in the DRAM cell.

CMOS imagers and charge-coupled devices (CCDs) collect light in a diode, creating a small amount of charge, which is amplified, creating an image. Unfortunately, the created charge is sensitive to atomic-level leakage, making imagers susceptible to damage.

The susceptibility of a CMOS structure to process damage depends on the size of the structures. Small structures can be more easily damaged by the electrical and thermal energy created during the process damage. For example, Finfets and silicon on insulator transistors have extremely small silicon areas and can be easily damaged. In addition, the continuous aggressive scaling of CMOS technologies has resulted in increased sensitivity to process damage, requiring more engineering diligence.

1.4 Survey of Process Damage Defects

Manufacturing CMOS chips requires extreme cleanliness and defect control as a large spectrum of defect mechanisms must be mitigated for successful production.

Table 1.2 shows a partial list of common CMOS technology defects grouped by general category. This book focuses on selected topics with strange electrical signatures rather than more physically oriented defects. The defects marked with a* are covered in this book, while ordinary physical defects are excluded. References are given in the table to direct the reader to information on topics not covered in this book.

Table 1.2 List of common CMOS process defects

Category	Defects	References
Particle	Foreign matter	Reinhardt and Kern (2018)
Patterning defect	Pattern collapse, resolution failure, proximity printing problem	Okoroanyanwu (2010)
Delamination	Adhesion failure	Murarka (1993)
Substrate defects	Dislocation, slip, loop, interstitials, strain defects, oxide precipitates, crystal-oriented precipitates	Kissinger and Pizzini (2015)
Chemical mechanical polishing	Chatter marks, scratches, electrostatic discharge*, topology	Babu (2016)
Process residuals	Fences, watermarks, residual resist, polymers	Reinhardt and Kern (2018)
Electrostatic* discharge	External ESD, process-induced ESD*, process arcing*	Voldman (2011)
Etching defects	Micro-trenching, edge notching, twisting	Nojiri (2012)
Metallurgy defects	Wormholes, salicidation defects, plug voids, top of plug defects, junction spiking, barrier failure, dendrites, missing films, etc.	Murarka (2000), Wolf (2002)
Plasma damage*	Antenna damage*, etc.	Cheung (1999)
Implantation defects*	End-of-range defects*, mask edge defects*, knock-on defects*, solid epitaxial growth defects*, damage through gate dielectrics*, etc.	McPherson (2002)
Surface damage*	Surface etch*, surface roughness*, contact etch damage*	Rimini (1995)
Low-k dielectric damage*	Multiple parameters*; see Chap. 8	Mack (1985)
Radiation damage*	External cosmic rays, processing radiation damage*, UV damage*	Romano et al. (2015)
Contamination*	Metal ions*, mobile ions*	Bentarzi (2011), Claeys and Simoen (2018)
Dielectric defects*	Pin holes, interface states*, traps*, vacancies*	Fleetwood et al. (2009), Schroder (2006)
Hydrogen defects*	See list in Chap. 14	Pankove and Johnson (1991)
Stressor and structural related stress defects*	Dislocations and slip lines related to the device structure on the wafer*	Fleetwood et al. (2009)
Fixed charge*	Fixed charge trapped in dielectrics*	Nicollian and Brews (1982)

References

Babu, S. (2016). *Advances in chemical mechanical planarization (CMP)*. Woodhead Publishing. https://doi.org/10.1016/C2014-0-01445-1

Bentarzi, H. (2011). *Transport in metal-oxide semiconductor structures—Mobile ions effects on the oxide properties*. Springer. https://doi.org/10.1007/978-3-642-16304-3

Cheung, K. P. (1999). *Plasma charging damage*. Springer. https://doi.org/10.1007/978-1-4471-0247-2

Claeys, C., & Simoen, E. (2018). *Metal impurities in silicon- and germanium-based technologies origin, characterization, control, and device impact*. Springer. https://doi.org/10.1007/978-3-319-93925-4

Fleetwood, D. M., Pantelides, S. T., & Schrimpf, R. D. (2009). *Defects in microelectronic materials and devices*. CRC Press Taylor and Francis Group. https://doi.org/10.1201/9781420043778

Kissinger, G., & Pizzini, S. (2015). *Silicon, germanium, and their alloys—Growth, defects, impurities, and nanocrystals*. CRC Press Taylor and Francis. https://doi.org/10.1201/b17868

Lee, S.-H. (2023). A noble design methodology to minimize plasma induced damage using a distributed network model in VNAND flash memory. In *International Electron Device Meeting*. https://doi.org/10.1109/IEDM45741.2023.10413668

Mack, M. E., et al. (1985). Wafer charging and beam interactions in ion implants. *Nuclear Instruments and Methods in Physics Research, 6*, 405–411. https://doi.org/10.1016/0168-583X(85)90665-2

McPherson, J. W., et al. (2002). Proposed universal relationship between dielectric breakdown and dielectric constant. In *IEEE IEDM*. https://doi.org/10.1109/IEDM.2002.1175919

Murarka, S. P. (1993). *Metallization—Theory and practice for VLSI and ULSI*. Butterworth-Heinemann.

Murarka, S. P., et al. (2000). *Copper—Fundamental mechanisms for microelectronic applications*. Wiley.

Nicollian, E. H., & Brews, J. R. (1982). *MOS (metal oxide semiconductor) physics and technology*. Wiley Interscience.

Nojiri, K. (2012). *Dry etching technology for semiconductors*. Springer. https://doi.org/10.1007/978-3-319-10295-5

Okoroanyanwu, U. (2010). *Chemistry and lithography*. SPIE Wiley.

Pankove, J. I., & Johnson, N. M. (1991). *Hydrogen in semiconductors*. Academic Press.

Park, D. (1998). *Plasma process-induced charging damage on thin gate oxides*. University of California, Berkeley.

Reinhardt, K. A., & Kern, W. (2018). *Handbook of silicon wafer cleaning technology* (3rd ed.). Elsevier William Andrew Applied Science Publishers. https://doi.org/10.1016/C2016-0-01001-X

Rimini, E. (1995). *Ion implantation basics to device fabrication*. Kluwer Academic Publishers.

Romano, L., Privitera, V., & Jagadish, C. (2015). *Defects in semiconductors*. Academic Press.

Schroder, D. K. (2006). *Semiconductor material and device characterization* (3rd ed.). Wiley Interscience. https://doi.org/10.1002/0471749095

Voldman, S. H. (2011). *ESD design and synthesis*. Wiley.

Wolf, S. (2002). *Silicon processing for the VLSI Era, Vol. 4: Deep-submicron process technology*. Lattice Press.

Zhong, W. (2001). *Plasma induced damage to Si and SiGe devices and materials new*. Jersey Institute of Technology.

Chapter 2
The Antenna Effect

2.1 Introduction

Process or plasma damage occurs when electrical charge accumulates on structures on the wafer surface during a process step or when electromagnetic energy couples into the wafer. The charge builds up, increasing the total energy on the surface of the wafer as the increasing charge raises the voltage on the structure. The accumulated energy can damage the wafer structures as the voltage and current move through the wafer. The dissipated energy causes electrical damage through multiple mechanisms. A small amount of damage breaks atomic bonds, and creates electron and hole traps, and interface states, shifting the electrical characteristics of the devices. A large amount of damage may destroy the structures and melt the structures on the wafer's surface, causing physical damage. Certain structures on the wafer may amplify these phenomena, increasing the severity of the problem. This problem is called the antenna effect, and it has nothing to do with traditional transmitting and receiving electromagnetic antennas. The word antenna in this book refers to conducting structures on the wafer that collect charge. The name antenna was adopted to describe this phenomenon.

Many different CMOS structures can create antennas, worsening damage. The antenna can be a rectangular conductor, a conductor with a high perimeter length, or an array of contacts etched into a dielectric on top of a conductor. The conductor can be metallization or gate conductors, junctions, or wells. Structures that can absorb charge (electrets), such as resist or masking layers and leaky dielectrics, can also cause plasma damage. Plasma damage can occur without an antenna, but it is usually much worse with an antenna.

K. Prall, *CMOS Plasma and Process Damage*,
https://doi.org/10.1007/978-3-031-89029-1_2

2.2 Direct Antenna Effect

The amount of charge and energy accumulated on the structures is proportional to the ratio of the structure collecting charge to the structure being damaged, with larger ratio structures receiving more damage. This phenomenon is called the antenna effect. Figure 2.1 shows a sketch of a CMOS antenna structure with a small transistor on an active area connected to a large antenna on the shallow trench isolation (STI). The charge collected by the antenna is dissipated into the transistor during the plasma damage event. However, the charge cannot dissipate from the structure on the isolation and flows into the leakier transistor gate oxide, worsening the damage because of the presence of the antenna.

The amount of charge collected on an antenna is proportional to the capacitance and area of the antenna. A larger antenna collects more charge than a small antenna, resulting in a higher current flowing through the small-area gate dielectric, increasing the gate dielectric damage. An extreme structure with a large antenna area and a small gate dielectric area creates the worst-case conditions for damage to the dielectric. The ratio of the antenna area to the gate dielectric is used as a figure of merit for the likelihood of damage given by Eq. (2.1).

$$\mathrm{AR} = \frac{\text{Area of antenna}}{\text{Area of gate dielectric}} \tag{2.1}$$

The antenna essentially amplifies the potential for damage by increased accumulated charge. For example, an antenna ratio of 1000:1 can increase the effective current per unit area of the receiving structure from milliamp levels to amperes of current, causing havoc on the wafer. The sensitivity of a specific process to

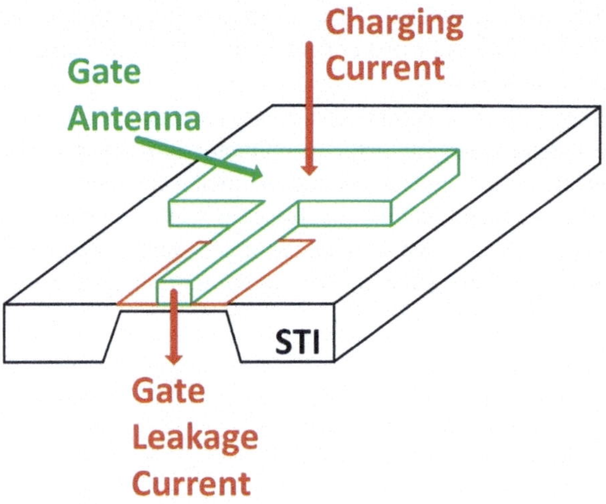

Fig. 2.1 Example of a simple gate antenna

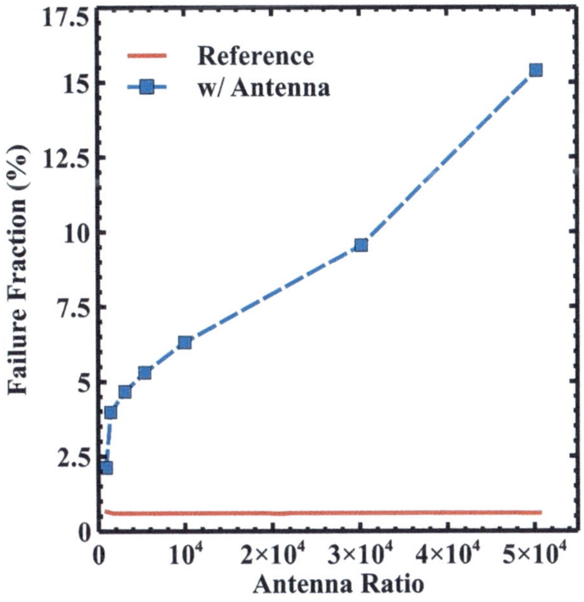

Fig. 2.2 Example comparing failure rates for structures with and without a metal finger antenna

antenna-enhanced damage can be determined by comparing the dielectric damage of multiple structures with different antenna areas to that of a structure with no antenna. Figure 2.2 compares an arbitrary dielectric failure criterion plotted against the antenna ratio for a spectrum of structures with increasing antenna area (Wang, 2004). The plot shows increasing dielectric damage as the antenna ratio increases. A reference device without an antenna is also shown to demonstrate the baseline performance with no antenna. The absence of an antenna does not guarantee that process damage will not occur, but it just makes damage less likely.

2.3 Indirect Antenna Effect

Direct process damage can occur when a structure susceptible to damage is directly exposed to CMOS processing of the structure. Process damage can also occur indirectly, caused by an unrelated process step occurring much later in the CMOS process flow. Figure 2.3 shows an example of indirect process damage (Choi et al., 2013). The plasma exposure of a metal line separated from the damaged transistor by many metal and dielectric layers can cause process damage. For example, exposure of the metal 4 (M4) layer to a plasma causes charge to flow through all the underlying layers of metals (M1–3) and damage the underlying transistors. This indirect damage needs only a low resistance path from the antenna to the damaged structure. Indirect damage can also occur without a conducting path through the

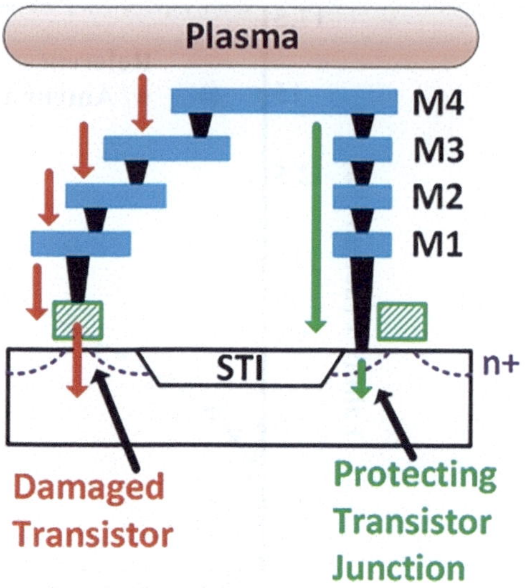

Fig. 2.3 Example of indirect plasma damage. The charge collected in antennas on metal 4 (M4) reaches the gate dielectric by conduction through metal lines, damaging the left transistor. This example shows that the left transistor is not protected by the junction diode of the right transistor until the metal 4 layer is formed, rerouting the damaging current through the right transistor junction, protecting the left transistor for process steps after metal 4 formation

metal. For example, wafer exposure to high-energy light (UV or X-ray) late in the process can damage underlying structures if the light can reach the structure. However, high-energy light damage does not show an antenna effect and does not correlate with antenna ratios.

The left transistor in Fig. 2.3 is protected after the metal 4 process steps when the right transistor junction diode is electrically connected to the gate of the left transistor through metallization. The junction connection of the right transistor provides an alternative leakage path, reducing the current through the gate dielectric and lessening the damage.

2.4 Source-Drain Antenna Effects

The most common plasma damage source is damage to transistors from antennas connected to the gate terminal. The source or drain of a CMOS transistor can also create a path for plasma damage through either the source-drain active area or through metallization connected to the source-drain. However, design rules generally forbid the formation of large active area antennas, as large source-drain regions usually have no circuit advantages. Charge injected into the source and drain terminals can damage the transistor by conduction through the gate dielectric in the source and drain overlap region (lightly doped drain), as shown in Fig. 2.4 (Yang et al., 2018). The transistor parasitic capacitance components shown in the figure can cause charging energy on the source and drain regions to be coupled to the gate conductor, creating a propensity for plasma damage. Increasing the source-drain

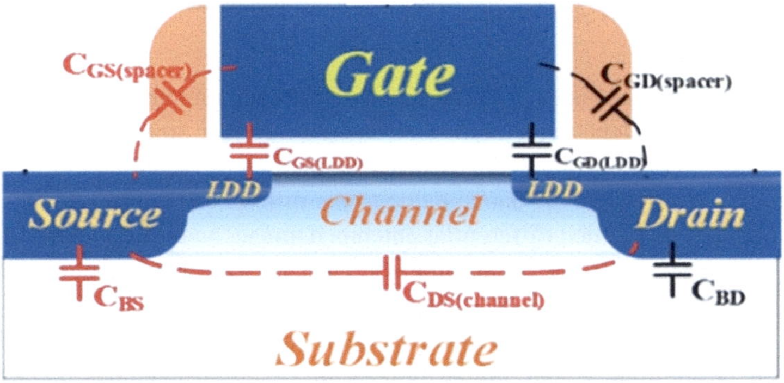

Fig. 2.4 Cross-section sketch of a CMOS transistor showing transistor related parasitic capacitance components. The spacer-lightly doped drain overlap capacitance is shown as $C_{GD(LDD)}$ and $C_{GS(LDD)}$. The spacer capacitance is $C_{GD(spacer)}$ and $C_{GS(spacer)}$. The source-drain to substrate capacitance is C_{BD} and C_{BS}. The source-drain capacitance is shown as $C_{DS(channel)}$

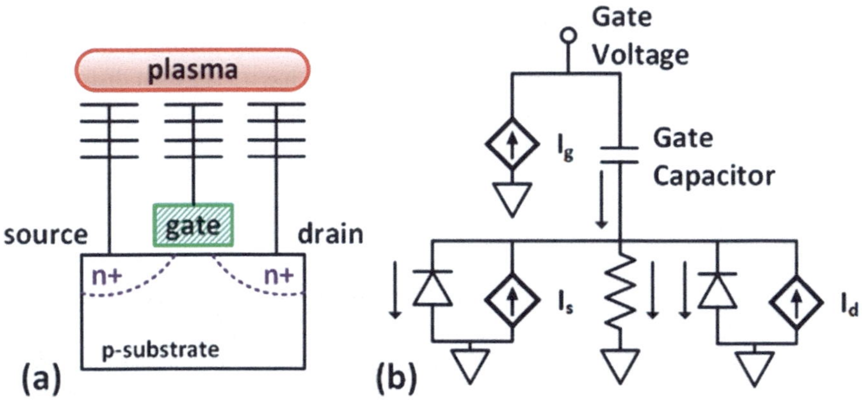

Fig. 2.5 Sketch showing a MOS transistor with antennas connected to the source, gate, and drain. (**a**) Cross-section sketch of transistors with antennas. (**b**) Equivalent circuit schematic of the transistor during plasma damage stress. See the reference for details on the circuit model

area creates a conducting antenna that can collect charge. The junction to substrate capacitance C_{BD} and C_{BS} increases with the source-drain area, storing increased energy.

This source-drain to gate overlap region tends to be sensitive to damage as it can easily modulate the transistor's behavior when damaged (see Chap. 5). Figure 2.5 shows a cross section, with antennas connected to the three terminals of the transistor (Lin & Sery, 2003). The plasma charging creates a current source flowing into the three terminals, damaging the transistor. Details on the damage are covered in the following chapters. Silicon-on-insulator (SOI) technology tends to be very susceptible to damage from the source and drain. This topic is covered in Chap. 8.

2.4.1 Differential Antennas

A transistor with antennas on both the gate and diffusion regions may have different amounts of damage depending on the size and difference in sizes of the antennas. A large antenna on one terminal, along with a small antenna on another terminal, can cause more extensive plasma damage due to the difference in charging conditions. This phenomenon is called differential antenna damage, and it has complex behaviors covered in Chap. 8. The differential antenna effect can be severe in silicon-on-insulator (SOI) technology, as described in Chap. 8.

The drain, source, and substrate connections can also impact plasma damage protection in addition to gate connections. Bond pads or antennas connected to these terminals can impact plasma damage. Figure 2.6 shows a sketch of the terminals of a MOS transistor undergoing plasma damage with antennas on all nodes (Lin, 2016a, b). The experiment performed in this figure connects various terminals to a diode with a temporary connection that can be removed with a fuse. The fuse is intact during process damage during the backend of line processing and is removed before electrical measurements. The plot in Fig. 2.6b shows a probability plot of threshold variation for different bond pad conditions during plasma damage. The configuration with all pads grounded through a diode has a very tight distribution, indicating that the grounded pads have successfully drained charge from the pad. Structures with floating pads that function as antennas have a degraded distribution, indicating increased variation due to plasma damage. It is helpful to have duplicate transistor layouts with standard bond pads and a minimum pad area that can be

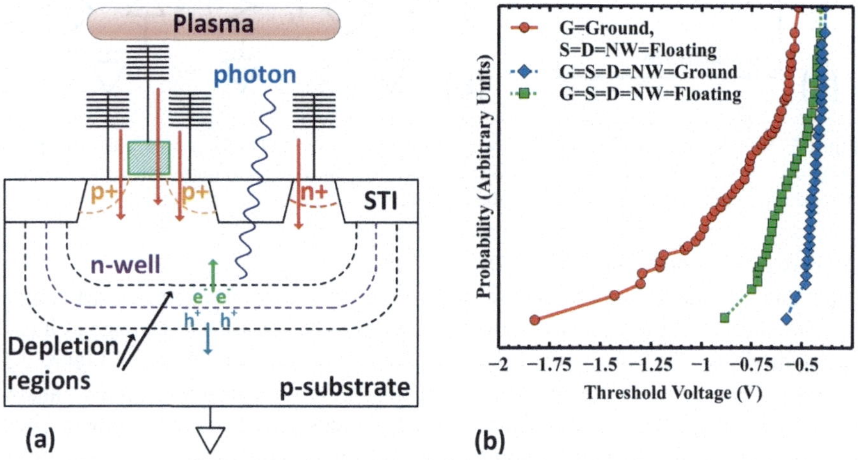

Fig. 2.6 The impact of bond pad electrical connections on the threshold voltage distribution of a PMOS transistor. (**a**) Shows a sketch of a PMOS transistor with bond pads under plasma damage conditions shown as antennas. Carrier generation by light exposure in the n-well is shown as a factor in the damage behavior. (**b**) Probability plot showing the threshold voltage variation with different bondpad connection configurations

probed. Comparing these two different structures can determine the impact of plasma damage on all the probe pads.

The source and drain areas can significantly impact the process damage. As the source and drain areas increase, the leakage increases, creating a high bias across the gate dielectric that increases damage. This effect typically dominates when the source and drain junctions are floating, making the junction less leaky as the area increases. Figure 2.7 shows an example plot comparing the gate dielectric leakage with different source-drain areas ranging from the design rule minimum to 100 times the minimum source-drain area (Lin & Sery, 2001). At the 100× area, more than 50% of the gate dielectrics tested leak, while there were no defective gate dielectrics at the design rule 1×.

Worst-case damage conditions for differential antenna plasma damage often occur when one antenna is much larger than the other, even if the overall antenna ratio calculated by the simple ratio of the antenna area to the gate area is lower. The likely cause of this phenomenon is that differential antennas with a slight difference charge the gate and drain to similar potentials. Conversely, significant differences in

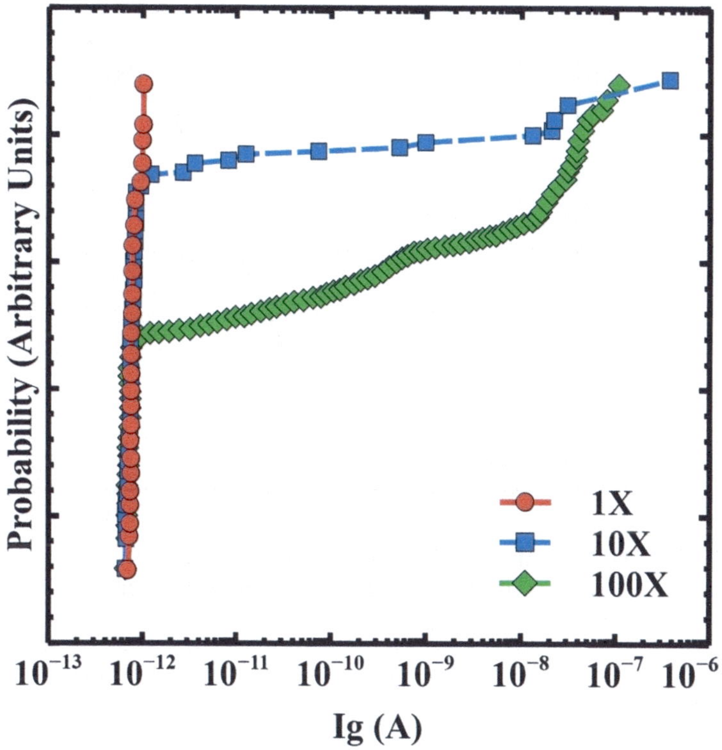

Fig. 2.7 Probability plot of gate leakage as a function of the source drain areas from a 1×, 10×, and 100× multiplier of the minimum design rules

antenna sizes cause large differences in charging from each terminal, resulting in considerable stress across the gate dielectric and increasing plasma damage. For example, Fig. 2.8 shows the results of an experiment varying the difference between the gate and drain antennas (Akbal et al., 2015b). The plot shows a high dielectric breakdown failure rate for a major difference (46,000) between a large gate antenna and a minimal drain antenna. The failure rate drops as the drain antenna increases, reaching zero as the difference reaches 100 and lower. For this case, the antenna ratio equations require subtraction of the antenna ratios.

The literature also shows cases where the differential antenna damage worsens when the gate and drain antenna areas are the same. For example, Fig. 2.9 shows a scatter plot of gate dielectric leakage for three different differential experiments (Lai et al., 2006). The number of extrinsic data points increases as the drain antenna area is increased until it matches the gate antenna ratio. The worst-case results occur for large antennas on the gate, drain, and source. This result is opposite to the results shown in Fig. 2.8.

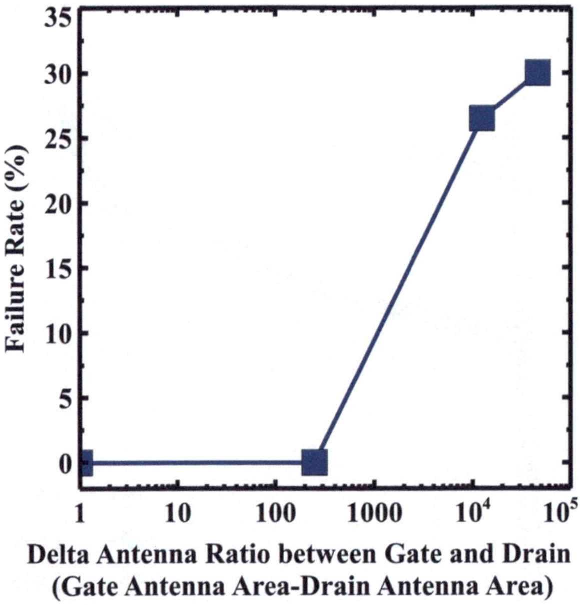

Fig. 2.8 Gate dielectric failure rate for differential antenna ratios between the gate and drain terminals for NMOS transistors, calculated by subtracting the drain antenna ratio from the gate antenna ratio

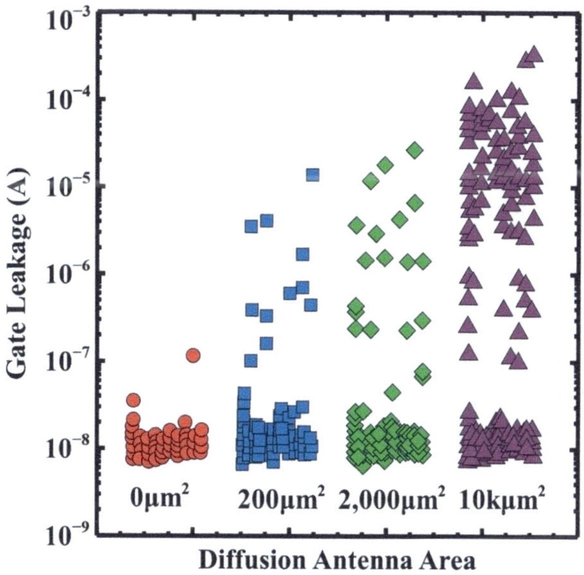

Fig. 2.9 NMOS gate dielectric leakage scatter plot for three different diffusion antenna ratios. Diffusion node antennas are placed on both the source and drain. The gate antennas are 10,000 μm^2 for each of the 3 experiments. The reference (0 μm^2) has no antennas

2.5 Capacitive-Coupled Antenna Effect

The direct and indirect antenna effects occur as a metal or doped semiconductor collects charge, and it flows to the weakest leakiest structure, causing damage. Process and plasma damage can also occur by capacitive coupling through dielectrics, where a voltage on one capacitor plate tries to increase the voltage of the other plate. Structures on the wafer can form intentional or unintentional capacitors. Charge collects on one plate, and the generated voltage is coupled onto the second plate, creating enough voltage to damage sensitive structures and devices.

Metal-insulator-metal (MIM) capacitors (also called metal oxide metal capacitors (MOM)) are commonly added to a CMOS process embedded into the backend of line metallization process. MIM capacitors provide an essential circuit element to the CMOS chip designer, increasing the circuit performance and capability by enabling high capacitance per unit chip area. MIM dielectrics can suffer from process damage, as described in Chap. 8, but they can also be a conduit to damage underlying devices such as CMOS transistors.

Figure 2.10 shows several possible MIM capacitor circuit configurations (Takeda et al., 1995; Weng et al., 2009). Figure 2.10a shows a MIM capacitor with the upper terminal connected to a metal antenna in the metal layer above the capacitor. The lower terminal is connected to another metal-routing line. Figure 2.10b shows a MIM capacitor with the upper terminal connected to a metal antenna in the metal layer above the capacitor. The lower terminal is connected to the substrate.

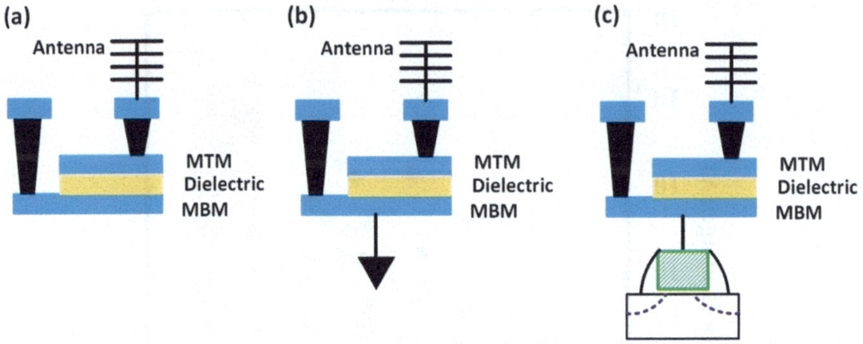

Fig. 2.10 Sketch and schematic of a metal insulator metal (MIM) capacitor in various configurations. (**a**) MIM is connected to a metal antenna and another metal line. (**b**) MIM capacitor is connected to an antenna and a ground in the semiconductor substrate. (**c**) MIM capacitor connected to an antenna and a CMOS transistor gate. MTM is defined as capacitor top metal. MBM is defined as capacitor bottom metal

Figure 2.10c shows a MIM capacitor with the upper terminal connected to a metal antenna in the metal layer above the capacitor. The lower terminal is connected to the gate of a CMOS transistor. The first two configurations only have the potential process damage to the MIM capacitor because there are no sensitive gate structures in the circuit.

The circuit connected to the CMOS gate shown in Fig. 2.10c can cause process damage. The charging current supplied to the antenna increases the voltage on the upper plate of the MIM capacitor. The voltage is split between the MIM capacitor and the gate capacitance. The voltage coupled to the gate may reach a high-enough voltage that can damage the gate dielectric and transistor. The MIM capacitor is also susceptible to damage during a plasma damage event.

The exact nature of the damage occurring in these configurations strongly depends on the electric fields created by the capacitive coupling. A large MIM capacitor connected to a small area gate creates the maximum potential for damage to the transistor because of the high gate dielectric electric field caused by these conditions. A high electric field occurs across the MIM capacitor for a large gate area and a small MIM capacitor area. The MIM capacitor is usually designed to withstand the operating voltages of the chip, which can be exceeded by plasma damage. The gate's dielectric thickness and reliability capability and MIM dielectric are critical factors in this case.

2.6 Virtual Antenna Effects

A coupling capacitor precipitating process damage does not necessarily have to be constructed from structures designed into a CMOS circuit, as in the MIM case. Instead, coupling capacitors can be formed from the plasma elements that occur

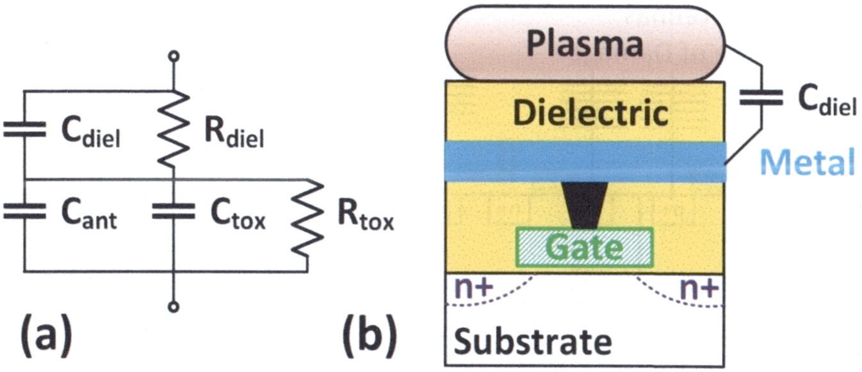

Fig. 2.11 Sketch of a dielectric on a metal feature on the surface of a wafer forming a capacitor with one capacitor terminal being the plasma and the metal being the second terminal by coupling to the plasma. (**a**) Equivalent circuit. (**b**) Cross-section process sketch showing the formation of a capacitor during plasma dielectric deposition

during processing in a fabrication tool. The wafer is separated from the plasma by a region with very few charge carriers called the sheath. The sheath can behave as a capacitor, coupling the wafer to the plasma. The plasma can serve as one terminal of the capacitor coupling to underlying conductors through a dielectric on the surface of the wafer. The capacitor consists of a plasma terminal and a conductor structure on the wafer surface separated by a dielectric that can couple enough voltage into the sensitive devices to damage them. Figure 2.11 shows a sketch showing a path for process damage through a capacitor formed by the plasma and a conducting feature on the surface of the wafer (Cheung, 1999).

2.7 Well Coupling-Related Damage

2.7.1 Bulk CMOS Well Coupling-Related Damage

The previous section discussed process damage caused by coupling through MIM capacitors to sensitive circuits. Other possibilities exist, including coupling through unintentional parasitic capacitances. Coupling through CMOS wells is an additional coupling mechanism that can cause process damage. CMOS wells are used to form doped regions to configure and separate NMOS and PMOS transistors for proper electrical operation. The wells also form junction diodes performing a critical isolation function.

The wells also form a junction capacitance at well interfaces and well-substrate interfaces. This junction capacitance can also couple voltages caused by plasma charging. The well capacitances can couple excess voltages, charging, damaging, and similarly destroying sensitive devices similar to the MIM capacitor problem

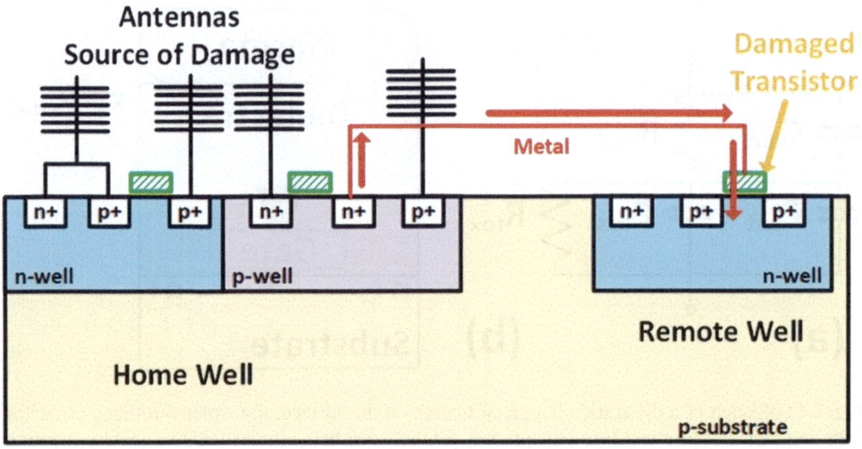

Fig. 2.12 Cross-section of a CMOS process showing transistors and wells demonstrating how plasma charging can couple from one well region to another well region

described previously. Figure 2.12 shows an example cross-section sketch of a CMOS circuit (Martin et al., 2011). Antennas connected to junctions in one well can change well potentials during processing to levels not seen during regular operation. The charged wells can couple voltages to electrically connected wells through metal lines or adjacent wells. CMOS devices in the connected well can undergo stress and damage, degrading circuit operation and reliability.

Due to large design rules at well boundaries, multiple transistors are placed in a well for layout efficiency for a typical CMOS layout. As a result, multiple connections to the wells from the source, drain, and well contacts create more propensity for process damage with the increasing number of well connections. However, wells also serve as diodes to protect dielectrics.

The coupling ratios for well-related coupling are more difficult to estimate than those of a simple metal or gate area antenna. This is because many factors impact process damage created by well coupling, including the following shown in Table 2.1 (Martin & Kamp, 2020). Complex computer analysis such as SPICE or technology computer-assisted design (TCAD) can provide the insight necessary to optimize design and design rules to minimize well coupling-related damage.

The well size, guard rings, and related rules change the resistance between the circuit elements and the substrate, impacting charge dissipation from the antenna. The CMOS well may float during plasma processing depending on the electrical connections of the well, or it may be shorted to other terminals. Figure 2.13 shows a plot of the gate leakage of a PMOS device from plasma damage for three different well configurations (Krishnan et al., 1998). Figure 2.13a shows the best results for a simple floating well connection with a small well. Figure 2.13b shows worse results with the well shorted to the substrate, creating a higher stress by bypassing the well diode. Figure 2.13c has a significant defect tail with a larger n-well, likely

Table 2.1 Factors affecting process damage by well coupling

Size and geometry of the metal antennas
Size, area, doping, breakdown, and electrical characteristics of the wells and junctions
Size and area of the degraded transistor
Oxide thickness and resistance to damage of the degraded transistor
The existence or absence and capability of discharge paths such as diodes
Resistivity of connections between wells and transistors
Antennas connected to the well where damage is occurring (reverse charging possible)

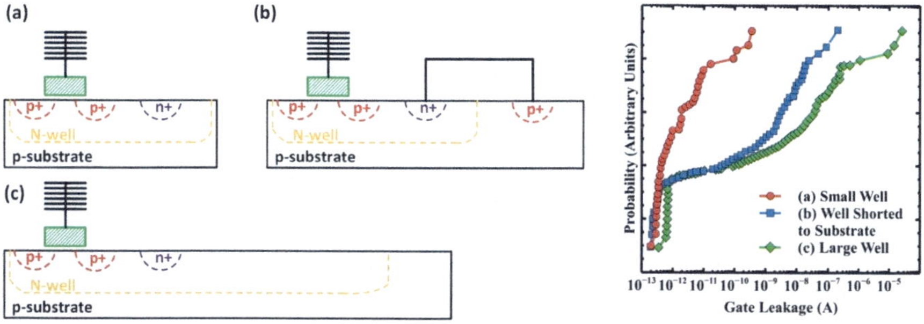

Fig. 2.13 Plot showing process damage gate leakage variation differences due to diode connection and well differences. (**a**) Small well. (**b**) Well shorted to substrate. (**c**) Large well

due to the increased well antenna area. The junction and well resistances can significantly influence well-related plasma damage.

The well coupling-related damage tends to be a function of both the antenna ratio and the well area, as shown in Fig. 2.14 (Martin & Kamp, 2020). The damage increases as the metal antenna area increases, coupling more charge into the well. The damage increases as the well area increases, coupling more voltage and energy into the gate dielectric. The upper right region of the plot has greater damage due to a large well antenna area and a large metal-to-well antenna ratio. The lower left region of the plot has no damage due to a small well and metal antenna.

2.7.2 *Triple Wells*

Basic CMOS technology consists of PMOS transistors placed in an n-well and NMOS transistors placed in a p-well. In addition, a third well is often added, typically consisting of a p-well surrounded by n-type regions. This structure is called triple-well technology, and the third well is called a deep n-well or triple well. NMOS transistors are formed in the p-well surrounded by the deep n-well. A CMOS

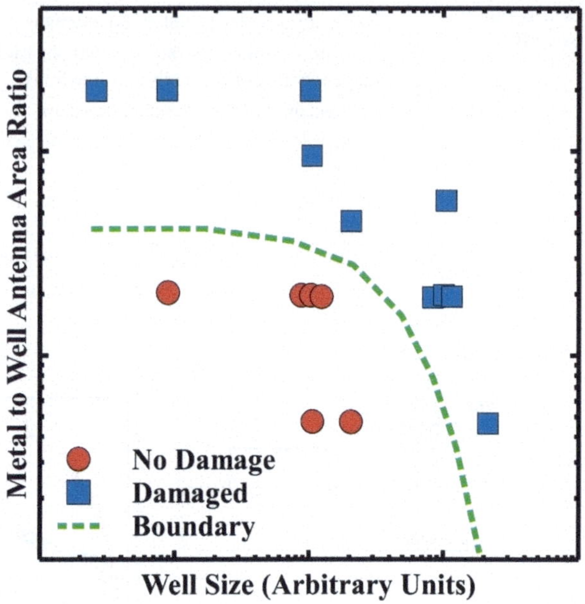

Fig. 2.14 Plot showing the two-dimensional behavior of well coupling related damage as a function of the metal antenna to well area ratio and the well size. The lower left region of the plot shows no damage, and the upper right region shows damage. The plot is assembled from various papers referenced by the author of the reference

triple well structure has several advantages: improved isolation, improved soft error rate, noise reduction, and circuit advantages. The circuit advantages include biasing the transistor body differently in the triple well compared to devices in the bulk p-well. The advantages for memories include improved refresh for DRAMs and optimization of the program and erase operations for nonvolatile memories.

Triple-well CMOS has disadvantages for plasma damage. An n+/p/n structure is formed by the source and drains, the p-triple well, and the deep n-well. Electrically, the n+/p/n structure forms two back-to-back diodes with better isolation and leakage properties than a single diode formed in a p-well in a p-substrate. The improved isolation reduces leakage, which can increase the degradation from process damage because the charge drained from the antennas is reduced.

There are different ways for damage to occur in triple-well structures, as shown in Fig. 2.15 (Martin & Kamp, 2020). First, the p-triple well can be charged from the antennas, and the voltage stress can propagate to an additional well (remote well), damaging the transistors in that well, as shown in Fig. 2.15a. Second, the isolated p-well can be charged from the plasma stress, and the remote well can increase the voltage stress on a transistor in the triple well (home well) if the remote p-well is near the ground, as shown in Fig. 2.15b.

As the well size increases, higher voltages can be coupled from the well as a function of the well capacitance ratios into the separate remote well, causing increasing damage to the transistors in the remote well. Figure 2.16 shows a Weibull

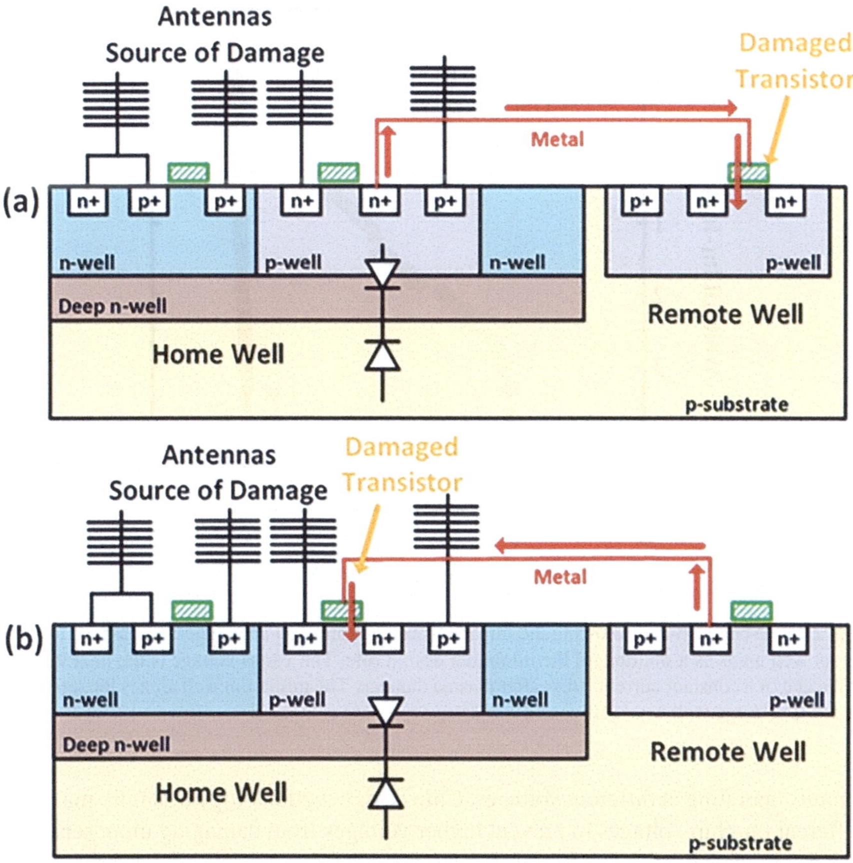

Fig. 2.15 Process cross-section sketches showing different possibilities for well damage. (**a**) Damage from a triple well (home well) with antennas into a transistor in a remote well. (**b**) Damage from a remote well into a triple well

probability plot of the gate leakage voltage after a current revealing stress for different well sizes (Martin, 2013). The reference curve has the minimum well area allowed by the design rules. Well sizes of 10 times and 100 times the minimum design rules are shown. The gate leakage significantly increases for the large well sizes.

2.8 Across Power Domains' Damage

Modern CMOS chips tend to operate with different supply source voltages connected to the chip externally or generated internally through voltage regulation or charge pumping. As a result, the chips have different power domains that separate

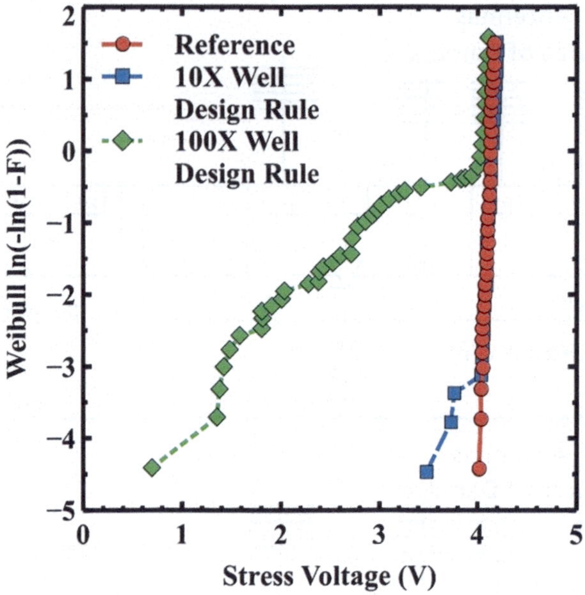

Fig. 2.16 Probability plot showing the impact of well coupling on gate dielectric damage for different well areas as a multiple of the minimum design rule. The *x*-axis voltage is the final voltage at the end of a constant current stress after plasma damage. The minimum well area is the reference group. Two large well sizes (10× and 100× design rule) are compared

circuits operating at different voltages. Chip design methodology carefully manages different on-chip voltages to prevent higher voltages from damaging more sensitive lower voltage circuits. The designed substrate bias also commonly varies between circuits for various reasons, including noise reduction, memory array operation, dynamic substrate biases, etc. Some CMOS circuits are typically placed in a triple well to isolate substrate voltages for transistors inside the triple well. Figure 2.17 shows an example schematic with two power domains (Vdd1 and Vdd2) (Chu et al., 2018). However, plasma damage can cause havoc across power domains exceeding designed conditions, with antenna damage propagating from different supply regions. Plasma damage can cross the power domains in either direction depending on multiple factors, including well configuration, gate dielectric thickness, supply voltages, etc. Damage can occur during processing when the current and voltage conditions exceed the design parameters.

Figure 2.18 shows a cross section of CMOS transistors in a circuit with different power domains. Antennas in the wells for one domain can collect process damage charge that transfers through transistors and metallization, damaging transistors in a different domain (Chu et al., 2017).

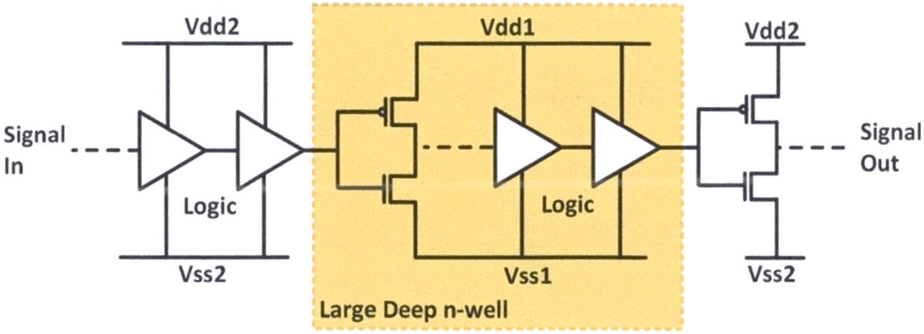

Fig. 2.17 Example circuit schematic showing across domain signal creating the potential for increased process damage. Vdd1 and Vdd2 are the supply voltages to the two power domains and Vss1 and Vss2 are the two grounds for the different power domains

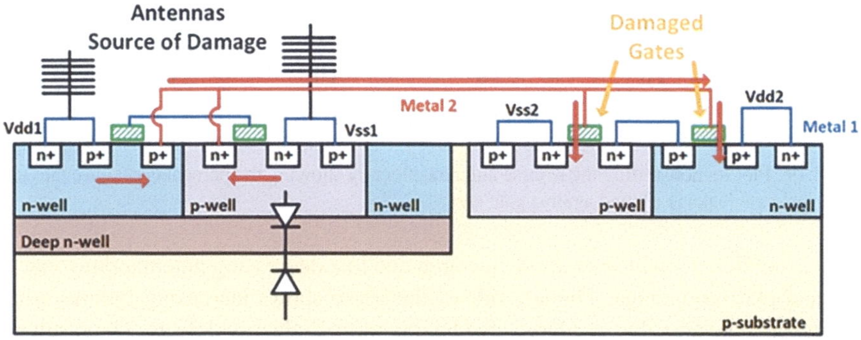

Fig. 2.18 Cross-section sketch showing antennas in one power domain coupling into a different power domain, causing plasma damage in a different domain following the path shown by the red arrows from the antennas in a triple well to transistors in different well with a different supply voltage

2.9 The Reverse Antenna Effect

Process damage normally increases as the antenna ratio increases, as shown in Fig. 2.2. Occasionally, different behaviors can be observed. Gate dielectric defects increase as the area of the capacitor increases. A large capacitor has a higher probability of being defective when compared to a small capacitor due to the statistical probability of defect density, which typically is constant per unit area. The process damage antenna behavior competes with increasing gate dielectric defect density, resulting in a peak failure rate at a specific gate dielectric area. The failure rate initially increases as the gate area increases, followed by a decreasing failure rate as the antenna ratio decreases. An example of this behavior is shown in Fig. 2.19 (Krishnan et al., 2002). A peak in the gate dielectric defect density is shown because of the two competing mechanisms. The right side of the bell-shaped curve shows

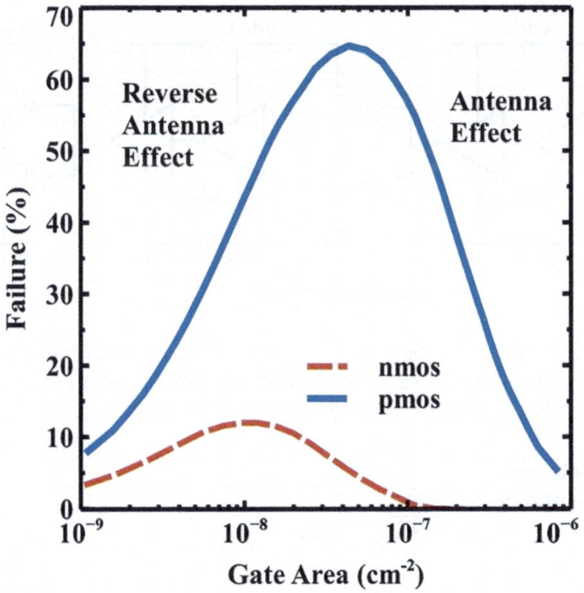

Fig. 2.19 Plot demonstrating the reverse antenna effect by showing the percentage failure rate for abnormal gate leakage plotted against gate area

decreasing defects with increased gate area due to a decreasing antenna ratio, causing less plasma damage. The left side of the curve shows increasing failures with increasing gate area due to a large area having more dielectric defects. The result of the two competing mechanisms is a worse-case condition with an intermediate gate area.

2.10 Remote Antennas

Antennas connected to a CMOS transistor can occur anywhere on the chip, even on the opposite side of the chip. Multiple antennas connected to the same transistor can also be on opposite sides of the chip. Separated antennas may be the worst-case situation for damage if there is any across chip or wafer variation in the plasma or processing conditions (Akbal et al., 2015a, b). The worst-case conditions occur when the plasma charges one antenna in a positive direction and another in a negative direction, creating a severe bias across the transistor (Martin, 2021). Figure 2.20 shows an antenna separation experiment (Lai et al., 2006). The gate antenna distance to the transistor is varied, and the gate dielectric damage is monitored. The source-drain antenna is placed close to the transistor in this experiment. Only the closely placed gate antenna (<100 μm) does not show evidence of gate damage,

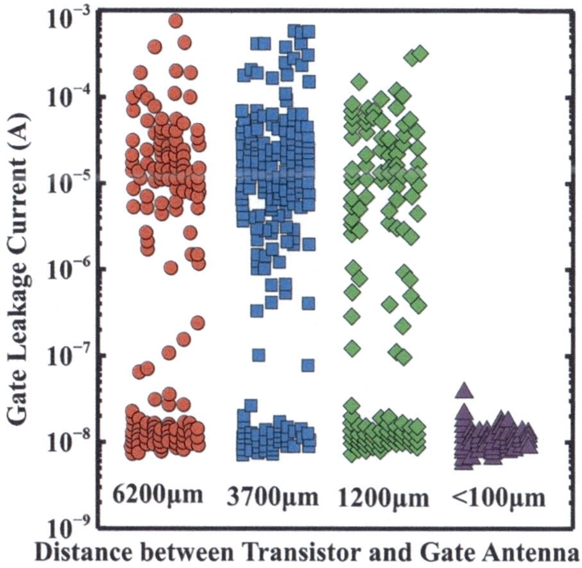

Fig. 2.20 Gate leakage for a remote antenna experiment with the distance to the gate antenna varied with the source-drain antenna distance held constant next to the transistor

while remote antennas have a significant extrinsic tail, indicating gate dielectric degradation.

References

Akbal, M., Ribes, G., & Vallier, L. (2015a). New insight in plasma charging impact on gate oxide breakdown in FDSOI technology. In *IEEE International Reliability Physics Symposium*. https://doi.org/10.1109/IRPS.2015.7112820

Akbal, M., et al. (2015b). Plasma induced damage investigation in the fully depleted SOI technology. In *International Conference of IC Design and Technology*. https://doi.org/10.1109/ICICDT.2015.7165900

Cheung, K. P. (1999). *Plasma charging damage*. Springer. https://doi.org/10.1007/978-1-4471-0247-2

Choi, W. H., et al. (2013). *An array-based circuit for characterizing latent plasma-induced damage*. In *International Reliability Physics Symposium*. https://doi.org/10.1109/IRPS.2013.6532005

Chu, Y.-L., et al. (2017). A CDM-like damage mechanism for multiple power domains fabricated with deep N-well processes. In *IEEE International Reliability Physics Symposium*. https://doi.org/10.1109/IRPS.2017.7936350

Chu, Y.-L., et al. (2018). A new mechanism of signal path charging damage across separated power domain deep N-well interface. In *IEEE International Reliability Physics Conference*. https://doi.org/10.1109/IRPS.2018.8353631

Krishnan, S., et al. (1998). Antenna device reliability for ULSI processing. In *IEEE IEDM*. https://doi.org/10.1109/IEDM.1998.746430

Krishnan, A. T., et al. (2002). Impact of gate area on plasma charging damage: The "reverse" antenna effect. In *IEEE IEDM*. https://doi.org/10.1109/IEDM.2002.1175895

Lai, W., et al. (2006). Ultra-thin gate dielectric plasma charging damage in SOI technology. In *IEEE International Reliability Physics Symposium*. https://doi.org/10.1109/RELPHY.2006.251246

Lin, W. (2016a). Bond-pad charging protection design for charging-free reference transistor test structures. *IEEE Transactions on Electron Devices, 63*(1), 272–279. https://doi.org/10.1109/TED.2015.2501418

Lin, W. (2016b). Plasma charging behavior of p-MOSFETs with floating bond pads and its application to charging-safe bond pad design. *IEEE Transactions on Electron Devices, 63*(7), 2722–2728. https://doi.org/10.1109/TED.2016.2569539

Lin, W., & Sery, G. (2001). Role of source/drain junction on plasma induced gate charging damage in N MOSFET. In *International Symposium on Plasma Process-Induced Damage*. https://doi.org/10.1109/PPID.2001.929991

Lin, W., & Sery, G. (2003). Multiple-terminal gate charging effect—Competing/compensating charging behavior. *IEEE Electron Device Letters, 24*(7), 481–483. https://doi.org/10.1109/led.2003.814997

Martin, A. (2013). Circuit relevant well charging from metal antenna and its degradation on digital MOS transistor reliability. In *IEEE International Integrated Reliability Workshop*. https://doi.org/10.1109/IIRW.2013.6804156

Martin, A. (2021). Reliability characterization of plasma induced charging damage (PID) for MOS transistors of a silicon on insulator (SOI) process using product relevant test structures and fWLR methods. In *IEEE International Integrated Reliability Workshop*. https://doi.org/10.1109/IIRW53245.2021.9635611

Martin, A., & Kamp, A. (2020). A new implementation approach for reliability design rules against plasma induced charging damage from well configurations of complex ICs. In *IEEE International Reliability Physics Symposium*. https://doi.org/10.1109/IRPS45951.2020.9128866

Martin, A., et al. (2011). Reliability degradation of MOS transistors originated from plasma process-induced charging of circuit blocks and detected with fWLR methods. In *International Integrated Reliability Workshop*. https://doi.org/10.1109/IIRW.2011.6142594

Takeda, T., et al. (1995). *Hot-carrier effects in MOS devices*. Elsevier.

Wang, Z. (2004). *Detection of and protection against plasma charging damage in modern IC technology*. Twente University.

Weng, W.-T., Lee, Y.-J., Lin, H.-C., & Huang, T.-Y. (2009). Effects of plasma damage on metal–insulator–metal capacitors and transistors for advanced mixed-signal/radio-frequency metal–oxide–semiconductor field-effect transistor technology. *Japanese Journal of Applied Physics, 48*(8), 086001. https://doi.org/10.1143/jjap.48.086001

Yang, T., et al. (2018). Improved performance of CMOS terahertz detectors by reducing MOSFET parasitic capacitance. *IEEE Access, 7*, 9783–9789. https://doi.org/10.1109/ACCESS.2018.2889154

Chapter 3
Diode and Transistor Protection

3.1 Introduction

Gate dielectrics are typically the most sensitive structures in a CMOS chip. However, other circuit elements such as diodes and transistors can protect gate dielectrics by draining charge away from the gate, reducing stress during process damage. Usually, for most CMOS circuits, every gate dielectric is naturally connected to a source or drain junction of the transistors driving the gate after processing is completed. The connected junctions partially protect the gate dielectrics from process damage from the point in the process flow where they are electrically connected to the end of the process flow. CMOS design practices commonly prevent the connection of transistor gate terminals without a gate connection to a drain or source. Design rules may prohibit connecting a gate terminal directly to a power supply to avoid unprotected transistor gates, avoiding worst-case circuits for process damage.

Transistors can be added to CMOS circuits solely to protect sensitive locations from plasma damage, placing junctions between the sensitive locations and the antennas. This strategy obviously can impact the chip area if used extensively. However, it can be useful for targeted specific critical circuits such as analog blocks where the protection transistors can degrade without affecting the circuit operation while protecting critical transistors from degradation. Figure 3.1 shows several example circuits (Martin et al., 2021; Zhao et al., 1996). Figure 3.1a shows an example where a CMOS transmission gate (pass gate) consisting of an NMOS and PMOS in parallel is placed between an antenna and a protected gate to protect transistors tied to the power supply and ground. Figure 3.1b shows a sacrifical transistor placed between the antenna and a protected transistor with the gate and drain tied to the antenna. The sacrifical transistor characteristics can degrade slightly without affecting the operation of the protected transistor. Careful analysis is needed to ensure that the added transistors improve process damage and not worsen the damage.

K. Prall, *CMOS Plasma and Process Damage*,
https://doi.org/10.1007/978-3-031-89029-1_3

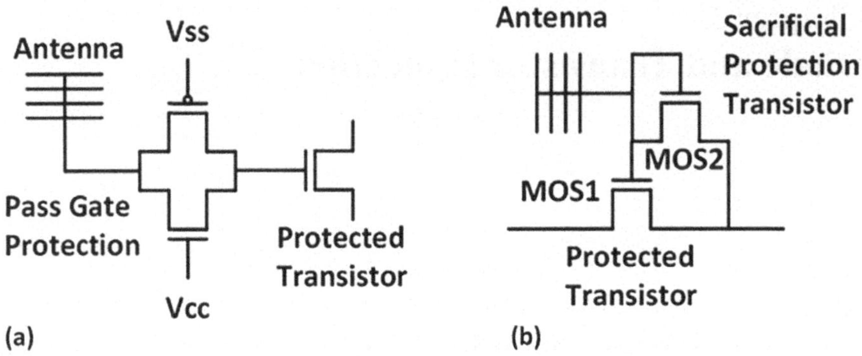

Fig. 3.1 Schematic sketches showing the use of CMOS transistors to protect sensitive locations. (**a**) Transmission (pass) gates to protect devices. (**b**) A sacrificial transistor (MOS2) is placed between the antenna and the gate of a protected transistor (MOS1)

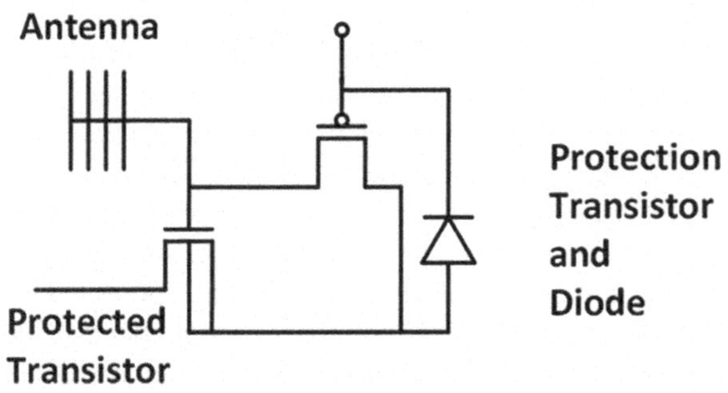

Fig. 3.2 Schematic showing a transistor-diode protection circuit to prevent plasma damage by placing the diode between the gate and substrate of the protected device

Circuits can be configured to protect devices from process damage by connecting a diode between the antenna and the substrate in parallel with the protected device. The diode creates a leakage path, draining the charge from the antenna, directing the voltage away, and limiting the stress from the protected device to minimize process damage. Diode leakage reduces the accumulation of charge on structures, limiting process damage. The diode must leak significantly more than the dielectrics, which is usually the case, enabling reduced dielectric stress by enabling the charging current to bypass the sensitive dielectric. Figure 3.2 shows a sketch of a schematic of a diode protection circuit (Hiura et al., 2020). The diode must be oriented so that it is reverse-biased during regular chip operation to avoid a forward bias condition where a large current may affect circuit operation.

Figure 3.3 shows a process cross-section sketch of a diode protecting a transistor gate from plasma damage. The diode must be connected to the gate as early in the process flow as possible. Typically, the diode is connected at the first metal level, as shown in the figure. A diode connected at the first metal level protects the gate during all subsequent processing steps, including the backend of line steps such as

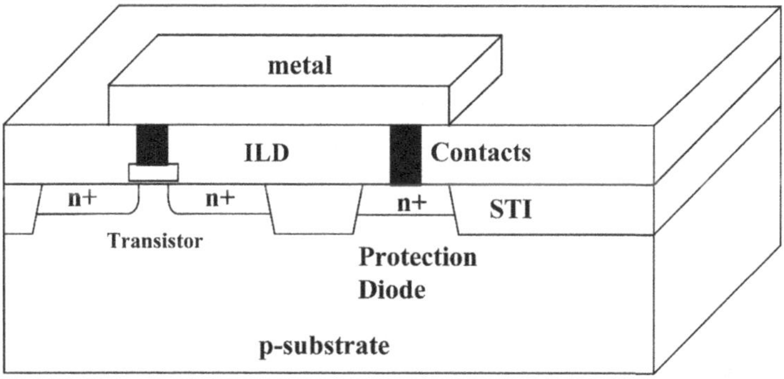

Fig. 3.3 Cross-section sketch of a gate protection diode

metal, vias, dielectric, bond pads, etc. However, the diode does not protect the CMOS gate dielectric during process steps before it is connected to the gate earlier in the process flow, such as the gate etch, implants, etc.

The addition of diode protection circuits to a CMOS design may create a significant layout and routing penalty, potentially increasing the chip area. Generally, it is not practical to add diodes to protect the gate of every transistor on a chip. Instead, specific sensitive circuits, such as matched transistors, or worse-case antennas can be protected as a part of the design process.

Parametric test structures isolate individual CMOS components, allowing detailed measurements and characterization of the isolated devices. Parametric structures commonly lack diodes on the gates to avoid measurement limitations. The protection diodes can interfere with some parametric measurements such as the gate current, gate-to-drain capacitances, etc.

Diodes can be added to parametric test structures to protect the parametric devices from process damage. Parametric structures are directly connected to bond pads, creating an antenna on most devices. Some structures should be repeated with and without a diode protecting the gate to allow easy diagnosis of plasma damage.

Diodes conduct current very asymmetrically, with the forward bias conducting a large amount of current and the reverse bias condition conducting a low current, which is many orders of magnitude lower. Therefore, the diode current during plasma damage conditions must be high enough for plasma charging in both directions to mitigate process damage effectively.

Plasma damage diode protection can significantly improve plasma damage compared to an unprotected device. Diode protection does not necessarily eliminate plasma damage but can significantly reduce the problem. Figure 3.4 shows some examples of successful plasma damage mitigation (Matsunaga et al., 2007; Sommer et al., 2011). Figure 3.4a shows improved gate leakage and reduced dielectric damage with diode protection for dielectrics damaged by multiple metal etches. Figure 3.4b shows a dramatic Vt shift and variability mitigated by diode protection during film deposition.

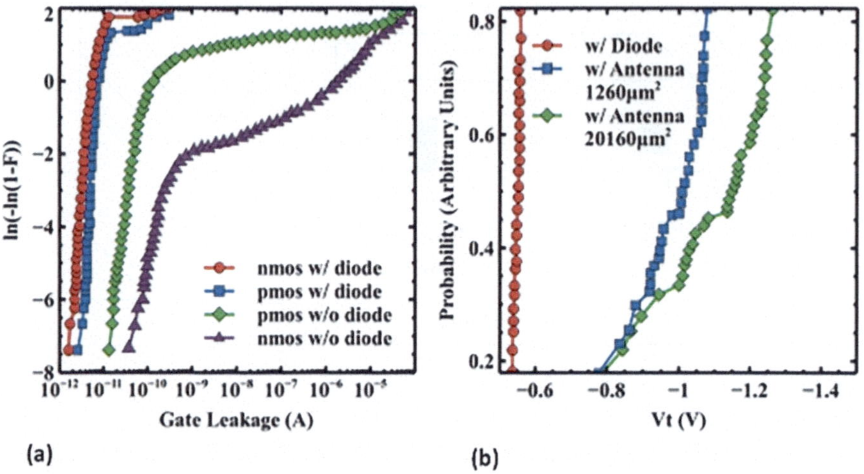

Fig. 3.4 Examples of the beneficial results of diode protection against plasma damage. (**a**) Probability plot of gate leakage for nmos and pmos transistors exposed to multiple metal etches with and without diode protection showing reduced gate leakage degradation with diodes. (**b**) Probability plot showing a pmos shifted threshold voltage and variability caused by a SiCN film deposition and a tight distribution with diode protection

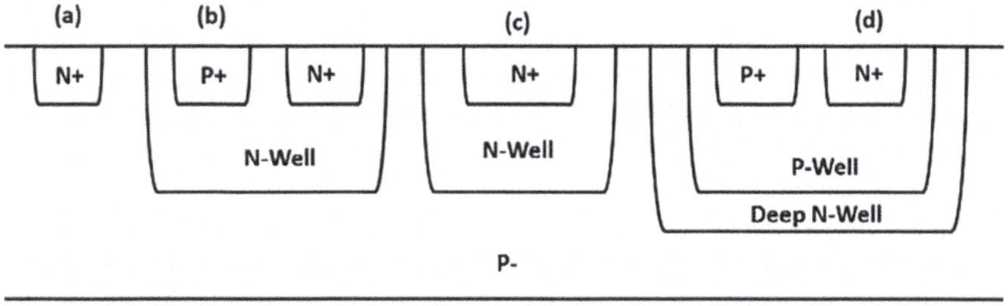

Fig. 3.5 Sketch showing different types of protection diodes in a CMOS process. (**a**) n+ to substrate diode. (**b**) p+ diode in an n-well. (**c**) n-well to substrate diode. (**d**) n+ diode in a triple well

A CMOS process has two complementary transistors: a NMOS transistor, usually placed in a p-type substrate, and a PMOS transistor in an n-well, requiring at least two types of diodes to protect all the transistors from plasma damage. Figure 3.5 shows a sketch of several types of ordinary diodes in a CMOS process. The characteristics of the different diodes affect the diodes' protection efficacy. Therefore, the protection diodes are selected based on the type of transistor to be protected and the electrical characteristics of the diode.

3.2 Diodes for Protection from Well Damage

Diodes can also protect from potential coupling-related well damage by preventing the well from charging during process damage or by draining the well charge, preventing damage to connected transistors. Figure 3.6 shows an example of methods to prevent damage from well coupling (Martin & Kamp, 2020). Figure 3.6a shows diode placement between the triple well and a downstream transistor, draining charge from the triple well and preventing damage to the downstream transistors. Figure 3.6b shows the placement of the diodes in the triple well, providing a path to drain the charge from the triple well, preventing damage. This protection strategy is more efficient as it uses one diode per well instead of one diode per transistor.

3.3 Transistor Clamp Protection

In addition to using a two-terminal diode as a clamp, CMOS transistors can be used as a process damage protection device, turning on during a plasma damage event, clamping the current, and protecting the circuit. These structures protect against well-related damage as they can shunt the damaging current to the ground, protecting circuits across power domains and outside the well. For example, Fig. 3.7a shows a NMOS transistor clamp used as a protection device for damage related to well charging for a CMOS process with a buried n-layer (Kuo et al., 2023). The clamp transistors are placed outside the wells and shunt the charging current to ground. Figure 3.7b shows a gate leakage probability plot of an input–output circuit connected to the damaging well. The gate leakage is substantially reduced with the addition of the NMOS clamp.

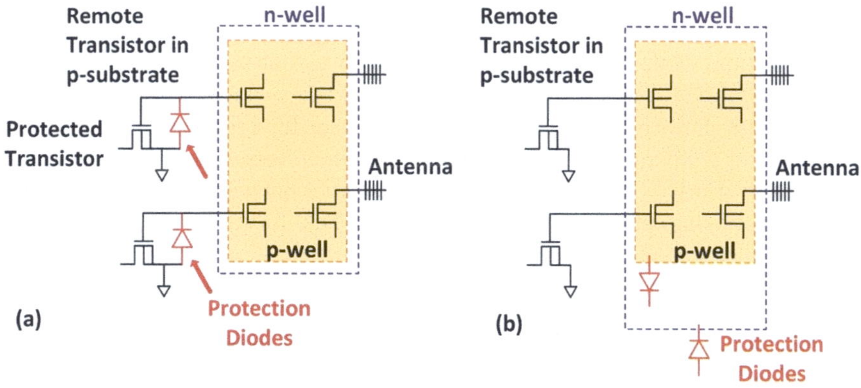

Fig. 3.6 Sketch showing diode protection strategies for protecting against well damage. (**a**) Diodes placed between the well and the downstream transistors. (**b**) Diodes placed in the charging well to drain charge from a triple well

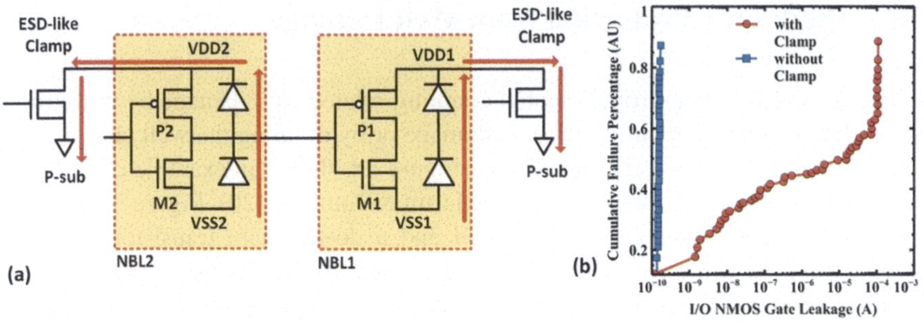

Fig. 3.7 Use of a CMOS transistor as a plasma damage protection device. (**a**) Circuit sketch showing nmos clamp transistors outside of a well with a n-type buried layer protecting circuits outside of the well. (**b**) Probability plot of the gate leakage current of an input-output circuit with and without the transistor clamp

3.4 Diode Operating Under Processing Conditions

CMOS processing temperatures can range from near room temperature to greater than 1000 °C for rapid thermal processing. Processing times can range from nanoseconds to milliseconds for rapid processing to hours for furnace processes. The protection diode characteristics change drastically throughout the range of CMOS processing conditions. During processing, the diode operating characteristics usually shift to a more favorable operating regime for protecting against plasma damage for processes with temperatures above room temperature.

The protection diodes during plasma damage operate under processing conditions that are vastly different from the normal chip operation. Plasma process conditions commonly reach temperatures of 400 °C during processing. In addition, wafers under processing conditions are often exposed to high-intensity, high-energy light (UV) that creates electron–hole pairs in the junction's depletion region, further increasing the current. The UV light may be blocked if the wafer region is protected by metal or UV-blocking films. The elevated temperature and light exposure cause the protection diode to become more conductive, enhancing plasma damage protection.

Figure 3.8 shows a simulated plot of diode leakage for a reverse-biased diode as a function of temperature (Wang, 2004, p. 79). The reverse-biased diode leakage increases by nearly 10 orders of magnitude between room temperature and 400 °C, greatly improving plasma damage protection. The forward bias current also increases at elevated temperatures (not shown).

Protection diodes exposed to light have increased electrical conduction if the photon energy is greater than the semiconductor band gap energy. For silicon, photoconduction begins at an energy of 1.1 eV or 1127 nm and varies as a function of wavelength. Plasma processes emit a large spectrum of wavelength as the ions and electrons recombine in the high-energy plasma, as described in Chap. 9, which provides an example of emitted wavelengths during plasma processing. The diode

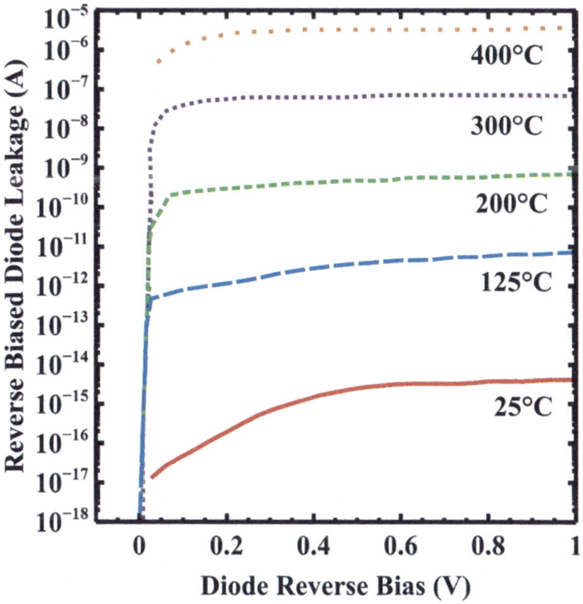

Fig. 3.8 Simulation of reverse diode leakage vs. temperature

photoconduction discharges the accumulating charge on the antenna, decreasing plasma damage.

Unfortunately, damage to dielectrics by high-energy photons can offset the beneficial photoinduced diode leakage. Photons with energy greater than the barrier energy of the metal–dielectric or dielectric–semiconductor barrier interfaces can inject photocurrents into the dielectric, causing electron damage and trapping (Snow et al., 1967). For the Si–SiO_2 barrier, this energy is about 3.2 eV. In addition, photons with energy greater than the band gap of the dielectric can create electron–hole pairs, creating severe damage to the dielectric. Figure 3.9 shows an example of a diode and gated diode with and without illumination, showing about three orders of magnitude increase in leakage with illumination (Krishnan & Amerasekera, 1998). The left side of the plot (0–2 V) shows gate-induced diode leakage due to band-to-band tunneling for the gated diode that is not affected by illumination, while leakage on the right side of the plot is dominated by junction leakage and is light-sensitive.

All the CMOS transistors leak at high rates when exposed to plasma processing, adding to the junction leakage. This leakage provides additional paths to drain charge from the antennas. Figure 3.10 shows an in situ transistor measurement during a downstream oxygen plasma resist strip process (Flounders et al., 1993). The wafer reaches about 300 °C, and there is a spike in the drain-to-source current.

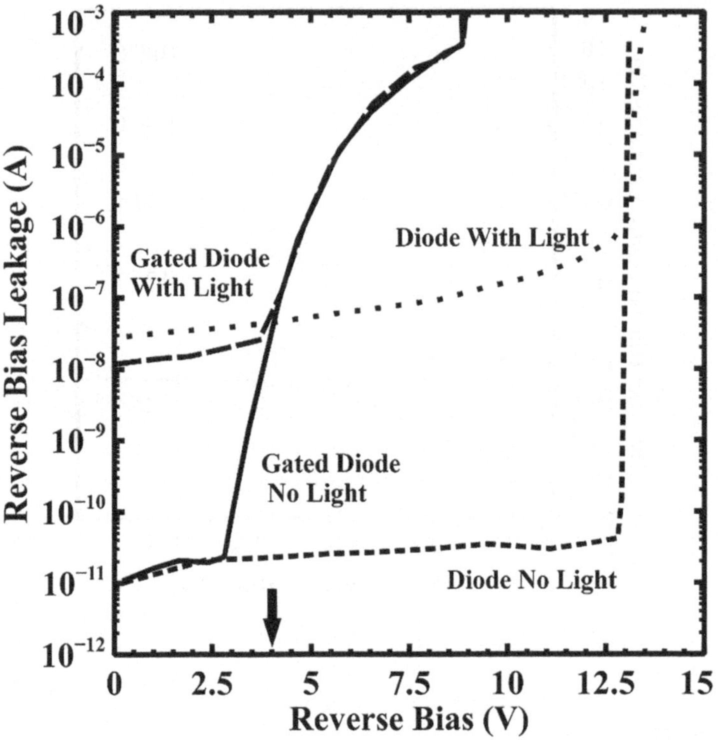

Fig. 3.9 Reverse-biased I-V plot of a diode and gated diode with and without illumination. The diode and gated diode currents are similar in the overlap region (left of the arrow). At higher biases, the leakage is higher due to gate-induced diode leakage. The diode area is 250 μm^2

3.5 Diode Protection for Silicon-on-Insulator Technology

Silicon-on-insulator (SOI) technology does not have the capability to add a diode in a straightforward manner like bulk CMOS (Fig. 3.3) because there is no substrate region accessible without a unique contact. Figure 3.11 shows a sketch of a lateral diode formed in an SOI process (Martin et al., 2021). As shown in the figure, two silicon regions are doped with opposite polarities, forming n+ and p+ regions. Additional contacts and interconnects are required, as shown in the figure. The lateral SOI diode requires increased area and interconnect, impacting the chip size if diodes are used frequently.

The special contacts shown in Fig. 3.12 can be added to an SOI chip, enabling a connection directly to the substrate (Pelella et al., 2007). The special contact is formed by etching a contact hole through the dielectric on top of the silicon regions (ILD), the isolation dielectric (STI), and the buried oxide. The contact etch is followed by implanting the p+ region and forming metallization. This p+ contact can drain the plasma charge from the local circuit after it is formed, reducing the

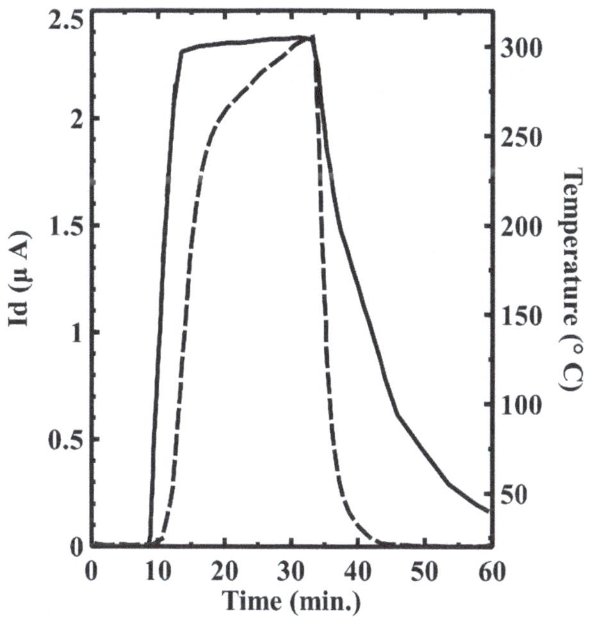

Fig. 3.10 Plot showing drain current during exposure to a plasma correlating to the wafer temperature

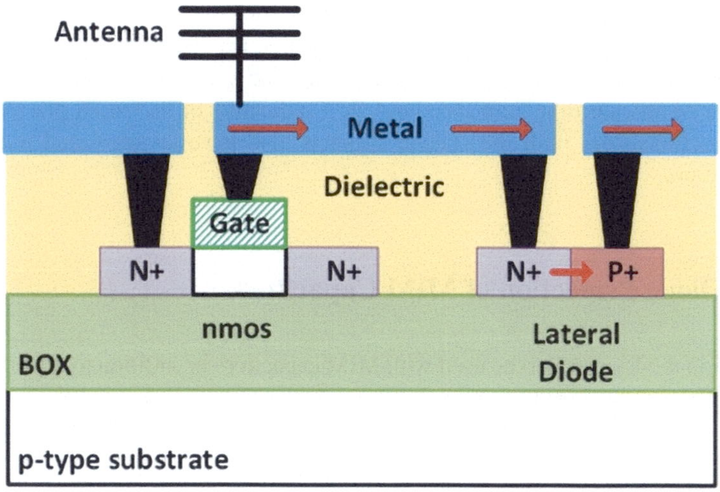

Fig. 3.11 Cross-section sketch of a silicon on insulator (SOI) CMOS wafer showing a lateral diode. The connection in red shows the conduction path of the diode

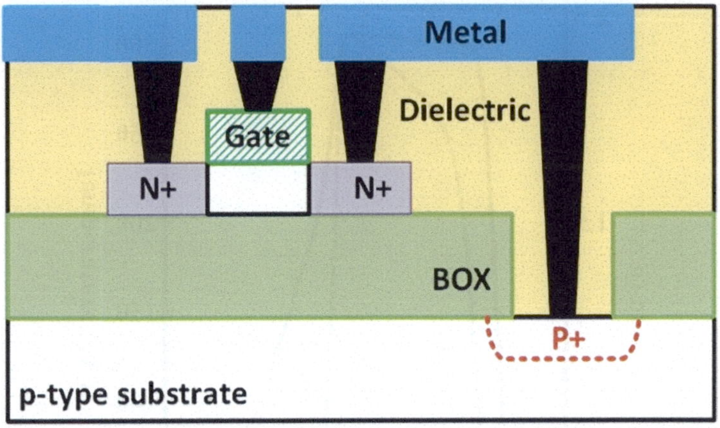

Fig. 3.12 SOI CMOS cross section showing a process damage protection device formed by a special p+ contact to the substrate through the buried oxide (BOX)

severity of the process damage. Alternatively, a n+ diode could be used. The reference claims these special contacts can be placed throughout the chip, serving as lighting rods and draining away plasma charge. This technique has the penalty of added process complexity and chip area.

The special p+ contact to the substrate or top-connected n+/p+ diode can actually increase process damage by providing a hard ground to the substrate for the floating body, increasing the stress across the gate dielectric during process damage. Figure 3.13 compares different antenna configurations for a sensitive SOI PMOS transistor (Martin et al., 2021). There is little difference between the reference and gate and drain antennas. However, adding a special substrate contact to the drain causes a significant time-zero threshold shift indicative of increased plasma damage. The transistors with the special contacts also have an increased reliability degradation.

3.6 Diode Protection of MIM Capacitors

Protection diodes can also be used with MIM capacitors in addition to other mitigation techniques, such as limiting antenna ratios. Figure 3.14 shows a process cross-section sketch of a MIM capacitor protected by a diode (Wang, 2004). The diode is connected to the MIM capacitor as early in the flow as possible, protecting the MIM structure from process damage at subsequent process steps. During operation, the diodes must be correctly biased to avoid forward biasing the diode and sinking large amounts of current into the substrate. Practically, diode protection of MIM capacitors in a logic chip may require increased area and interconnect. However, the diodes may be placed under the MIM capacitor, minimizing the area penalty.

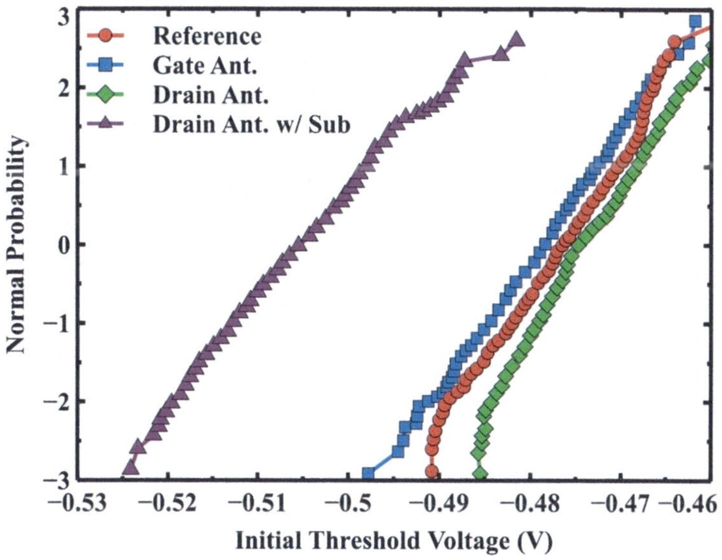

Fig. 3.13 Time-zero threshold voltage shifts for a SOI pmos transistor with various antenna configurations including, a no-diode reference, gate antenna, drain antennas, and drain antenna with a special substrate contact

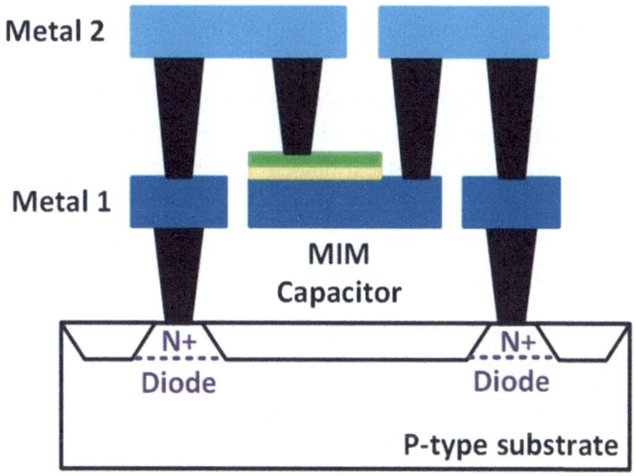

Fig. 3.14 Sketch showing diode protection for a MIM capacitor

3.7 Protection of Memory Wordlines

Wordlines in a memory chip are common gates extended across a memory array, connecting the gates of many individual memory cells to reduce the array area. As a result, the wordlines do not have any leakage path to drain the injected charge

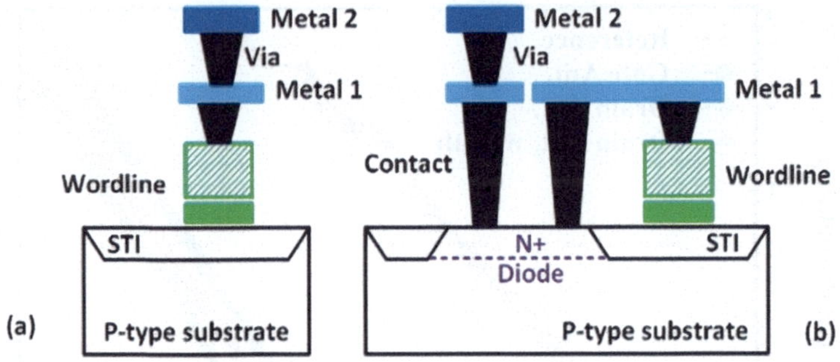

Fig. 3.15 Process cross-section sketch of a row driver to flash wordline connection. (**a**) Direct connection. (**b**) Diode protection by routing through a n+ jumper

from process damage. In addition, for nonvolatile memories, there may be two dielectrics in series consisting of the tunnel oxide and blocking or interpoly dielectric, further reducing the leakage and allowing increased voltage that can overstress the array transistors connected to the wordline.

The wordline connects the control gates of many NVM cells running across a memory array. NVM cells are very susceptible to plasma damage, as described in Chap. 8. A diode can protect the cells along a wordline by placing a diode at the location where the wordline connects to the row driver circuit. Figure 3.15 shows a process cross-section sketch for a wordline connection (Takebuchi et al., 1996). Figure 3.15a shows a direct connection without a diode. Figure 3.15b shows a wordline connection through a protection diode. In this example, the unprotected wordline had multiple defective NVM cells on each wordline. The protected wordline had no failures. Wordlines that are biased with both positive and negative voltages cannot use a simple diode and require a back-to-back diode structure.

However, adding wordline diodes to a high-density NVM chip may be impractical. Figure 3.16 shows a transmission electron microscopy (TEM) photograph of the wordline contacts for a high-density 3D NAND device (Choe, 2018). The wordline contacts are labeled 3 in TEM. Adding diodes to a very dense array of connections would significantly increase the chip size. Therefore, the process must be highly optimized for high-density arrays because diode protection is not viable. The contact depth increases from left to right in the TEM image. The leftmost contact receives more plasma damage during the time necessary to clear the dielectric from the rightmost contact during the contact etch. Additionally, 3D NAND wordline structures are susceptible to damage, as the varying dielectric distance to the wordline shown in Fig. 3.16 results in a large overetch for the higher wordlines. Process damage to wordline structures typically manifests as excess electrical damage that varies with the wordline location.

Other memories, including DRAMs, can also experience process damage through the wordline interconnects. For DRAMs, the most common symptoms of wordline-related damage are single-atom defects, as described in Chap. 8.

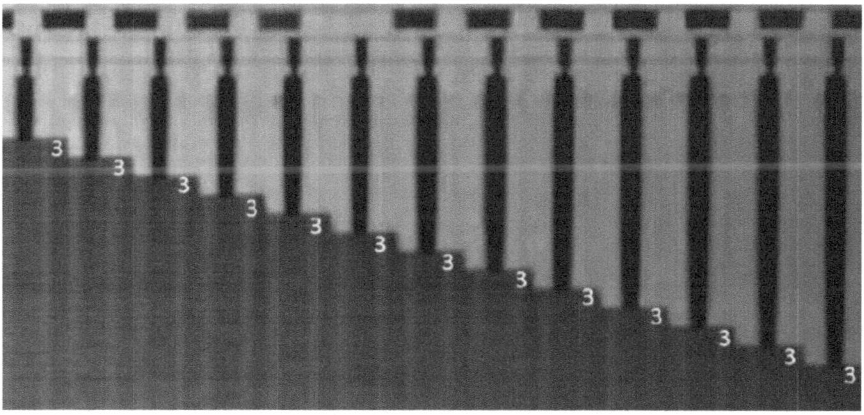

Fig. 3.16 TEM cross-section showing high density wordline contacts typical of a 3D NAND chip. (Source Techinsights)

References

Choe, J. (2018). Memory process, design and architecture. In *IEEE IEDM*.

Flounders, A. W., et al. (1993). In situ monitoring of radiation damage to thermal silicon dioxide films exposed to downstream oxygen plasmas. *Journal of the Electrochemical Society, 140*(5), 1414–1424. https://doi.org/10.1149/1.2221571

Hiura, Y., et al. (2020). Study of lower voltage protection against plasma process induced damage by quantitative prediction technique. In *International Reliability Physics Symposium*. https://doi.org/10.1109/IRPS45951.2020.9128947

Krishnan K., & Amerasekera, A. (1998). Antenna protection strategy for ultra-thin gate MOSFETs. In *IEEE International Reliability Physics Symposium*. https://doi.org/10.1109/RELPHY.1998.670660

Kuo, H.-Y., et al. (2023). Protection schemes for plasma induced damage from well-side antennas. In *International Reliability Physics Symposium*. https://doi.org/10.1109/IRPS48203.2023.10117947

Martin, A., & Kamp, A. (2020). A new implementation approach for reliability design rules against plasma induced charging damage from well configurations of complex ICs. In *IEEE International Reliability Physics Symposium*. https://doi.org/10.1109/IRPS45951.2020.9128866

Martin, A., Kamp, A., & Berger, J. (2021). Reliability characterization of plasma induced charging damage (PID) for MOS transistors of a silicon on insulator (SOI) process using product relevant test structures and fWLR methods. In *2021 IEEE International Integrated Reliability Workshop (IIRW)*. https://doi.org/10.1109/IIRW53245.2021.9635611

Matsunaga, N., Yamaguchi, H., & Shibata, H. (2007). Spreading antenna effect of PID in damascene interconnect process. *IEEE Transactions on Electron Devices, 54*(6), 1486–1491. https://doi.org/10.1109/ted.2007.896360

Pelella, M. M., et al. (2007). A plasma damage mitigation concept for SOI technologies: Lightning rods. In *IEEE International SOI Conference*. https://doi.org/10.1109/SOI.2007.4357860

Snow, E. H., Grove, A. S., & Fitzgerald, D. J. (1967). Effects of ionizing radiation on oxidized silicon surfaces and planar devices. *Proceedings of the IEEE, 55*(7), 1168–1185. https://doi.org/10.1109/PROC.1967.5776

Sommer, S. P., Paschen, U., Figge, M., & Vogt, H. (2011). Light switched plasma charging protection device for high-field characterization and flash memory protection. *IEEE Transactions on Device and Materials Reliability, 11*(1), 81–85. https://doi.org/10.1109/tdmr.2010.2088399

Takebuchi, M., et al. (1996). New interconnect plasma induced damage analyzed by flash memory cell array. In *IEEE IEDM*. https://doi.org/10.1109/IEDM.1996.553150

Wang, Z. (2004). *Detection of and protection against plasma charging damage in modern IC technology*. Twente University.

Zhao, J., et al. (1996). Investigation of charging damage induced mismatch for submicron mixed-signal technology. In *IEEE IRPS*. https://doi.org/10.1109/RELPHY.1996.492058

Chapter 4
Signatures of Process Damage

4.1 Introduction

Plasma and process damage can have many different signatures that are observable in multiple ways in the massive data taken during a CMOS manufacturing process. Some signals may be observed in global signals, such as chip yield, burn-in yield, and qualification failure. Parametric data is often the primary means of detecting damage problems. The damage parametric data can take multiple statistical forms, including a shift in the mean value, the development of a defect subpopulation showing as kurtosis, creating a defect tail, or a significant increase in the variation of a parameter. Plasma and process damage can show up in statistical process control data, such as parametric trend charts correlating with tool excursions, preventative maintenance, etc. Damage may be cumulative as a wafer is processed through the CMOS process flow, with multiple steps damaging the chip, each damaging step adding to the degradation of the chip. This chapter covers these topics, showing examples of actual data.

4.2 Signatures of Process Damage

Plasma damage can be challenging to diagnose, with varying symptoms and many different electrical effects. Plasma damage can show random or systematic patterns across a chip, region, or wafer. Virtually, any electrical chip problem may be caused or aggravated by process damage. Plasma damage can worsen preexisting electrical problems and may worsen electrical problems due to unrelated causes such as non-uniformity. The electrical symptoms of plasma damage can be consistent or widely vary depending on the specific electrical parameters affecting the chip operation.

K. Prall, *CMOS Plasma and Process Damage*,
https://doi.org/10.1007/978-3-031-89029-1_4

Some common symptoms of plasma damage occur across the full spectrum of potential plasma damage problems. Table 4.1 shows a list of these common symptoms. These mostly unique signatures allow plasma damage to be diagnosed correctly, separating plasma damage from other sources of process problems.

The defining signature for plasma damage is the correlation with antenna structures, which is not observed for other causes of process damage. Circuits and layouts with antennas typically fail first. This signature is unique for plasma damage, isolating the cause of a process problem. The improvement in plasma damage with a diode connected to the gate is also a unique signature of plasma damage.

Plasma damage can cause a specific spatial signature ranging from everything being dead to only a single device on every chip being dead. Unfortunately, baffling spatial signatures are a common theme for plasma damage, where the damage occurs on specific devices, leaving similar devices undamaged as if the defect can differentiate between wafer locations. As a result, the ability of plasma damage to appear intelligent can be befuddling.

Plasma damage usually correlates with a specific process tool or process step, showing a good correlation with the tool or recipe parameters that change the charge fluence or energy of charge carriers reaching the wafer.

The final thermal processes and alloys can partially recover plasma damage, resulting in partial healing with time after damage. High-energy UV light can also partially reverse plasma damage.

Table 4.1 Common symptoms of plasma damage

Typically, it is much worse on structures with antennas
Structures with diodes on the gate may have less damage
Damage occurs on specific wafer locations, design rules, and circuits. Slightly different structures are not affected. It may affect only one of several identical structures on the chip
Damage varies with process conditions, especially parameters related to power, bias, throughput, voltage, etc.
The problem correlates with specific processing tools
Sensitive to final alloy conditions with more time, temperature, and hydrogen typically improving symptoms
Recovery with time at room and elevated temperatures may occur
It may improve with post-testing thermal bakes (200–500 °C)
It may improve with high-temperature processing (500+ °C)
Exposure to UV light may improve symptoms
Parametric shifts between inline and final testing may be observed
Damage can be detected with parametric test structures. It can be worse on parametric test structures depending on design rules. Parametric data can be bad, but product results good
Only one or some of identical structures may be affected
Damage is detected with a diagnostic stress. See Chap. 13
Across wafer variation that may or may not correlate with wafer physical characteristics
Reliability degradation in structures with antennas may occur, while structures without antennas are better

Plasma damage is often first detected in parametric testing, measured on test structures placed in the scribeline, showing poor electrical behavior. Poor parametric electrical results can sometimes occur on a wafer with reasonable product chip results due to the increased sensitivity of parametric structures.

Finally, plasma damage can degrade the chip's reliability, creating latent defects that are difficult to detect without extensive reliability testing. In addition, different designs fabricated in the same process can have different reliability characteristics due to the interaction of the design and layout with process damage. The reliability and qualification of different designs in the same process can be affected by design-specific plasma damage.

4.3 Statistical Signature of Process Damage

Plasma damage has a myriad number of potential symptoms that are manifested in many different statistical distributions of electrical or physical parameters. Process damage can modify a normal electrical distribution by causing a shift in the mean or widening the distribution, increasing the standard deviation, and degrading the process control capability.

Process damage can also create non-normal distributions where the parameter distribution is a mixture of intrinsic and defective extrinsic data. Additionally, the distribution can become bimodal, as some devices are degraded or destroyed while others are undamaged.

Figure 4.1 shows some examples of finfet parameter distributions created by plasma damage (Hiblot et al., 2019). Figure 4.1a shows that plasma damage created an additional unique distribution with reduced gate leakage current (due to trapping). Figure 4.1b shows a plasma damage distribution with a long extrinsic tail of high gate leakage. Figure 4.1c shows plasma damage, shifting 30% of the measurements, leaving the rest of the data unmodified. Figure 4.1d shows that the mean and standard deviation are increased by plasma damage, while the distribution remains normal.

4.4 Cumulative Damage

Process damage can occur at many steps in a CMOS process. Similar or identical processes are often repeated many times in a CMOS process flow, creating the potential for repeated process damage. Each time a wafer is processed through a repeated step, process damage can repeatedly occur, with device degradation worsening with each additional exposure to the degrading process step. For example, a CMOS process usually has many layers of metal. The processing for each layer of metal is typically identical or very similar. Figure 4.2 shows a sketch of example

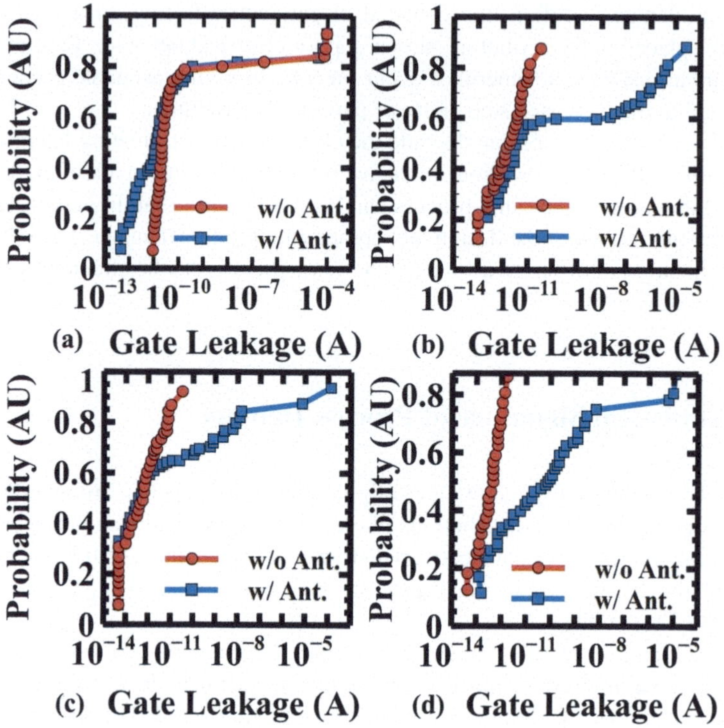

Fig. 4.1 Examples of different finfet gate leakage statistical distributions caused by plasma damage with an antenna compared to distributions without an antenna and without plasma damage. (**a**) Bimodal distribution after damage. (**b**) Defect tail (kurtosis) after damage. (**c**) Defect tail widening distribution after damage. (**d**) Large increase in mean and variation after damage

structures constructed with repeated process steps and can be subjected to repeated process damage (Martin, 2020).

Accumulated damage can occur as a summation of processing damage that a CMOS wafer incurs through multiple process steps. The accumulation of process damage can occur as the wafer is processed through many different equipment types and process recipes. The following chapters describe the most common steps of the process that often cause plasma damage. The accumulated damage can be quantified by antenna test structures that combine antennas on different layers into one structure, as shown in Fig. 4.2. The damage is measured inline after each antenna has completed processing throughout the process flow. Performing parametric measurements inline after specific process steps is crucial to isolating the exact process step causing damage. Figure 4.3 shows an example of the cumulative damage caused by processing through multiple process modules, including a conductive gate metal and five interconnect layers (Simon et al., 2000).

Process damage from multiple damaging steps can repeatedly damage CMOS structures, causing increasing damage as the wafer progresses through the entire

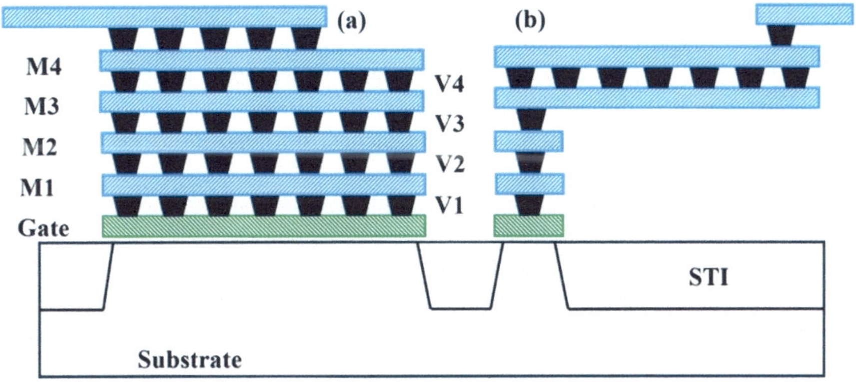

Fig. 4.2 Examples of CMOS structures which can accumulate process damage. (**a**) Repeated metal (M1–M4) and via (V1–V4) antenna structures on multiple layers that can cause cumulative damage. (**b**) Isolated antenna limited to one process region for detection of plasma damage in a limited process region

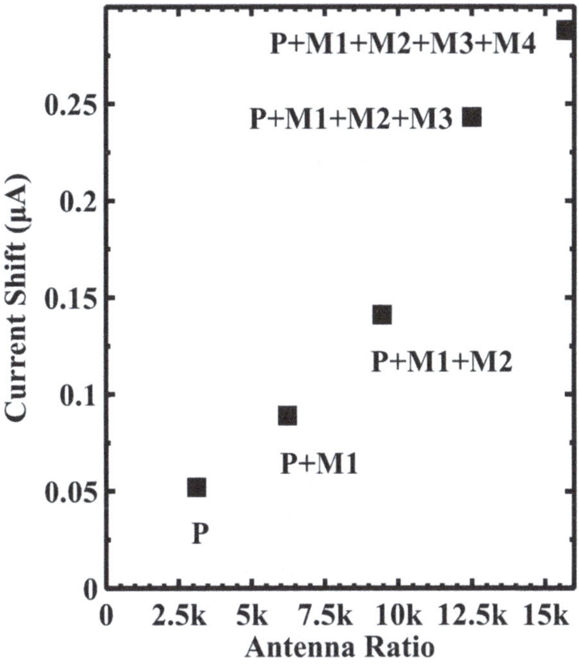

Fig. 4.3 Example of accumulated damage resulting in transistor drive current shift on a test structure with multiple antennas on four conductors in a CMOS process flow. The antennas are Gate Metal (P), Metal 1 (M1), Metal 2 (M2), Metal 3 (M3), and Metal 4 (M4)

process flow. Some regular process steps, such as a thermal treatment, may partially recover the damage, but the subsequent process damage steps degrade the device further. The damage tends to be additive throughout the process flow. Process steps tend to be repeated throughout the process flow, creating an environment for repeated damage. Steps such as resist ashing, etches, film depositions, etc., are often repeated.

Figure 4.4 shows another example of cumulative process damage causing gate leakage (Wang et al., 2001). Identical structures are compared with one antenna at each process module, and multiple combined antennas for all modules are monitored. This comparison allows the examination of cumulative damage effects. The structures with a single antenna isolate the process damage at one process region, with the individual failure rate ranging from 0% to 20%. The failure rate for structures with multiple antennas on multiple layers ranges from 18% to 85%. The cumulative antenna structures suffer from process damage at multiple steps, resulting in ever-increasing damage as the wafer is processed.

Figure 4.5 shows a sketch of a plasma damage test structure created with a finfet (Wu et al., 2015). A finfet gate with no contact serves as a floating gate. The floating gate is capacitively coupled through contacts adjacent to the floating gate. The

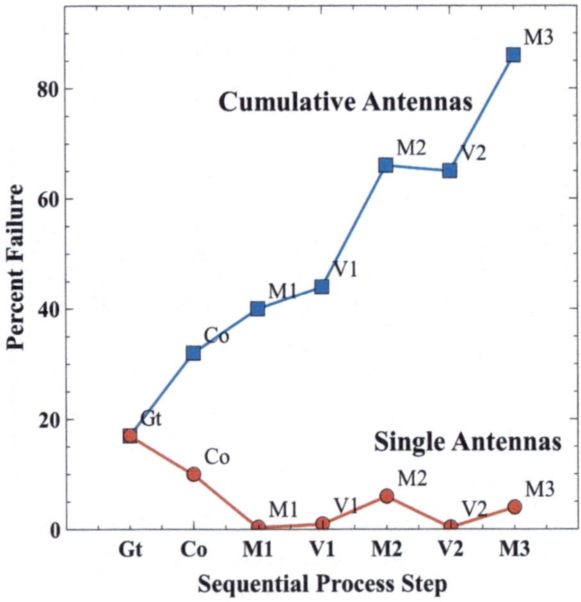

Fig. 4.4 Plot showing the percentage failure for gate leakage current failures with antenna structures at multiple levels in a CMOS process. The bottom curve shows the failure rate for an antenna only at one level. The top curve shows the failure rate for combined antennas on all levels monitored with antennas. The process steps shown are Gt-gate etch, Co-contact etch, M1-metal 1 etch, V1-via 1 etch, M2-metal 2 etch, V2-via 2 etch, M3-metal 3 etch

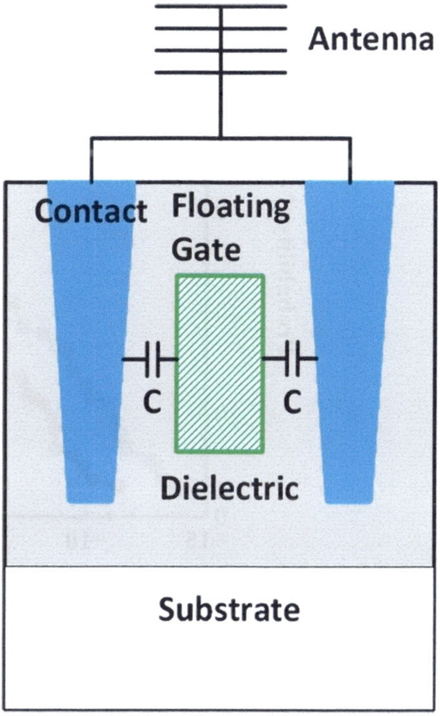

Fig. 4.5 Cross-section sketch showing a finfet-based floating gate structure used for plasma damage measurements. The contacts next to the floating gate capacitively couple to the floating gate, creating a floating gate device. Antennas are contacted to the contacts. Plasma damage stores charge on the floating gate. The charge is read by using the antenna as the gate and reading the drain current with a finfet source-drain contact (not shown)

contacts are connected to various types of antennas, allowing for the thorough characterization of plasma damage in the CMOS finfet process.

Figure 4.6 shows another example of cumulative damage across multiple process steps measured with the finfet floating gate structure (Wu et al., 2015). In this case, the variability of the threshold voltage V_t of a specialized finfet floating gate test structure is degraded by multiple process steps, causing process damage. V_t in this example shifts in both positive and negative directions, indicating that the antennas on different wafer regions collect electrons and ions or holes, respectively. The positive and negative differences change as a function of the metal layer, with less ion charging at higher metal layers.

4.5 Process Damage by Excursions

Semiconductor manufacturing is heavily controlled by statistical process control, electrical monitoring of tool performance, and other techniques. Process excursions are typically quickly detected and remedied with a limited loss of wafers inline in the manufacturing process. Process damage can be subtle as it may not be detected inline during processing or after the wafer is completed during electrical testing at

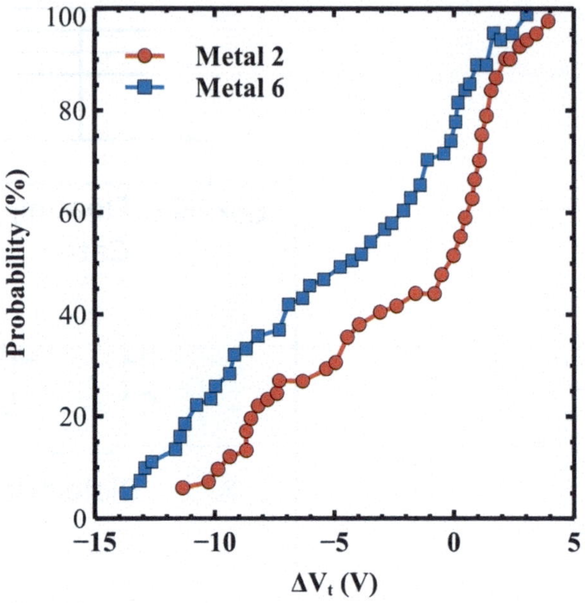

Fig. 4.6 Probability plot showing V_t distribution for a finfet flash process damage test structure showing cumulative damage throughout the metallization section of a CMOS process from metal 2 through metal 6. The damage is measured at metal 2 and again at metal 6. Both positive and negative threshold voltage shifts indicate that the plasma is not uniform across the wafer. The cumulative damage shifts the V_t more negative as the wafer processing progresses

time zero. Latent damage can be undetected without reliability testing. Process damage excursions can create an elevated risk of shipping unreliable products when an excursion occurs. Figure 4.7 shows an example of a process damage reliability excursion caused by a deviation of a dielectric deposition chamber (Martin et al., 2009). The excursion was undetectable by parametric and electrical testing. The excursion was detected by performing a two second revealing gate current stress of 2 C/cm^2 (see Chap. 13). The excursion was detected by measuring the threshold voltage before and after the current stress was applied and analyzing the difference. This example highlights the need for fast production reliability monitoring of a production process.

Also, plasma damage can vary throughout the maintenance cycles of fabrication equipment as the system undergoes periodic cleaning and part replacement. For example, Fig. 4.8 shows a factor of 2 variation in plasma damage before and after periodic maintenance of a high-current ion implanter as measured by a specialized test wafer using nonvolatile memory (CHARM) (Gonzalez et al., 1988; Siu et al., 2002).

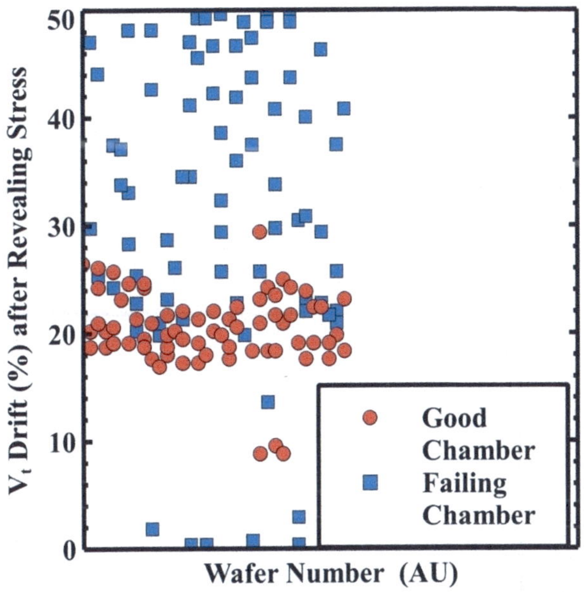

Fig. 4.7 Plot of threshold voltage drift versus wafer number comparing a good chamber and a failing chamber of backend of line dielectric deposition chambers after a constant current reliability stress. The problem was not detected with normal parametric data collection and required a revealing stress to detect it (see Chap. 13)

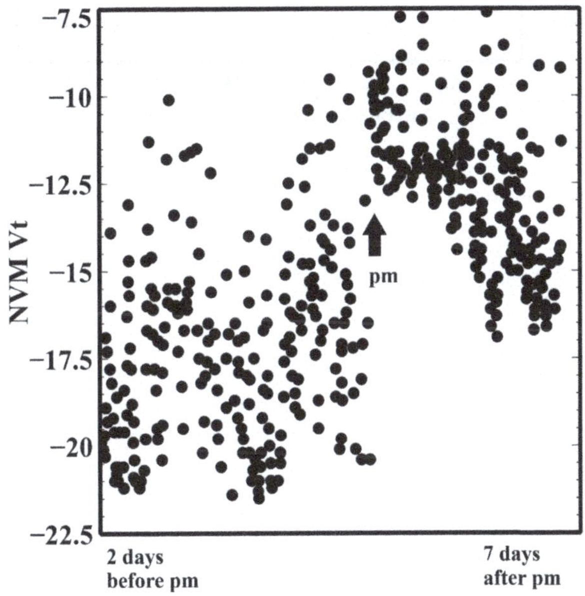

Fig. 4.8 Plot showing a large shift in plasma damage measured as a threshold voltage shift before and after preventative maintenance (pm) for an ion implantation system measured with a special nonvolatile test wafer

4.6 Burn-In Yield Correlation

Plasma and process damage often causes reliability problems, as described in Chap. 6. The damage observed with time-zero parametric yield can sometimes be correlated with chip-level reliability testing. For example, Fig. 4.9 shows a correlation between a parametric antenna test structure and product-level burn-in yield across multiple manufacturing sites (Mason et al., 2000).

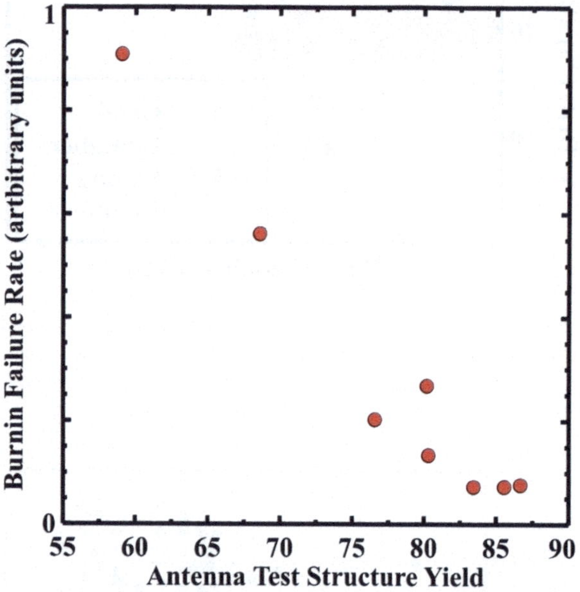

Fig. 4.9 Plot showing burn-in failure rate (arbitrary units) plotted against the yield of a parametric antenna structures for multiple fabrication facilities

References

Gonzalez, H., et al. (1988). Monitoring in high current ion implanters yields optimum preventive maintenance schedules and procedures. In *Integrated Reliability Workshop*. https://doi.org/10.1109/IIT.1999.812176

Hiblot, G., et al. (2019). Comparative analysis of the degradation mechanisms in logic and I/O finfet devices induced by plasma damage. In *IEEE IRPS*. https://doi.org/10.1109/IRPS.2019.8720525

Martin, A. (2020). Plasma induced charging damage: From appropriate MOS test structures to antenna design rules, a comprehensive process qualification procedure. In *2020 IEEE International Integrated Reliability Workshop (IIRW)*. https://doi.org/10.1109/IIRW49815.2020.9312854

Martin, A., Bukethal, C., & Ryden, K. H. (2009). Fast wafer level reliability monitoring: Quantification of plasma-induced damage detected on productive hardware. *IEEE Transactions on Device and Materials Reliability, 9*(2), 135–144. https://doi.org/10.1109/tdmr.2009.2020601

Mason, P. W., et al. (2000). Relationship between yield and reliability impact of plasma damage to gate oxide. In *International Symposium on Plasma Process-Induced Damage*. https://doi.org/10.1109/PPID.2000.870560.

Simon, P., et al. (2000). Identification of plasma-induced damage conditions in VLSI designs. *IEEE Transactions on Semiconductor Manufacturing, 13*(2), 136–144. https://doi.org/10.1109/66.843628

Siu, S., et al. (2002). Utility of CHARM-2 in diagnosing sources of plasma charging damage in high density etchers and in assisting hardware development. In *7th International Symposium on Plasma and Process-Induced Damage*. https://doi.org/10.1109/PPID.2002.1042614

Wang, Z., et al. (2001). *Plasma process-induced latent damage on gate oxide—Demonstrated by single-layer and multi-layer antenna structures*. IPFA. https://doi.org/10.1109/IPFA.2001.941490

Wu, C. H., et al. (2015). Mapping of wafer-level plasma induced charge contour by novel on-chip in-situ recorders in advanced FinFET technologies. In *IEEE IEDM*. https://doi.org/10.1109/IEDM.2015.7409644

Chapter 5
Electrical Signatures of Process Damage

5.1 Introduction

Process damage can wreak havoc with CMOS chips, causing a wide range of electrical symptoms and damage. The damage can create time-zero electrical problems, which can be observed immediately after wafer processing, resulting in probe and parametric electrical changes that cause multiple problems, including yield loss, margin loss, functionality, etc. Process damage can also cause chip reliability problems by partially wounding the CMOS devices, resulting in reduced chip lifetime and failure in the field (see Chaps. 4 and 6).

5.2 Similarity of Process Damage and Electrical Stress

The electrical effects of process damage tend to be remarkably similar to damage caused by electrical gate stress by applying voltage or current stress to a CMOS structure. This similarity provides some insights into the actual process damage, which similarly stresses the devices. Electrical stress results can be used as a proxy for modeling process damage in simulations. Figure 5.1 compares the interface state density D_{it} caused by plasma exposure to a constant current electrical stress (Wang, 2004). The constant current stress conditions are selected to match the plasma damage conditions. This comparison provides some insights into the harmful conditions caused by plasma damage during processing. The concept shown here is that the process damage creates similar voltage and current conditions on the surface of the wafer as the electrical stress does, allowing an estimation of the processing damage conditions that cannot be directly measured in situ during processing.

K. Prall, *CMOS Plasma and Process Damage*,
https://doi.org/10.1007/978-3-031-89029-1_5

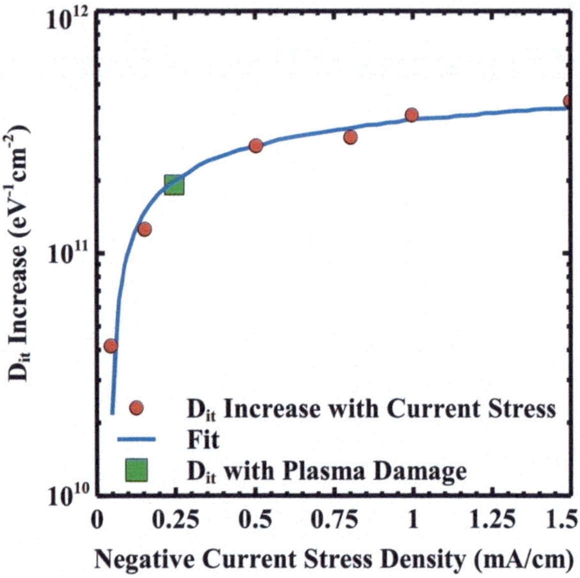

Fig. 5.1 Example plot showing similar degradation of plasma damage and a constant current electrical stress

5.3 Similarity of Process Damage and Radiation Damage

Some of the process damage electrical characteristics described in this chapter are very similar to those caused by exposing CMOS chips to ionizing radiation such as cosmic rays, neutrons, X-rays, etc. Radiation hardness is an essential property of CMOS devices, particularly in space and nuclear environments. Additionally, commercial CMOS chips are assessed for radiation hardness to ensure problem-free operation in everyday environments. The following sections describe common process damage electrical effects caused by process damage. Some of the process damage electrical effects described in Table 5.1 and Chap. 6 can be caused by exposure to radiation, with the possible exception of structural damage caused by electrostatic discharge, metallization degradation, etc. The electrical symptoms of process damage and radiation damage are very similar because the fundamental damage mechanisms are similar, caused by the creation of traps, interface states, etc. Additionally, process damage can degrade the radiation hardness of CMOS technology, as described in Sect. 5.9.6.

Table 5.1 Common electrical signatures of time-zero process damage

Damage signature	Affected devices	Chapter/ Section
Electrostatic discharge physical damage	All	Section 5.4.1
Dielectric-related parameters		
Dielectric breakdown	All fets including transistors, MOS capacitors, isolation, nonvolatile memories, DRAM, imagers, MIM capacitors	Section 5.4.3
Increased dielectric trapping	All fets, including transistors, MOS capacitors, isolation, nonvolatile memories, DRAM, and imagers	Section 5.6
Change in capacitance-voltage characteristics	All fets including transistors, MOS capacitors, isolation, nonvolatile memories, DRAM, imagers, MIM capacitors	Section 5.4.2 Section 5.9.2
Increased interface states	All fets including transistors, MOS capacitors, isolation, nonvolatile memories, DRAM, imagers, MIM capacitors	Section 5.5
Gate leakage-stress-induced leakage current (SILC)	All fets including transistors, MOS capacitors, isolation, nonvolatile memories, DRAM, imagers, MIM capacitors	Chapter 7
Degraded low-k dielectrics	Low-k dielectrics	Chapter 8
Metal-insulator-metal (MIM) capacitor breakdown and degradation	MIM capacitor	Chapter 8
Transistor-related parameters		
Threshold voltage shifts	All fets, including transistors, MOS capacitors, isolation, nonvolatile memories, DRAM, and imagers	Section 5.6
Subthreshold slope degradation Subthreshold conduction degradation	All fets, including transistors, MOS capacitors, isolation, nonvolatile memories, DRAM, and imagers	Section 5.7.2
Transconductance (g_m) degradation Transistor drive degradation	All fets, including transistors, MOS capacitors, nonvolatile memories, DRAM, and imagers	Section 5.5.1
Source-drain resistance degradation Channel length shift	All fets, including transistors, MOS capacitors, nonvolatile memories, DRAM, and imagers	Section 5.7.4
Gate-induced diode leakage	All fets, including transistors, MOS capacitors, isolation, nonvolatile memories, DRAM, imagers, MIM capacitors	Section 5.7.3
Increase in variation and mismatch	All fets including transistors, MOS capacitors, isolation, nonvolatile memories, DRAM, imagers, MIM capacitors	Section 5.8.1
Increased noise generation	All fets including transistors, MOS capacitors, isolation, nonvolatile memories, DRAM, imagers, MIM capacitors	Section 5.8.2

(continued)

Table 5.1 (continued)

Damage signature	Affected devices	Chapter/ Section
Junction leakage	All fets, including transistors, MOS capacitors, isolation, nonvolatile memories, DRAM, and imagers	Section 5.7.3
Source-drain leakage Punch-through Isolation leakage	All fets, including transistors, MOS capacitors, isolation, nonvolatile memories, DRAM, and imagers	Section 5.9.4
Giant random telegraph signal	All fets, including transistors, MOS capacitors, isolation, nonvolatile memories, DRAM, and imagers	Chapters 7 and 8
Ring-oscillator speed degradation	All CMOS logic fets	Section 5.8.3
Contact resistance degradation	All fets, including transistors, MOS capacitors, isolation, nonvolatile memories, DRAM, and imagers	Section 5.9.1
Nonvolatile memory-related parameters		
Cycling-related stress-induced leakage current	Nonvolatile memories	Chapter 8
Read disturb	Nonvolatile memories	Chapter 8
Program disturb	Nonvolatile memories	Chapter 8
Overerase	Nonvolatile memories	Chapter 8
Cycling window closure or shift	Nonvolatile memories	Chapter 8
Endurance	Nonvolatile memories	Chapter 8
Data retention (high and low temperatures)	Nonvolatile memories	Chapter 8
Charge gain	Nonvolatile memories	Chapter 8
Over programming	Nonvolatile memories	Chapter 8
Erratic programming	Nonvolatile memories	Chapter 8
Parasitic charge trapping	Nonvolatile memories	Chapter 8
DRAM-related parameters		
Degraded refresh	DRAM	Chapter 8
Row hammer	DRAM	Chapter 8
Variable retention time	DRAM	Chapter 8
CMOS imager and CCD-related parameters		
Increased dark current	CCDs, imagers	Chapter 8
Increased white current and defects	CCDs, imagers	Chapter 8
Other parameters		
Decreased radiation hardness	All	Section 5.9.6
Degraded electrostatic discharge protection	All	Section 5.9.8

5.4 Time-Zero Process Damage Electrical Effects

Process damage can be observed through the degradation of many electrical characteristics. The various structures commonly damaged by process damage can have slightly different electrical signatures, as shown in Table 5.1. Any device can simultaneously have multiple damage signatures consistent with the typical device abnormalities possible for that device. The potentially affected devices are shown in the second column. This table does not include higher-order metrics such as yield, chip failure rate, etc., and shows only degradation modes that can be measured parametrically on test structures. Latent reliability process damage problems are described in Chap. 6.

The process damage correlation with the structure layout consisting of different antenna ratios is considered the definitive proof of plasma damage. This antenna correlation with electrical effects is often used to isolate plasma damage from other types of process damage.

5.4.1 Electrical Static Discharge Process Damage (ESD)

CMOS processes and equipment have the potential to generate an electrical arc or electrostatic discharge, which can leave visible physical damage on the wafer, as shown in Fig. 5.2 (Cantin et al., 1999; Lee et al., 2008; Ma, 2004). The amount of damage is proportional to the energy discharged in the damaging process. The energy discharged in the damaging process typically causes local heating sufficient to melt the materials on the surface of the wafer. The discharge often results in lightning-like marks on the surface of the wafer. An arc typically destroys an affected CMOS chip but can possibly wound nearby chips, creating latent damage. The defects can be random across the chip and wafer or have a specific pattern occurring on only certain locations and structures based on the chip and the local layout of the wafer.

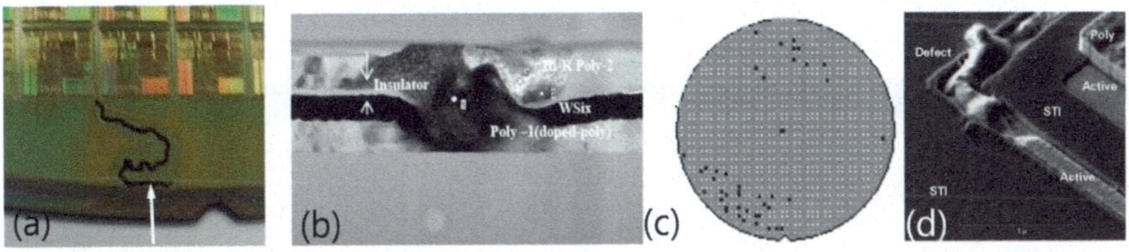

Fig. 5.2 Example of electrostatic or arcing damage to a wafer. (**a**) Arc mark from a chip to the edge of a wafer. (**b**) Cross-section TEM of metal and silicon melted together due to electrostatic discharge in a scrubber clean process. (**c**) Wafer map of defects from defect in (**d**). (**d**) Defect showing melted silicon active area due to ion implantation damage

Electrostatic discharge can leave unique damage signatures in the wafer, giving convincing evidence that an electrostatic discharge occurred. Commonly, tendrils can be observed in the electrostatically damaged region, such as the defect shown in Fig. 5.2a. This tendril type is commonly observed even in lighting damage, as shown in Fig. 5.3.

Additionally, structural damage can occur during an electrostatic event, as shown in the more extreme examples in Fig. 5.2. For small amounts of damage, it is often difficult to differentiate between ESD damage and other origins of damage, such as dielectric breakdown. The physical destruction is a signature of ESD damage, while dielectric breakdown can have many causes.

5.4.2 Dielectric Damage Caused by Process Damage

The process damage charge flowing through gate oxide and other dielectrics can damage the dielectric's atomic structure, creating broken bonds and atomic-level locations that trap electrons or holes. These atomic-level defects can affect many CMOS transistor device parameters described in this section. Numerous methods exist to characterize and quantify these defects, including capacitance-voltage techniques, charge pumping, etc. (Schroder, 2006). Parametric structures with antennas can be characterized with these techniques to measure these defects with and without process damage.

Fig. 5.3 Photograph of a lightning strike on a sidewalk showing tendrils. (Source unknown)

5.4.3 Dielectric Breakdown

Dielectric breakdown is the catastrophic failure of a dielectric caused by excessive voltage or current, exacerbated by thermal stress. Dielectric breakdown is commonly observed in process damage situations. Dielectric breakdown can leave catastrophic physical damage similar to arcing, or it can leave subtle minor damage that is difficult to detect or no observable defect at all (Selvarajoo et al., 2005; Tang et al., 2004). The dielectric damage is also proportional to the amount of energy dissipated in the damage event. Dielectric damage and physical arcing damage can be closely related or caused by the same process damage event. Figure 5.4 shows TEM images of various types of structural damage due to dielectric breakdown (Lombardo et al., 2005; Raghavan, 2009; Rider, 2021; Shubhakar et al., 2015; Tang et al., 2004). The temperature increases as more energy is dissipated during the breakdown event, creating more damage. The rough temperature of the event can be estimated depending on what materials are melted or mixed. Low-energy dielectric breakdown can result in minute amounts of damage.

Figure 5.4a shows a TEM image showing slight damage at the gate dielectric interface forming subnanometer-scale defects formed during a low energy damage event. Figure 5.4b shows a hard dielectric breakdown (HBD) with dielectric breakdown-induced epitaxy (DBIE), where the silicon regrows from the damaged region, forming a quasi-single crystal region grown from the substrate. Figure 5.4c shows a TEM image, with the gate silicide and polysilicon intermixed because the high temperature was above the melting point. Figure 5.4d shows damage to the underlying silicon and tungsten plug contact heated to around the melting point of tungsten.

Dielectric damage can vary from a slight increase in leakage through stress-induced leakage, higher leakage with soft breakdown, and a high level of leakage with a hard short. Soft dielectric breakdown may result in variable leakage through time or multiple measurements. Soft breakdown typically occurs in low-energy

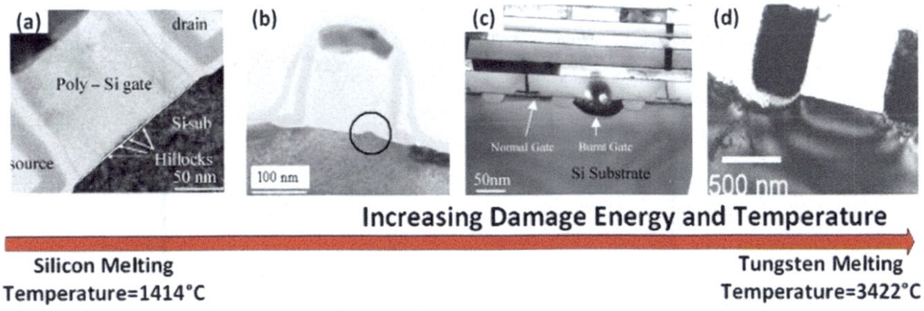

Fig. 5.4 TEM Photographs showing different amounts of structural damage due to dielectric breakdown as a function of local temperature during the breakdown event. (**a**) Low energy breakdown forming hillocks. (**b**) Dielectric breakdown and breakdown-induced epitaxy (DBIE) of silicon forming a silicon bump. (**c**) High energy breakdown event forming a melted gate (burnt) region. (**d**) High energy event resulting in the melting of tungsten plug contact

breakdown events. The electrical resistance of a defect can vary with time and partially recover toward its pre-breakdown value. As a result, soft breakdown can behave erratically, resulting in sporadic confusing electrical results. Figure 5.5 shows a plot of a high-k dielectric breakdown for a constant voltage stress progressing from soft breakdown to a lower resistance hard breakdown state (Okada et al., 1999).

Dielectric damage can occur in any dielectric used in a CMOS process, including gate dielectrics, intermetal dielectrics, isolation dielectrics, etc. Dielectric breakdown destroys the material's extremely high electrical resistance, creating a leaky material or an electrical short. Figure 5.6 shows some examples of various dielectric breakdown and leakage levels compared to the initial undamaged I-V curve dominated by direct or Fowler-Nordheim tunneling. A dielectric may leak by stress-induced leakage current (SILC) for low amounts of damage. SILC leakage is a conduction mechanism caused by charge carriers moving through the dielectric hopping between the damage-created defect traps (Ielmini et al., 2004). Figure 5.6a shows an I-V curve where SILC leakage overwhelms the undamaged Fowler-Nordheim or direct tunneling current (Okada et al., 1999). As the damage increases, the leakage transitions from SILC to soft breakdown, finally ending up in the low-resistance hard breakdown state. Figure 5.6b shows this progression, transitioning from the right side of the plot to the left side of the plot (Strong et al., 2009, p. 300).

Dielectrics subjected to process damage are degraded by the charge flowing through them. The charge damages the dielectric by breaking the atomic bonds in the dielectric and at the dielectric-semiconductor interface, creating interface states

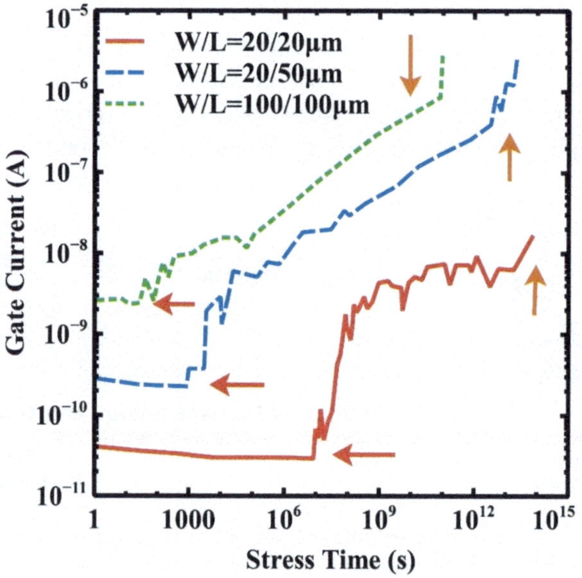

Fig. 5.5 Plot of gate current versus time showing high-k dielectric breakdown for various size transistors progressing from the initial soft breakdown (red arrow) to hard breakdown (orange arrow)

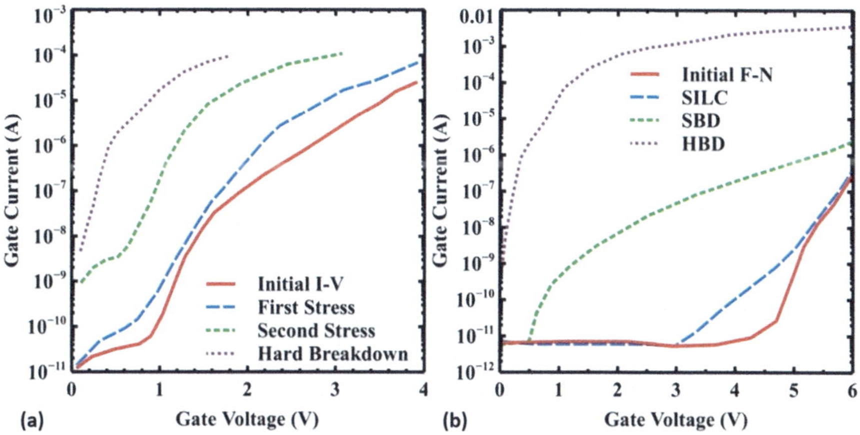

Fig. 5.6 Dielectric breakdown I-V curves showing various amounts of damage. (**a**) A 2.4 nm oxide dominated by direct tunneling showing initial leakage and stress-induced leakage current and breakdown. The unstressed intial curve is shown, along with various amounts of increasing stress induced leakage current after current stress. (**b**) A thicker 4.35 nm dielectric dominated by Fowler-Nordheim tunneling showing initial Fowler-Nordheim leakage and various levels of damage, including stress induced-leakage current (SILC), soft breakdown (SBD), and hard breakdown (HBD)

and carrier traps. This degradation creates electrical changes or makes the dielectric more susceptible to further damage. Characterization techniques such as charge pumping (Barbottin & Vapaille, 1999), capacitance-voltage (C-V) analysis, trapping measurements, etc., allow for direct or indirect measurements of the traps and interface states (Barbottin & Vapaille, 1989; Nicollian & Brews, 1982). Interface states and traps are the primary causes of the parametric shifts in many of the following sections.

5.4.4 Enhanced Breakdown Due to Radius of Curvature

The electric field across a dielectric is increased at the corners of conductors or semiconductors, enhancing conduction. The corner field enhancement increases conduction by field emission and increases degradation caused by plasma damage in structures such as the finfet gate corners, metal interconnect corners, gate edges, and shallow trench isolation edges (Yamabe & Imai, 1987). Figure 5.7a shows a sketch of a conductor corner covered by a dielectric. The corner radius is denoted by r, and the dielectric thickness is denoted by t_{ox}. The location of the enhanced field is denoted by E. Figure 5.7b shows the electric field enhancement as a ratio of the corner field divided by the field in a flat region as a function of the radius of curvature divided by the dielectric thickness.

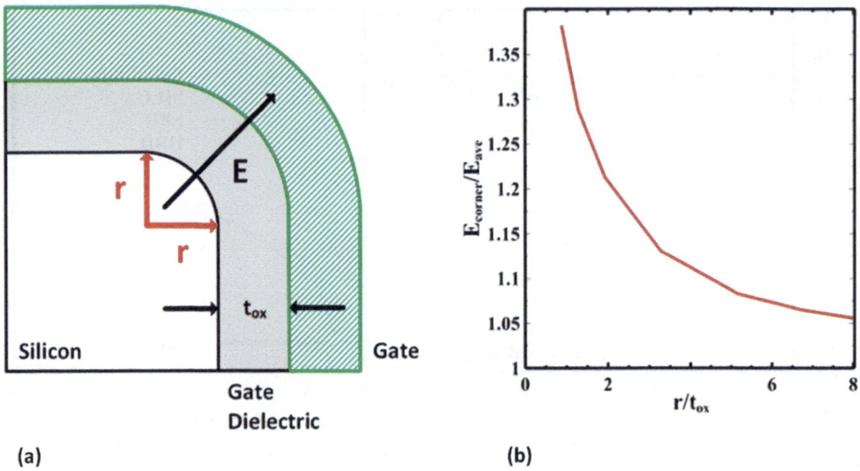

Fig. 5.7 Sketch and plot showing increased electric field at a structural corner. (**a**) Sketch of corner showing radius r and dielectric thickness t_{ox}. (**b**) Plot showing the ratio of the increased electric field at a corner to the average electric field in a flat region as a function of the ratio of the radius to the dielectric thickness for a silicon dioxide dielectric. The peak electric field is at the corner shown with the arrow labeled E

5.5 Interface States

Interface states are electrically active defects at the semiconductor surface-dielectric interface that can interact with charge carriers flowing in the transistor inversion region at the interface. Charge carriers in the inversion region provide most of the current flowing through a MOS transistor. Interface states interact with the inversion charge, temporarily trapping the channel current carriers and lowering the gain and drive of the transistors. Interface states are generally attributed to dangling bonds between the gate dielectric and semiconductor at the interface. Figure 5.8 shows a quasistatic C-V plot of a MOS capacitor with various metal antennas (Shin & Hu, 1993). The flat band and inversion region of the plot are distorted due to the formation of interface states. This distortion can degrade transistors and capacitors. Interface states can also create reliability hazards for the CMOS chip. Figure 5.9 shows a plot of the interface state density due to back-end-of-line (BEOL) plasma damage with and without an antenna (Hiblot et al., 2019).

5.5.1 Gain (g_m) Degradation

The gain (increase in signal strength) or transconductance (g_m) is a CMOS transistor metric used to measure a transistor's ability to amplify a gate signal into a drain-to-source current. The transistor g_m is defined as the ratio of the partial derivative of the

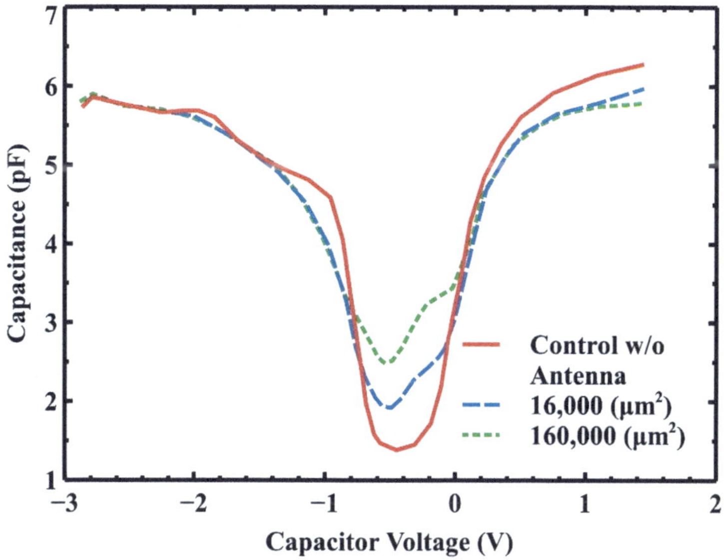

Fig. 5.8 Example of interface state generation caused by plasma damage shown in a quasistatic CV plot for a 60 s metal etch. A wet etched control capacitor is shown. Two different metal antenna sizes are shown with increasing interface state generation shown in the dip of the C-V curve

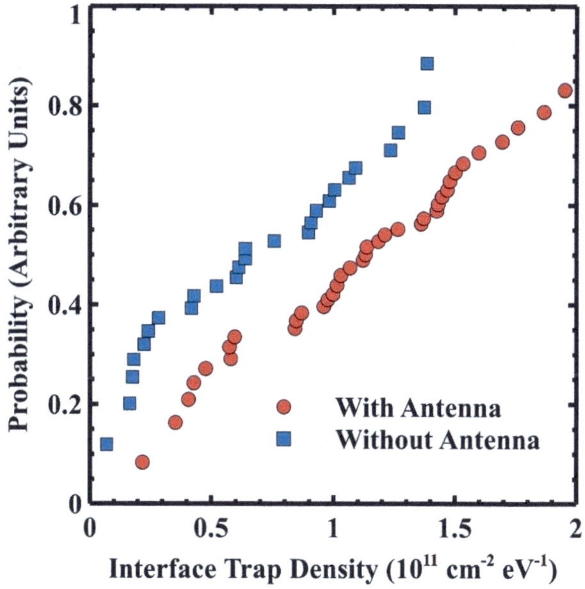

Fig. 5.9 Probability plot of interface state density with and without a backend of line antenna in local interconnect, vias, and metal 1 layers

drain current to the partial derivative of the gate voltage, as shown in Eq. (5.1). The transconductance gives the slope of the drain current with the change in gate voltage.

$$g_m = \frac{\partial I_d}{\partial V_g} \tag{5.1}$$

The electron trapping, trapped charge, and interface states generated by process damage impede the flow of carriers in the transistor channel, resulting in a reduced gain.

Figure 5.10 shows an example of g_m degradation for a finfet technology due to back-end-of-line process damage during interconnect processing with and without an antenna (Hiblot et al., 2019). Degradation of the g_m characteristics can shift the circuit-operating parameters or cause circuit failure.

Interface states are a primary cause of conductance generation, as the states interfere with the movement of charge carriers at the gate dielectric-silicon surface. Interface states can be degraded by plasma and process damage, as shown in Fig. 5.11. The exposure of gate dielectrics to an argon plasma is shown to increase the number of interface states as a function of the energy level in the silicon bandgap.

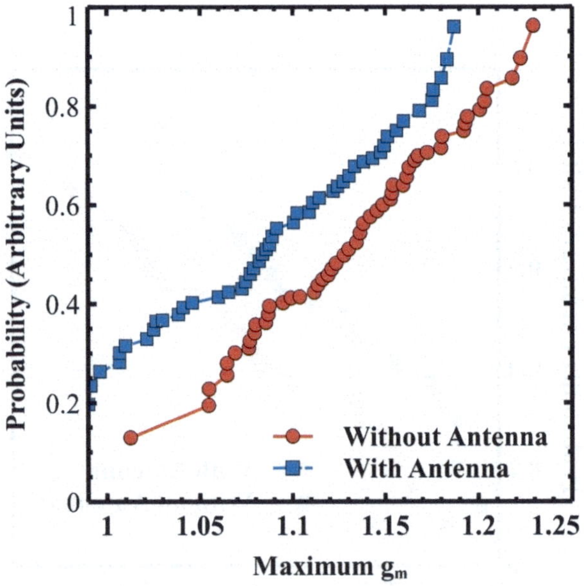

Fig. 5.10 Example of conductance degradation due to process damage in bulk finfet technology

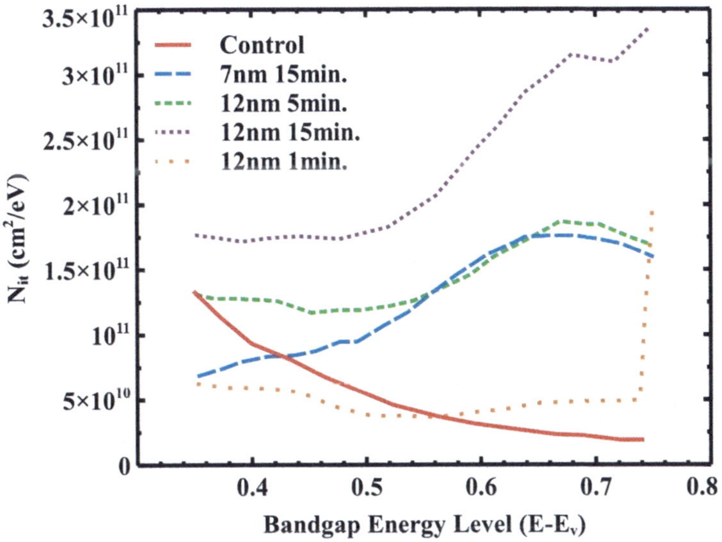

Fig. 5.11 Interface state density for different argon inductively coupled plasma exposures for a 7 and 12 nm gate dielectric

5.6 Trapping

Traps can degrade devices by capturing charge carriers for short time periods, extended periods, or permanently. Trapping can shift several device parameters, including the threshold voltage, conductance, and dielectric conduction, and create reliability hazards. Traps can be located in the dielectric bulk or at the interface. In addition, process damage can create traps as current flows through dielectrics or through high-energy radiation damage. The traps can be created without trapping a carrier, remaining neutral and unobservable. However, after process damage, transistor operation or stress can populate the traps, causing electrical shifts and reliability degradation. The gate dielectric trapping behavior is a strong function of the dielectric thickness, with thicker dielectrics having a high propensity for trapping.

Traps can be detected by forcing current through a dielectric and monitoring the voltage across the capacitor through time. As the traps fill with charge carriers, the Fowler-Nordheim or direct tunneling current flowing through the dielectric increases or decreases as the trapped carriers modify the electric field across the dielectric by changing the shape of the quantum mechanical barrier. Figure 5.12 shows an example of a trapping curve where the voltage across the capacitor is monitored as a function of time while the current is being forced (Cheung, 1996). As shown in the figure, both hole and electron trapping can occur and shift the voltage across the capacitor in opposite directions. The green line shows a curve where the electron and hole current can be observed through time as the shape of the barrier changes,

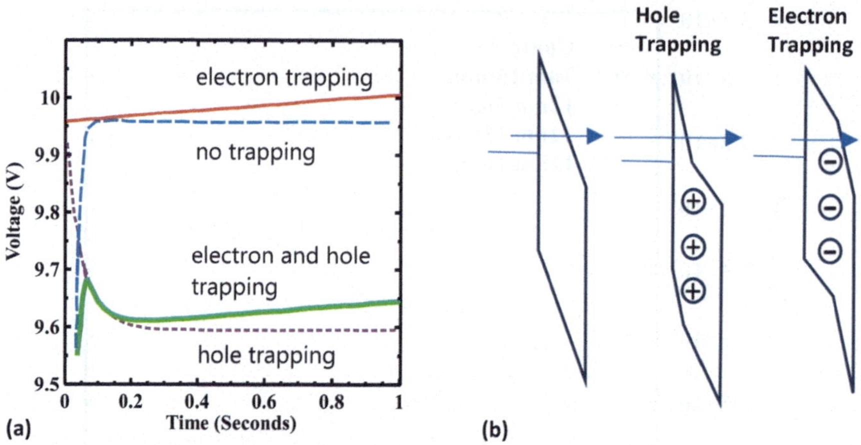

Fig. 5.12 Voltage versus time sketch of the voltage measured across a gate dielectric for a forced constant current. (**a**) Electron and hole trapping curves are shown along with the combined curve. A no trapping curve is shown as a reference. (**b**) Band diagram sketch showing the change in the barrier thickness and shape with electron and hold trapping

as shown in Fig. 5.12b. As the holes are trapped in the oxide, the effective E-field increases at the emitting interface, leading to increased current. As electrons are trapped, the effective E-field decreases at the emitting interface, thus decreasing the current. The trapping characteristics may vary widely, depending on the dielectric, damage, and measurement conditions. The location of the traps in the dielectric is also important in the behavior of the trapping curve, modifying the electric field and changing the conduction.

The shape of the combined trapping curve also depends on the current level being forced. High current levels can prevent the observation of either hole or electron trapping, as the signal from the opposite carrier type of trapping overwhelms the smaller signal. Figure 5.13 shows the capacitor gate voltage versus time plot for a constant current stress with different antenna ratios (Brozek et al., 1996). The control curve in Fig. 5.13a with no antenna shows a slight observable hole trapping and slow electron trapping. Figure 5.13b, c show rapid electron trapping with no observable hole trapping at the stress conditions. Figure 5.13d shows a curve for the largest antenna. The curve shows initial hole trapping and a large amount of electron trapping.

5.6.1 *Threshold Voltage Shifts*

Process damage can shift the threshold voltage (V_t) of the transistor, causing havoc with the circuit operation, resulting in loss of yield, loss of electrical margin, etc. The threshold voltage may shift positively or negatively during process damage depending on the current direction and the trapping rates of holes and electrons. The

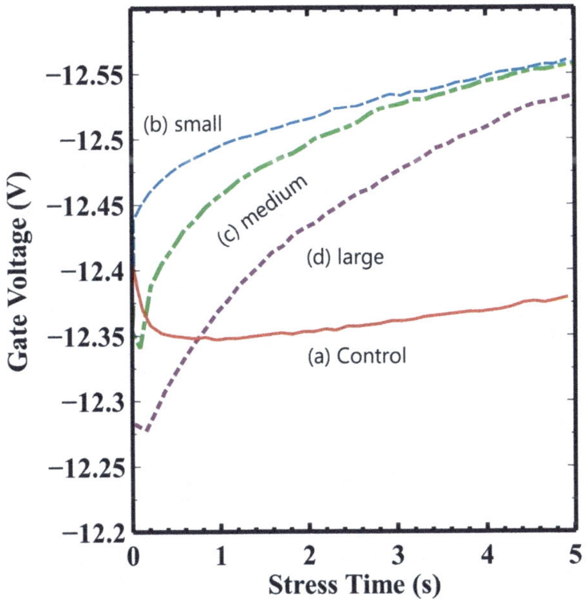

Fig. 5.13 Process damage creating dielectric traps observed by Fowler-Nordheim current stress on PMOS transistors versus time with back end of line antennas. Different antenna sizes are shown: (a) control transistor without antenna, (b) small, (c) medium, (d) large. The forced current is 0.1 A/cm^2. The transistor size is 0.5 ×5 .0 μm with a 9 nm dielectric

threshold voltage can also shift in one direction and then change, shifting in the opposite direction as damage progresses, following a similar behavior to the gate voltage shown in Fig. 5.13. For an NMOS transistor, positive trapped charge decreases V_t, and negative trapped charge raises V_t. For a PMOS transistor, the results are the opposite.

Figure 5.14a shows a plot of the threshold voltage for a finfet transistor array with and without an antenna, showing a negative shift caused by process damage in the metal and interconnect process module (Hiblot et al., 2019). Figure 5.14b shows a plot of the threshold voltage shift with and without damage due to electron beam exposure (Pyo & Kim, 2013; Vici et al., 2020).

The amount of threshold shift can be a function of the design rules in addition to the antenna area or perimeter ratios to the gate area. Figure 5.15 shows the threshold voltage plotted against the antenna area for different antenna structures and design rules (Matsunaga et al., 2007). There is a variation in V_t that is dependent on the finger spacing, indicating a dependence on the space design rule, as described in Chap. 9.

The threshold voltage due to process damage can also shift due to the active area and gate design rules. Figure 5.16 shows a plot of the PMOS threshold voltage as a function of the gate and active area layout (Li et al., 2007). Subtle changes such as the number and position of dummy gates and the active area layout can cause significant changes in the threshold voltage shift due to plasma damage. The figure

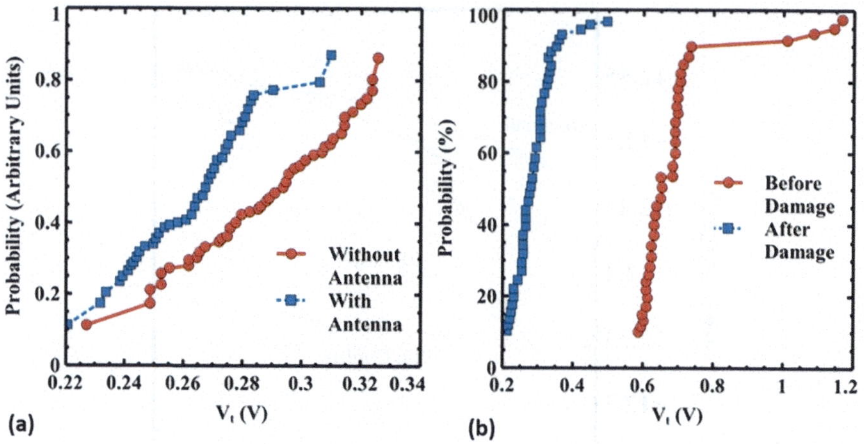

Fig. 5.14 Plots showing threshold voltage shifts due to plasma damage on a NMOS transistor. (**a**) Finfet NMOS V_t shift caused by back end of line via and metal processing. (**b**) Planar NMOS V_t shift caused by process damage due to electron beam spin on glass curing. All antennas in this experiment (right plot) shifted an identical amount, indicating charging was not involved, and light damage was the cause of the V_t shift

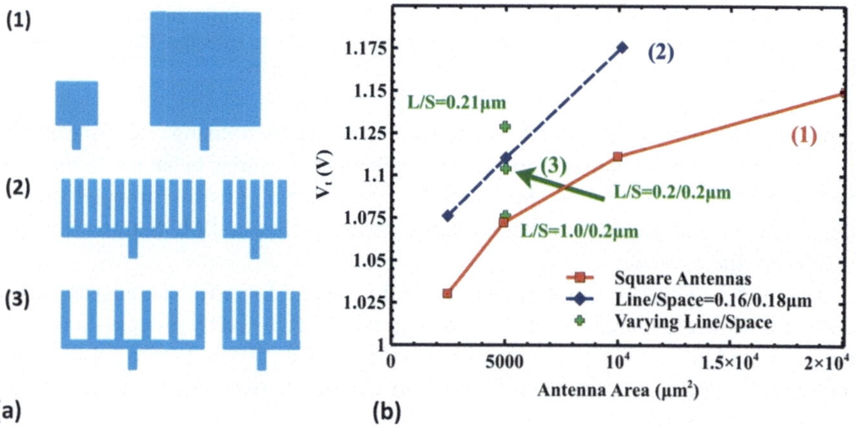

Fig. 5.15 Change in V_t caused by process damage during back end of line plasma enhanced chemical vapor deposition of a silicon carbon nitride film. (**a**) Sketch showing the different experimental structures. (1) Square metal area antenna with varying area. (2) Line and space antenna with varying area. (3) Constant area and varying pitch (**b**) threshold voltage shift as a function of antenna area and type of structure. The damage is design rule dependent, varying with the structure and the design rules

shows the V_t shift as a function of the hydrogen alloy time, showing an increasing shift for some transistors with the alloy time. The hydrogen alloy passivates interface states, indicating that the layout with a small active area (AA) and small gate is independent of the alloy time, and it has a much lower interface state density than

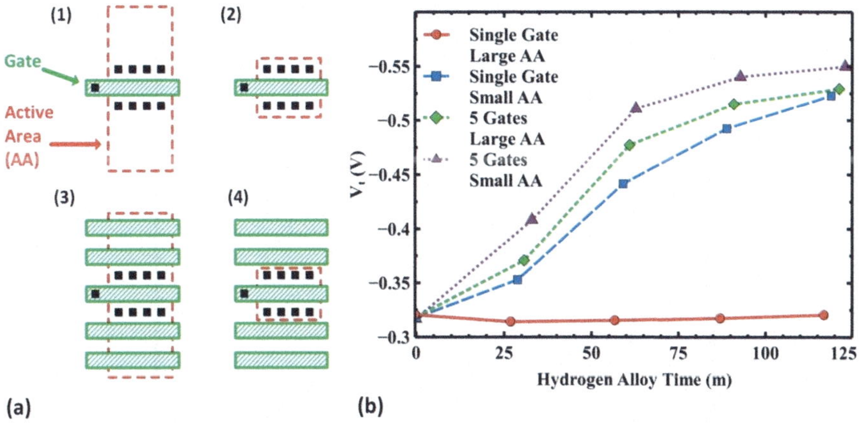

Fig. 5.16 Process damage V_t shift with different design rules as a function of the hydrogen alloy time caused by mechanical stress and selective plasma etching of a silicide block dielectric layer. (**a**) The layout experiment varies the active area (AA) size and the number of gates. (**b**) The layout dependent V_t shift is attributed to mechanical stress in the shallow trench isolation or the salicide block and stressor dielectric

the other devices. The cause of the plasma damage is attributed to the dry etch removal of the silicide block dielectric layer. Removal of the dielectric layer by wet etching shows no shift as a function of the layout. The stressor layer also plays a significant role in the threshold voltage shift in this experiment.

Plasma damage can vary widely depending on the transistor channel length. Figure 5.17 shows a plot of the threshold voltage shift versus the gate length, with the largest channel length showing a negligible shift and the shortest channel length showing a significant shift. This phenomenon may be due to spacer or channel edge plasma damage. It also may be due to charge trapping in the nitride composite spacer (Li et al., 2007). There is a slight change in the antenna ratio as the channel length changes but not enough to explain the significant shift in the threshold voltage.

Contact and via processing can also cause a V_t shift. Figure 5.18 shows a plot of threshold voltage variation as a function of the number of contacts connected to the transistor gate metallization (Tsui, 2000).

5.7 Transistor Leakage Mechanisms

5.7.1 Introduction

A CMOS transistor's ability to prevent unwanted leakage between all its terminals is an essential characteristic of the device, enabling chip operation and low power capability. Figure 5.19 shows a sketch of the MOS transistor leakage mechanisms

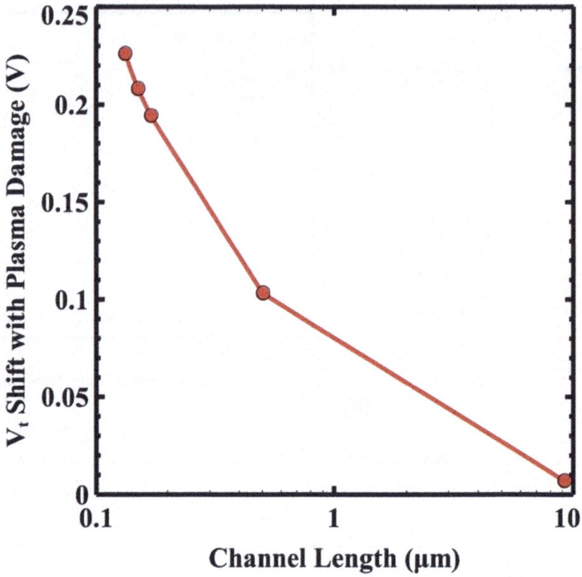

Fig. 5.17 Plot showing the threshold voltage shift due to plasma damage plotted against channel length

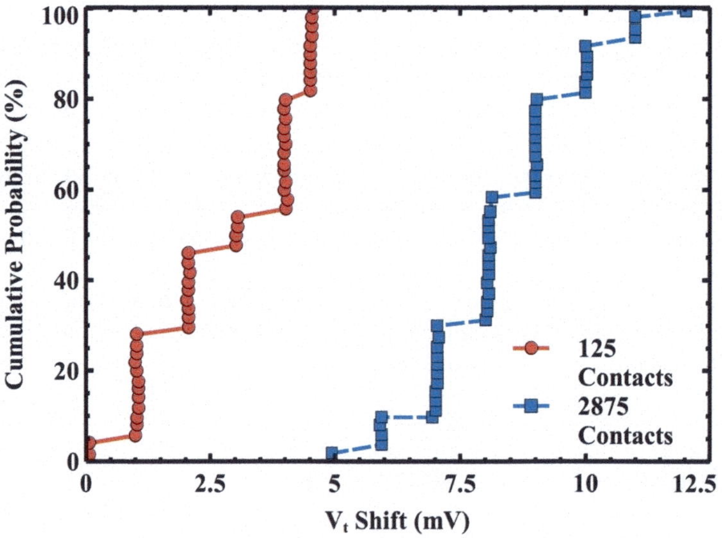

Fig. 5.18 Probability plot showing the threshold voltage shift as a function of the number of contacts connected to the gate of the transistor

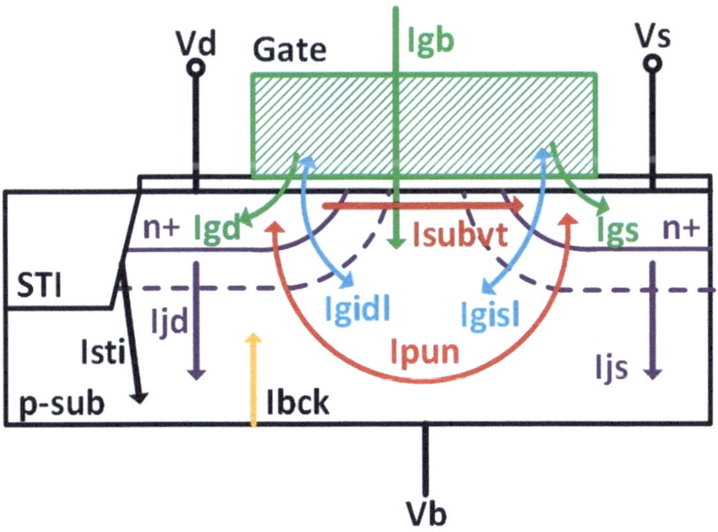

Fig. 5.19 Cross-section sketch of a MOS transistor showing different leakage mechanisms. I_{jd} and I_{js} are the diode leakage mechanism from the drain and source junctions to the substrate. I_{subvt} is the subthreshold gate-modulated leakage current between drain and source that occurs when the transistor is off. I_{pun} is a leakage between the source drain junctions that occurs at high drain biases when the gate is off. Gate dielectric leakage is denoted by I_{gb} for leakage through the channel. I_{gd} and I_{sd} is the dielectric leakage from the gate to the source and drain in the source or drain to the gate overlap region. The gate induced diode leakage caused by band to band tunneling is labeled I_{gidl} and I_{gisl}. Also, leakage can occur at the STI edge (I_{sti}). For thin wafers, generation from the backside can reach the active transistor (I_{bck})

(Calimera et al., 2012). Process damage can potentially worsen most of these leakage mechanisms. This section briefly describes the leakage mechanisms and degradation caused by process damage.

Gate dielectric leakage from the gate to channel (I_{gb}) or source and drain (I_{gd} and I_{gs}) intrinsically occurs through Fowler-Nordheim tunneling or direct tunneling, as described in Sect. 5.9.3. Plasma damage can degrade the insulating properties of the gate dielectric, causing dielectric breakdown discussed previously or stress-induced leakage current (SILC) (see Chap. 7).

Leakage can occur from the drain and source to the substrate by junction leakage (I_{jd} and I_{js}) or gate-induced diode leakage (I_{gidl} and I_{gisl}). Drain-to-substrate diode leakage usually requires severe process damage (e.g., arcing) or other structural damage, such as the creation dislocations (Yamaguchi et al., 1996). However, drain-to-substate leakage can occur for other reasons besides plasma damage, including salicide processing, metal contamination, structural defects, contact leakage, etch damage, etc. (Claeys & Simoen, 2018). In addition, process damage to the gate dielectric can degrade the edge damage as shown in Sect. 5.7.3.

Leakage can occur from drain to source due to punch-through (I_{pun}) or subthreshold conduction (I_{subvt}). Drain-to-source leakage is usually caused by the process damage-induced transistor parameter shifts (e.g., V_t shift, subthreshold degradation) described in this chapter. The punch-through path shown by I_{pun} in Fig. 5.19 is less sensitive to plasma damage as the deep channel doping dominates the bulk leakage path. Additionally, other mechanisms can occur, such as STI edge leakage or carrier generation from the back of the wafer for thinned wafers.

5.7.2 *Subthreshold Conduction Degradation*

Subthreshold conduction is an essential property of CMOS devices. A MOS transistor in the off-state still conducts a small amount of current that can severely degrade the power dissipation of a CMOS chip (Hu, 2010). Figure 5.20 shows a sketch of the log of the drain current plotted against the gate voltage. The log plot allows the subthreshold conduction to be easily observed as a straight line at gate voltages below the on-state of the transistor in this example. Two points on the plot are used to calculate the subthreshold slope S from the plot in units of mV per decade of current. The subthreshold slope is given by Eq. (5.2) and is a strong function of temperature (T). The parameter η is a technology-dependent parameter specific to an individual technology and transistor.

$$S\left(\frac{\text{mV}}{\text{decade}}\right) = \eta\, 60\text{mV} \frac{T}{300\,\text{K}} \tag{5.2}$$

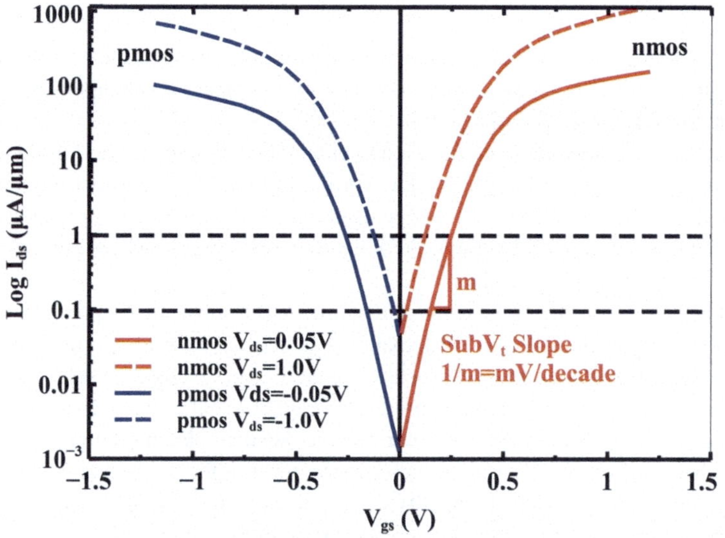

Fig. 5.20 Log I_{ds} versus V_{gs} plot showing subthreshold conduction. The sketch shows the definition of subthreshold slope as the inverses slope of I_{ds}/V_{gs} measured in the subthreshold region (curve is straight) expressed as mV per decade of current

An ideal transistor would have a perfect vertical infinite slope in the subthreshold plot, turning on instantly at the threshold voltage, eliminating the leakage for digital circuits. Fundamental physics limits the subthreshold slope to 60 mV/decade at room temperature (300 K). The theoretical limit is for $\eta = 1$ and is given by Eq. (5.3), where k is Boltzmann's constant (in eV), and q is the electron charge. The minimum theoretical limit is a function of temperature and fundamental constants and contains no parameters that can be optimized.

$$\frac{60\text{mV}}{\text{decade}} = \frac{kT}{q}\ln(10) \tag{5.3}$$

Process damage can also degrade the subthreshold conduction region of transistor operation (Kamei et al., 2007). For example, a V_t shift described previously can degrade the threshold region by shifting the NMOS transistor threshold voltage left in Fig. 5.20a toward zero, increasing the off-state leakage. Additionally, the subthreshold slope can be degraded by process damage due to carrier traps, interface states, and fixed charge by creating a path for charge carriers across the channel. Figure 5.21 shows an example plot showing degradation in the subthreshold regions of CMOS transistors with high-k and silicon dioxide dielectrics. V_t is shifted toward zero in the lower-left plot. The flat curves near zero V_{gs} show drain-to-source leakage where the gate control of the transistor is lost. The transistor leaks severely as V_t approaches zero voltage as the channel is always on. Junction leakage and subthreshold conduction can be differentiated by determining if the drain current flows to the substrate or the source.

Figure 5.22 shows an example of subthreshold slope degradation caused by a metal etch as a function of gate oxide thickness with and without diode protection (Park & Hu, 1998). Diode-protected transistors show no degradation as a function of antenna length, while unprotected transistors show degradation as a function of dielectric thickness. Process damage strongly depends on the gate oxide thickness, with the thicker dielectrics experiencing a more significant threshold shift, indicative of increased trapping and interface state generation.

5.7.3 Gate-Induced Diode Leakage Degradation

Gate-induced diode leakage (GIDL) is one of the leakage mechanisms shown in Fig. 5.19 that can be degraded by process damage (Chen et al., 1987). GIDL is a band-to-band tunneling phenomenon that occurs in the transistor region where the gate overlaps the source and drain junctions, as shown in Fig. 5.23. A high electric field across the gate dielectric in this region is generated by the difference in the gate and source or drain voltages, increasing with increasing difference in source/drain to gate potentials. The high gate electric field bends the energy bands in the source-drain region to the point where the diode depletion region barrier width is thin enough that carriers can tunnel through the junction barrier. Both electrons and

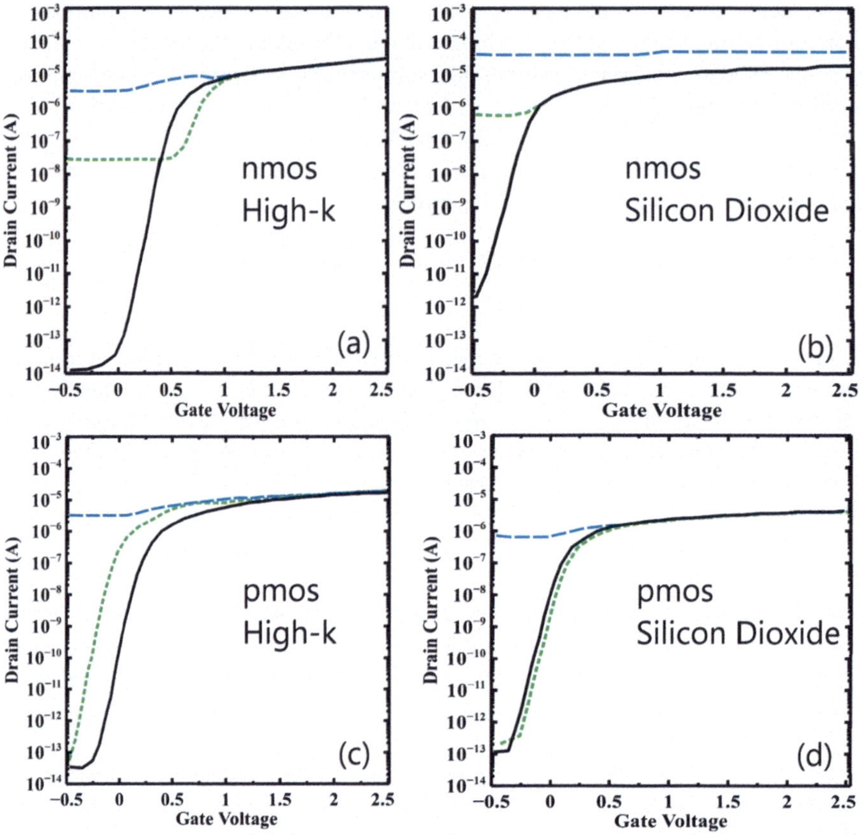

Fig. 5.21 Example showing subthreshold degradation due to Ar/O_2 plasma damage for high-k and SiO_2 NMOS and PMOS transistors measured at a drain voltage of 0.05 V. (**a**) NMOS high-k dielectric. (**b**) NMOS SiO_2 dielectric. (**c**) PMOS high-k dielectric. (**d**) PMOS SiO_2. Three levels of damage are shown: undamaged (solid line), moderate damage (green dotted line), heavy damage (blue dashed line)

holes are involved in the GIDL mechanism. For an NMOS device, the tunneling electrons are collected by the source and drain. Holes created through the GIDL process are injected into the substrate and flow toward the substrate contacts for bulk CMOS. The direction of electron and hole flow is shown in the figure. Plasma-damage-created defects such as traps and interface states may cause an increase in the GIDL leakage current (Kim, 2015).

GIDL can be degraded by process damage, creating traps and interface states, increasing the GIDL current (Chen et al., 1989). Figure 5.24 shows an example of GIDL degradation caused by plasma damage from a metal antenna. GIDL has a unique electrical signature that eliminates ambiguity. The GIDL current increases as the off-state gate voltage is biased further away from the threshold voltage, as shown in the figure. The degradation shown correlates with an increase in the interface

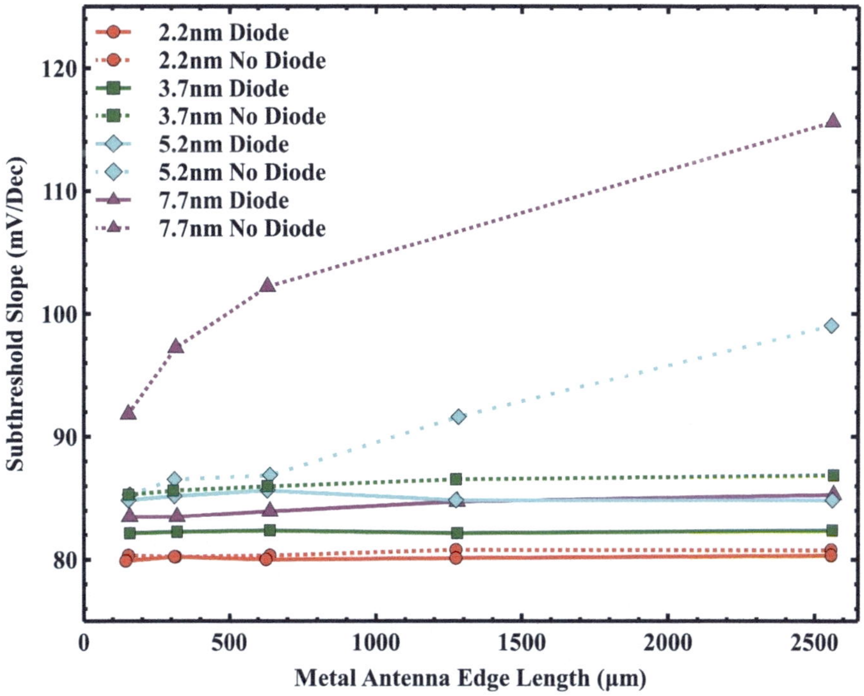

Fig. 5.22 Plot showing the subthreshold slope as a function of metal antenna perimeter for several different dielectric thicknesses with and without diode protection

state density. The GIDL current is sensitive to source-drain damage in the gate overlap region, as described in Sect. 5.7.4. Process-damage-related defects, such as traps, etc., can increase the GIDL through trap-assisted SILC by adding a dielectric leakage component, as shown in Chap. 7 (Kim, 2015; Ma et al., 2001).

5.7.4 Lightly Doped Drain (LDD) Region Damage

The region where the gate overlaps the doped source or drain junction is called the source-drain overlap region. Charge flowing into the gate dielectric can reach the substrate through conduction over the channel or over the source-drain regions. This overlap region is sensitive to process damage. There can be many process sources for damage to the source-drain region. For example, direct physical damage to the gate edge and spacer edge can occur from exposure to the plasma, causing chemical contamination, kinetic damage, etc. Direct exposure to the plasma can generate electrically active defects, such as traps, etc.

The source-drain regions may have higher current flow depending on several factors, including floating potentials of the well, plasma charging from the source or

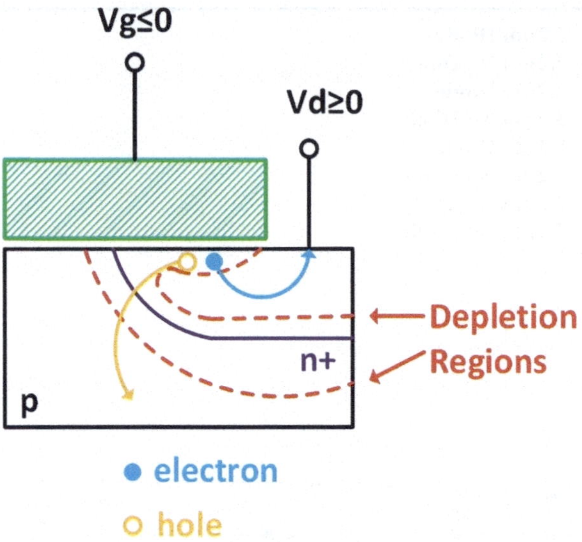

Fig. 5.23 Sketch of a NMOS transistor source drain to gate overlap region showing GIDL current. The depletion regions are shown by the dashed lines and show a depletion region created in the heavily doped n+ source drain region due to the high electric field between the gate and the drain. The depletion region is not to scale and is exaggerated under the gate

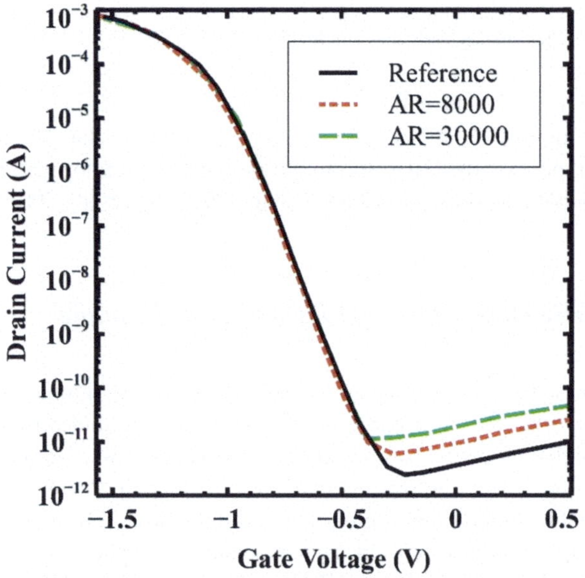

Fig. 5.24 Subthreshold plot of log I_{ds} versus V_{gs} showing gate induced diode leakage (GIDL) on the bottom right side of the plot for two different antenna ratios and a reference without an antenna. The subthreshold plot for a PMOS transistor with a 5 nm gate dielectric. The GIDL current is increased by plasma damage from metal 1 processing and correlates with metal antenna ratio

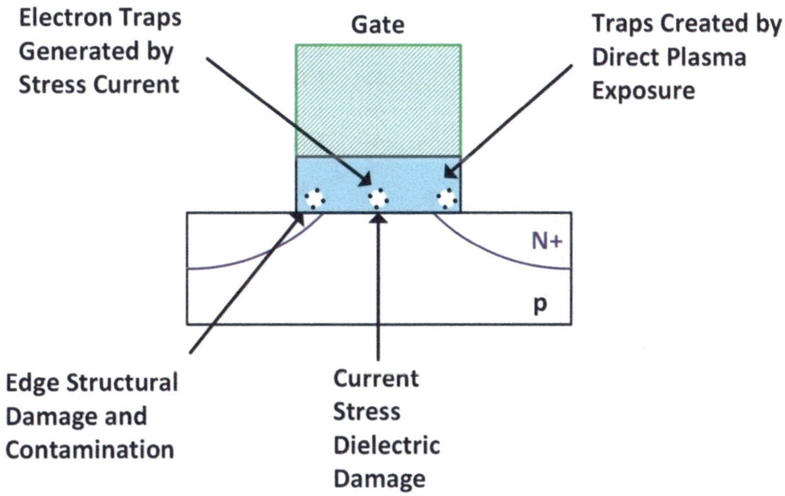

Fig. 5.25 Sketch comparing process damage in the channel with process damage over the source-drain regions

drain, potentials of the source-drain regions, etc. In addition, the source-drain region usually has a very tiny area relative to the channel area. Therefore, the current density in the source-drain overlap region during plasma damage can be much higher than the current density in the channel, creating a propensity for more damage in a small region. The source-drain overlap region can be susceptible to trapped charge, interface states, or dielectric leakage. The lightly doped drain (LDD) region resistance can be modulated, changing the source-drain resistance and impacting the transistor electrical parameters. The modulation of the LDD region can change the transistor's channel length and mobility, impacting the transistor's current drive. This region of the transistor is critical to its reliability performance. For example, the hot-carrier lifetime, etc., is controlled by this region and can be degraded by damage. Figure 5.25 shows a sketch highlighting the differences between damage in the channel and the source-drain overlap region (Brozek et al., 1995).

Modulation of the source-drain resistance with damage can easily occur. Figure 5.26 shows the source-drain resistance plotted against the gate voltage for two different etchers under two overetch conditions (Brozek et al., 1995). In this example, the resistance shifts by a factor of 2, creating a significant shift in the electrical device parameters.

The effective electrical channel length is a metric describing the distance between the source and drain in an MOS transistor measured from the junction edges. Many CMOS parameters are a strong function of the channel length, affecting the drive current and many other parameters. The resistance modulation of the source-drain resistance can create a significant shift in the effective electrical channel length, measured across several devices with different channel lengths. Figure 5.27 shows a plot of the change in the effective channel length (L_{eff}) plotted against the gate

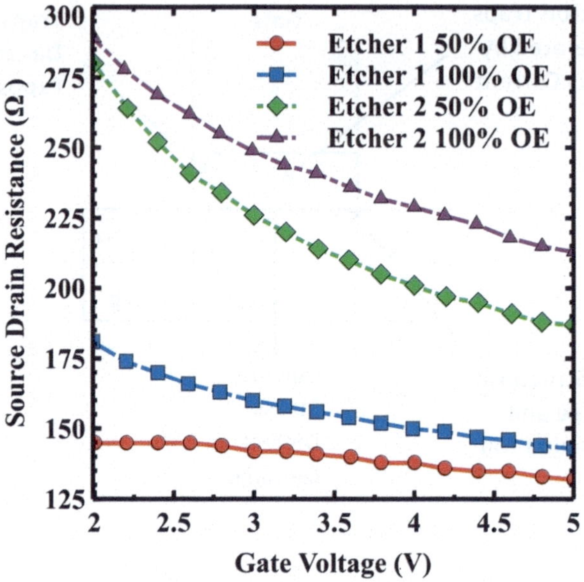

Fig. 5.26 Plot of source-drain resistance versus gate voltage for two different gate etchers at 50% and 100% over etch. The damage occurred during the gate metallization etch

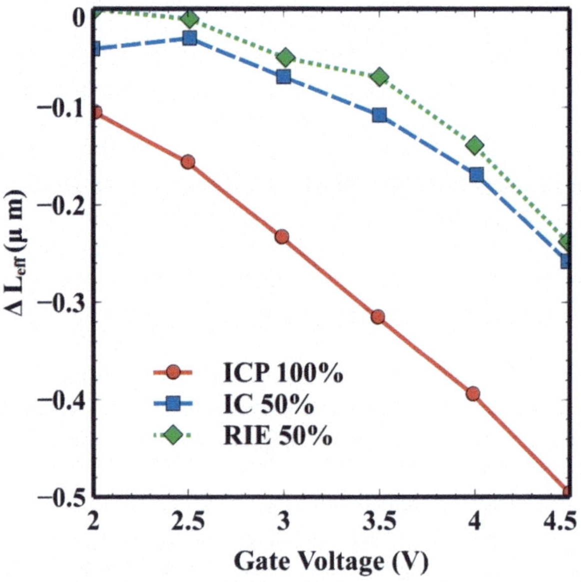

Fig. 5.27 Plot showing the change in electrical effective channel length for a 90 Å gate dielectric transistor for two different gate etchers. A reactive ion etch (RIE) at 50% over etch, and an inductively coupled plasma (ICP) are compared with two different over etches for the ICP etcher (50% and 100%). The negative ΔL_{eff} is attributed to an increase in resistance in the lightly doped drain region

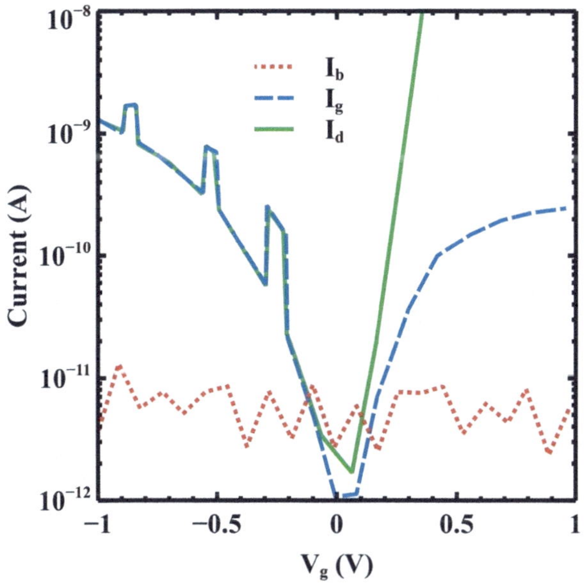

Fig. 5.28 Drain, gate, and substrate current for a metal gate high-k CMOS transistor damaged by the gate etch. The I-V curves show a high amount of noise. The drain bias is 50 mV

voltage for two different types of gate etchers (Li et al., 1994). L_{eff} is strongly modulated by the gate etch conditions.

Damage to the source-drain region can have other effects than parametric and reliability shifts. Damage to the gate edge and dielectric can cause excess noise through the creation of interface states, traps, and leakage. Figure 5.28 shows transistor current-voltage currents with significant noise caused by source-drain edge damage for a high-k dielectric transistor (Cho et al., 2011).

5.7.5 Punch-Through Degradation

Punch-through is another CMOS transistor leakage mechanism shown in Fig. 5.19. In this case, the leakage current can flow between the source and the drain. A CMOS transistor is designed to have a strong electrical barrier between the source and drain, created by tailoring the doping in the channel and source-drain edges. Plasma damage can inject charge in various channel locations, including an asymmetric charge distribution that can weaken the punch-through resistance or cause punch-through-related reliability problems. For example, contact etch damage was shown to degrade a punch-through-related PMOS reliability problem called hot-electron-induced punch-through (HEIP) (Koyangi et al., 1987).

5.8 Transistor Degradation Mechanisms

5.8.1 Process Damage-Induced Transistor Mismatch

Process damage commonly degrades the variability of many transistor parameters, increasing the mismatch of matched transistors in critical circuits in CMOS chips. CMOS wafers exposed to excessive process damage can be expected to have increased variation in the affected device parameters in this section. Degraded mismatch of critical circuits can cause yield, margin, performance, and reliability problems. Process damage can often be mistaken for process variability, poor circuit design, testing problems, etc. Design rules and layout methodology (see Chap. 12) are critical to controlling process damage for matched transistors. All antenna structures connected to matched transistors of similar critical circuits should have an identical layout matching any potential antenna structures to achieve identical damage. Design modifications to protect the critical circuits with diodes, pass transistors, etc., can also mitigate process damage (Zhao et al., 1996).

Figure 5.29 shows an example of process damage creating matching issues (Shin et al., 1993). The circuit diagram shown in the figure is for a differential pair amplifier (Baker, 2019, p. 735). This amplifier detects low-level signals and magnifies them, providing a useful signal. It is commonly used in critical applications such as sense amplifiers for memory arrays. The transistors in the circuit must be carefully matched for the amplifier to work correctly. This type of circuit is sensitive to plasma damage. Different antenna sizes are connected to the inputs A1 and A2 by using different pad sizes. The output V_0 can be monitored to determine the offset voltage error caused by the damage. Figure 5.29b shows the offset voltage error caused by plasma ashing. The error shown as tens of millivolts is significant because

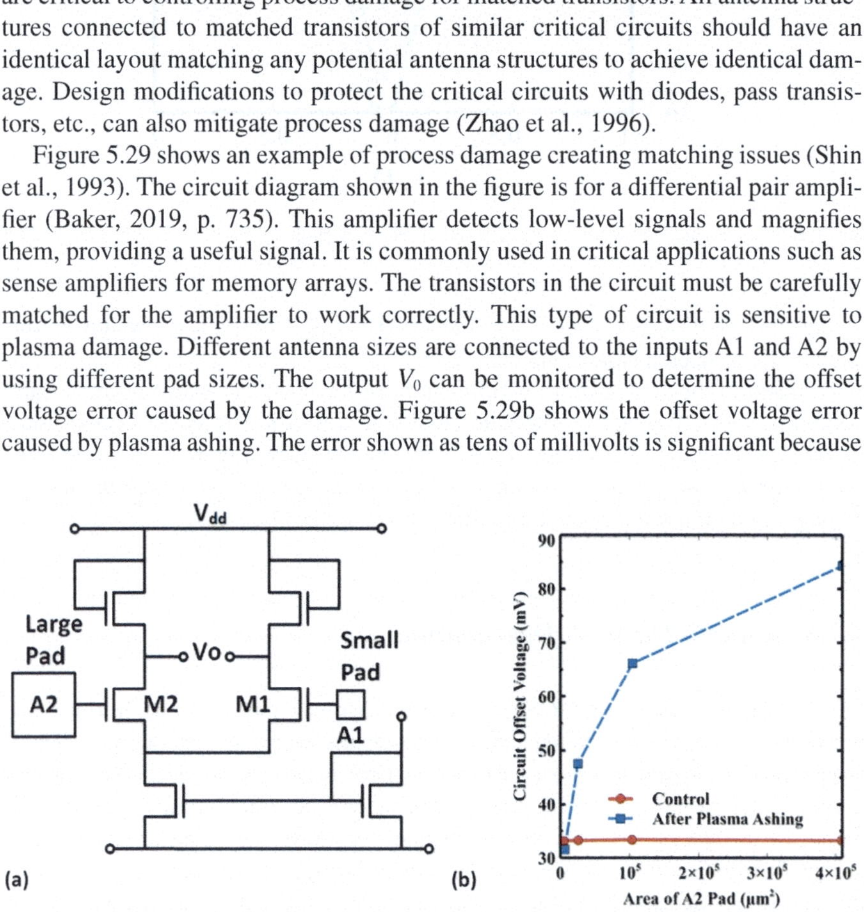

Fig. 5.29 Example of degradation in transistor matching with plasma damage with a metal antenna. (**a**) Schematic of a differential pair amplifier with different antenna sizes on transistors M1 and M2. (**b**) Offset voltage between the transistors M1 and M2 with different antenna sizes on transistor M2. The area of the small pad is 6400 μm^2

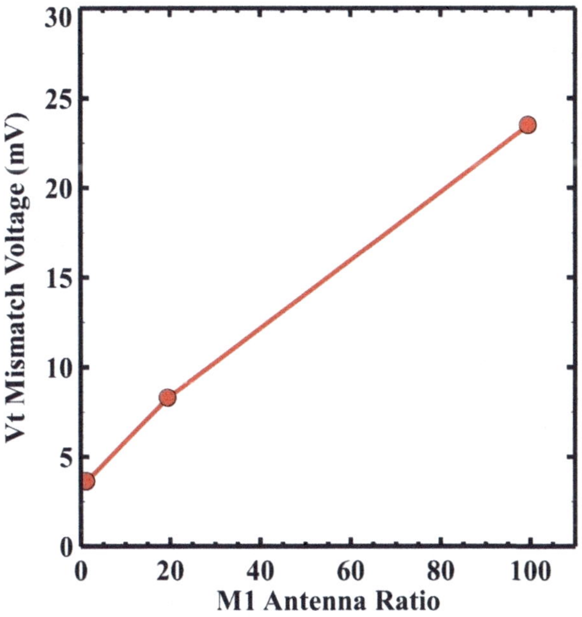

Fig. 5.30 Transistor V_t mismatch for different ratios of gate antennas due to damage from metal etching

it is of the same order of magnitude as signals typically amplified by the circuit and would cause a severe malfunction. The control line in the figure uses a wet metal etch to avoid plasma damage, proving the cause of the mismatch. An additional example is shown in Fig. 5.30, where metal patterning causes a significant V_t mismatch, highlighting the need for an identical layout (Zhao et al., 1996).

5.8.2 Process Damage Noise Degradation

Noise is ever present in electronic systems and is universally generated by thermal processes at temperatures above absolute zero. Additionally, electronic and defect processes in semiconductors generate noise that can be significantly increased by process damage. Some sources of noise are fundamental and not related to process damage. These sources include thermal noise caused by the Brownian motion of electrons or shot noise caused by individual electrons moving across a junction. Other noise sources are less fundamental and can be modulated or degraded by process damage. These sources include 1/f noise that dominates at low frequencies caused by mobility variation, fluctuations in carrier numbers, and random thermal signal (RTS) noise caused by carrier capture and emission. Process damage typically modulates 1/f and RTS noise. These two noise sources tend to be confounded and interact.

Figure 5.31 shows an example of this interaction observed in CMOS imagers (Wang, 2008). Figure 5.31a is a histogram of the digital output of a CMOS imager pixel as a function of frequency, showing a 1/f noise characteristic. Figure 5.31b–d shows the pixel noise for three pixels taken at different points on the 1/f histogram with various levels of noise. Multimodal noise can be observed in some pixels (Wang, 2008).

RTS noise in CMOS transistors strongly depends on the transistor area, with the noise increasing for smaller transistors with a 1/WL (length and width) dependence. Scaling makes the noise worse, and process damage degrades the noise in a scaled transistor more severely. The RTS noise for a CMOS transistor can be roughly approximated as a voltage source in series with the gate by Eq. (5.4), where K is a process-dependent constant, C_{ox} is the gate capacitance, W and L are the transistor width and length, f is the frequency of the noise, and C is a constant between 0.7 and 1.2 (Tsividis & McAndrew, 2011, p. 453).

$$V_n^2 = \frac{K(V_{GS})}{C_{ox}WL}\frac{1}{f^C} \tag{5.4}$$

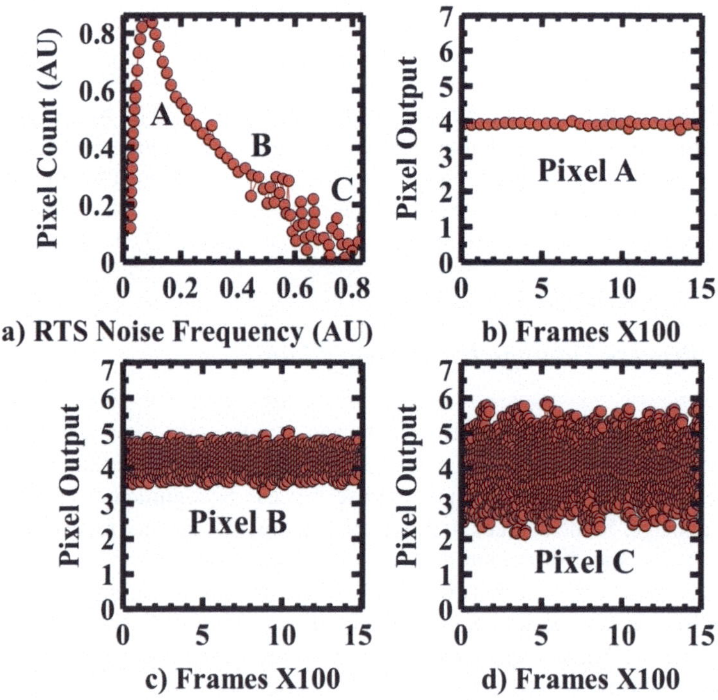

Fig. 5.31 Example of 1/f and RTS noise in a CMOS imager across 15,000 frames (pictures). (**a**) 1/f noise histogram with arbitrary units. (**b**) Pixel A, low noise pixel output. (**c**) Pixel B, medium noise pixel. (**d**) Pixel C, high noise pixel

RTS problems can cause severe problems in highly scaled devices such as NVM, DRAM, imagers, finfets, etc., with process damage degrading the devices further.

The damaged dielectrics and semiconductor interfaces caused by process damage can increase the transistor noise. The increased noise is caused by the increased number of traps and interface states caused by the damage. Charge carriers populate and depopulate these defects, creating noise as the carriers are added or subtracted from the operating transistor channel current. In addition, damage to the gate dielectric can create a noise interaction between the gate and drain caused by the defects modulating the tunneling current between the gate and drain. Damage to the source-drain overlap region is described in Sect. 5.7.4.

Transistor noise can degrade the chip operating margin (speed, voltage, etc.), causing a loss in operation robustness, manifested as a testing problem, yield problem, design problem, process variation, etc. Figure 5.32 shows an example of increased noise with plasma damage with variations in antenna sizes. First, the transistor drain current, etc., is measured continuously as rapidly as possible to obtain many measurements under the same conditions. Then, the current is plotted against the elapsed time of the measurements to generate plots of the variation.

Figure 5.33a shows a noise plot for two different antenna sizes for transistors damaged by an inductively coupled plasma (ICP) metal etch. The larger antenna size increases noise with the noise spikes caused by the traps filling and detrapping, yielding current conduction at different levels. The spikes have a time constant associated with the time required for the trap to emit or absorb a charge carrier.

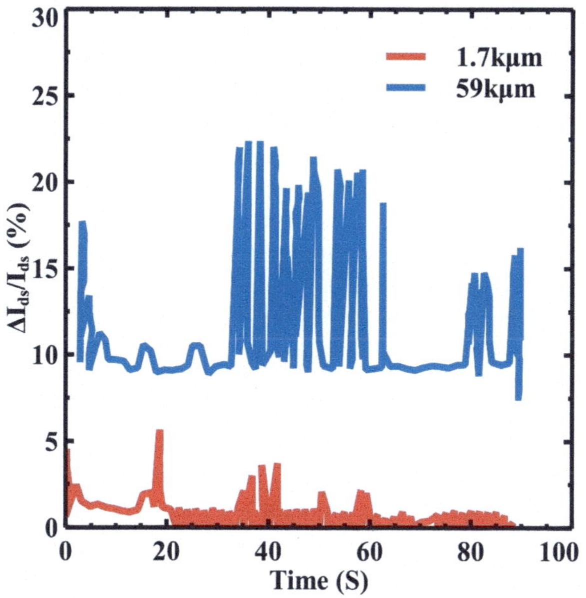

Fig. 5.32 Drain current noise characteristic measured as a function of time for two different metal antenna sizes

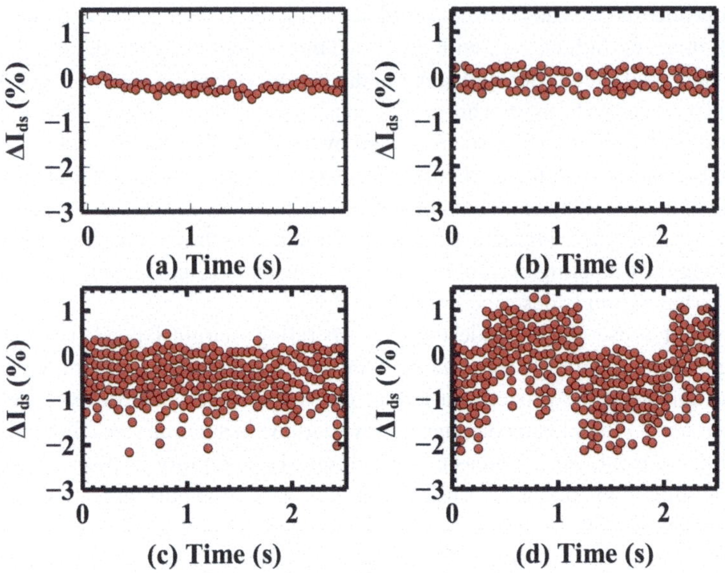

Fig. 5.33 Drain current noise characteristics. (**a**) Drain current noise plasma damage for a SiO_2 dielectric before damage. (**b**) SiO_2 dielectric after exposure to an inductively coupled plasma with a metal antenna. (**c**) High-k dielectric before damage. (**d**) High-k dielectric after exposure to an inductively coupled plasma with a metal antenna. The current is normalized to an average reading

Figure 5.33b shows the noise signal for a transistor with a metal antenna before and after exposure to an inductively coupled plasma. The identical structure is compared with a conventional SiO_2 dielectric and a high-k dielectric normalized by dividing by the mobility. The high-k dielectric has increased noise in this case due to a higher propensity for traps and interface states.

Noise behavior in CMOS devices is a complex topic (Bonani & Ghione, 2001). Several different metrics are commonly used to express noise behavior. For example, Fig. 5.34 shows the noise metrics drain current noise (A^2/Hz) and the noise power spectrum (dB/Hz) demonstrating noise degradation with plasma damage (Shin et al., 1993; Tzeng et al., 2002).

Noise can sometimes be observed during routine I-V measurements. Figure 5.28 (shown previously) shows plots of the drain, gate, and substrate currents for a high-k gate transistor damaged by the plasma gate etch. The noise can be observed as abnormal noisy I-V curves, as shown in the figure.

The RTS noise signatures in the previous figures oscillate between two or more metastable states. The oscillation frequency between states depends on the type of defect trapping and detrapping the charge carrier. The energy level of the trap determines the difficulty of trapping or ejecting a charge carrier, determining the characteristic frequency of the RTS defect. Figure 5.35 shows a plot of the characteristic frequency for a few RTS defects as a function of temperature (Janesick, 2001). The time constant is a strong function of temperature, changing by many orders of

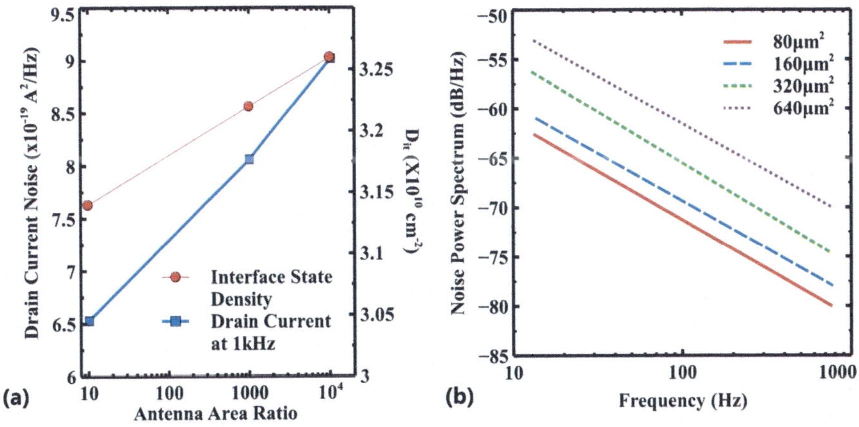

Fig. 5.34 Noise plots showing degradation due to plasma damage. (**a**) Drain current noise as a function of metal antenna ratio damaged during metal etch and resist strip and correlation with interface state density. (**b**) Noise power spectrum versus frequency for different antenna sizes during metal etch

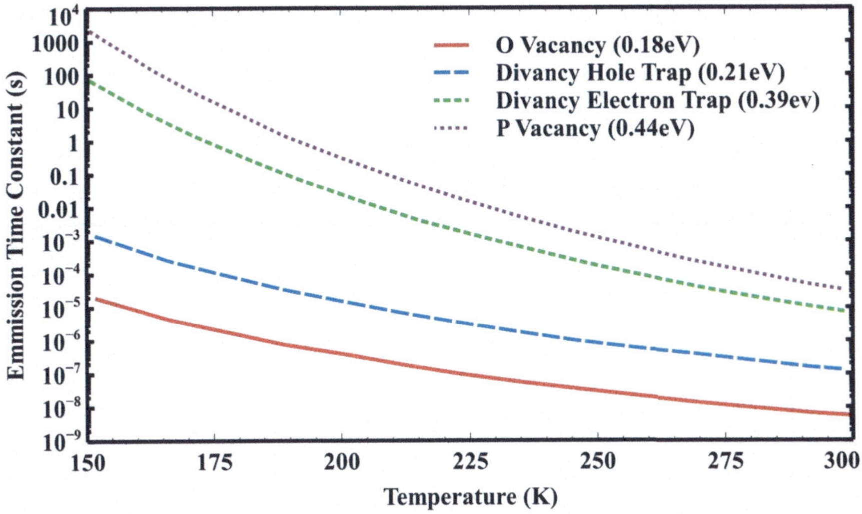

Fig. 5.35 Emission time constant for various types of RTS defects as a function of temperature with the trap energy

magnitude. Traps with low energy levels have a low time constant, while traps with high energy levels have a high time constant. The extracted energy level is somewhat helpful in identifying the exact type of defect, as shown in the figure. Most RTS traps tend to have a characteristic energy near half the bandgap (0.55 eV for silicon) because most defects are Schockley-Read-Hall (SRH) defects that dominate the carrier lifetime in silicon.

Long-time constant SRH defects are challenging to screen while testing the chip, requiring extensive testing and burn-in at different temperatures, adding significant cost. The defects described here can easily escape testing, reaching the customer in the field and degrading quality and reliability. Customer-specific processes, such as mounting chips on boards, expose the chip to a low-temperature thermal cycle, which can uncover latent defects such as RTS noise sites. As shipped from the factory, a good chip can become defective through customer processes, requiring some mitigation to replace or recover the chip. For example, DRAM datasheets contain methods for built-in self-testing (BIST) and repair techniques, replacing defective cells with redundant cells (Park et al., 2021). This allows for repair after the DRAM is mounted to a circuit board.

5.8.3 Ring Oscillator Degradation

Ring oscillators consisting of an odd-numbered series of inverters connected serially are used to provide a relatively simple circuit to measure the performance of CMOS technology, providing relevant metrics for actual circuit operation. As expected, ring oscillators can degrade with process damage, given that many transistor parameters are affected by process damage. Therefore, the complex circuits in a CMOS chip can be expected to have degradation similar to that of ring oscillators. Figure 5.36 shows an example of ring oscillator degradation with gate etch processing and hot electron damage (Li et al., 1995). Both hot-electron stress and plasma

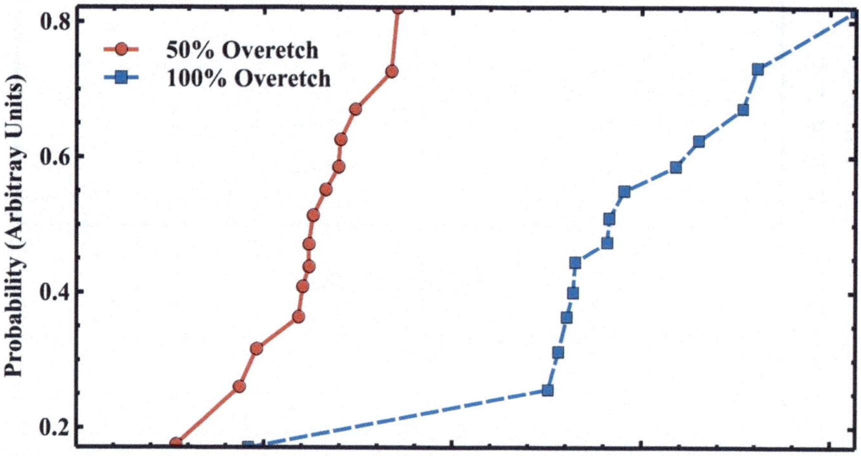

Fig. 5.36 Probability plot showing the plasma damage degradation in ring oscillator performance after hot electron stress with differences in gate etch

damage to the same transistors are needed to observe the damage shown. This plot is an excellent example of latent damage that manifests degradation after reliability testing.

5.9 Other Signatures

5.9.1 Contact Resistance Degradation

Contact etch processes remove a dielectric, creating a hole for a metal contact to junctions and ohmic contacts in the substrate. The etch must remove the dielectric and stop on the surface of the wafer without significant damage to the substrate, salicide, etc. Contact etches are high-energy processes forming a vertical anisotropic contact hole. The energetic etch can damage the bottom of the contact hole, creating a noncrystalline damaged region or an interfacial layer that can degrade the contact resistance or cause nonohmic behavior. CMOS processes are heavily optimized to clean and remove damaged layers at the bottom of the contact to minimize the electrical resistance. High-energy ions from the etch plasma can be implanted into the substrate beneath the contact. Often, a dielectric layer is placed over the silicon substrate to allow the contact etch to stop or partially stop in the film, allowing a transition to a less damaging etch. Figure 5.37 compares three different contact etch process options, showing an extreme variation in contact diode electrical characteristics (Yang, 1998).

5.9.2 Degradation in MOS Capacitance-Voltage Characteristics

The capacitance-voltage (C-V) characteristics of a MOS capacitor provide essential information about a CMOS process. Also, as MOS capacitors are used as a circuit element, it is crucial that the C-V characteristics are well-controlled and match the design models. Plasma and process damage can significantly modify a MOS capacitor's characteristics. As a worst-case example, Fig. 5.38 shows C-V plots of the native oxide dielectric after direct exposure to an argon plasma (Nakakubo, 2015). The undamaged reference C-V is abnormal due to the low-quality and thin native oxide. Different argon plasma treatments change the C-V characteristics toward a more normal behavior as the native oxide becomes thicker than the reference. In less severe cases, plasma damage can shift the threshold voltage due to fixed charge, change the shape due to interface states, etc.

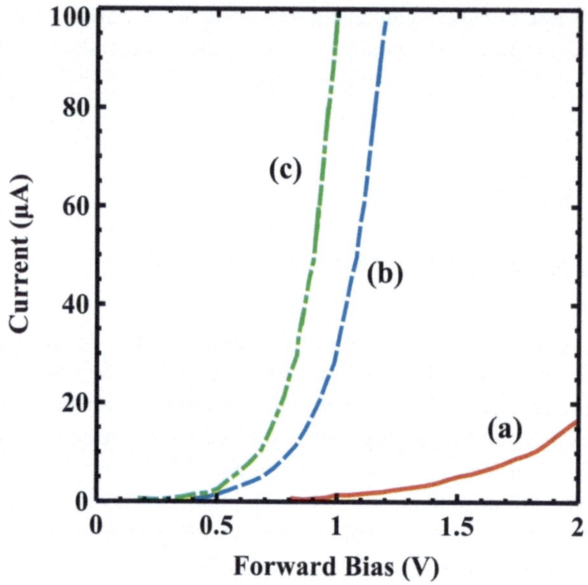

Fig. 5.37 Contact diode I-V characteristics of a contact formed with three different etch processes. (a) A tetraethylorthosilicate (TEOS) glass is used as a bottom layer above the substrate etched with a standard low power contact etch process. (b) A nitride layer was used as the bottom layer etched with the same low power etch as (a). (c) The same etch process as (1) with an added damage removal soft etch

5.9.3 Impact of the Gate Dielectric on Process Damage

The gate dielectric is the material most commonly damaged by process damage. The entire wafer surface is covered with transistors containing gate dielectric, creating many structures susceptible to process damage. In addition, the dielectric properties can determine how sensitive the dielectric is to process damage. Figure 5.39 shows an example gate leakage plot of process damage as a function of the gate oxide thickness for different damaging process steps (Krishnan et al., 1998). Note that the worst-case damage occurs at around 3–4 nm dielectric thickness. The reference does not explain why the metal etch experiment in Fig. 5.39 increases monotonically instead of showing a peak at 3–4 nm.

The gate dielectric thickness dependence is a complicated mixture of electrical and physical parameters and may differ for different CMOS technologies and processes. Some insights can be gained by performing analytical or technology computer-aided design of the antenna damage circuit. Figure 5.40 shows a plot of the peak electrical field during plasma damage for a range of gate oxide thicknesses and antenna ratios, indicating the peak plasma damage occurs in the 35–50 Å range (Weng, 2007).

The gate dielectric thickness can determine the severity of process damage. Thin dielectrics less than about 3 nm leak more than thicker dielectrics due to direct

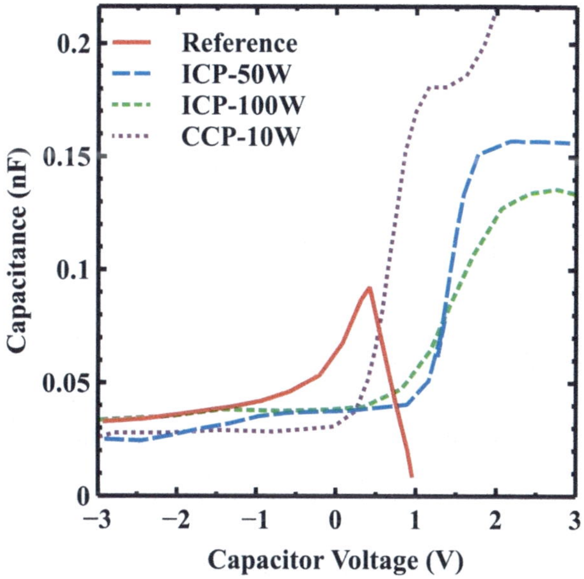

Fig. 5.38 Capacitance-voltage plots of a mos capacitor damaged by the exposure of a silicon surface exposed to an argon plasma with three different plasma conditions, power, and a reference. The C-V characteristics are measured on the surface native oxide with a mercury probe. The abbreviations are as follows: *ICP* inductively coupled plasma, *CCP* capacitively coupled plasma

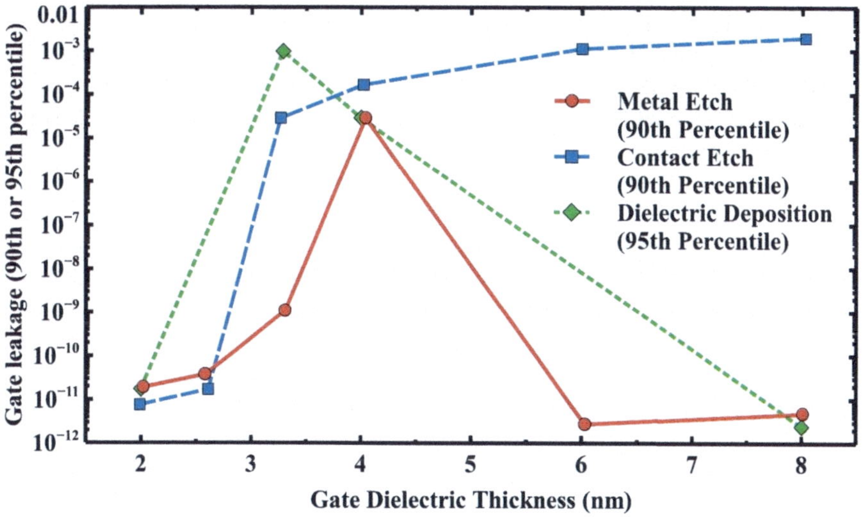

Fig. 5.39 Process damage gate dielectric leakage plots showing the 90% or 95% leakage current measured at 5 MV/cm plotted against gate oxide thickness for a contact etch, metal etch, and dielectric deposition. The dielectric deposition was performed on top of a gate antenna

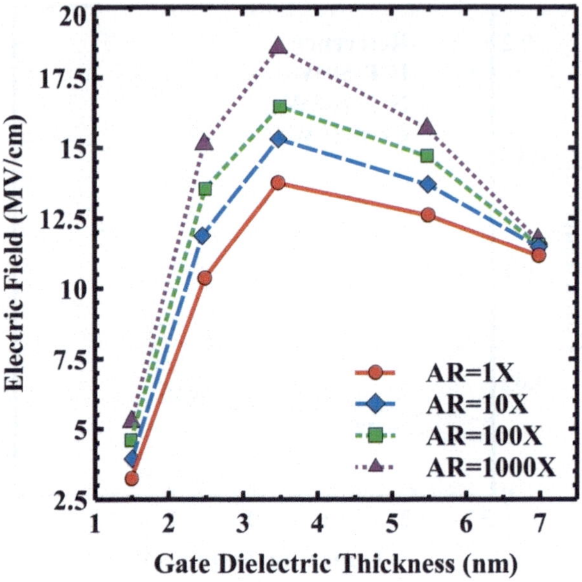

Fig. 5.40 Calculated peak electric field as a function of gate dielectric thickness and antenna ratio from an analytical model. The model combines the oxide leakage, plasma current, and the antenna ratio. See the reference for details

tunneling conduction, making them more resistant to damage due to more charge being drained from the gate during the damage event. Thicker dielectrics conduct less because the thicker dielectric prevents direct tunneling. Thicker dielectrics conduct by Fowler-Nordheim (F-N) tunneling and leak less than thin dielectrics, making them more susceptible to damage. Figure 5.41 shows the gate dielectric leakage for different dielectric thicknesses. Direct tunneling causes thin dielectrics to leak at low voltages, which improves process damage by draining the charge that accumulates on antennas, reducing the severity of process damage. Thin dielectrics may perform better because the charge to break down is higher as electrons can pass through the dielectric with a low probability of damage. Thicker dielectrics have less leakage and more propensity for damage by carrier conduction. The worst dielectric thickness for plasma damage may be for an intermediate thickness, as shown in Fig. 5.41 (Ranuárez et al., 2006). Thicker dielectrics also trap more tunneling charge carriers, resulting in increased parametric shifts.

In addition to the different tunneling mechanisms shown in Fig. 5.41, slight variations in the gate dielectric thickness in the different tunneling regions can significantly impact the tunneling. Figure 5.42 shows a similar plot with smaller increments in gate dielectric leakage, clearly distinguishing between the direct tunneling and Fowler-Nordheim tunneling regions (Strong et al., 2009).

Process damage is a strong function of the dielectric thickness, as shown in Fig. 5.39. Slight changes in the dielectric thickness of a few angstroms can shift the

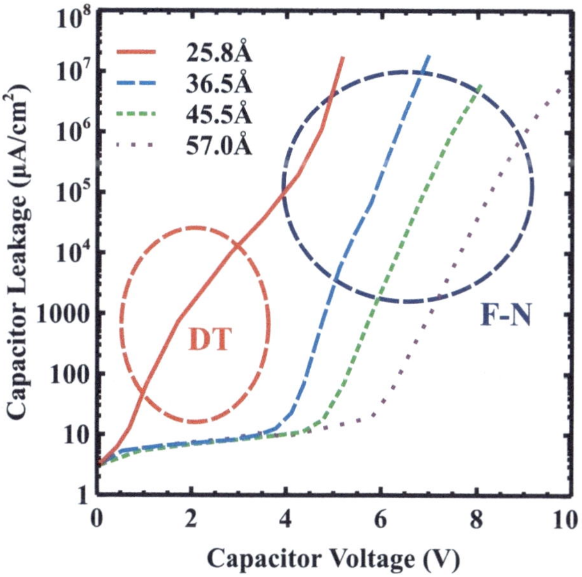

Fig. 5.41 Gate dielectric leakage plotted against gate voltage for several dielectric thicknesses. The Fowler-Nordheim (F-N) tunneling and the direct tunneling regions are shown. The direct tunneling region only occurs for dielectrics less than around 30 Å

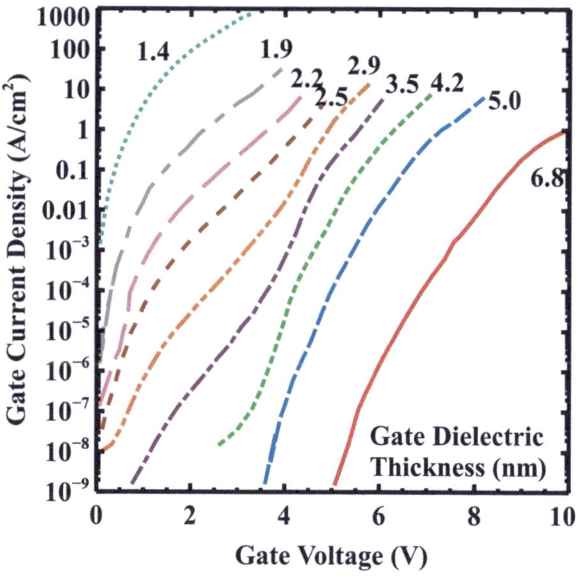

Fig. 5.42 Gate leakage plot as function of gate voltage and dielectric thickness shows a clear distinction between direct tunneling on the left side and Fowler-Nordheim tunneling on the right side

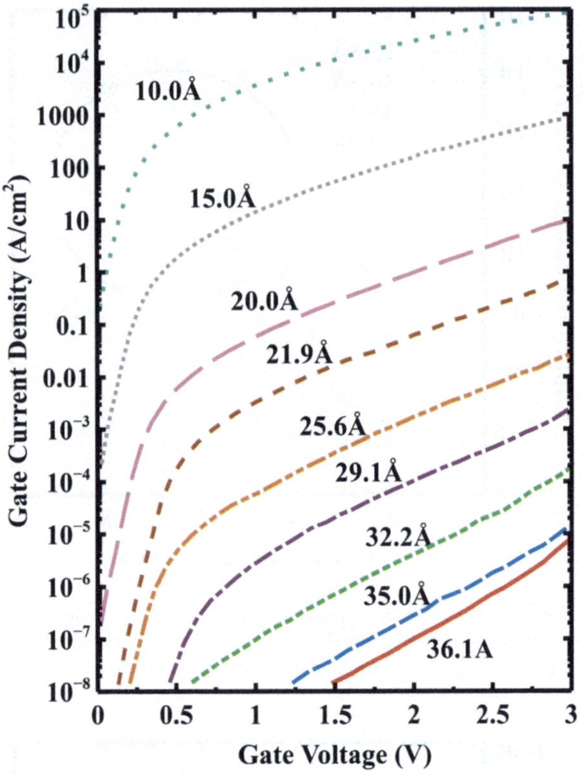

Fig. 5.43 Simulated gate dielectric leakage as a function of thickness for a NMOS transistor with SiO_2 dielectric

gate leakage orders of magnitude, easily changing the delicate balance between the charge flowing onto the antenna from the processing conditions and charge flowing off the antenna through dielectric leakage. A modification of this balance can create a significant change in the process damage. Figure 5.43 shows the gate dielectric leakage plotted against the voltage and thickness (Strong et al., 2009). A change in the thickness of 1 Å can change the dielectric leakage by an order of magnitude. The process control capability for a gate dielectric is usually a few angstroms, requiring process damage characterization across the possible thickness ranges in manufacturing.

In addition to gate leakage, other process damage-related parametric shifts are strong functions of the dielectric thickness. The severity of damage varies with the dielectric thickness, ranging from no shift to a significant shift across a range of gate dielectric thicknesses. Figure 5.44 shows an example of a V_t shift across a wafer due to metal etch damage (Park & Hu, 1998). The thicker dielectrics show a significant shift, while the shift for thin dielectrics is negligible. Thicker dielectrics have a higher propensity for charge trapping because a carrier transiting a thicker dielectric has a higher probability of damaging the dielectric, creating a defect, and trapping

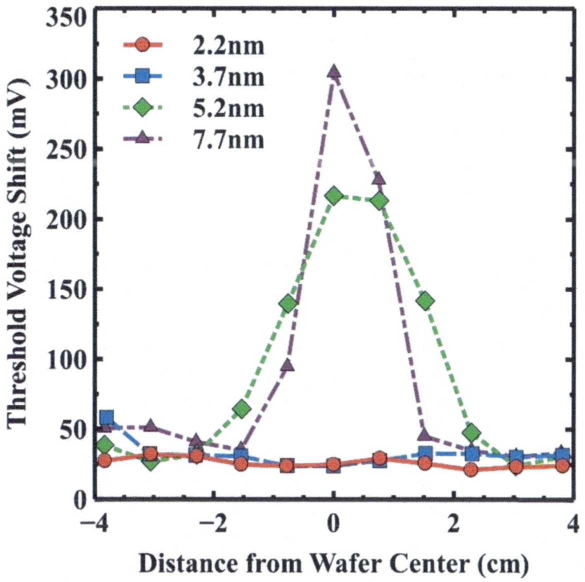

Fig. 5.44 Plot of a NMOS threshold voltage shift across a wafer for several different dielectric thicknesses damaged by a metal etch of antenna. The antenna area is 640 × 640 μm^2, and the transistor size is 10 × 1.5 μm^2

a carrier. Also, a charge trapped in a thicker dielectric has a lower probability of escaping the trap due to a longer tunneling distance. The stress-induced leakage current (SILC) is reduced in thick dielectrics as the probability of trap-to-trap tunneling across the thick dielectric is less likely. Conversely, thinner dielectrics have a low propensity for trapping, having a low probability of encountering an atom in the material and a high probability of a carrier escaping the trap through tunneling.

5.9.4 Isolation Edge Leakage

The edge of the channel where the edge of the active transistor meets the shallow trench isolation (STI) region can be susceptible to process damage. Process damage to this region can create a parasitic channel parallel to the active transistor channel. The parasitic channel can create a leakage path, significantly deviating from the designed transistor behavior, impacting chip performance and leakage. Figure 5.45 shows a sketch of a transistor with the field edge leakage path highlighted (Lee & Lee, 2013). The leakage path occurs where the channel meets the STI edge.

The susceptibility is caused by the lack of gate control in this region and the difficulty in doping the edge. Fixed charge or interface states created in the parasitic region by process damage can lower the threshold of the parasitic transistor forming the leakage path. Fixed charge can cause a much larger threshold voltage shift in the

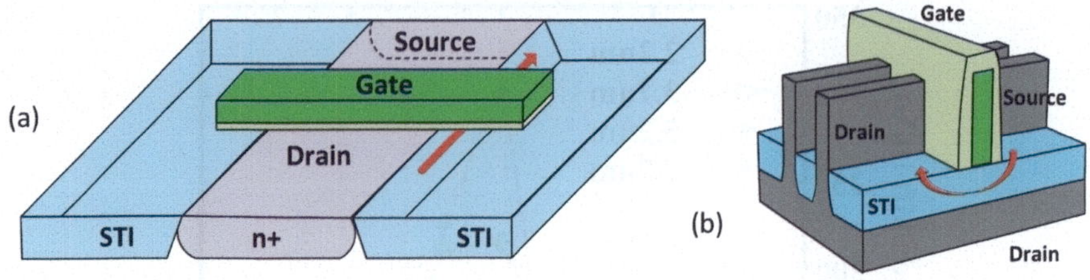

Fig. 5.45 Sketch showing a CMOS transistor with a field or STI edge leakage path. (**a**) Planar leakage path. (**b**) Finfet leakage path

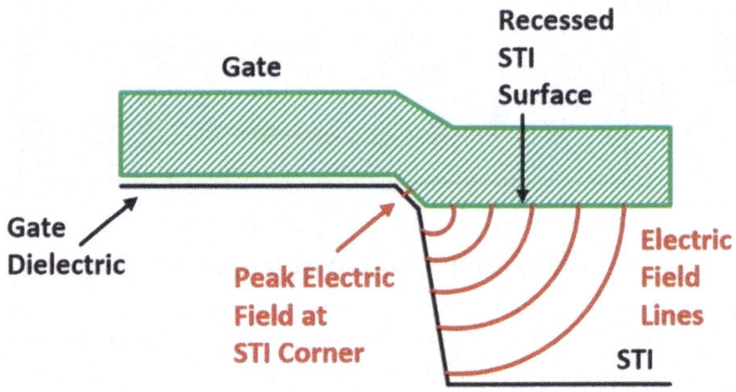

Fig. 5.46 Sketch showing enhanced electric field enhancement at the STI-channel edge where the top of the shallow trench isolation is recessed below the silicon surface, enhancing the electric field

parasitic leakage region because of the thick isolation dielectric compared to the active CMOS fet. The threshold shift V_t as a function of the fixed oxide charge Q_f is given by Eq. (5.5), where t_{ox} is the dielectric thickness, and ε is the dielectric permittivity. This relationship indicates that the thick field edge region with a large t_{ox} has the propensity to leak with a smaller amount of fixed charge than the active transistors. Thus, it is not unusual for the field edge to fail first for smaller amounts of process damage.

$$V_t \sim -\frac{Q_f t_{ox}}{\varepsilon} \tag{5.5}$$

The geometry of the field edge affects the electrical field lines. The electric field enhancement occurs as a function of the radius of curvature at the corners, as described in Sect. 5.4.4. Figure 5.46 shows a sketch showing the enhanced field lines at the edge and corners of the isolation, making leakage worse (McLain et al., 2017). The STI dielectric in the sketch is recessed below the silicon surface, enhancing the corner electric field. The shallow trench isolation in a CMOS process is optimized to lower the corner electric field.

The parasitic transistor at the isolation field edge normally never conducts over the normal operating voltage range of the active CMOS transistors. However,

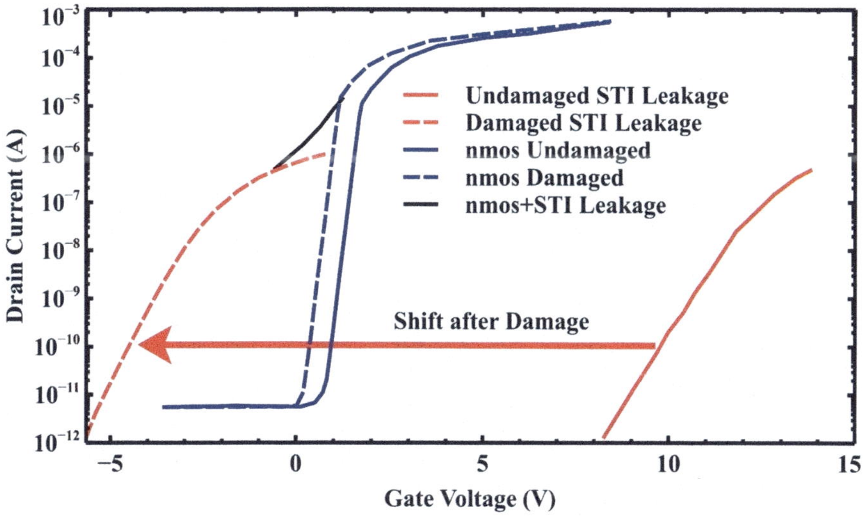

Fig. 5.47 Example plot of a sub-threshold kink showing the NMOS field (STI) edge parasitic transistor turning on at a different gate voltage than the mosfet. The undamaged parasitic STI transistor leakage on the right shifts to the left after plasma damage (red lines). The blue lines shows the I-V active NMOS transistor I-V curve before and after damage. The black curve shows the combined active NMOS FET and the damaged parasitic isolation transistor current

process damage can shift the parasitic device's operation into the transistor's operating range, creating two devices in parallel and causing havoc on the chip operation. The parasitic transistor current-voltage characteristics can be observed in the characteristics of the active transistors, as shown in Fig. 5.47 (Schwank et al., 2008; Teng et al., 2014). Typically, the threshold voltage of the undamaged parasitic isolation transistor is far higher than that of the CMOS transistors, as shown on the right side of the plot, and never affects the I-V characteristics of the active CMOS devices. As the isolation edge is damaged, the threshold voltage of the parasitic transistor drops. As a result, it can conduct at the same voltages as the CMOS transistor, severely distorting the I-V characteristics. Both conduction paths can be observed in the combined I-V curve. The ordinary transistor NMOS I-V characteristic shifts to the left with plasma damage. The damaged parasitic transistor is shown on the left side of the figure with a lower V_t and low current due to the limited width of the STI edge parasitic. The total transistor current is the sum of the two transistors, typically causing catastrophic failure of the CMOS chip. The NMOS or PMOS isolation may fail depending on the polarity of the charge, with NMOS failing with positive charge and PMOS failing with negative charge. The parametric properties, such as diode characteristics, etc., can also be modified by trapped charge.

Modifications to the gate layout to reduce field isolation failure are possible in some CMOS processes. The channel length at the edge of the shallow trench can be extended, as shown in Fig. 5.48a (Esqueda et al., 2005). This dog-bone layout can provide experimental evidence of enhanced leakage at the field edge due to process

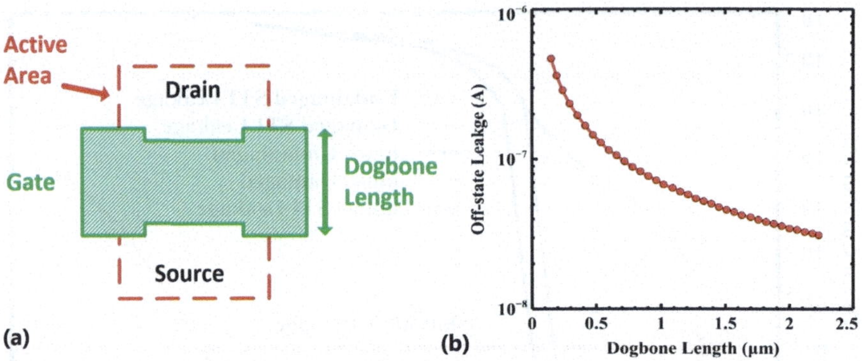

Fig. 5.48 Dog-bone transistor layout and electrical simulation showing symptoms of leakage along the shallow trench isolation. (**a**) Top down layout showing extended gate length along the shallow trench isolation to vary STI edge leakage. The center gate length is 0.18 μm. The sketch is not to scale (**b**) simulated gate leakage as a function as dog-bone length showing reduced leakage with an increased dog-bone length

damage. Extending the channel length at the STI edge can reduce damage-induced leakage. Figure 5.48b shows a simulation of the off-state leakage as a function of dog-bone length for a fixed amount of STI edge charge density. This technique is useful for longer-channel CMOS devices such as planar devices and may not be viable for finfet devices. Other layout variations shown in Chap. 11, such as donut devices, can form a transistor, eliminating the STI edge leakage and providing proof of STI leakage (Lacoe, 2008). These options are likely impractical in advanced CMOS technologies.

The isolation process can impact the susceptibility to isolation failure. For example, a thin nitride liner is sometimes used in shallow trench isolation (STI) to protect the silicon during oxidation. The nitride liner can trap electrons or holes, changing the electrical behavior along the vertical regions near the surface. Figure 5.49 shows an STI cross-section sketch where the holes can become trapped, affecting the leakage in the field edge regions in red (Teng et al., 2014). The trapped holes lower the threshold voltage locally at the shallow trench isolation (STI) edge, creating a leakage path at the boundary of the STI edge and the gate dielectric.

5.9.5 *High-k Dielectric Dependence*

CMOS technology transitioned to high-k dielectrics, moving away from silicon dioxide and silicon nitride-based dielectrics to reduce dielectric leakage and enable continued scaling. High-k dielectrics behave differently under process damage conditions. Significant differences in the band structure, such as reduced bandgap and band offsets, may cause increased dielectric leakage under process damage

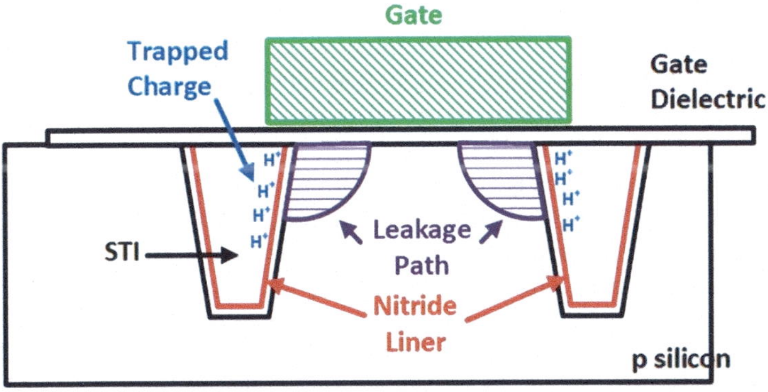

Fig. 5.49 Sketch showing hole trapping in the nitride liner in shallow trench isolation

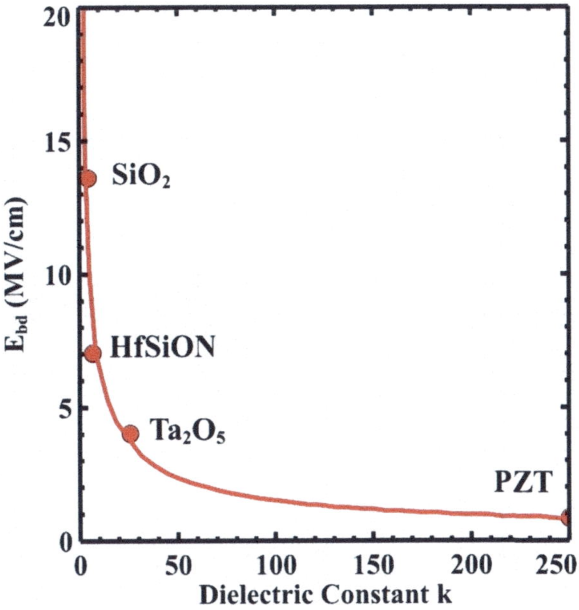

Fig. 5.50 Dielectric breakdown field versus dielectric constant

conditions, creating the propensity for increased or reduced damage due to draining the damaging charge. High-k dielectrics tend to have lower bandgaps and offsets when compared to the traditional dielectric SiO_2 (McPherson et al., 2002). Differences in bond strength change how resistant high-k dielectrics are to damage. High-k dielectrics can be fundamentally weaker than silicon dioxide and silicon nitride. Figure 5.50 shows the breakdown field plotted against the dielectric constant, indicating that higher dielectric constant materials with weaker breakdown fields may be more susceptible to process damage (McPherson et al., 2003). The ionic bonds in the high-k dielectric shift much farther than SiO_2 under the presence of an electric field, making the bonds easier to break. This is a fundamental property of high-k dielectrics.

High-k dielectrics, typically consisting of transition-metal oxides, are vastly differ from traditional SiO_2-based dielectrics, with a large spectrum of process options. Many fundamental material properties are quite different for high-k dielectrics, including bandgap, band offsets, bond strength, etc. This spectrum of differences can cause high-k dielectrics to occasionally be significantly better than SiO_2. For example, the trapping characteristics can cause the dielectric to be more resilient to stress. Figure 5.51 compares constant current stress measurements for a high-k and silicon dioxide dielectric with the same electrical thickness (Kamei et al., 2014). The high-k results without process damage survive approximately an order of magnitude longer than the SiO_2 dielectric, while the results are similar to SiO_2 with plasma damage. The difference in the lifetime is attributed to trapping characteristic differences in this experiment.

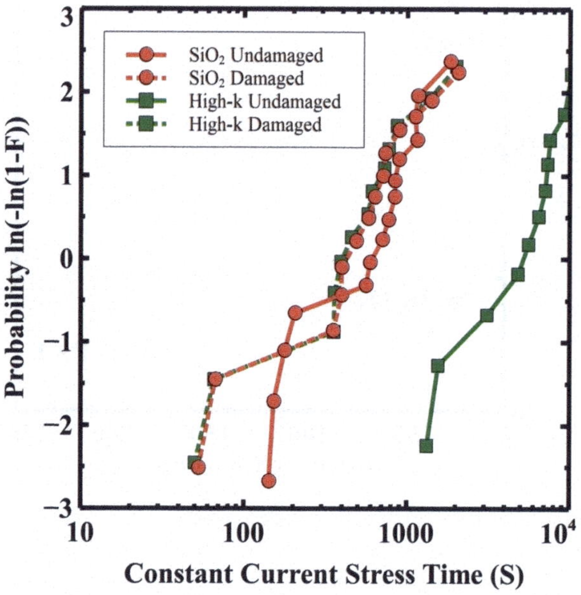

Fig. 5.51 Probability plot comparison of SiO_2 and high-k dielectric constant current (QBD) stress failure times with and without plasma damage

5.9.6 Degradation of Radiation Hardness

The radiation hardness of CMOS chips is vital for space and military applications, as well as for commercial applications. The electrical signatures of process damage and radiation damage can be remarkably similar, if not indistinguishable. Plasma damage can significantly degrade the radiation hardness of CMOS transistors. Figure 5.52 shows an example of radiation hardness degradation by plasma damage where the transistors are exposed to X-rays after process damage (Yue & Sinha, 1998). The time-zero differences (before radiation exposure) in V_t and subthreshold slope with and without a gate antenna are minor. After exposure to X-ray radiation, the latent damage from process damage can be seen, increasing the shifts in the transistor characteristics.

5.9.7 Increased Low-Radiation Dose Rate Sensitivity

There is a common phenomenon in radiation hardness testing called enhanced low dose rate sensitivity (ELDRS), where radiating a device with a slower amount of radiation per second actually causes more radiation damage (Fleetwood et al., 2024; Tsetseris et al., 2005). This phenomenon may interact with plasma damage, changing the behavior and possibly resulting in radiation hardness degradation. The enhancement can be factors of 5 or more. The cause of this phenomenon was attributed to increased H_2 molecule formation at high dose rates, which affected

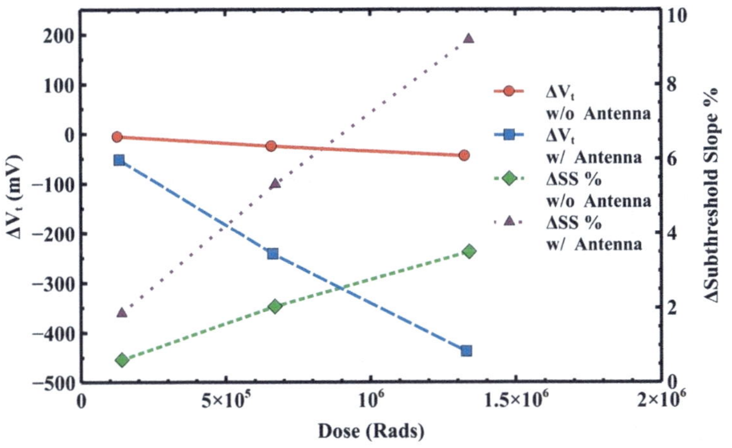

Fig. 5.52 Example of degradation of radiation hardness degradation by process damage demonstrated by comparing a gate antenna and no antenna before and after X-ray exposure. The plot shows threshold voltage shift and subthreshold slope degradation (ΔS)

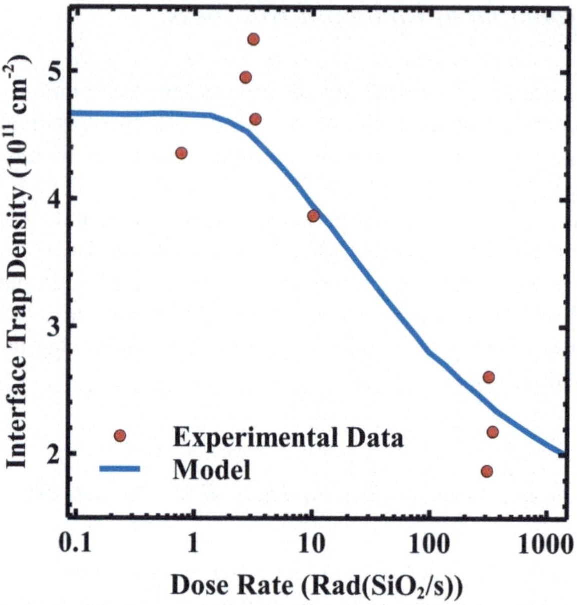

Fig. 5.53 Plot showing enhanced low dose rate sensitivity (ELDRS) of interface state generation for a constant total dose radiation exposure (10 Kev X-rays) at different dose rates

hydrogen kinetics as the higher dose rates saturated the dielectric with hydrogen. As an example, Fig. 5.53 shows the interface state density as a function of the dose rate for a capacitor irradiated to 200 krad (SiO_2). The interface state density is higher as the dose rate decreases (Fleetwood, 2013). Process damage can potentially degrade ELDRS (Cellere et al., 2002) and the radiation hardness of CMOS chips.

5.9.8 Degradation of Electrostatic Discharge Circuits

Electrostatic discharge (ESD) circuits protect the chip's bond pads from electrostatic discharge during chip and package handling or during the product life. These circuits are designed to meet specific discharge models, such as the human body model or machine models (Liou, 2016). Although these circuits are not designed to protect against plasma damage, they serve this function. Plasma damage conditions operate far outside the designed conditions for the common ESD models. Plasma damage can degrade or damage the ESD circuits if the plasma damage conditions cause damage in electrical operation regions where the ESD circuits fail.

5.9.9 Diode Leakage

Diode leakage can be degraded by plasma and process damage through many mechanisms. The most common mechanisms are listed in Table 5.2. Degradation can occur through metal-related mechanisms such as contamination in the junction and metal filaments or other features spiking (shorting) the junction, which may occur in salicide and other silicon surface metallization processes. The silicon surface can be damaged through kinetic ion damage, removal of the surface, and contamination by plasma components. Charge injected through plasma damage into the dielectrics surrounding the diode can change the electrical fields, affecting the junction diode behavior. Diodes are stressed through plasma damage and can be physically degraded by excessive currents, resulting in thermal damage. Ion implantation can create several types of defects, including defects caused by local amorphization of the substrate and slip and dislocations during post-implant crystal regrowth (see Chap. 9).

The investigation of diode leakage usually involves determining if the leakage is area or STI edge-dependent, which can reveal if the defect is located at the STI edge. If the leakage tracks the area of the diode, then the leakage is presumed to be uniform. If the leakage does not track the area of the diode, then the leakage is presumed to be a point defect, such as a spiked junction.

For example, Fig. 5.54 shows a plot of diode reverse current leakage as a function of polysilicon etch time on the surface of a wafer (Miyawaki et al., 1991). The reverse bias current leakage increases with increasing etch time until the surface is exposed. The damage slope increases at this point, showing a higher damage rate per unit time as the silicon surface is directly exposed to the plasma. The problem was solved by reducing the plasma ion energy to minimize the damage.

Table 5.2 Common process and plasma-damage-induced mechanisms for diode leakage

Leakage mechanism	References
Metal contamination	Claeys and Simoen (2018)
Junction spiking by metallization	Claeys and Simoen (2018)
Surface damage	Yang (1997)
Excessive silicon loss	Yang (1997)
Trapped charge in surrounding dielectrics	
Knocked on contaminants	Nakakubo (2015)
Excessive electrical stress during damage	
End-of-range ion implantation defects	Wolf and Tauber (1986)
Slip and dislocations after high-dose implantation	Tung et al. (2003)

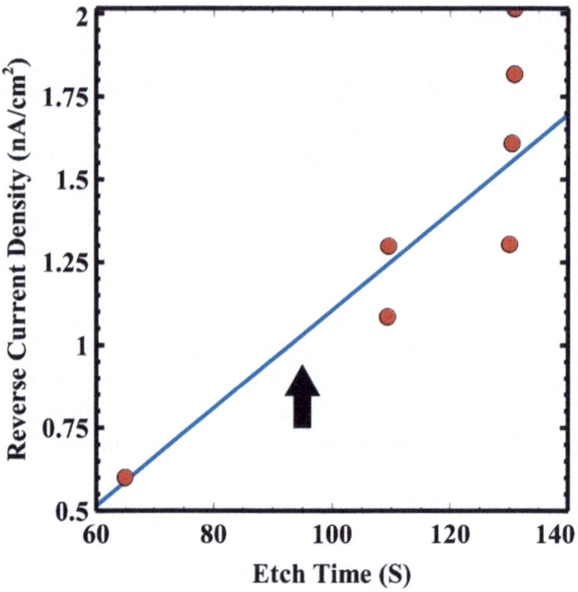

Fig. 5.54 Reverse bias diode leakage as a function of etch time for a gate polysilicon etch as the polysilicon is removed and the silicon surface is exposed. The point marked with an arrow shows the point where the silicon is exposed

References

Baker, R. J. (2019). *CMOS circuit design, layout, and simulation* (4th ed.). IEEE Press Wiley. https://doi.org/10.1002/9780470891179

Barbottin, G., & Vapaille, A. (1989). *Instabilities in silicon devices silicon passivation and related instabilities* (Vol. 2). Elsevier Science Publishers B.V.

Barbottin, G., & Vapaille, A. (1999). *Instabilities in silicon devices new insulators and radiation effects* (Vol. 3). Elsevier Science Publishers.

Bonani, F., & Ghione, G. (2001). *Noise in semiconductor devices modeling and simulation.* Springer. https://doi.org/10.1007/978-3-662-04530-5

Brozek, T., et al. (1995). Assessment of uniform and non-uniform damage in plasma etched submicron transistors. In *International Conference on VLSI Technology, Systems, and Applications.* https://doi.org/10.1109/VTSA.1995.524633

Brozek, T., et al. (1996). Hole trap generation in the gate oxide due to plasma-induced charging. *IEEE Electron Device Letters, 17*(9), 440–442. https://doi.org/10.1109/55.536286

Calimera, A., et al. (2012). Design techniques and architectures for low-leakage SRAMs. *IEEE Transactions of Circuits and Systems, 59*(9), 1992–2007. https://doi.org/10.1109/TCSI.2012.2185303

Cantin, C., Laviron, C., & Gove, G. (1999). Charging control on high energy implanters. *Microelectronics Reliability, 49*(2), 209–214. https://doi.org/10.1016/j.microrel.2008.11.012

Cellere, G., et al. (2002). Radiation-induced depassivation of latent plasma damage. *Microelectronics Engineering, 60*, 439–450. https://doi.org/10.1016/S0167-9317(01)00704-3

Chen, J., et al. (1987). Subbreakdown drain leakage current in MOSFET. *IEEE Electron Device Letters, 8*(11), 515–517. https://doi.org/10.1109/RELPHY.1997.584241

Chen, I.-C., et al. (1989). Interface-trap enhanced gate-induced leakage current in MOSFET. *IEEE Electron Device Letters, 10*(5), 216–218. https://doi.org/10.1109/55.31725

Cheung, K. P. (1996). On the use of Fowler-Nordheim stress to reveal plasma-charging damage. In *International Symposium on Plasma Process-Induced Damage*. https://doi.org/10.1109/PPID.1996.715192

Cho, H.-J., Son, Y., Oh, B., Jang, S., Lee, J.-H., Park, B.-G., & Shin, H. (2011). Investigation of gate etch damage at metal/high-k gate dielectric stack through random telegraph noise in gate edge direct tunneling current. *IEEE Electron Device Letters, 32*(4), 569–571. https://doi.org/10.1109/led.2011.2106108

Claeys, C., & Simoen, E. (2018). *Metal impurities in silicon- and germanium-based technologies origin, characterization, control, and device impact*. Springer. https://doi.org/10.1007/978-3-319-93925-4

Esqueda, I. S., et al. (2005). Two-dimensional methodology for modeling radiation-induced off-state leakage in CMOS technologies. *IEEE Transactions on Nuclear Science, 52*(6), 2259–2264. https://doi.org/10.1109/TNS.2005.860671

Fleetwood, D. M. (2013). Total ionizing dose effects in MOS and low-dose-rate-sensitive linear-bipolar devices. *IEEE Transactions on Nuclear Science, 60*(3), 1706–1730. https://doi.org/10.1109/TNS.2013.2259260

Fleetwood, D. M., et al. (2024). Effects of interface traps and hydrogen on the low-frequency noise of irradiated MOS devices. *IEEE Transactions on Nuclear Science, 71*(4), 555–568. https://doi.org/10.1109/TNS.2023.3323548

Hiblot, G., Subirats, A., Liu, Y., & Van Der Plas, G. (2019). Electrical characterization of BEOL plasma-induced damage in bulk FinFET technology. *IEEE Transactions on Device and Materials Reliability, 19*(1), 84–89. https://doi.org/10.1109/TDMR.2018.2881740

Hu, C. (2010). *Modern semiconductor devices for integrated circuits*. Prentice Hall.

Ielmini, D., et al. (2004). Impact of correlated generation of oxide defects on SILC and breakdown distributions. *IEEE Transactions on Electron Devices, 51*(8), 1281–1297. https://doi.org/10.1109/TED.2004.831959

Janesick, J. R. (2001). *Scientific charge-coupled devices*. The International Society for Optical Engineering.

Kamei, M., et al. (2007). Mechanisms for junction degradation of advanced MOSFETs induced by plasma processing. In *IEEE International Conference on Integrated Circuit Design and Technology*. https://doi.org/10.1109/ICICDT.2007.4299553

Kamei, M., et al. (2014). TDDB lifetime enhancement of high-K MOSFETs damaged by plasma processing—Conflicting results in plasma charging damage evaluation. In *IEEE IIRW*. https://doi.org/10.1109/IIRW.2014.7049506

Kim, D.-H. (2015). AVERT: An elaborate model for simulating variable retention time in DRAMs. *Microelectronic Engineering, 55*, 1313. https://doi.org/10.1016/j.microrel.2015.06.064

Koyangi, M., et al. (1987). Hot-electron-induced punchthrough (HEIP) effect in submicrometer PMOSFET's. *IEEE Transactions on Electron Devices, 34*(4), 839–844.

Krishnan, S., et al. (1998). Antenna device reliability for ULSI processing. In *IEEE IEDM*. https://doi.org/10.1109/IEDM.1998.746430

Lacoe, R. C. (2008). Improved integrated circuit performance through the application of hardness by design methodology. *IEEE Transactions on Nuclear Science, 55*(4), 1903–1925. https://doi.org/10.1109/TNS.2008.2000480

Lee, M. S., & Lee, H. C. (2013). Dummy gate-assisted n-MOSFET layout for a radiation-tolerant integrated circuit. *IEEE Transactions on Nuclear Science, 60*(4), 3084–3091. https://doi.org/10.1109/tns.2013.2268390

Lee, J.-H., et al. (2008). Scrubber clean process induced CDM ESD-like: CSM (charged surface model) event caused by dummy patterns. In *IEEE IRPS*. https://doi.org/10.1109/RELPHY.2008.4558904

Li, X., et al. (1994). Effect of plasma poly etch on effective channel length and hot carrier reliability in submicron transistors. *IEEE Electron Device Letters, 15*(4), 140–141. https://doi.org/10.1109/55.285403

Li, X.-Y., et al. (1995). Degraded CMOS hot carrier life time—Role of plasma etching induced charging damage and edge damage. In *IEEE IRPS*. https://doi.org/10.1109/RELPHY.1995.513689

Li, R., Kong, W.-R., Tao, K., Yu, L.-J., Huang, K., Ning, J., Geng, C.-Q., & Wang, C.-D. (2007). Threshold voltage shift due to mechanical stress-enhanced plasma process-induced damage in 0.13 micron PMOSFET. *IEEE Electron Device Letters, 28*(5), 360–362. https://doi.org/10.1109/led.2007.894644

Liou, J. J. (2016). *Electrostatic discharge protection advances and applications*. CRC Press. https://doi.org/10.1201/b18976

Lombardo, S., et al. (2005). Dielectric breakdown mechanisms in gate oxides. *Applied Physics Reviews, 98*, 121301. https://doi.org/10.1063/1.2147714

Ma, S. (2004, July 8). *Plasma induced charging damage: From a semiconductor equipment vendor point of view*. Plasma Etch User Group (PEUG).

Ma, S., et al. (2001). Gate-induced leakage enhanced by plasma charging damage. *IEEE Transactions on Electron Devices, 48*(5), 1006–1008. https://doi.org/10.1109/16.918252

Matsunaga, N., Yamaguchi, H., & Shibata, H. (2007). Spreading antenna effect of PID in damascene interconnect process. *IEEE Transactions on Electron Devices, 54*(6), 1486–1491. https://doi.org/10.1109/ted.2007.896360

McLain, M. L., Barnaby, H., & Schlenvogt, G. (2017). Effects of channel implant variation on radiation-induced edge leakage currents in n-channel MOSFETs. *IEEE Transactions on Nuclear Science, 65*, 2235–2241. https://doi.org/10.1109/tns.2017.2705118

McPherson, J. W., et al. (2002). Proposed universal relationship between dielectric breakdown and dielectric constant. In *IEEE IEDM*. https://doi.org/10.1109/IEDM.2002.1175919

McPherson, J. W., Jinyoung, K., Shanware, A., Mogul, H., & Rodriguez, J. (2003). Trends in the ultimate breakdown strength of high dielectric-constant materials. *IEEE Transactions on Electron Devices, 50*(8), 1771–1778. https://doi.org/10.1109/ted.2003.815141

Miyawaki, M., et al. (1991). Reduction of fixed-pattern noise of BASIS due to low kinetic energy reactive ion and native-oxide-free processing. *IEEE Transations on Electron Devices, 38*(5), 1037–1043. https://doi.org/10.1109/16.78376

Nakakubo, Y. (2015). *A study of defect generation phenomena in single crystalline silicon substrate during plasma processing and the characterization techniques*. Kyoto University.

Nicollian, E. H., & Brews, J. R. (1982). *MOS (metal oxide semiconductor) physics and technology*. Wiley Interscience.

Okada, K., et al. (1999). A concept of gate oxide lifetime limited by B-mode stress induced leakage currents in direct tunneling regime. In *IEEE Symposium on VLSI Technology*. https://doi.org/10.1109/VLSIT.1999.799338

Park, D., & Hu, C. (1998). Plasma charging damage on ultrathin gate oxides. *IEEE Electron Device Letters, 19*(1), 1–3. https://doi.org/10.1109/55.650333

Park, J., et al. (2021). An in-DRAM BIST for 16 Gb DDR4 DRAM in the 2nd 10-nm-class DRAM process. *IEEE Access, 9*, 33487–33497. https://doi.org/10.1109/ACCESS.2021.3061349

Pyo, S. G., & Kim, S. (2013). Plasma damage mechanism of electron beam curing process for spin on dielectrics. *IEEE Electron Device Letters, 34*(5), 575–577. https://doi.org/10.1109/led.2013.2250246

Raghavan, N. (2009). Post breakdown reliability enhancement of ULSI circuits with novel gate dielectric stacks. In *ISIC 2009*.

Ranuárez, J. C., Deen, M. J., & Chen, C.-H. (2006). A review of gate tunneling current in MOS devices. *Microelectronics Reliability, 46*(12), 1939–1956. https://doi.org/10.1016/j.microrel.2005.12.006

Rider, G. (2021). Myths, misconceptions and mistakes in the electrostatic protection of field-sensitive items—Why it's time to re-visit device protection. *Global Journal of Researches in Engineering: J General Engineering, 21*(3).

Schroder, D. K. (2006). *Semiconductor material and device characterization* (3rd ed.). Wiley Interscience. https://doi.org/10.1002/0471749095

Schwank, J. R., et al. (2008). Radiation effects in MOS oxides. *IEEE Transactions on Nuclear Science, 55*(4), 1833–1853. https://doi.org/10.1109/TNS.2008.2001040

Selvarajoo, T. A. L., Ranjan, R., Kin-Leong, P., Lei-Jun, T., Chih Hang, T., & Wenhe, L. (2005). Dielectric-breakdown-induced epitaxy: A universal breakdown defect in ultrathin gate dielectrics. *IEEE Transactions on Device and Materials Reliability, 5*(2), 190–197. https://doi.org/10.1109/tdmr.2005.846674

Shin, H., & Hu, C. (1993). Monitoring plasma-process induced damage in thin oxide. *IEEE Transactions on Semiconductor Manufacturing., 6*(2), 96–102. https://doi.org/10.1109/66.216927

Shin, H., et al. (1993). Impact of plasma charging damage and diode protection on scaled thin oxide. In *IEEE IEDM*. https://doi.org/10.1109/IEDM.1993.347309

Shubhakar, K., et al. (2015). An SEM/STM based nanoprobing and TEM study of breakdown locations in HfO2 dielectric stacks for failure analysis. *Microelectronics Reliability, 55*, 1450. https://doi.org/10.1016/j.microrel.2015.07.027

Strong, A. W., et al. (2009). *Reliability wearout mechanisms in advanced CMOS technologies*. Wiley. https://doi.org/10.1002/9780470455265

Tang, L. J., Pey, K. L., Tung, C. H., Radhakrishnan, M. K., & Lin, W. H. (2004). Gate dielectric-breakdown-induced microstructural damage in MOSFETs. *IEEE Transactions on Device and Materials Reliability, 4*(1), 38–45. https://doi.org/10.1109/tdmr.2004.824374

Teng, A. S., et al. (2014). Gate bias temperature stress-induced off-state leakage in NMOSFETs: Mechanism, lifetime model and circuit design consideration. In *IEEE IRPS*. https://doi.org/10.1109/IRPS.2014.6861182

Tsetseris, L., et al. (2005). Common origin for enhanced low-dose-rate sensitivity and bias temperature instability under negative bias. *IEEE Transactions on Nuclear Science, 52*(6), 2265–2271. https://doi.org/10.1109/TNS.2005.860670

Tsividis, Y., & McAndrew, C. (2011). *Operation and modeling on of the MOS transistor* (3rd ed.). Oxford University Press.

Tsui, B. Y. (2000). Plasma charging damage during contact hole etch in high-density plasma etcher. *Microelectronics Reliability, 40*, 2039–2046. https://doi.org/10.1016/S0026-2714(00)00023-8

Tung, C. H., Sheng, G. T. T., & Lu, C. Y. (2003). *ULSI semiconductor technology atlas*. Wiley-Interscience. https://doi.org/10.1002/0471668796

Tzeng, P.-J., et al. (2002). Plasma charging effects on submicron MOS devices. *IEEE Transactions on Electron Devices, 49*(7), 1151–1157. https://doi.org/10.1109/TED.2002.1013270

Vici, A., Russo, F., Lovisi, N., & Irrera, F. (2020). On border traps in back-side-illuminated CMOS image sensor oxides. *IEEE Transactions on Electron Devices, 67*(5), 2022–2027. https://doi.org/10.1109/ted.2020.2983039

Wang, Z. (2004). *Detection of and protection against plasma charging damage in modern IC technology*. Twente University.

Wang, X. (2008). *Noise in sub-micron CMOS image sensors* [PhD, University of Southampton, UK].

Weng, W.-T. (2007). A comprehensive model for plasma damage enhanced transistor reliability degradation. In *IEEE International Reliability Physics Symposium*. https://doi.org/10.1109/RELPHY.2007.369916

Wolf, S., & Tauber, R. N. (1986). *Silicon processing for the VLSI era, Vol. 1: Process technology*. Lattice Press.

Yamabe, K., & Imai, K. (1987). Nonplanar oxidation and reduction of oxide leakage currents at silicon corners by rounding-off oxidation. *IEEE Transactions on Electron Devices, 34*(8), 1681–1687. https://doi.org/10.1109/T-ED.1987.23137

Yamaguchi, T., Sasaki S., & Nikoh, H. (1996). Control of RIE induced surface damage and junction leakage current in field isolation process. In *1st International Symposium on Plasma Process-Induced Damage*. https://doi.org/10.1109/PPID.1996.715237

Yang, M. (1997). Silicon damage mechanism in oxide etch. In *International Symposium on Plasma Process-Induced Damage*. https://doi.org/10.1109/PPID.1997.596694

Yang, M. (1998). Plasma damage and contact resistance. In *1998 5th International Conference on Solid-State and Integrated Circuit Technology*. https://doi.org/10.1109/ICSICT.1998.785809

Yue, J., & Sinha, S. (1998). Enhanced hole and interface trap generation in plasma damage bulk and SOI MOSFETs under total dose irradiation. In *3rd International Symposium on Plasma Process-Induced Damage*. https://doi.org/10.1109/PPID.1998.725565

Zhao, J., et al. (1996). Investigation of charging damage induced mismatch for submicron mixed-signal technology. In *IEEE IRPS*. https://doi.org/10.1109/RELPHY.1996.492058

Chapter 6
Latent Damage and Reliability Degradation

6.1 Introduction

Process damage can cause latent damage in CMOS chips where the chip's long-term reliability is degraded, while time-zero degradation is minimal or undetectable. Latent damage can be exceedingly difficult or impossible to detect at the end of the process flow, where the chips are tested and screened before shipping to the customer. Latent damage can be a severe field reliability hazard if it escapes into the field. Process damage can be detected electrically by monitoring parametric and product-level performance after the completion of processing. If the spectrum of parametric testing described in Chap. 5 does not detect plasma damage, latent damage may exist, creating a reliability hazard. Common latent damage reliability problems are described in the next section.

Figure 6.1 shows an example of latent damage (Kim et al., 1999). There is no correlation between the transistor parameters and the antenna ratios at time zero, with all antenna sizes overlayed on the same line as the reference on the left side of the plot. However, after a dielectric reliability stress test or revealing stress, the parameters shift as a function of the antenna ratio. This result indicates that the gate dielectric and interface have been degraded with increased traps and other defects. This defect can cause severe reliability problems as the chip performance shifts with time. Defects and traps created during plasma damage can create electrically neutral traps that are not populated with trapped carriers. These electrically neutral traps are not detectable with routine time-zero testing. Additional testing with carefully designed methods is needed to detect latent damage.

Figure 6.2 shows an example of hot electron lifetime degradation caused by a plasma-enhanced intermetal dielectric deposition process (Cheung, 1995). The dielectric deposition damaged the gate dielectric, creating oxide defects through trapping, as described in Chap. 5. The initial trapping rate is measured from the slope of the curve obtained by forcing a current in a capacitor and measuring the

K. Prall, *CMOS Plasma and Process Damage*,
https://doi.org/10.1007/978-3-031-89029-1_6

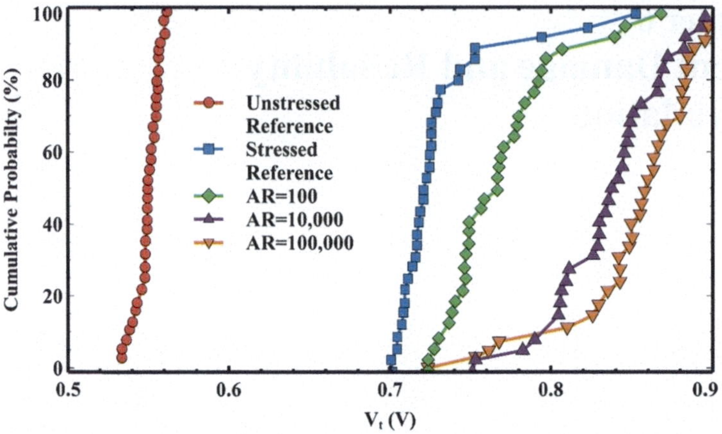

Fig. 6.1 Example of latent process damage caused by multiple process steps. Initial parametric data shows an insignificant deviation from the unstressed reference. The probability plot shows threshold voltage distribution as a function of antenna ratio after a gate dielectric reliability stress of 1 C/cm^2 charge forced through the dielectric

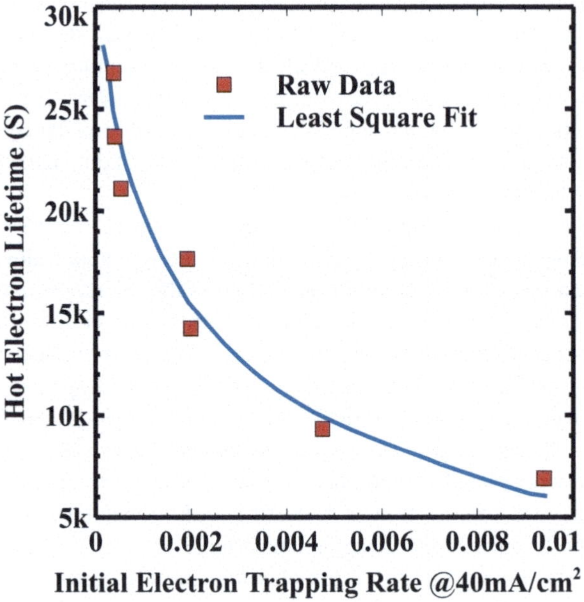

Fig. 6.2 Hot electron lifetime as a function of the initial electron trapping rate caused by dielectric damage during an interlevel dielectric deposition

voltage (Chap. 5) (Cheung, 1994). The initial electron trapping rate correlates to the hot-electron lifetime.

6.2 Damage Recovery

Following a process damage step, process steps later in the manufacturing flow may partially recover the damage, healing the damaged dielectrics by detrapping trapped charge, rebonding broken bonds, etc. Typically, thermal process steps can provide enough thermal energy to rebond broken bonds by thermal excitation of the atoms in the material. This process is called thermal annealing. Alloying processes treat the wafer in a hydrogen (or deuterium) ambient (see Chap. 14). An alloy step repairs damage by hydrogen bonding to dielectric defects, neutralizing charge, and replacing broken bonds. Annealing in a non-hydrogen environment such as N2 or Ar can recover damage primarily through thermal energy without the reactive hydrogen. Healing steps can partially recover parametric shifts caused by process damage, reduce latent damage, and improve reliability degradation. Healing is a process that makes latent damage challenging to detect by recovering observable parametric shifts, leaving difficult to detect hidden damage.

Thermal annealing results improve with increasing time and temperature. Figure 6.3 shows the partial process damage recovery of interface state defects

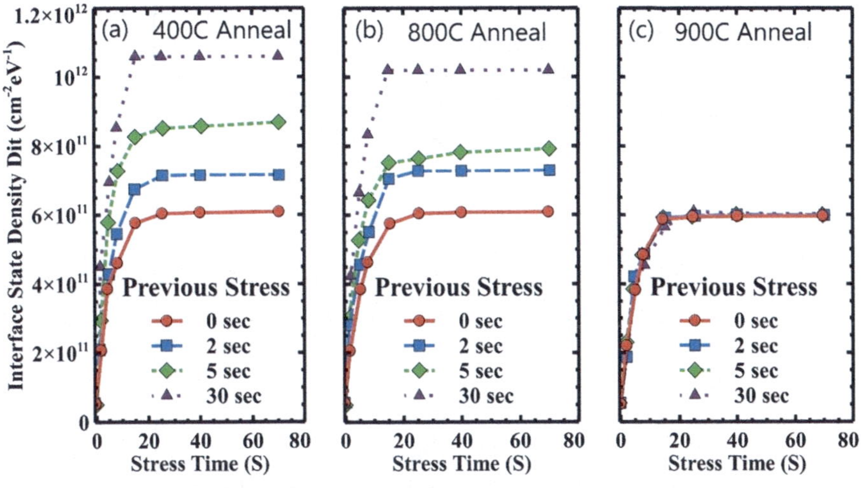

Fig. 6.3 Interface state density recovery after voltage stress with an anneal. (**a**) 400 °C in H_2/N_2 (**b**) 800 °C in N_2. (**c**) 900 °C in N_2. Polysilicon capacitors with an 8.5 nm SiO_2 dielectric are stressed at 9 V for 0, 2, 5, and 30 s. After the stress, the capacitors are annealed at 400 °C 20 min in N_2/H_2, 800 °C 20 min in N_2, and 900 °C in N_2. After the anneal, the capacitors are again stressed at 9 V, with the interface state density measured by high frequency and quasistatic capacitance-voltage measurements at various times

caused by a DC gate dielectric stress that simulates process damage (King & Hu, 1994). The experiment consists of a gate stress, followed by different thermal anneals to partially repair the damage and hide the degradation. Next, a second electric stress is used to determine whether the thermal anneal healed or hid the damage. The unstressed control is the curve labeled 0 s. The stressed curves deviate from the unstressed curve, generating increased interface states. The high-temperature anneal (900 °C) experiments show all the stressed groups merging to the unstressed curve, indicating that the damage was mostly healed with the high-temperature anneal. The DC stress is followed by thermal annealing and a second DC dielectric stress simulating a reliability stress measurement. The second stress causes an increasing interface state density that saturates at different levels depending on the first stress step time. The latent damage from the first step can be observed with the second stress. The latent damage is not observable in the time-zero interface state density. Annealing at increasing temperatures is more effective at eliminating the latent damage. The damage appears to be recovered entirely with an anneal at 900 °C, while residual latent damage remains at lower temperatures. However, the damage remains after the 900 °C anneal but is not detected in this measurement.

Partial recovery occurs at lower temperatures. Figure 6.4 shows the shift in the gate voltage for a constant current stress on a gate dielectric simulating dielectric damage (King & Hu, 1994). The gate voltage increases to maintain a constant

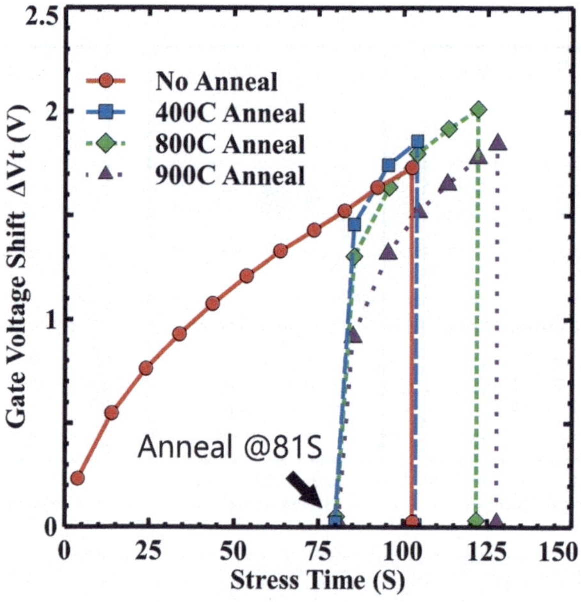

Fig. 6.4 Gate voltage shift as a function of electrical constant voltage stress. An anneal at different temperatures is inserted at about 81 s and the stress restarted after the anneal. Note that the anneal slightly improves the trapping

current during the stress as charge is trapped opposing the voltage. Midway through the stress, the wafers are annealed at various temperatures. The gate voltage shift decreases after the thermal annealing, indicating that charge was removed or compensated for during the anneal. Temperatures from 400 to 900 °C show this effect, partially healing the simulated process damage. The 900 °C anneal appears to mostly remove the damage, with lower temperatures showing partial recovery. Recovery can occur slowly at temperatures less than 400 °C, even at room temperature or below over hours, days, or weeks.

Recovering CMOS device damage is partially effective in alleviating device damage, but the residual degradation remains, resulting in significantly inferior devices compared to undamaged devices. In addition, the compromised devices usually trap charge faster and degrade faster when subjected to additional damage. After the 900 °C anneal, it appears that the damage is partially annealed, but it takes less time to obtain the same gate voltage shift compared to the initial stress results. This finding indicates that even the 900 °C anneal is insufficient to heal the damage completely.

Another example of plasma damage recovery is shown in Fig. 6.5, where the silicon surface is exposed to a polysilicon etch that damages the silicon surface of a diode (Miyawaki et al., 1991). The plasma degraded the reverse diode leakage, as shown in Chap. 5. Different anneals were used to recover the damage partially. A 1000 °C 30-min high-temperature anneal recovers the plasma damage. Also, a 400 °C 30-min hydrogen alloy recovers the damage.

Dark current noise in a CMOS imager is the background noise observed when a pixel is in the dark (Wang, 2008), as described in Chap. 8. Recovery of noise from process damage can also occur in CMOS imagers. Figure 6.6 shows a plot of damage recovery of the dark current in a CCD imager after exposure to deep UV light simulating a deep UV scanner (Li et al., 2004). The dark current partial recovery is

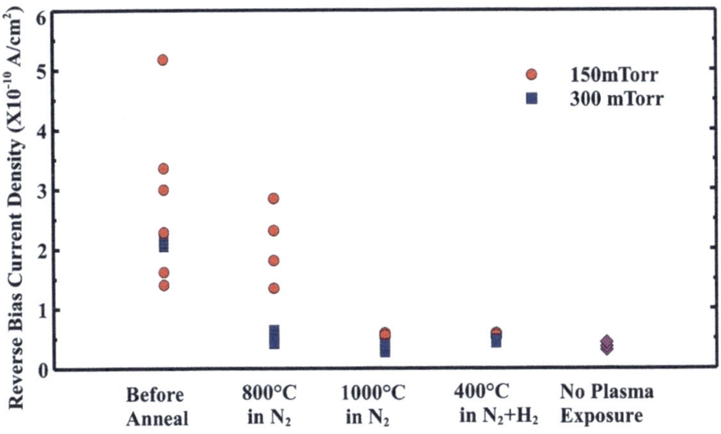

Fig. 6.5 Recovery of reverse biased diode leakage with annealing after exposure to a polysilicon plasma etch at two different pressures at 100 W over the exposed silicon substrate

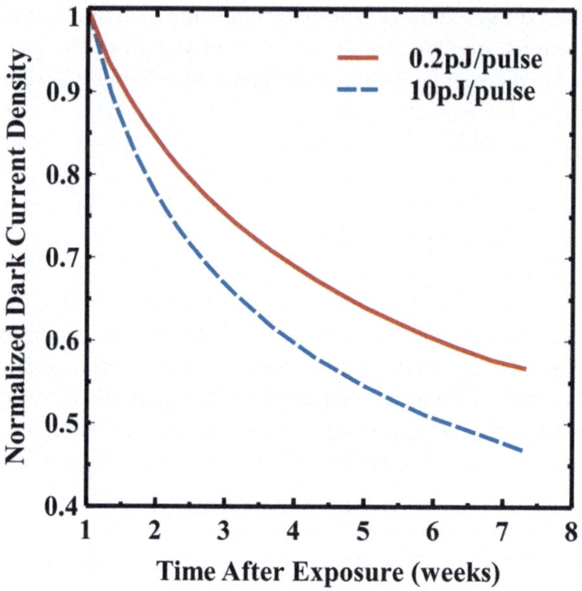

Fig. 6.6 Damage recovery in dark current to a charge-coupled device (CCD) image sensor exposed to a UV laser stored at room temperature

observed over several weeks as defects disappear at a low rate from the room-temperature thermal energy. The room-temperature thermal energy causes the repair of broken bonds, detrapping of trapped carriers, etc.

Each defect type has different annealing characteristics. For example, Fig. 6.7 shows a plot of defect annealing versus approximate temperature for vacancy and dopant-related defects based on measurements and first-order modeling (Hull, 1999, p. 646). Each defect anneals at different temperatures, transitioning from a specific concentration to a negligible concentration. This plot shows the annealing temperature for a 15–30 min anneal. Defects can be annealed at lower temperatures than the points shown in the plot for longer times (hours and days). Defects that anneal near the maximum temperatures where CMOS chips operate or are tested can result in a change in circuit behavior with time. The defects shown to anneal in the 100–300° region may be a threat to the stability of CMOS chips.

6.3 Reliability Degradation by Process Damage

6.3.1 Introduction

The presence of antennas on a CMOS chip may greatly increase plasma and process damage, causing havoc with yield, reliability, and the economic success of a semiconductor company. For example, Fig. 6.8 shows the chip burn-in failure rate

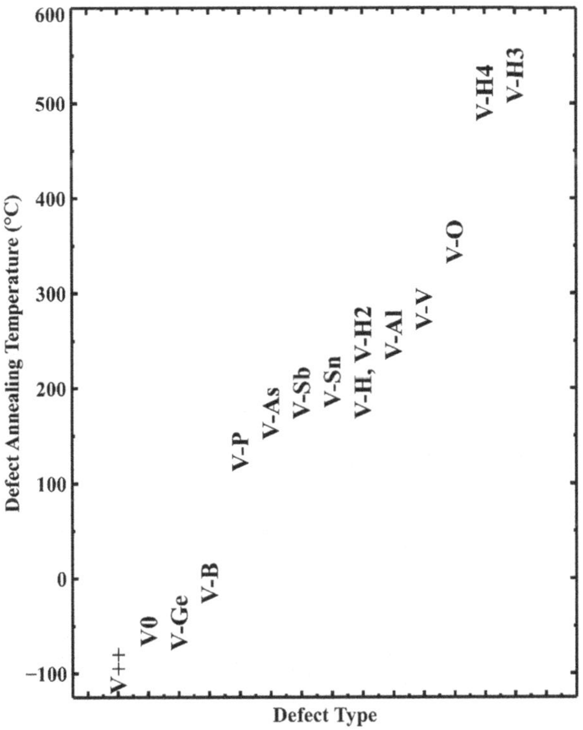

Fig. 6.7 Plot showing defect annealing as a function of temperature for common vacancy, interstitial, and impurity interaction related silicon defects. The annealing time was 15–30 min. The vacancy related defects are denoted with a V. The dopants are denoted by their atomic abbreviation

plotted against the yield of a parametric antenna test structure across different fabs (Mason et al., 2000). The plot strongly correlates increased burn-in failures with low antenna structure yield. The root cause of this problem was determined to be caused by a variation in the plasma power, temperature, and chamber impedance for a radio-frequency argon sputter process used as a back-sputtering via cleaning process. Additionally, product yield also correlated with the antenna test structure yield.

Process damage can degrade chip reliability, worsening many common CMOS reliability mechanisms. Many common CMOS potential failure mechanisms can be modulated by process damage. A partial list of reliability failure modes or degradation mechanisms caused or worsened by process damage is shown in Table 6.1. The degradation of time-zero parameters is often followed by reliability degradation related to the initial damage. For example, time-zero degradation in oxide parameters, such as Vt shifts, SILC, etc., correlates with degraded dielectric reliability.

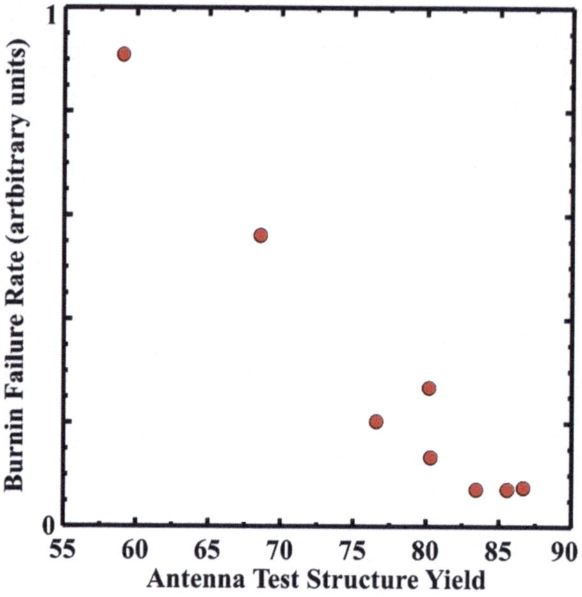

Fig. 6.8 Plot showing the burn-in failure rate (arbitrary units) plotted against the yield of a parametric test antenna structure for multiple fabrication facilities

6.3.2 CMOS Transistor Failure Modes

CMOS transistors fail through multiple mechanisms, all of which can be degraded and accelerated by process damage. Figure 6.9 shows transistor sketches with the different operating conditions for each common failure mode. p-channel and n-channel transistors can behave quite differently for different stress conditions due to differences in mobility, etc. All these failure modes interact with hydrogen, as described in Chap. 14.

Hot carriers are generated by high electric fields during transistor operation, causing the carriers to have sufficient energy to damage the interface or become trapped in the dielectrics shifting the parameters and causing device failure. In addition, hot carriers can be generated in the transistor channel during on-state stress, particularly when biased in the saturation region of operation near the peak substrate current. The two primary modes of channel hot carrier degradation, as shown in Fig. 6.9a, b, are channel hot carrier (CHE) and channel-initiated secondary electron (CHISEL) modes. For CHE stress, both electrons and holes can be trapped depending on the bias, with a high gate voltage favoring electron injection and a low gate voltage favoring hole injection in NMOS transistors. CHISEL is a mode caused by substrate bias accelerating carriers, creating secondary carriers that can be injected into the dielectrics (Takeda, 1995).

High-energy carriers can be generated during off-state stress (OSS) with a high drain bias. This mechanism is also called nonconducting stress (NCS) or

Table 6.1 Partial list of CMOS reliability failure mechanisms possibly degraded by process damage

Reliability mechanism	Affected devices	References
Reduced dielectric lifetime	All Fets including transistors, MOS capacitors, isolation, nonvolatile memories, DRAM, imagers, MIM capacitors, and intermetal dielectrics	Eriguchi et al. (1997)
Hot carrier resistance degradation	All Fets including transistors, MOS capacitors, isolation, nonvolatile memories, DRAM, and imagers	Li (1995)
Negative or positive bias temperature instability	All Fets including transistors, MOS capacitors, isolation, nonvolatile memories, DRAM, and imagers	Matsunaga (2002)
Cycling-related stress-induced leakage currents	Nonvolatile memories	Beug et al. (2011)
Read disturb	Nonvolatile memories	Lee (2002)
Program disturb	Nonvolatile memories	Oh (2008)
Overerase	Nonvolatile memories	Lee (2002)
Cycling window closure or shift	Nonvolatile memories	Beug et al. (2011)
Data retention (high and low temperatures)	Nonvolatile memories	Cai (2015)
Endurance	Nonvolatile memories	Lee (2002)
Charge gain	Nonvolatile memories	Kameyama (2000)
Erratic programming	Nonvolatile memories	Chimenton (2011)
Parasitic charge trapping	Nonvolatile memories	Prall (2007)
Degraded refresh	DRAM	Yi (2004)
Row hammer	DRAM	Park (2014)
Variable retention time	DRAM	Lee et al. (2012)
Increased dark current, white pixels, and defects	CCDs, imagers	Carrere et al. (2014)
Degraded metallization reliability	All	Tsai et al. (2018a, b)
Decreased radiation hardness	All	Yue and Sinha (1998)

hot-electron-induced punch-through (HEIP). The mechanism is shown in Fig. 6.9c. The high drain bias can generate carriers in the drain depletion region that are accelerated into the dielectric, becoming trapped. HEIP is caused when carriers are trapped in the channel, causing a decreased electrical channel length and reduced punch-through resistance. The carriers can also become trapped at the isolation edge, creating a parasitic leakage path.

Additionally, uniform gate stress can also cause parametric shifts, as shown in Fig. 6.9d, e. This stress is called negative bias instability (NBTI) or positive bias instability (PBTI). The physics of NBTI and PBTI is complicated, including traps, dissociated bonds with a hydrogen interaction, etc. (Mahapatra, 2016).

Finally, the time-dependent dielectric breakdown (TDDB) described in Chap. 5 and shown in Fig. 6.9f is a failure mechanism.

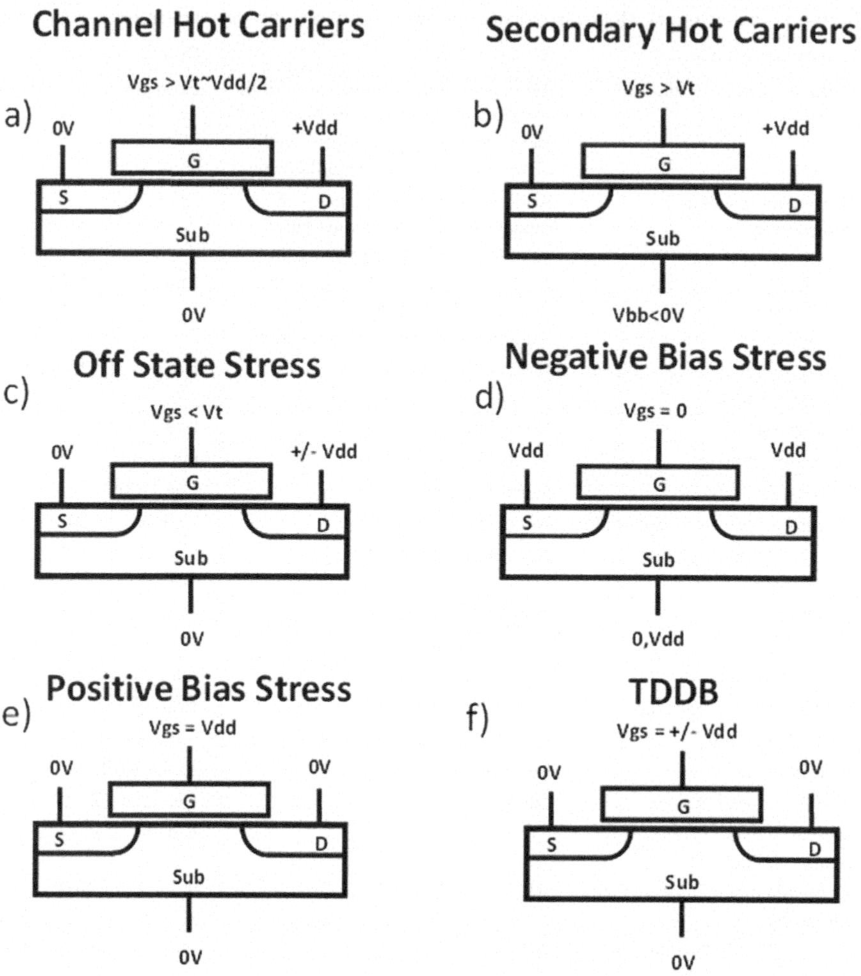

Fig. 6.9 Sketch of NMOS transistors showing common failure modes and their associated stress conditions. The voltage conditions are reversed for a PMOS transistor. (**a**) Channel hot carrier stress (CHE). (**b**) Channel initiated secondary electron (CHISEL). (**c**) Off-state stress (OSS), non-conducting stress (NCS), hot electron induced punch-through (HEIP). (**d**) Negative bias temperature instability (NBTI). (**e**) Positive bias temperature instability (PBTI). (**f**) Time dependent dielectric breakdown (TDDB) and charge trapping

6.3.3 Degradation in Dielectric Lifetime

A dielectric may survive the initial process damage event with no noticeable shift in dielectric leakage and no impact on product yield or function. However, the dielectric must be capable of surviving the required chip lifetime without failure or degradation in function. The process damage may degrade the dielectric's useful lifetime, resulting in a reliability hazard for chip failure in the field during its operating lifetime. This effect is called time-dependent dielectric breakdown (TDDB). Dielectrics are characterized by test structures to predict the dielectric and product lifetimes (Strong et al., 2009). Test structures matching the antenna ratio required by the design rules specified for the CMOS process can be compared to test structures without antennas to determine the degradation due to the antenna effect.

Dielectrics are typically characterized by accelerated voltage, current, and temperature stress beyond normal operating conditions. The lifetimes under accelerated conditions are extrapolated to the operating conditions to predict the lifetime under the chip operating conditions. Lifetime-accelerated testing commonly uses two methods: constant voltage or constant current stress. A constant voltage test forces a constant voltage while monitoring the current flowing through the dielectric. A current criterion determines the time of failure when the increasing leakage reaches a specified point, and the resulting voltage is recorded (VBD test). The second commonly used test forces a constant current on a dielectric test structure while monitoring the voltage. This test is also called charge to breakdown (QBD). A voltage criterion is established where the measured voltage is monitored over time. Dielectric failure is defined when the voltage across the dielectric crosses the criterion. Figure 6.10 shows examples of VBD and QBD dielectric breakdown tests (Strong et al., 2009, pp. 297, 300).

Figure 6.11 shows an example of charge to breakdown and time to breakdown degradation caused by process damage during a plasma etch process versus the antenna ratio (gate metal area on isolation divided by the gate metal area on the active area) for a 7 nm gate dielectric (Kosako et al., 1998).

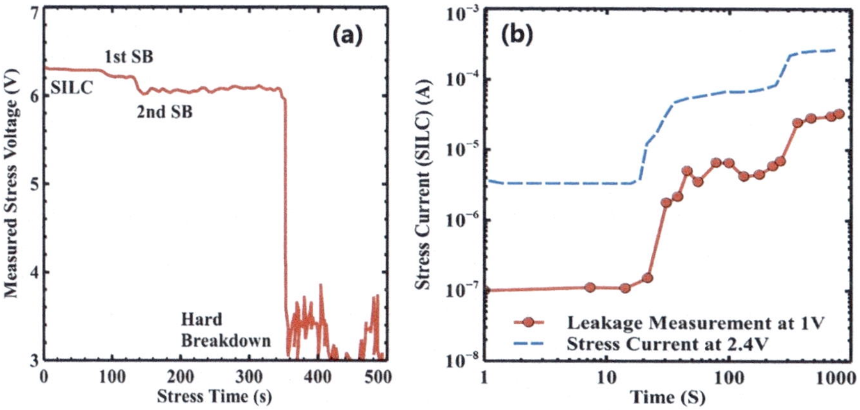

Fig. 6.10 Example plots of common dielectric lifetime tests. (**a**) Charge to breakdown (QBD) with a constant forced current while monitoring the voltage for a 4.2 nm silicon dioxide stressed at 1 μA. The different labels refer to stress induced leakage (SILC), first and second soft breakdown (SB), followed by hard breakdown (HB). (**b**) Voltage to breakdown (VBD) with a forced voltage while measuring a current. The step function increase in current is due to the dielectric breakdown. The upper curve is the forced voltage at 2.4 V across a 1 nm thick dielectric. The lower curve is a periodic measurement at 1 V

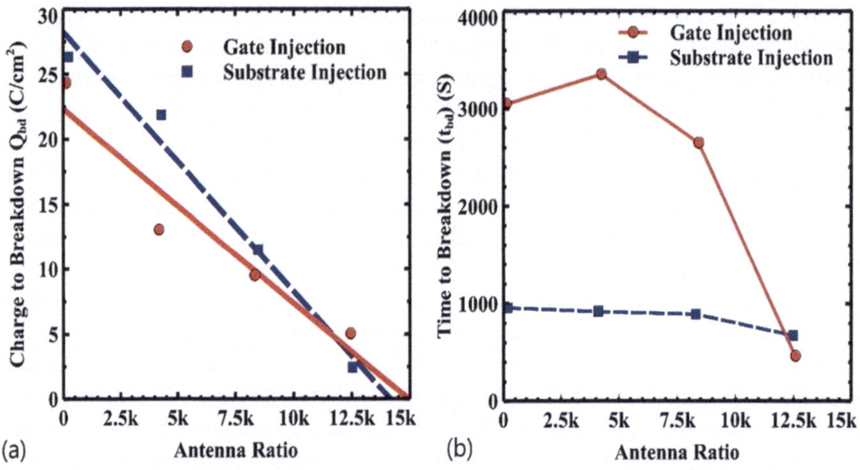

Fig. 6.11 (**a**) Example plot of charge to breakdown versus antenna ratio measured with both gate and substrate injection. (**b**) Degradation in time to breakdown with increasing antenna ratio

6.3.4 Hot Carrier Lifetime Degradation

CMOS transistors are designed to resist damage from high-energy hot carriers generated during operation using techniques such as the lightly doped drain. Constant voltage scaling of the transistor increases the electrical field in the channel caused by aggressively scaling transistors with limited voltage scaling, increasing the energy of channel carriers. Process damage can degrade hot carrier resistance, creating potential reliability hazards for CMOS chips operating in the field. Transistors are tested with accelerated voltages and temperatures (usually lower temperatures where the channel mobility is higher) to obtain hot carrier data quickly. This data is extrapolated to normal operating conditions. Process damage degrading hot carrier lifetime may be mistaken for inadequate transistor design, requiring added process cost to recover the reliability.

Figure 6.12 shows some examples of hot carrier lifetime degradation caused by process damage (Rangan et al., 1999; Sachdeva et al., 2006; Tzeng et al., 2002; Vici et al., 2020). Figure 6.12a shows the degradation of conductance or gain (g_m) with different antenna lengths for two different dielectric thicknesses. Figure 6.12b shows a plot of drain current reduction as a function of antenna length for two different dielectric thicknesses. The thinner dielectric (2.5 nm) has little or no degradation, while a 4 nm dielectric shows significant degradation. Figure 6.12c shows a PMOS hot carrier lifetime metric consisting of a function of the substrate current plotted against the antenna ratio. Figure 6.12d shows the same metric for an NMOS transistor.

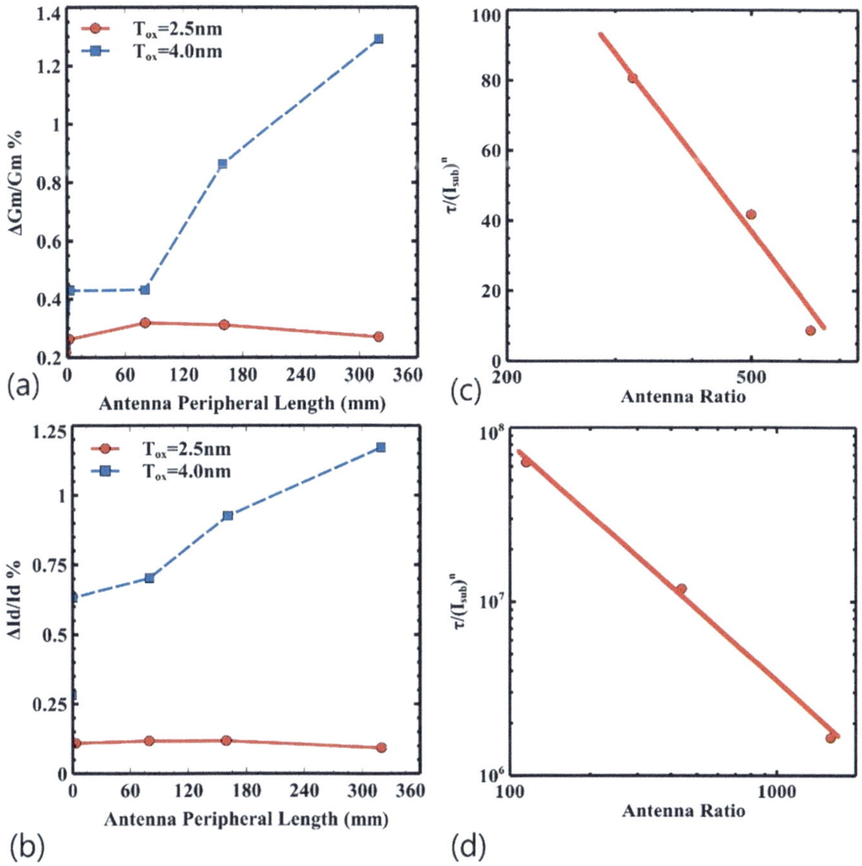

Fig. 6.12 Examples of hot carrier reliability degradation by process damage. (**a**) NMOS transistor gain (gm) degradation for two different oxide thicknesses plotted as a function of antenna length for a gate metal antenna. (**b**) Change in NMOS drain current with process damage versus antenna length for a gate metal antenna. (**c**) A hot carrier reliability metric plotted against antenna ratio for a PMOS transistor damaged during metal 1 (M1) processing. (**d**) A hot carrier reliability metric plotted versus antenna ratio for a NMOS transistor damaged during metal 1 (M1) processing. The metric $\tau/(I_{sub})^n$ is a metric for estimating the transistor lifetime

6.3.5 Negative Bias Temperature Instability (NBTI) Degradation

Negative bias temperature instability (NBTI) is a reliability failure mechanism affecting mainly PMOS devices due to boron–hydrogen effects, etc. (Fleetwood et al., 2009; Strong et al., 2009). A PMOS transistor biased in inversion (V_{gs} <0) operated at higher temperatures (30–200 °C) accumulates positively charged defects and interface states. Process damage also creates the same defects and can accelerate NBTI degradation, shortening the transistor's lifetime. The NBTI lifetime of a transistor can be reduced by orders of magnitude by process damage. Figure 6.13 shows an example of PMOS NBTI lifetime reduction caused by process damage (Krishnan et al., 2001). NMOS transistors can suffer from a similar reliability problem called positive bias temperature instability (PBTI). Also, see Chap. 8 for an additional example of NBTI degradation caused by dielectric deposition on top of a MIM capacitor.

6.4 Metallization Reliability Degradation

6.4.1 Introduction

The chip specification in the datasheet limits the energy delivered to the chip by the system to prevent damage. Additionally, CMOS chips are designed to carefully manage power and energy during operation to prevent damage to the chip. For example, the maximum current density for metallization is limited to avoid electromigration, which destroys the metal lines by slow migration of metal atoms, resulting in an open circuit. Under plasma or process damage conditions, the voltages and current flowing through the metal interconnect are uncontrolled, possibly exceeding the capability of metallization. Excessive current can result in time-zero metallization failure or latent degradation in the long-term reliability that can fail in the field during the lifetime of the chip.

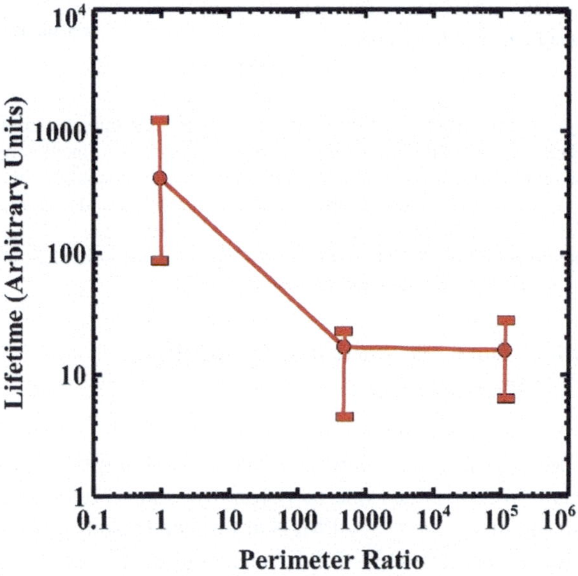

Fig. 6.13 PMOS with a 4 nm gate dielectric transistor NBTI lifetime (arbitrary units) versus the perimeter ratio for process damage during Al metal 1 processing

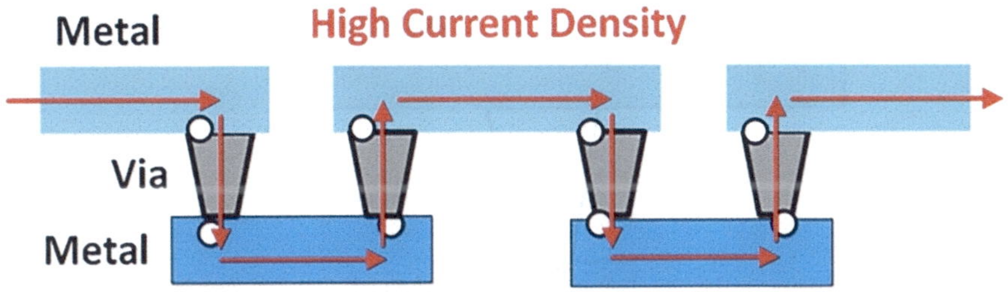

Fig. 6.14 Cross-section sketch showing a via chain with initial damage and plasma damage caused voiding that degrades abnormally fast under electromigration testing

Furthermore, the intermetal dielectrics may be damaged or degraded, causing reliability problems. In addition, excessive chip current can accelerate chip corrosion, stress migration, etc., during processing (Tsai et al., 2018a, b). Worst-case chip locations containing large antennas, etc., tend to fail first for this problem. Low-k dielectrics are covered separately in Chap. 8. Figure 6.14 shows a cross-section sketch of a via chain damaged by plasma damage and subjected to electromigration reliability testing (Tsai et al., 2018a, b). The damaged chain may fail abnormally early because of the damage, as the electromigration lifetime is partially used up by high current density during plasma damage. Any voids that grow during the plasma damage stress increase the metallization resistance. The voids tend to occur at interfaces between metals.

6.4.2 *Intermetal Dielectric Degradation*

The large amounts of intermetal dielectrics that separate the metal layers in a CMOS chip are potential sources of chip yield or reliability failure. The intermetal dielectric may be excessively stressed during a process damage event, resulting in damage very similar to the process damage to gate dielectrics covered in Chap. 5. Figure 6.15 shows a sketch of a CMOS wafer cross section, highlighting potential locations for dielectric process damage at the gate dielectric and the intermetal dielectric (Su et al., 2020). Depending on the plasma, electrical, and structural conditions, the damage may occur at either location or both.

The electric field at the corners in metal features is enhanced due to the radius of curvature effect, as described in Chap. 5, resulting in a higher electric field compared to a simple parallel plate capacitor. The electric field is increased proportionally to the inverse of the radius of curvature squared. Due to this effect, process damage can be expected to worsen at the corners of metal lines. Tighter design rules with constant voltage scaling result in higher fields, increasing the effect.

The stress on intermetal dielectrics and the resultant damage depends on the metal lines' electrical connections. Figure 6.16 shows an example with two different connections to the lower metal lines (m1) (Su et al., 2020). The charging event

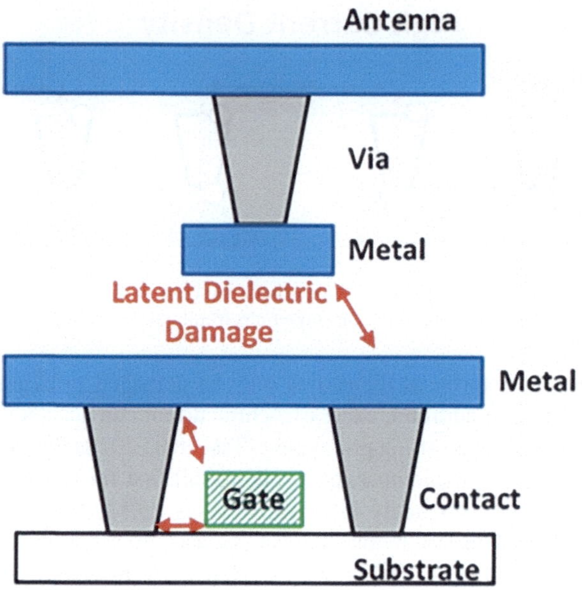

Fig. 6.15 CMOS process cross-section sketch showing multiple locations of dielectric damage to the gate dielectric and the intermetal dielectric

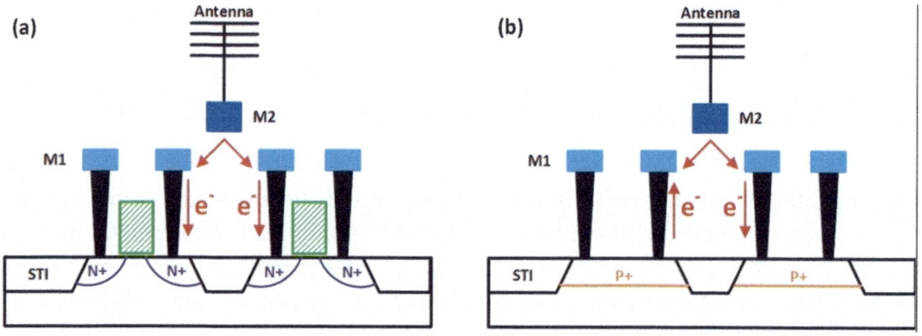

Fig. 6.16 CMOS cross-section sketch showing different configurations for intermetal dielectric damage between metal 1 (M1) and metal 2 (M2). (**a**) Metal 1 (M1) is isolated from the substrate by the np junction. (**b**) Metal 1 is directly connected to the p-type substrate through a p+ region

from the metal 2 (m2) antenna stresses the dielectric between metal 1 (m1) and metal 2 (m2). If the metal 1 line is completely floating (no connection to the substrate), reduced damage to the dielectric occurs because the metal 1 line capacitively couples to metal 2, raising the metal 1 potential. However, if the metal 1 line is connected to a diode, as shown in Fig. 6.16a, the metal 1 line is partially isolated from the substrate by the reverse-biased junction, resulting in less damage to the intermetal dielectric compared to Fig. 6.16b. Finally, if the metal 1 line is hard-wired to the substrate, as shown in Fig. 6.16b, the substrate connection serves as a

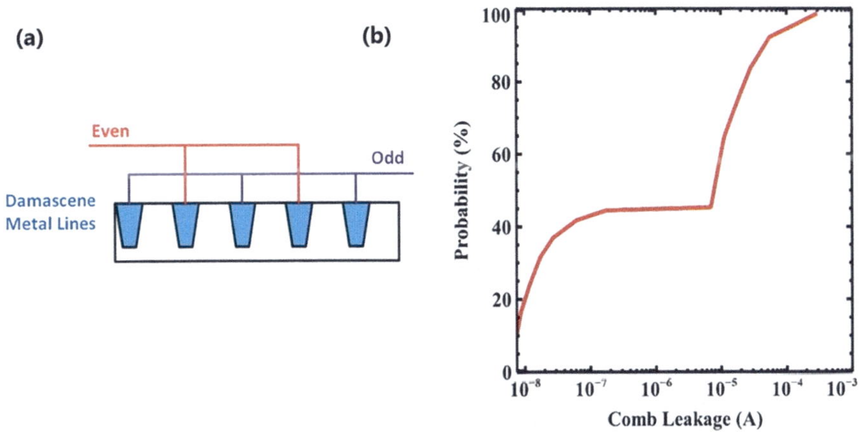

Fig. 6.17 SILC damage in intermetal dielectrics. (**a**) Sketch of an intermetal dielectric test structure consisting of an interdigitated comb of even and odd metal lines. (**b**) Plot of intermetal dielectric SILC leakage caused by plasma damage showing bimodal leakage

pseudo-ground, creating the worst-case damage conditions by allowing bidirectional stress with no field-sharing.

In addition to the dielectric breakdown of the intermetal dielectric, process damage can cause stress-induced leakage (SILC) in an identical manner to SILC in gate dielectrics, as described in Chap. 7. Process damage degrades the leakage properties of the intermetal dielectric, causing a reliability hazard and potentially affecting time-zero circuit operation. Figure 6.17a shows an intermetal comb process damage test structure sketch (Tsai et al., 2018a, b). An antenna (not shown) may be connected to one terminal of the comb. A distribution plot of the SILC leakage caused by process damage is shown in Fig. 6.17b, where 60% of the finger capacitors leak after the damage.

6.4.3 *Corrosion*

Figure 6.18 shows an example of this phenomenon where increased plasma intensity across the wafer accelerates wafer corrosion (Tsai et al., 2018a, b). The plasma intensity varies across the wafer and is measured by nonvolatile memory cells, creating a plasma damage intensity metric. Figure 6.19 shows an example TEM image of a voided via connection to a metal line interface (Irrera et al., 2010; Tsai et al., 2018a, b). The plasma damage electrical stress can accelerate corrosion, stress, and electromigration degrading reliability (Bothra et al., 1998).

One factor associated with corrosion is the residual byproducts left on the wafer after processing, particularly after plasma etch. Etch byproducts left on the wafer, including reactive species such as fluorine or chlorine, can continue etching after the

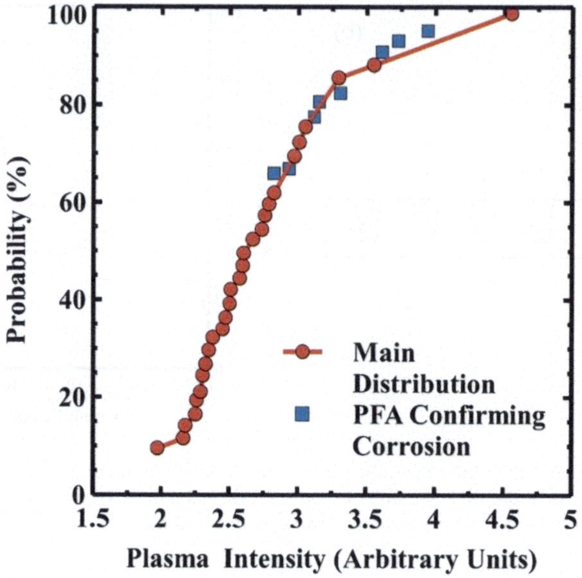

Fig. 6.18 Probability showing metal corrosion correlation with regions of the wafer near the center that have increased plasma damage. Plasma intensity was measured with a nonvolatile memory test structure. The corrosion was confirmed with physical failure analysis on sites with increased damage

etch is completed, attacking metal structures such as tungsten vias. In addition, exposure to moisture in room air can accelerate reactions. Postprocessing strips and cleans are heavily optimized to eliminate byproducts and corrosion. An in situ strip in the etch chamber is commonly used to minimize corrosion (Fatkhotdinov et al., 2002). Across wafer damage variation can be related to across wafer residual variation.

6.4.4 Electromigration and Stress Migration

Electromigration and stress migration are critical reliability risks for CMOS metallization. Backend of line processes are heavily optimized to ensure electromigration, and stress migration does not cause product failures over the product life. Process and plasma damage can potentially degrade these reliability risks. For example, Fig. 6.20 shows a plot of differences in electromigration lifetime with different plasma treatments before low-k dielectric deposition (Wang et al., 2002).

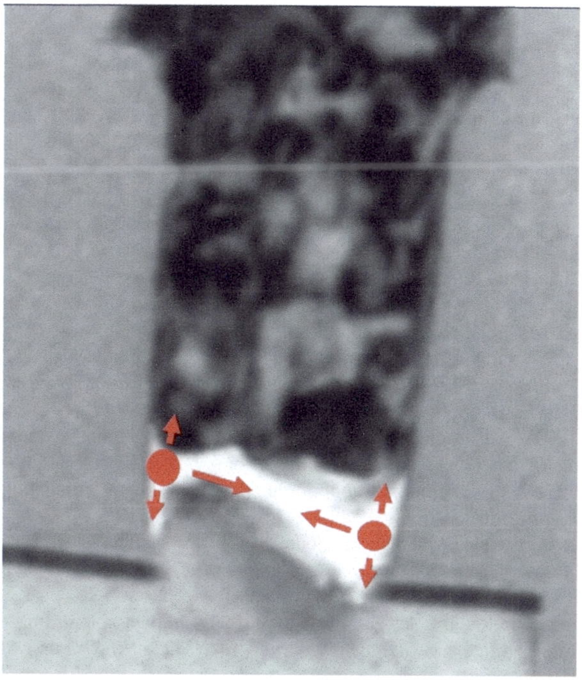

Fig. 6.19 Example cross-section transmission electron microscope picture showing a via void at a metal interface

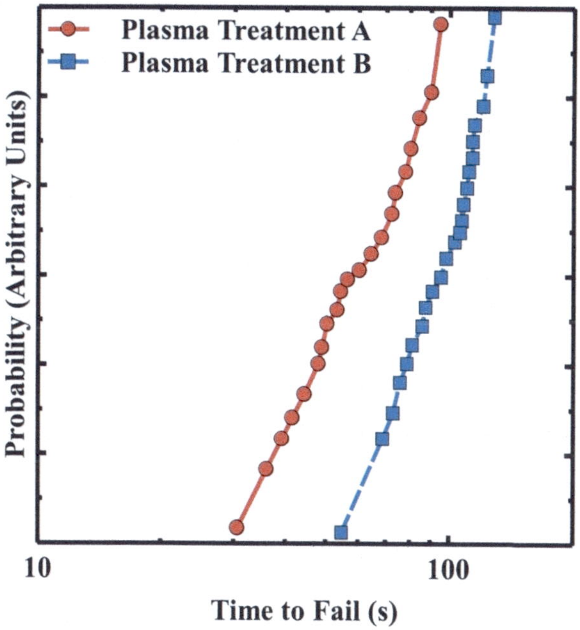

Fig. 6.20 Probability plot showing differences in electromigration lifetime caused by different plasma treatments before low-k dielectric deposition

References

Beug, M. F., Melde, T., Paul, J., & Knoefler, R. (2011). TaN and AL2O3 sidewall gate-etch damage influence on program, erase, and retention of sub-50-nm TANOS nand flash memory cells. *IEEE Transactions on Electron Devices, 58*(6), 1728–1734. https://doi.org/10.1109/ted.2011.2121070

Bothra, W., Sur, H., & Liang, V. (1998). A new failure mechanism by corrosion of tungsten in a tungsten plug process. In *IEEE International Reliability Physics Symposium*. https://doi.org/10.1109/RELPHY.1998.670473

Cai, Y., et al. (2015). Data retention in MLC NAND flash memory: Characterization, optimization, and recovery. In *International Symposium on High-Performance Computer Architecture*. https://doi.org/10.1109/HPCA.2015.7056062

Carrere, J. P., Place, S., Oddou, J. P., Benoit, D., & Roy, F. (2014). CMOS image sensor: Process impact on dark current. In *2014 IEEE International Reliability Physics Symposium*. https://doi.org/10.1109/IRPS.2014.6860620

Cheung, K. P. (1994). An efficient method for plasma-charging damage measurement. *IEEE Electron Device Letters, 15*(11), 460–462. https://doi.org/10.1109/55.334667

Cheung, K. P. (1995). A new method to monitor gate oxide reliability degradation. In *Symposium on VLSI Technology*. https://doi.org/10.1109/VLSIT.1995.520869

Chimenton, A., et al. (2011). A statistical model of erratic behaviors in flash memory arrays. *IEEE Transactions on Electron Devices, 58*(11), 3707–3711. https://doi.org/10.1109/TED.2011.2165722

Eriguchi, K., Yamada T, Kosaka, Y., & Niwa, M. (1997). Impacts of plasma process induced damage on ultra thin gate oxide reliability. In *IEEE IRPS*. https://doi.org/10.1109/RELPHY.1997.584257

Fatkhotdinov, M., et al. (2002). Impact of post metal etch resist strip in plasma on plasma charge-induced erosion of tungsten-plugs. In *ESSDERC*.

Fleetwood, D. M., Pantelides, S. T., & Schrimpf, R. D. (2009). *Defects in microelectronic materials and devices*. CRC Press Taylor and Francis Group. https://doi.org/10.1201/9781420043778

Hull, R. (1999). *Properties of crystalline silicon*. INSPEC.

Irrera, F., et al. (2010). Model of stress induced voiding in the back-end-of-line of a CMOS process. In *MOS-AK 2010 Workshop*.

Kameyama, H., et al. (2000). A new data retention mechanism after endurance stress on flash memory. In *IEEE IRPS*. https://doi.org/10.1109/RELPHY.2000.843914

Kim, Y.-K., et al. (1999). Application of the deuterium sintering process to improve the device design rule in reducing plasma induced damages. In *4th International Symposium on Plasma Process-Induced Damage*. https://doi.org/10.1109/PPID.1999.798814

King, J. C., & Hu, C. (1994). Effect of low and high temperature anneal on process-induced damage of gate oxide. *IEEE Electron Device Letters, 15*(11), 475–476. https://doi.org/10.1109/55.334672

Kosako, Y., et al. (1998). Stress mode of gate oxide charging during the MERIE and the ICP processing and its effect on the gate oxide reliability. In *3rd International Conference on Plasma and Process Induced Damage*. https://doi.org/10.1109/PPID.1998.725611

Krishnan, A. T., et al. (2001). Impact of charging damage on negative bias temperature instability. In *IEEE IEDM*. https://doi.org/10.1109/IEDM.2001.979650

Lee, D.-K., et al. (2002). Influence of plasma edge damage on erase characteristics of NOR flash EEPROM using channel erase method. In *IEEE IRPS*. https://doi.org/10.1109/RELPHY.2002.996659

Lee, C.-Y., Lai, C.-S., Yang, C.-M., & Wang, D. H. L. (2012). DRAM data retention and cell transistor threshold voltage reliability improved by passivation annealing prior to the deposition of plasma nitride layer. *IEEE Transactions on Device and Materials Reliability, 12*(2), 406–412. https://doi.org/10.1109/TDMR.2012.2188895

Li, X.-Y., et al. (1995). Degraded CMOS hot carrier life time—Role of plasma etching induced charging damage and edge damage. In *IEEE IRPS*. https://doi.org/10.1109/RELPHY.1995.513689
Li, F. M., Nixon, O., & Nathan, A. (2004). Degradation behavior and damage mechanisms of CCD image sensor with deep-UV laser radiation. *IEEE Transactions on Electron Devices, 51*(12), 2229–2236. https://doi.org/10.1109/ted.2004.839758
Mahapatra, S. (2016). *Fundamentals of bias temperature instability in MOS transistors*. Springer. https://doi.org/10.1007/978-81-322-2508-9
Mason, P. W., et al. (2000). Relationship between yield and reliability impact of plasma damage to gate oxide. In *International Symposium on Plasma Process-Induced Damage*. https://doi.org/10.1109/PPID.2000.870560
Matsunaga, N., et al. (2002). NBTI analysis of antenna pMOSFET with thermally recovered plasma-induced damage. In *7th International Symposium on Plasma and Process Induced Damage*. https://doi.org/10.1109/PPID.2002.1042629
Miyawaki, M., et al. (1991). Reduction of fixed-pattern noise of BASIS due to low kinetic energy reactive ion and native-oxide-free processing. *IEEE Transactions on Electron Devices, 38*(5), 1037–1043. https://doi.org/10.1109/16.78376
Oh, D., et al. (2008). Program disturb phenomenon by DIBL in MLC NAND flash device. In *Joint Nonvolatile Semiconductor Memory Workshop and International Conference on Memory Technology*. https://doi.org/10.1109/NVSMW.2008.7
Park, K., et al. (2014). Active-precharge hammering on a row induced failure in DDR3 SDRAMs under 3x nm technology. In *IEEE International Integrated Reliability Workshop*. https://doi.org/10.1109/IIRW.2014.7049516
Prall, K. (2007). Scaling nonvolatile memory below 30nm. In *IEEE Nonvolatile Semiconductor Memory Workshop*. https://doi.org/10.1109/NVSMW.2007.4290561
Rangan, S., et al. (1999). A model for channel hot carrier reliability degradation due to plasma damage in MOS devices. In *IEEE IRPS*. https://doi.org/10.1109/RELPHY.1999.761641
Sachdeva, R., et al. (2006). Electrical properties and diffusion behavior of hafnium in single crystal silicon. *Applied Physics A, 84*, 351–357. https://doi.org/10.1007/s00339-006-3637-6
Strong, A. W., et al. (2009). *Reliability wearout mechanisms in advanced CMOS technologies*. Wiley. https://doi.org/10.1002/9780470455265
Su, C., Tsai, Y. P., Lin, C. J., & King, Y. C. (2020). Test pattern design for plasma induced damage on inter-metal dielectric in FinFET Cu BEOL processes. *Nanoscale Research Letters, 15*(1), 96. https://doi.org/10.1186/s11671-020-03328-7
Takeda, T., et al. (1995). *Hot-carrier effects in MOS devices*. Elsevier.
Tsai, Y.-P., Liou, P.-C., Lin, C. J., & King, Y.-C. (2018a). Plasma charging effect on the reliability of copper BEOL structures in advanced FinFET technologies. *IEEE Journal of the Electron Devices Society, 6*, 875–883. https://doi.org/10.1109/JEDS.2018.2859365
Tsai, Y. P., et al. (2018b). 7nm FinFET plasma charge recording device. In *IEEE IRPS*. https://doi.org/10.1109/IEDM.2018.8614520
Tzeng, P.-J., et al. (2002). Plasma charging effects on submicron MOS devices. *IEEE Transactions on Electron Devices, 49*(7), 1151–1157. https://doi.org/10.1109/TED.2002.1013270
Vici, A., Russo, F., Lovisi, N., & Irrera, F. (2020). On border traps in back-side-illuminated CMOS image sensor oxides. *IEEE Transactions on Electron Devices, 67*(5), 2022–2027. https://doi.org/10.1109/ted.2020.2983039
Wang, X. (2008). *Noise in sub-micron CMOS image sensors* [PhD, University of Southampton, UK].
Wang, R. C. J., et al. (2002). Investigation of electromigration properties and plasma charging damages for plasma treatment process in Cu interconnects. In *International Symposium on Plasma and Process-Induced Damage*. https://doi.org/10.1109/PPID.2002.1042635
Yi, J. Y. (2004). *A study on the retention time distribution with plasma damage effect*. ICEIC.
Yue, J., & Sinha, S. (1998). Enhanced hole and interface trap generation in plasma damage bulk and SOI MOSFETs under total dose irradiation. In *3rd International Symposium on Plasma Process Induced Damage*. https://doi.org/10.1109/PPID.1998.725565

Chapter 7
Atomic-Level Defects and Electrical Effects

7.1 Introduction

CMOS technology's relentless scaling and advancement have resulted in ever-smaller devices and dimensions. Scaled small devices have become more susceptible to smaller and smaller defects reaching the scale where a single atomic-level defect can cause device failure. Additionally, a few atomic defects can link electrically, creating potential chip failure mechanisms. The large spectrum of possible atomic defects is challenging to summarize. This chapter attempts to describe the atomic defects in advance of a detailed description of the technology-specific electrical behavior caused by these defects (Chap. 8).

Atomic-level defects frequently impact CMOS devices in various degrees, depending on the type of device and the device dimensions. In addition, some technologies are inherently more susceptible because of the underlying physics of operation, allowing the detection of atomic-level defects. Table 7.1 shows a partial list of single-atom electrical defects by the technology type. Details on the electrical behavior are described in Chaps. 5 and 8. Atomic-level defects can be caused by process damage. These defects may occur at minimal levels of damage and can be observed before other types of damage occur. Typically, these defects can only be observed in devices like memory arrays, imager arrays, etc. Large-area devices such as parametric capacitors usually fail to observe these effects because the atomic-level defects are swamped by large-area effects such as tunneling. This section introduces some of the single defect mechanisms.

K. Prall, *CMOS Plasma and Process Damage*,
https://doi.org/10.1007/978-3-031-89029-1_7

Table 7.1 Partial list of atomic-level electrical defects affecting CMOS devices

Technology
All advanced CMOS technologies: giant RTS noise, diode leakage
Nonvolatile memories, including data retention, charge gain, over erase, over programming, endurance, cycling window shift or closure, erratic programming and erasing, electrical disturb mechanisms
DRAM including refresh, row hammer, variable retention time
CCD and imagers, including dark current and white pixels

7.2 SiO_2-Silicon Atomic-Level Defects

At the atomic level, the surface of a nearly perfect silicon wafer contains atomic-level imperfections that can be electrically active, resulting in the defects described in this section. An atomic force microscope image of a silicon surface is shown in Fig. 7.1, highlighting the atomic-level irregularity of the silicon surface (Bowler, 1997). The silicon surface can contain a step in the surface caused by the end of an atomic layer crystalline plane, as shown in the lower right region of the image. Missing or irregularly placed atoms in the silicon can be observed in the image and may cause electrically active defects, dimers, etc. (Werner, 2006). Single-atom defects and dimers consisting of pairs of electrically linked atoms of the silicon defects can be observed on the silicon surface. Common silicon defects are sketched in the figure.

CMOS process technologists carefully optimize the processing conditions to create the best possible silicon-dielectric interface and eliminate as many silicon interface defects as possible. Typical processes optimized include the dielectric deposition preclean, dielectric deposition, and post-deposition annealing. Additionally, plasma and process damage must be mitigated.

One of the key advantages of using silicon for integrated circuits is the use of silicon dioxide as a dielectric and its electrical properties in conjunction with silicon, forming an electrically and chemically stable interface (Huff & Gilmer, 2005). The discovery of this system was a breakthrough that enabled the exponential growth of the semiconductor industry for decades. The advantages of this system include electrical and thermal stability, reliability, control of surface electrical defects, dopant control, junction leakage, etc. These beneficial properties are due to the strong covalent bonding of the SiO_2 dielectric, resulting in a higher melting point of 1740 °C, which is higher than silicon's (1410 °C). As a result, silicon dioxide is resistant to damage, requiring high kinetic energy to damage the dielectric. However, much weaker defects exist in the dielectrics, causing many reliability and electrical problems.

The Si-SiO_2 system has been one of the most heavily studied material systems for many years, and numerous defect types have been discovered and researched. Figure 7.2 shows a diagram of common defects in this system (Fleetwood et al., 2009). The defects are shown roughly positioned where they exist in the system.

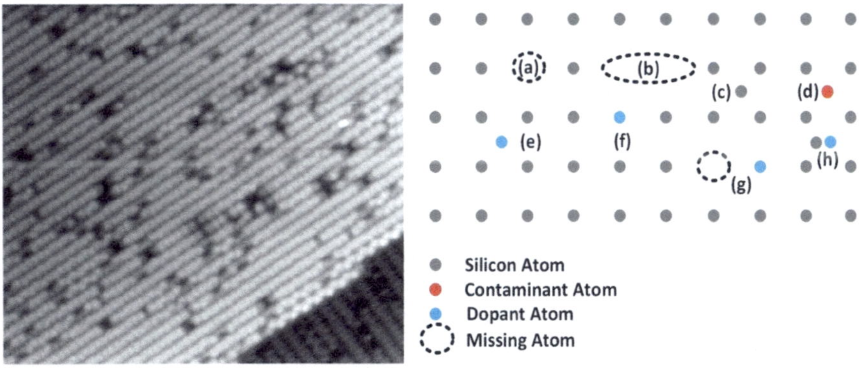

Fig. 7.1 Scanning tunneling microscope image of a silicon <001> surface shown on the left side. Sketch of different defect types. Schematic of different defect types shown on the right side, i.e., (a) vacancy, (b) di-vacancy, (c) self-interstitial, (d) contaminant interstitial, (e) dopant interstitial, (f) substitutional impurity, (g) impurity vacancy pair, (h) impurity self-interstitial pair

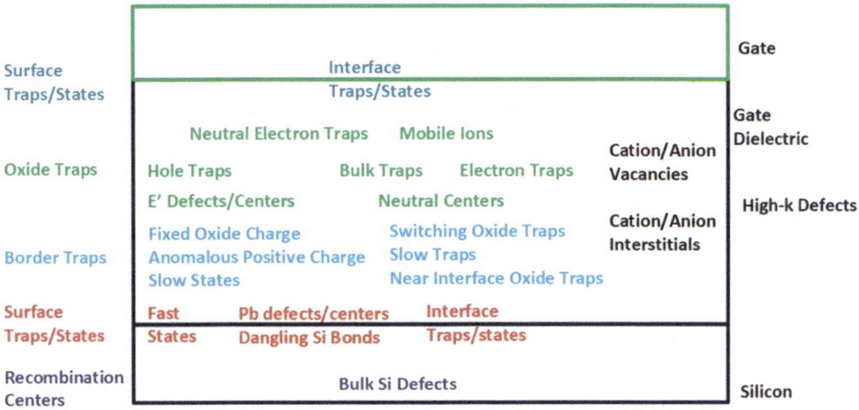

Fig. 7.2 Diagram showing common silicon dioxide-silicon defects with their rough position in the MOS system along with high-k defects

Defects deep in the silicon, such as recombination centers and bulk silicon defects, affect bulk and surface properties, degrading electrical properties such as minority carrier lifetime and junction leakage. Si-SiO_2 surface defects such as fast states and interface traps affect surface electrical properties such as transistor mobility and subthreshold leakage. Near-surface defects, called border traps, slow states, and fixed charges, impact surface electric and dielectric properties. Bulk dielectric oxide traps such as electron and hole traps can shift critical transistor parameters such as the threshold voltage. Finally, defects at the top of the SiO_2-gate metal interface can modify electrical properties by pinning the work function and shifting the threshold voltage.

7.3 High-k Dielectric Single-Atom Defects

The dielectric SiO_2, with minor variations, has served the semiconductor industry for decades. High-dielectric (high-k) constant materials have recently replaced silicon dioxide in some applications, such as logic. High-k dielectrics have numerous advantages, including reduced gate leakage, better electrical control of the MOS channel, etc. (Wong, 2012).

High-k dielectrics commonly use transition-metal oxides, including HfO, HfSiO, ZrO, etc., because of their compatibility with silicon and electrical properties. These types of dielectrics have a strong ionic bond characteristic with the metal cation and oxide anion bonding, which is much different from the covalent silicon dioxide bonding. As a result, transition-metal oxides are quite different from silicon dioxide dielectrics and have different dominant defects affecting the electrical behavior, failure, and degradation. The dominant defects in transition-metal oxides tend to be vacancy and interstitial defects. Vacancy defects are defects where an anion or cation atom is missing from the dielectric. Interstitial defects are defects where an extra atom exists in the dielectric. Both the metal (cation) and oxide (anion) atoms can occur in vacancy and interstitial defects. In addition, the defects can trap an electron or hole, creating a charged defect. The defects can also trap two electrons or two holes, creating a doubly charged defect. Table 7.2 lists these types of defects in transition-metal oxides. High-k dielectrics can also have the defects shown in Fig. 7.2, creating a complex spectrum of potential defects (Maier, 2004; Wong, 2012, p. 88).

A crucial factor in high-k dielectrics is their compatibility with silicon. A metal oxide gate dielectric can interact with the silicon substrate with desirable or undesirable results. The hafnia-related material systems form a stable oxide with silicon, resulting in a viable CMOS technology. Undesirable effects include intermixing, causing silicate formation, formation of an unstable oxide, or the reduction of the dielectric by the silicon. The undesirable effects can result in a nonviable dielectric. Therefore, the interface between the high-k dielectric and the silicon substrate is critical to obtain a reliable dielectric with the capabilities for advanced CMOS

Table 7.2 Types of defects in transition-metal oxides

Anion vacancy
Anion on interstitial site
Anion on regular lattice site
Cation vacancy
Cation on interstitial site
Cation on regular lattice site
Charged anion vacancy
Charged cation vacancy
Charged interstitial anion
Charged interstitial cation

technology. An extremely thin SiO_2 layer often exists under the high-k dielectric layer, serving an essential function at the silicon interface and improving atomic layer deposition and the electrical quality of the interface. Figure 7.3 shows an example TEM image of a high-k dielectric on top of silicon dioxide at the silicon interface (Robertson, 2004).

The presence of the SiO_2 dielectric interfacial layer can have electrical effects, and it is heavily optimized in a CMOS process. Figure 7.4 shows a sketch of a dielectric-silicon interface showing the impact on the dielectric defects (Fleetwood et al., 2009, p. 128). The diagram shows that the oxide charge and fixed interface states occur in two distinct locations and materials, affecting the electrical characteristics of the film and the transistor. The electrical characteristics of the two regions are typically different.

High-k dielectrics tend to be more nanocrystalline and less amorphous than silicon dioxide, creating the propensity for defects at the grain boundaries that can cause electrical defects. Figure 7.5 shows a plane-view TEM image showing an example of a nanocrystalline structure for a hafnia and silicon dioxide mixture (Robertson, 2004).

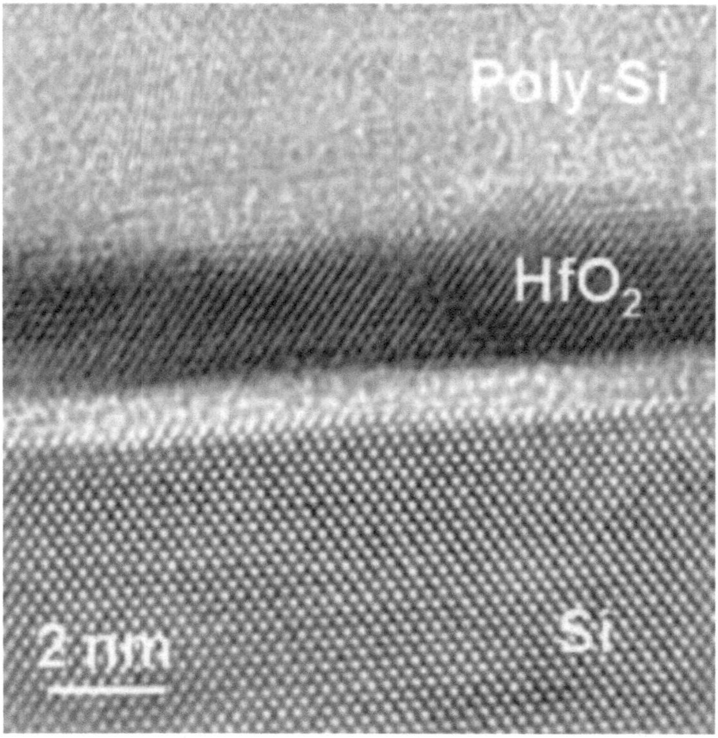

Fig. 7.3 High resolution TEM showing a high-k HfO_2 dielectric on top of an interfacial SiO_2 dielectric. The bright layer beneath the HfO_2 layers is the interfacial SiO_2 layer

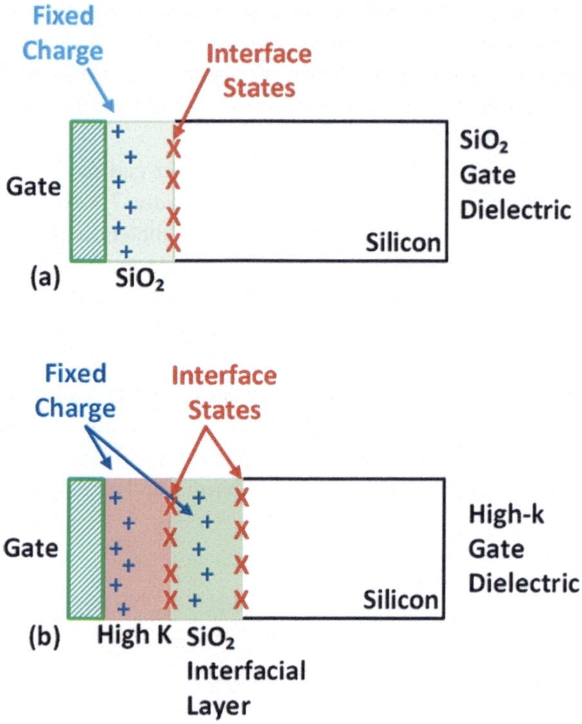

Fig. 7.4 Sketch showing the impact of a SiO_2 interfacial layer under a high-k dielectric. (**a**) SiO_2 dielectric without an interfacial layer. (**b**) High-k dielectric with an interfacial SiO_2 layer showing two separate interface and fixed charge regions in the SiO_2 and high-k layers

7.4 Stress-Induced Leakage Current

Stress-induced leakage current (SILC) is a common problem in MOS dielectrics after electric stress or process damage, providing an excellent example of atomic-level defects (Fleetwood et al., 2009). SILC is the degradation of the insulating properties of the dielectric caused by atomic-level defects created by charge carriers flowing through the dielectric during process damage. The charge carriers have enough kinetic energy to break the bonds in the dielectric, creating traps and interface states. The created defects serve as tunneling sites, allowing charge carriers to tunnel between defects, creating an abnormally high current between the gate and substrate. A single atomic-level defect can cause a significant increase in the dielectric current in dielectrics less than 90 Å. Thicker dielectrics can leak through multiple defects that are aligned physically and by energy levels in the band structure. Figure 7.6 shows a sketch of SILC conduction through a dielectric (Ielmini et al., 2004; Okuyama et al., 1998). A single or double trap can cause significant current to flow through the dielectric, affecting sensitive devices.

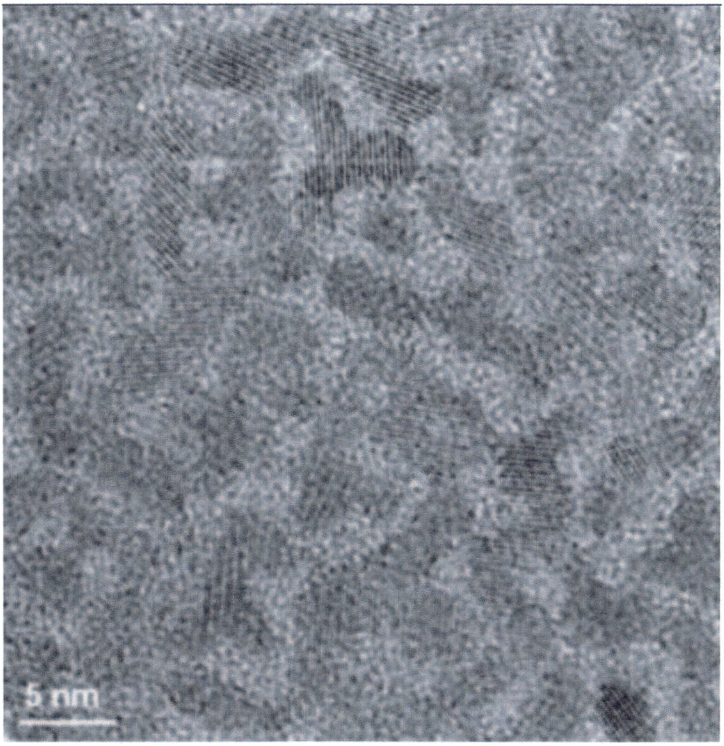

Fig. 7.5 Plan view TEM showing the nanocrystalline structure of the HfO_2/SiO_2 alloy system

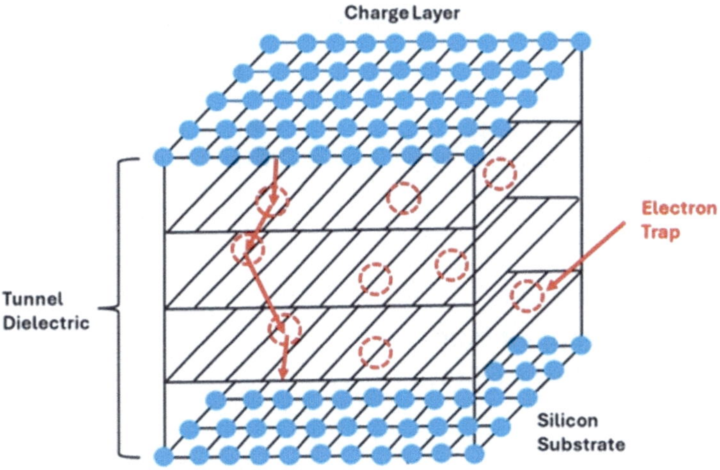

Fig. 7.6 Sketch showing stress induced leakage current (SILC) flowing through multiple trap defects in a dielectric

Multiple conduction mechanisms can occur with SILC (Fleetwood et al., 2009, p. 500). Transient leakage can occur as a positive trapped charge or trap is annihilated by an electron, as shown in Fig. 7.7a. Additionally, the trapped charge can contribute to the SILC current as carriers flow into or out of the interface states, as shown in Fig. 7.7b. The trap-assisted SILC current can tunnel through traps, as shown in Fig. 7.7c, d. Due to the transient nature of some of the SILC mechanisms, some of the SILC current is temporary, changing with time.

SILC current is a strong function of the dielectric thickness, as more trapping sites are needed for thicker dielectrics, requiring traps to be aligned spatially between the silicon and the gate, allowing carriers to tunnel trap-to-trap across the dielectric.

Nonvolatile memories are very sensitive to SILC currents because of the very small numbers of electrons involved in the cell operation, as described in Chap. 8. Figure 7.8 shows a probability plot of tunnel oxide leakage for a flash memory array with different dielectric thicknesses after 10^4 cycles (Ielmini et al., 2002). Single atomic-level defects can play a significant role in the reliability of sensitive devices such as flash memories and other devices, limiting the scaling of the tunnel oxides. Thicker tunnel dielectrics are required to limit SILC conduction by statistically reducing the impact of tunneling through a few defects. Two bits in the distribution

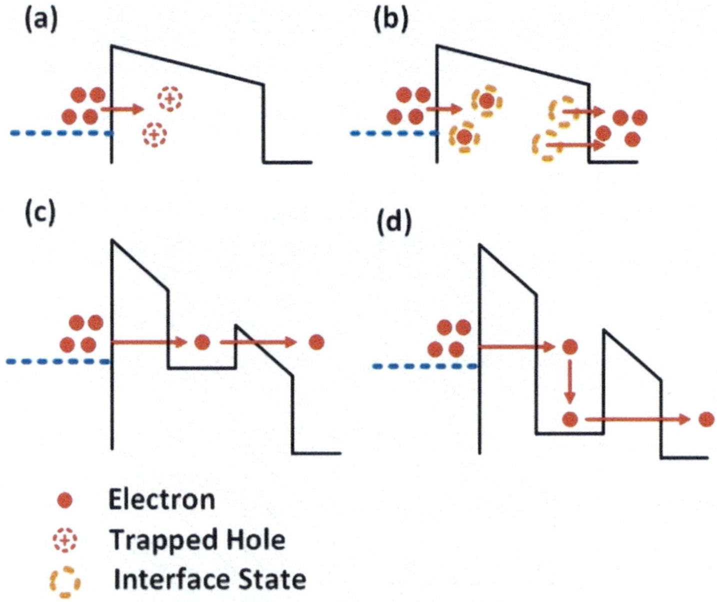

Fig. 7.7 Band energy sketch of various SILC conduction mechanisms in a dielectric. (**a**) Irreproducible component due to annihilating positive trapped charge with an electron. (**b**) Transient component due to charging and/or discharging of interface traps. (**c**) Component, attributed to elastic trap assisted tunneling. (**d**) Component, attributed to inelastic trap assisted tunneling (energy loss)

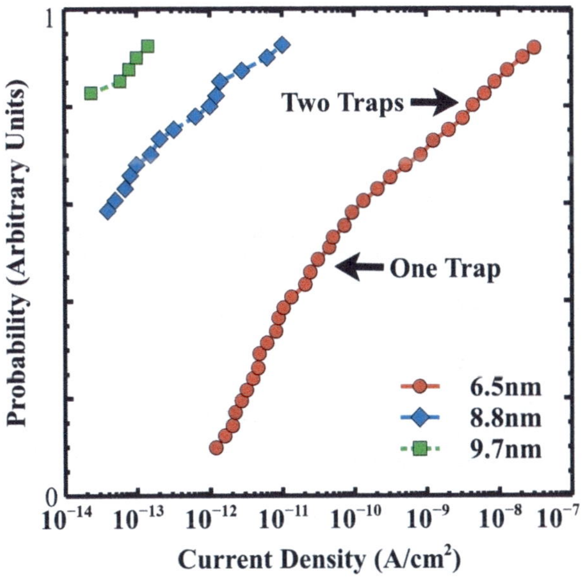

Fig. 7.8 Probability plot showing SILC dielectric leakage as a function of dielectric thickness for a flash memory array showing dramatically reduced leakage with increasing dielectric thickness. Two cells are highlighted, one with the leakage caused by a single atomic level trap and another with two atomic level traps

were determined to be caused by a single trap and a double trap. The higher leakage cell was determined to be caused by SILC leakage through two traps, and the lower level leakage was determined to be caused by leakage through one trap, explaining the higher leakage in the two-trap bit.

7.5 Giant RTS

Single-atomic defects can have an outsized impact on the electrical behavior of a CMOS transistor when the defect occurs in an unlucky location in a small device. Figure 7.9 shows a sketch representing a three-dimensional simulation of an NMOS transistor, including traps and acceptor dopants in the channel (Bukhori et al., 2010). For an advanced CMOS transistor, the presence of dopant atoms can locally reduce the channel current by repelling the channel carriers, resulting in a filamentary conducting path (in blue) around the dopants. Traps in the channel can interact with the dopant atoms, modulating the current in the filamentary conduction. A single trap can modulate the remaining filament, creating an anomalously giant random telegraph signal as the trap is filled or emptied.

Figure 7.10 shows a plot of the drain current versus the gate voltage for a flash memory cell with a giant RTS defect (Fantini et al., 2007). As the gate voltage is

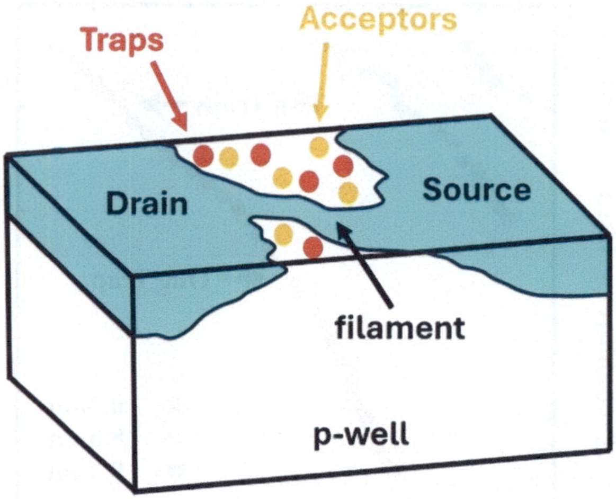

Fig. 7.9 Sketch showing filamentary conduction based on a 3D computer simulation of a NMOS transistor with traps and dopant atoms modulating the current conduction path in blue

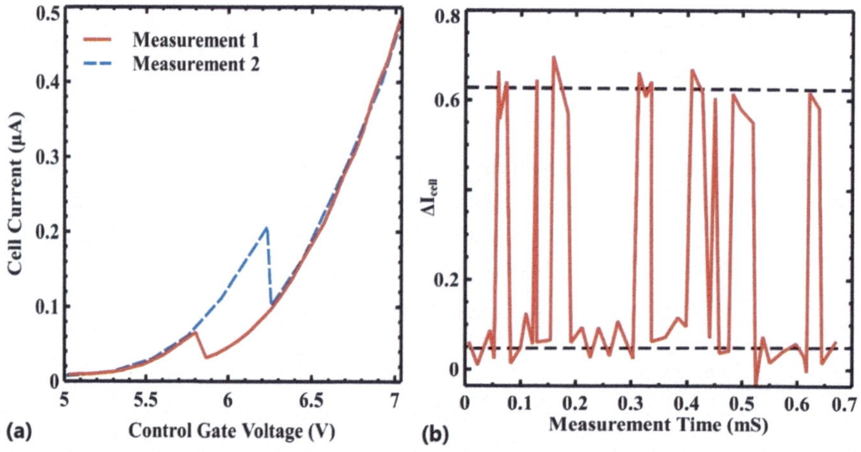

Fig. 7.10 Giant RTS in a flash cell. (**a**) Lot showing giant RTS in a flash memory cell as the control gate is ramped. (**b**) The current variation as a function of time

ramped, the drain current oscillates between two levels as the trap fills, blocking the current or empties, removing the current block. The binary current levels are shown in the figure. The change in drain current is the same order of magnitude as the typical drain current without RTS, which can have a significant impact on the NVM cell reliability.

References

Bowler, D. R. (1997). *A theoretical investigation of gas source growth of the Si (001) surface*. University College London - Wolfson College Trinity. https://www.ucl.ac.uk/~ucapdrb/Thesis/Masterch4.html#x14-440032

Bukhori, M. F., et al. (2010). Simulation of statistical aspects of charge trapping and related degradation in bulk MOSFETs in the presence of random discrete dopants. *IEEE Transactions on Electron Devices, 57*(4), 795–803. https://doi.org/10.1109/TED.2010.2041859

Fantini, P., et al. (2007). Giant random telegraph signals in nanoscale floating-gate devices. *IEEE Electron Device Letters, 28*(12), 1114–1116. https://doi.org/10.1109/LED.2007.909835

Fleetwood, D. M., Pantelides, S. T., & Schrimpf, R. D. (2009). *Defects in microelectronic materials and devices*. CRC Press Taylor and Francis Group. https://doi.org/10.1201/9781420043778

Huff, H. R., & Gilmer, D. C. (2005). *High dielectric constant materials VLSI MOSFET applications*. Springer. https://doi.org/10.1007/b137574

Ielmini, D., et al. (2002). A statistical model for SILC in flash memories. *IEEE Transactions on Electron Devices, 49*(1), 1955–1961. https://doi.org/10.1109/TED.2002.804730

Ielmini, D., et al. (2004). Impact of correlated generation of oxide defects on SILC and breakdown distributions. *IEEE Transactions on Electron Devices, 51*(8), 1281–1297. https://doi.org/10.1109/TED.2004.831959

Maier, J. (2004). *Physical chemistry of ionic materials—Ions and electrons in solids*. John Wiley and Sons. https://doi.org/10.1002/0470020229

Okuyama, Y., et al. (1998). Monte Carlo simulation of stress-induced leakage current by hopping conduction via multi-traps in oxide. In *IEEE IEDM*. https://doi.org/10.1109/IEDM.1998.746501

Robertson, J. (2004). High dielectric constant oxides. *European Physical Journal Applied Physics, 28*, 265–291. https://doi.org/10.1051/epjap:2004206

Werner, M. (2006). *Damage formation and annealing studies of low energy ion implants in silicon using medium energy ion scattering*. The University of Salford.

Wong, H. (2012). *Nano-CMOS gate dielectric engineering*. Taylor and Francis Group. https://doi.org/10.1201/b11270

Chapter 8
Technology-Specific Process Damage

8.1 Introduction

Some types of CMOS technology are more sensitive to process and plasma damage. As CMOS devices are scaled, the transistors become more sensitive to defects. Planar nonvolatile memory cells had the smallest CMOS dimensions for a while, making them very susceptible to atomic-level defects. Additionally, flash memory cells typically store more than one bit per cell, requiring finer discernment of the threshold voltage (or current), substantially adding to the sensitivity.

Some CMOS devices operate with very few charge carriers, which decrease with each generation of scaling. These devices include nonvolatile memories, DRAMs, and CMOS imagers. Planar nonvolatile memories were scaled until just a few electrons were used for each bit (Prall, 2007). DRAMs have been scaled down to the fundamental limits of the storage capacitor, reaching around less than 10,000 electrons per bit. CMOS imagers have been optimized to detect a few photons, sometimes even a single photon, making them very sensitive to single-atom defects. These technologies tend to be the most sensitive to low levels of defects.

This chapter includes other types of devices sensitive to process damage. Metal-insulator-metal capacitors are sensitive to damage because they use thin dielectrics in the plasma discharge path. Low-k dielectric films are sensitive because the materials used are easily damaged.

Advanced logic devices, such as silicon-on-insulators (SOIs) and finfets, are sensitive to damage due to their very small dimensions, which can be damaged with low energy levels.

K. Prall, *CMOS Plasma and Process Damage*,
https://doi.org/10.1007/978-3-031-89029-1_8

8.2 Process Damage to Nonvolatile Memory Cells

Nonvolatile memory (NVM) cells are particularly susceptible to process damage as they are designed to collect charge to store the memory state. These transistors are optimized to collect charge and, therefore, are unintentionally optimized for maximum process damage. NVM transistors contain a region to trap charge. For a floating gate cell, a small polysilicon gate collects the charge. A silicon oxide nitride oxide silicon (SONOS) cell and related technologies contain a charge-trapping dielectric layer. The charge flowing due to process damage flows through the NVM cell, causing damage to the dielectrics and resulting in degraded function and reliability problems.

NVM cells are constructed with the trapping layer sandwiched between a tunneling dielectric and a blocking dielectric with a gate on one side and a semiconductor on the other, as shown in Fig. 8.1. The NVM cell's gate, source, or drain collects charge during process damage situations, increasing the voltage on the terminals until a current starts to flow through the transistor. For example, the current can flow through the blocking dielectric into the trapping layer and then through the tunnel oxide into the semiconductor. Alternately, the current could flow in the opposite direction. The current flowing through the cell causes dielectric damage to the tunnel and blocking dielectrics in a similar manner as described in the previous chapters. An NVM cell manifests process damage in additional ways due to the functional requirements of the memories beyond the mechanisms observed in a simple CMOS transistor.

A sketch of an NVM cell is shown in Fig. 8.1. During plasma damage, charge collects in the trapping layer, and the transistor V_t shifts proportionally to the charge in the trapping layer. The trapping layer gains or loses charge through tunneling driven by the electric field until the current through the blocking oxide is equal to the current through the tunnel oxide. At this point, the V_t stops shifting unless the conditions change. The charge in the trapping layer may remain in the trapping layer during subsequent processing. Some process steps may cause the charge to discharge from the trapping layer, shifting the cell's threshold voltage toward its

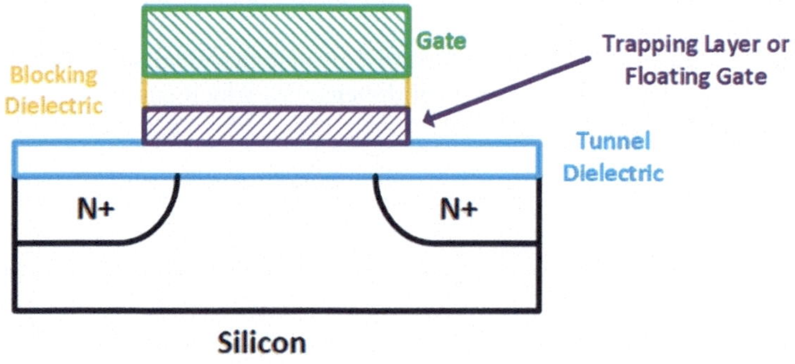

Fig. 8.1 Sketch showing a nonvolatile memory cell

native uncharged V_t. The steps that can shift the NVM V_t require imparting enough energy to the trapped charge to eject it from the trapping layer. High-energy exposure to thermal energy (temperatures greater than 200–400 °C) during processing or high-frequency light (UV or X-ray) can provide sufficient energy to detrap the carriers, erasing the cell. This exposure can eliminate the observable V_t shift, but latent damage may remain.

Figure 8.2 shows a plot of NVM cell behavior where positive and negative voltage stresses (simulating plasma damage) are applied to the gate for 1 min, and the resulting threshold voltage caused by the stress is plotted on the *y*-axis (Cheung, 1999). V_t does not shift at low voltages because there is not enough field to inject or remove the charge from the trapping layer, resulting in zero threshold voltage shifts for these conditions. At high-stress voltages, V_t shifts positively or negatively depending on the polarity of the stress. Process damage similarly shifts V_t like the voltage stress. Observation of cells with threshold voltages shifted from the native V_t (no trapped charge) clearly indicates that the cell has been exposed to process damage. Parametric measurements of NVM transistors during or after processing can be used to diagnose process damage. Specially designed wafers with NVM cells have been used to analyze process damage, but these are limited to process conditions that do not change or damage the wafer. Special finfet structures have been used to analyze plasma damage (Wu et al., 2015).

The threshold voltage of an NVM cell can be calculated from a simple capacitance model of the transistor and trapping layer. Figure 8.3 shows a sketch of a

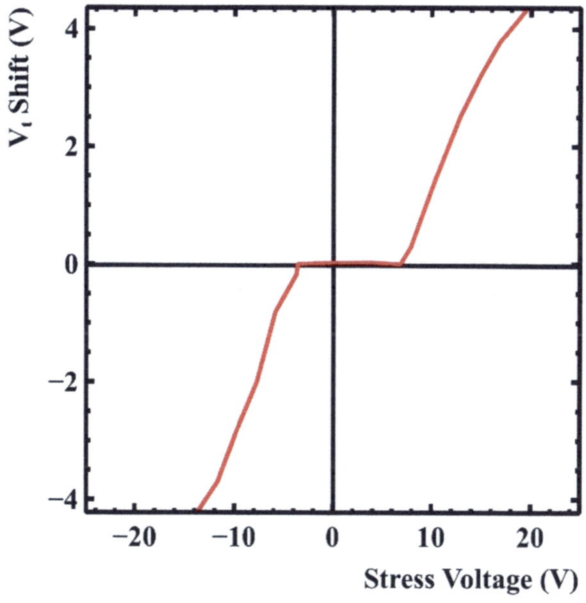

Fig. 8.2 V_t shift of a metal nitride oxide semiconductor (MNOS) NVM cell exposed to a 1 min voltage stress used to simulate plasma damage

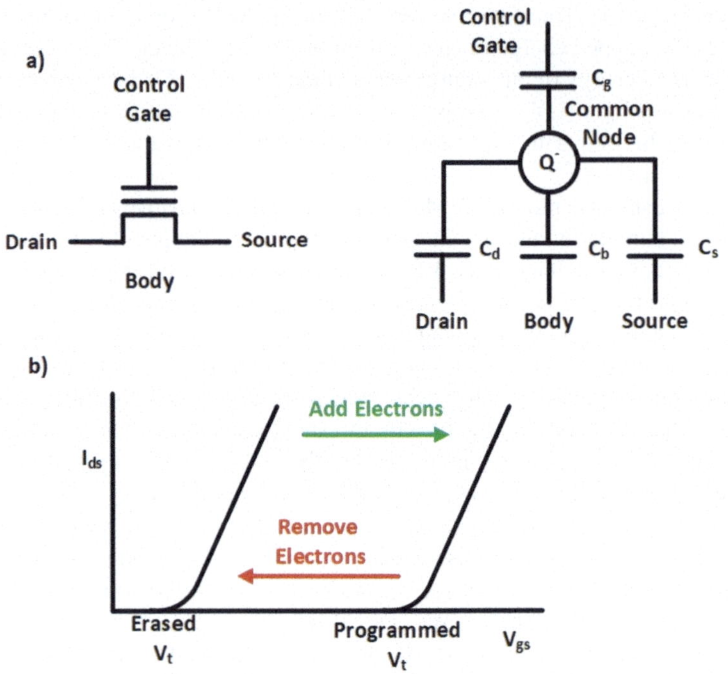

Fig. 8.3 Sketch showing a capacitive coupling model of a nmos nonvolatile memory cell. (**a**) Capacitive components connected to the trapping layer or floating gate. (**b**) Drain current versus gate voltage showing the threshold shift by adding electrons to or removing electrons from the charge storage layer

circuit schematic for an NVM cell (Brewer & Gill, 2008). Capacitors connect each transistor terminal to a common node: the trapping layer or floating gate. The capacitor values are typically calculated from the cell's technology computer (TCAD) simulations. A negative or positive charge is added to the common node, shifting the curve in Fig. 8.3b right or left, respectively. The floating gate voltage or trapping layer is calculated by a simple circuit analysis, giving Eq. (8.1), where V_{cg}, V_d, V_s, V_b are the voltages applied to the control gate, drain, source, and substrate, respectively. The common node voltage is defined as V_{fg}. The constant α in front of the voltage is the capacitive coupling ratio, where the terminal capacitance is divided by the total capacitance C_t as shown in Eq. (8.2), giving the coupling ratio equations shown in Eq. (8.3). This simple capacitor model does not include the capacitance from adjacent memory cells (Prall, 2007).

$$V_{fg} = \propto_g V_{cg} + \propto_d V_d + \propto_s V_s + \propto_b V_b \tag{8.1}$$

$$C_t = C_g + C_d + C_b + C_s \tag{8.2}$$

$$\propto_g = \frac{C_g}{C_t}; \quad \propto_d = \frac{C_d}{C_t}; \quad \propto_s = \frac{C_s}{C_t}; \quad \propto_b = \frac{C_b}{C_t} \tag{8.3}$$

These equations allow the calculation of the floating gate or trapping layer voltages V_{fg}. The current flowing on or off the floating gate can be calculated from the Fowler-Nordheim or direct tunneling equations (Brewer & Gill, 2008; Ranuárez et al., 2006).

The exposure of NVM cells to a plasma can shift V_t positively or negatively during a process damage event as charge is added or removed from the trapping layer, with V_t controlled by the net charge on the common node. Figure 8.4 shows an example of threshold voltage shifts caused by exposure to an Ar plasma (Alba et al., 1997). The shift increases with increasing plasma exposure time. In addition, the shift has an across-the-wafer variation, with the wafer center increasing in a positive direction and the wafer edge decreasing in a negative direction caused by plasma nonuniformity across the wafer.

Process and plasma damage can shift the initial threshold voltage measured when the wafer is tested after the completion of processing, giving erratic and confusing parametric information. Knowing the actual uncharged threshold voltage is essential for process control and engineering reasons. The completed wafer NVM threshold voltage shift from the native V_t indicates the amount of plasma damage the

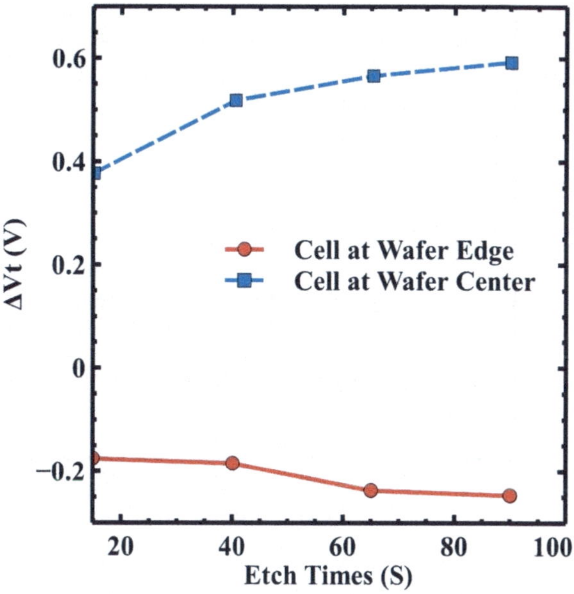

Fig. 8.4 Change in a flash cell V_t with exposure to an Ar plasma. Cells on the edge of the wafer lose electrons, shifting the V_t negative. Cells in the center of the wafer gain electrons, increasing the V_t in the positive direction

cell received during wafer fabrication, with more significant shifts indicating more damage. The direction of the V_t shift indicates the polarity of the damage.

Exposing the wafer to UV light to remove the charge from the NVM cell is sometimes performed before electrical testing to obtain the native threshold voltage. Charge injected onto a NVM cell during process damage can be neutralized later in the process by thermal process steps or steps exposing the wafer to high-energy light, masking the severity of the process damage. Figure 8.5 shows an example of process damage affecting an NVM array after processing (Roizin et al., 2003). The cells with protection diodes on the gates (see Chap. 3) show a tight distribution at the native V_t, while unprotected cells show a V_t shift of more than 1 V caused by plasma damage with a large amount of variation. This example highlights the sensitivity of NVM chips to process damage.

The diodes connected to the gates of nonvolatile memories affect the behavior during UV exposure by modifying the electric fields across the dielectrics, changing the ejection and injection behaviors of carriers operating in the trapping region.

Nonvolatile memories are highly susceptible to process damage due to their very small cell sizes and small cell capacitances, resulting in cell operation with very few charge carriers (Prall, 2007). Some NVM cells operate with as few as ten electrons per bit. Therefore, single electrons can be observed in the operation of an NVM cell.

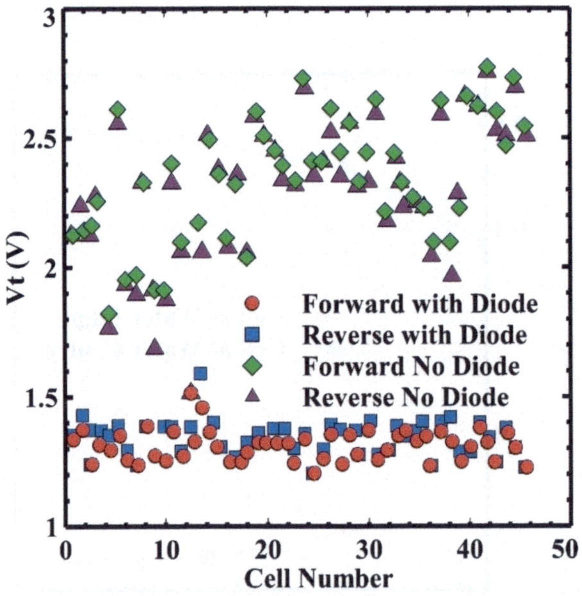

Fig. 8.5 Plot of the V_t of NVM SONOS NROM cells exposed to a plasma, plotted against cell number with and without diode protection. Forward and reverse denote forward and reverse directions for the two bits in the cell

Figure 8.6 shows a plot demonstrating observations of individual electrons in a NAND memory cell's operation, indicating the extreme sensitivity to atomic-level defects (Nicosia et al., 2015).

The sensitivity of NVM cells to single charge carriers and single atomic-level defects in dielectrics causes a host of operational and reliability problems requiring extensive optimization to achieve a successful memory. Process damage can worsen these problems. Process damage can be manifested either as an intrinsic defect affecting all cells or an extrinsic defect creating a few defective bits in a memory array with abnormal behavior. NVM suffers from technology-specific problems in addition to the CMOS parametric problems described previously (Chap. 5). Table 8.1 shows some NVM problems that can be caused or aggravated by process damage. The following sections present examples of some of these issues.

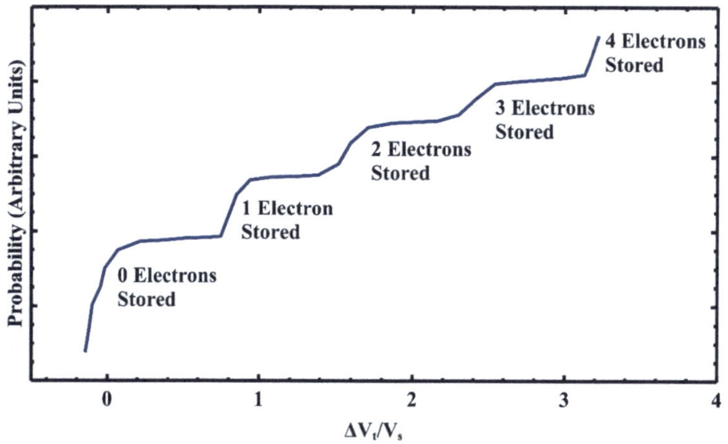

Fig. 8.6 Observation of single electrons in the operation of NAND flash memory cells. V_t measurements are averaged by 20 measurements. V_s is the stepped gate voltage applied to the cell during programming

Table 8.1 NVM operational and reliability problems caused or degraded by plasma damage

Cycling-related stress-induced leakage current (Ielmini et al., 2002)
Data retention (Beug et al., 2011; Gilmer et al., 2010; Takebuchi et al., 1996, 1998)
Charge gain (Takebuchi et al., 1996)
Over-erase and over-programming (Beug et al., 2011)
Endurance, cycling window closure, or shift (Gilmer et al., 2010; Park et al., 1995)
Erratic programming or erase (Beug et al., 2011; Cha et al., 1998; Chimenton et al., 2011)
Electrical disturb mechanisms (Read Disturb, Erase Disturb, Program Disturb) (Chiu et al., 2016; Oh et al., 2008; Zambelli et al., 2020)
Giant random telegraph signal (Fantini et al., 2007)
Parasitic charge trapping (Prall, 2007)

8.2.1 Stress-Induced Leakage in Nonvolatile Memories

The stress-induced leakage (SILC) problem described previously in Chap. 7 plays a dominant role in nonvolatile memories (NVMs), causing or contributing to many reliability problems observed in NVMs. The problems caused by SILC are described in the following sections. NVMs have unique operational physics requiring electrons or holes to intentionally transit through dielectrics by tunneling, hot electron injection, etc., to and from the trapping layer or floating gate. For every carrier transiting the dielectric, there is a finite probability that the carrier can damage the dielectric at the atomic level. The damage probability depends on the carrier energy, position, dielectric thickness, etc. Therefore, NVMs are heavily optimized to minimize SILC damage.

The atomic-level defects caused by SILC damage create a local electrically active defect that can serve as a leakage path, allowing carriers to tunnel or hop between defects, causing dielectric leakage. SILC contributes to all the mechanisms shown in Table 8.1. NVMs are designed so that a single atomic-level defect in the dielectric does not cause bit failure. Mitigating SILC requires that the tunnel dielectrics are thick enough that multiple defects are required for the cell to fail. In addition, the defects must align physically within a small distance, allowing tunneling or hopping. Defects outside of this small range do not contribute to the leakage current. The energy levels of the defects in the bandgap must also align to allow leakage. Figure 8.7 shows a sketch of the local leakage path in the dielectric caused by aligned defects (Okuyama et al., 1998).

Critical optimization and targeting of the tunnel oxide process and thickness are necessary to mitigate SILC-related reliability problems in nonvolatile memories. Figure 8.8 shows an example highlighting the importance of tunnel oxide thickness

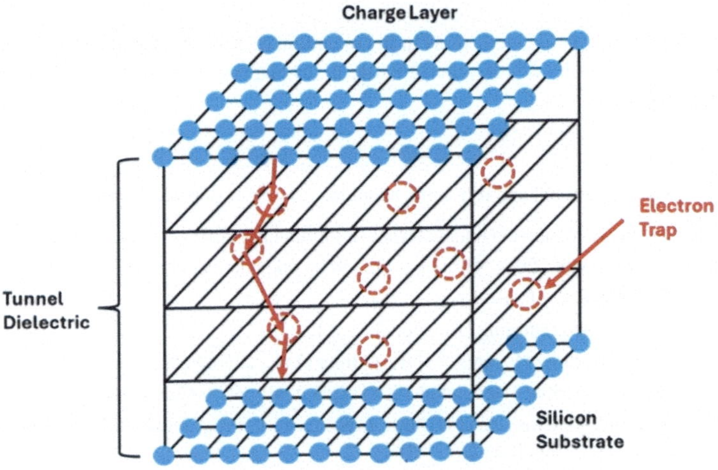

Fig. 8.7 Sketch showing a stress induced local leakage conduction path caused by defects aligned spatially and energetically

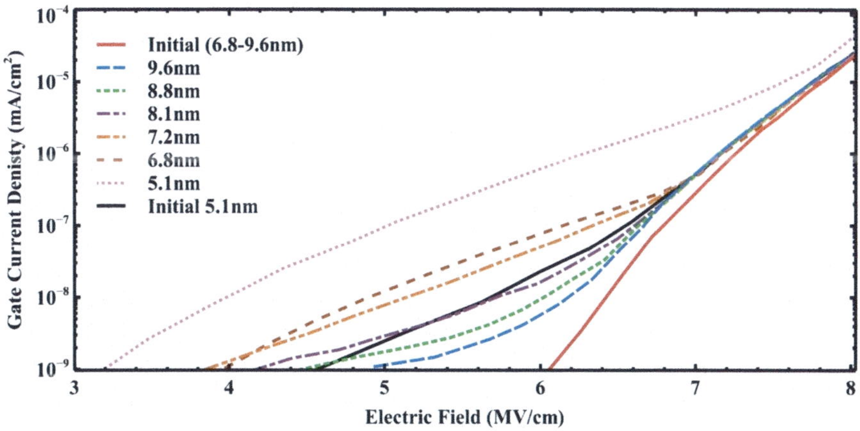

Fig. 8.8 Plot of tunnel oxide current density versus electric field for multiple tunnel oxide thicknesses after current stress. The prestress curves are shown as a solid line and the post-stress curves are shown as a dashed line

with a plot of tunnel oxide leakage plotted against the electric field for a range of tunnel oxide thicknesses after a current stress (Brewer & Gill, 2008). The stress-induced leakage current increases with decreasing dielectric thickness. The initial unstressed SILC leakage current is identical for all thicknesses except for the thinnest 51 Å group. The thicker dielectric groups show the lowest increase with the current stress, partially mitigating the effects of SILC.

Silicon oxide nitride oxide silicon (SONOS)-based nonvolatile memory cells are the basis for 3D NAND technology. The SONOS cell is optimized quite differently from the SILC examples shown here. The tunnel dielectric is very thin, around 20 Å. This mitigates SILC damage as the tunneling electrons are less likely to cause damage as they pass through the thin dielectric. SONOS technology has the advantage of the insulating trapping layer (the nitride), which reduces SILC as the electrons in the trapping layer cannot flow easily to the SILC defect locations like they can in a conducting floating gate.

8.2.2 Data Retention

Data retention is the critical parameter that makes a nonvolatile memory nonvolatile. The data must stay in the NVM cell semipermanently, as specified by the data sheet. Ideally, all charge carriers stay precisely where they are placed, providing a permanent data storage mechanism. The data retention capabilities in the real world allow an NVM data retention specification of typically 1–10 years uncycled or as low as a few months cycled (Cai et al., 2015).

Data retention is measured by programming a data pattern into an NVM chip, waiting for a specific time at a specific temperature, and rereading the chip to determine if the pattern remains. First-order data retention is determined by counting the number of bits in the array that fail after a specific time. NVM chips are designed with test modes that measure the cells' V_t in the product array. Sometimes, varying the supply or internal voltages can be used to estimate V_t. This capability allows all or part of the memory array to be scanned. As a result, detailed data can be collected to accurately determine the data retention capability of every bit in the array. In addition, the shift in threshold voltage with time can be used to predict the lifetime of bits under specific data storage conditions.

Data retention must be guaranteed across the NVMs' operating and storage temperature range. High-temperature retention tests are typically performed and extrapolated to lower temperatures. However, the retention behavior can be complex due to multiple data retention mechanisms that can be degraded by process damage. The multiple mechanisms include thermionic emission, trap-to-trap tunneling, stress-induced leakage current, etc. The loss of charge from the trapping layer at elevated temperatures can be due to thermionic emission, where the charge obtains sufficient thermal energy to be ejected from the trapping layer. This mechanism is not generally degraded by plasma damage. The retention loss at lower temperatures (25–100 °C) can be caused by stress-induced leakage. Multiple competing mechanisms for charge loss yield a complex behavior for NVM retention. Figure 8.9

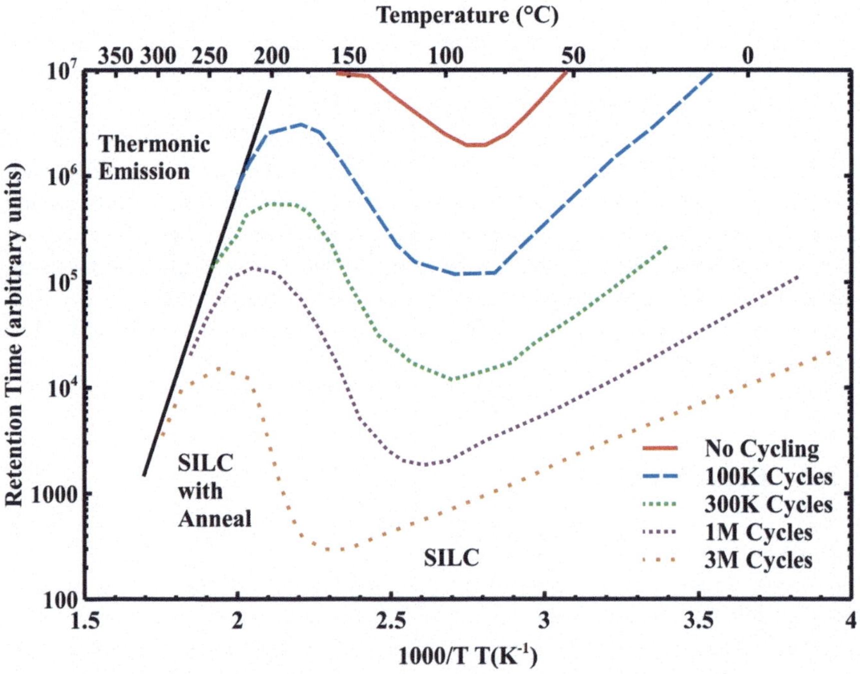

Fig. 8.9 NVM data retention plotted against temperature showing complex behavior with multiple mechanisms without cycling and through various cycling counts. At high temperatures, thermionic emission (TE) from the trapping or floating gate layer dominates. At lower temperatures, stress-induced leakage current (SILC) dominates. At temperatures around 200 °C, retention improves as the SILC damage is partially annealed

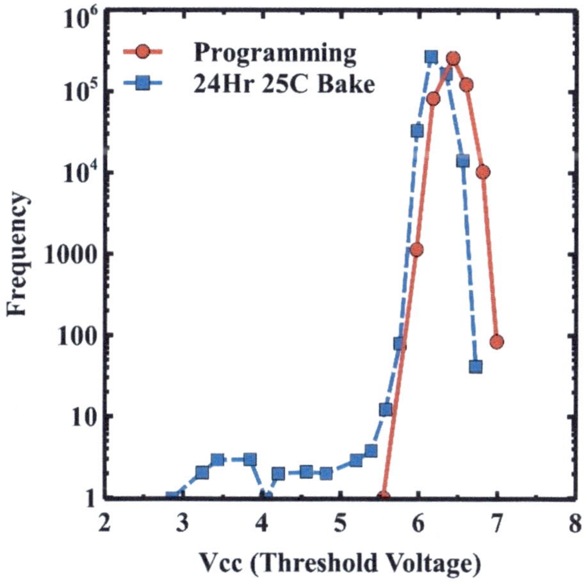

Fig. 8.10 V_t distribution of a chip-level 1 Mbit memory array damaged by a via etch, measured by varying V_{cc}. (a) After programming is shown in red. (b) After 24 h storage at room temperature is shown in blue. The V_{cc} provides an estimate of the threshold voltage

shows an example plot of the retention lifetime plotted against the temperature (Kameyama et al., 2000). A mechanism-based model consisting of two mechanisms, thermionic emission (TE) and stress-induced leakage (SILC), is shown. The low-temperature retention loss can occur at room temperature (Arai et al., 1998). In addition, thermal annealing may occur, as shown in Chap. 6. The worst-case retention is at around 85 °C in this example. The thermal energy can anneal the SILC leakage at higher temperatures, reducing the current.

Additionally, 3D NAND memories generally use silicon oxide nitride oxide silicon (SONOS)-based cells that have specific additional failure modes. These modes include charge migration and charge redistribution in the nitride storage layer (Fan et al., 2022). These data retention loss modes can be degraded by plasma damage.

Process damage causing stress-induced leakage current typically degrades retention at lower temperatures. An NVM cell that loses charge carriers from its trapping region may also lose the data state if sufficient carriers are lost. Figure 8.10 shows an extrinsic data retention problem caused by plasma damage for a room-temperature data retention test caused by a via etch process (Takebuchi et al., 1996). A few defective bits rapidly lose charge and become visible at low threshold voltages, dropping out of the main intrinsic distribution due to abnormal dielectric leakage. The rest of the bits are undamaged and behave normally.

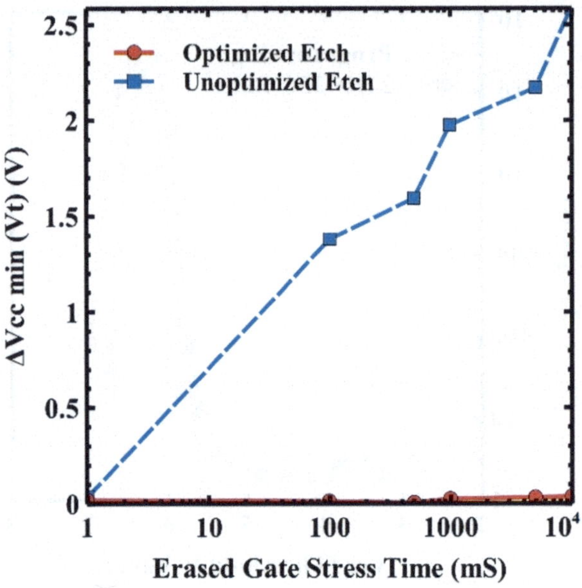

Fig. 8.11 Charge gain from an erased state for a gate stress in a 1 Mbit flash array with two different via etch processes. V_{cc} variation is used to indirectly measure the threshold voltage in the array

8.2.3 Charge Gain

Dielectrics degraded by plasma damage can inadvertently cause charge gain in addition to charge loss, causing erased state retention failures, with charge flowing into the trapping layer by the same defects causing charge loss. Charge gain can cause the failure of the NVM chip by changing the logical state of the memory cell. The charge gain from an erased state due to process damage is shown in Fig. 8.11 with a gate stress (Takebuchi et al., 1996). One via etch recipe caused defects in the NVM cell, creating leakage in the tunnel dielectric. This defect allows charge to flow onto the trapping layer or floating gate, increasing the threshold voltage. A second via etch recipe is free from this defect.

8.2.4 Over-Erase and Over-Programming

For NVM, it is crucial for the program and erase behaviors of all cells in a memory chip to behave uniformly through time and the required number of cycles. Deviant bits that program and erase differently from the majority bits can cause chip failure by shifting the bit to a different logical state.

NVM chips typically use algorithms to force reduced variation in the V_t distribution to ensure that all bits in the array reach the correct logical state. The algorithms can partially compensate for the deviant bits, but there are limits to the algorithm's ability to force deviant bits into the main distribution.

Figure 8.12 shows an example of over-erase in a NOR flash memory for two different stacked gate etch steps (Lee et al., 2002b). One recipe (split B) yields a normal distribution, while a second recipe (Split A) has an abnormal distribution caused by defective bits erasing abnormally fast, creating a defect kurtosis tail at lower threshold voltages (over-erase). Process damage has degraded the tunnel dielectric, causing excessive leakage from the floating gate.

Figure 8.13 shows the variation in erase speed between four different NOR flash cells caused by damage to the source-drain regions by the polysilicon floating gate etch and ion implantation (Lee et al., 2002a). Careful measurements of a large number of bits are needed to determine if the erase speed change is due to an intrinsic shift or a defect mechanism causing erratic behavior.

Defective bits created by plasma damage can over-program as well as over-erase. An abnormal current through the tunnel or blocking dielectric injects or depletes charge from the trapping layer, causing the abnormal functioning of the defective bit. Figure 8.14 shows the distributions of 16 Mbit NAND chips before and after 1 M write-erase cycles (Park et al., 1995). The deposition of the final NAND passivation layer degrades the variation in the programming speed for a fraction of the bits, increasing the deviation of the overall distribution. The three plots show the increased deviation after cycling caused by adding passivation to the wafer.

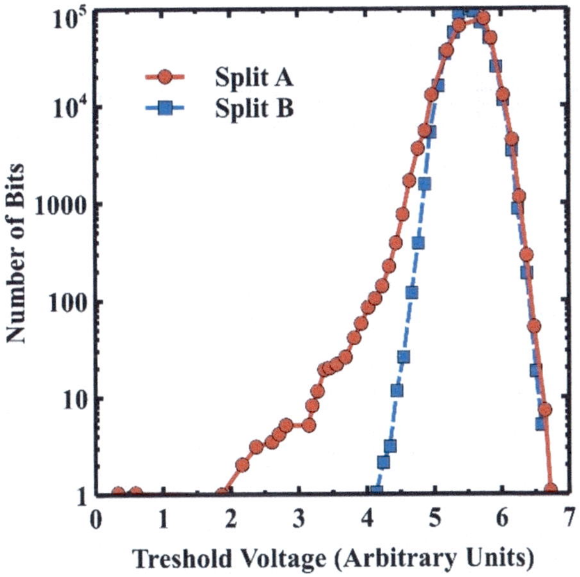

Fig. 8.12 NOR flash erase distribution comparing two different stacked gate etch processes and tools showing over erase with one of the etches. Split A is a magnetically enhanced reactive ion etch tool using HBr/Cl_2/He-O_2 process. Split B is based on a transfer coupled plasma tool using Cl_2 and O_2

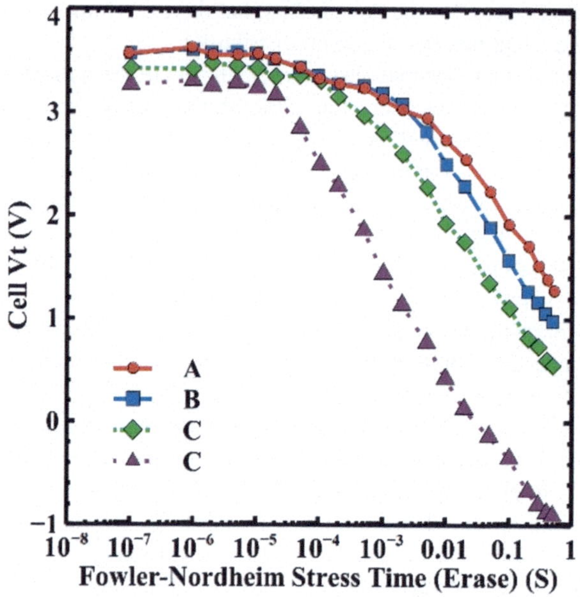

Fig. 8.13 Variation in erase speed for four NOR flash cells caused by the floating gate etch under channel erase conditions. Group A used a magnetically enhanced reactive ion etch tool (MERIE) with a 7° angled ion implant. Group B used a magnetically enhanced reactive ion etch tool (MERIE) with a 0° angled ion implant. Group C used a angled ion implant tool with a 7° angled ion implant. Group D used a transfer coupled plasma tool with a 7° angled ion implant

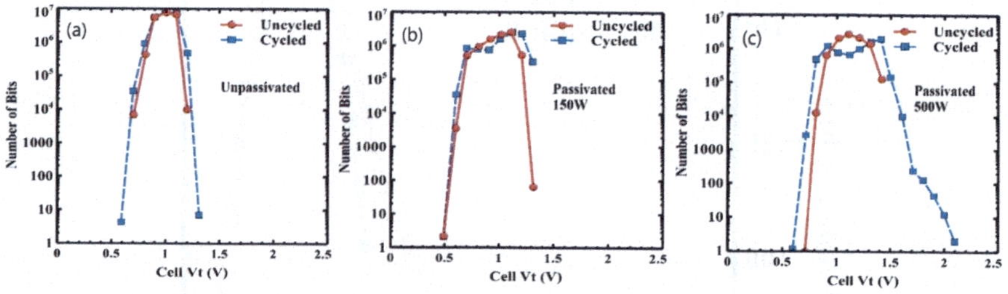

Fig. 8.14 Over programming caused by damage during the deposition of the nitride passivation layer in a NAND flash process. (**a**) Distribution of threshold voltages for an unpassivated 16 Mbit array. (**b**) Distribution of threshold voltages for a nitride passivated 16 Mbit array deposited at 150 W. (**c**) Distribution of threshold voltages for a nitride passivated 16 Mbit array deposited at 500 W

Figure 8.14b shows a slight increase in variation compared to the unpassivated wafer distribution shown in Fig. 8.14a. Figure 8.14c shows further variation degradation caused by increased plasma power during the deposition. After cycling, the highest power deposition group has a fast programming subset of bits with a high threshold voltage. The PECVD deposition process created latent defects in the tunnel oxide, which degraded with cycling.

8.2.5 *Endurance, Cycling Window Shift, or Closure*

The ideal behavior of an array of NVM memory cells through the cycling life of the chip needs to be consistent with the program and erase V_t reaching the same values through cycling. Unfortunately, actual NVM cells change their behavior through cycling as charge becomes trapped in the dielectrics. In addition to a shift in behavior, cells may catastrophically fail, no longer functioning, programming, erasing, or retaining data.

Figure 8.15 shows a plot of the program and erase threshold voltages as a function of the number of cycles for an array of NAND flash memory cells (Park et al., 1995). Ideally, the plot would be perfectly flat throughout cycling instead of turning up at a high cycle count. However, the absolute value of the threshold voltages, in this case, increases during cycling, and the difference between the program and erase states decreases, closing the operating window of the array. The erase slowdown and program speed increase degrades the overall chip performance. The NVM chip typically manages this shift by modifying the program and erase waveforms to force the threshold voltage to be consistent through cycling. Figure 8.15 shows degradation in the window behavior caused by process damage from the nitride plasma deposition at the end of the process. The passivation process step shifts the program-erase window as the recipe power changes with improved results for

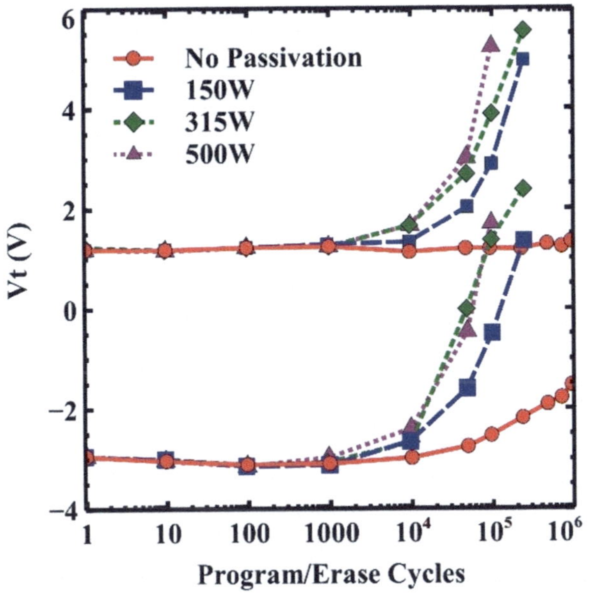

Fig. 8.15 16Mbit NAND Flash single cell program-erase window shift and closure as a function of plasma power during silicon nitride deposition. An unpassivated control group shows better performance without plasma damage

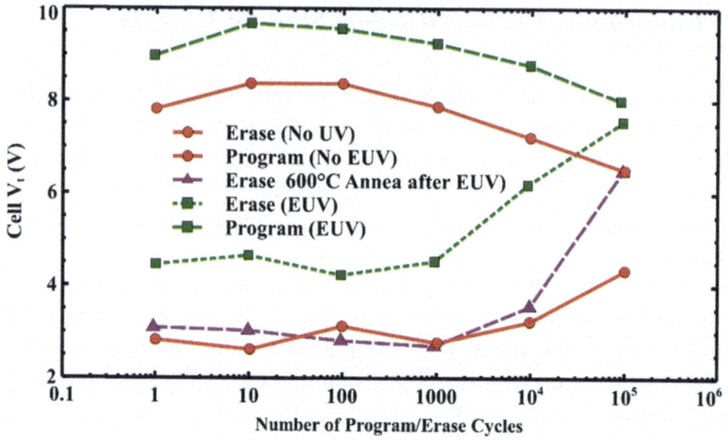

Fig. 8.16 Program-erase window closure and shift characteristics for a polysilicon silicon-oxide-nitride-oxide-silicon (SONOS) after exposure to extreme ultraviolet light (EUV) simulating EUV lithography. The thermal anneal partially recovers the cycling window

reduced power. The unpassivated experimental group has much better behavior, indicating that the deposition caused the damage.

Figure 8.16 shows the program-erase windows for a control process and a cell exposed to extreme ultraviolet light (EUV) to simulate the use of EUV lithography (Tsui et al., 2011). Exposure to EUV light shifts the entire program-erase window in a positive direction. This experiment is a worse-case test of damage by EUV light as the wafer is exposed to EUV light without resist and other layers that would partially or fully block the EUV light.

8.2.6 *Erratic Programming and Erasing*

NVM cells should behave consistently throughout the cycling life of the product, as described previously. The gradual, well-behaved shifts shown previously constitute one nonideal cycling behavior of an NVM memory. A second nonideal behavior is an erratic program and erase phenomenon. Erratic behavior occurs when an NVM cell behavior varies from cycle to cycle instead of gradually over time. Figure 8.17 shows some example plots of erratic programming (Chimenton et al., 2011). The program or erase V_t undergoes a quantum shift at random cycles, shifting from one value to another. The threshold can remain at the new value for a few or many cycles, sometimes reverting to its original value. This erratic behavior can cause bits to fail intermittently throughout the life of the NVM product.

The cause of this type of erratic behavior is attributed to erratic electron or hole tunneling modulated by trapped charge in the dielectrics. The trapped charge can enhance or inhibit tunneling locally through the dielectric (see Chap. 7). The trap

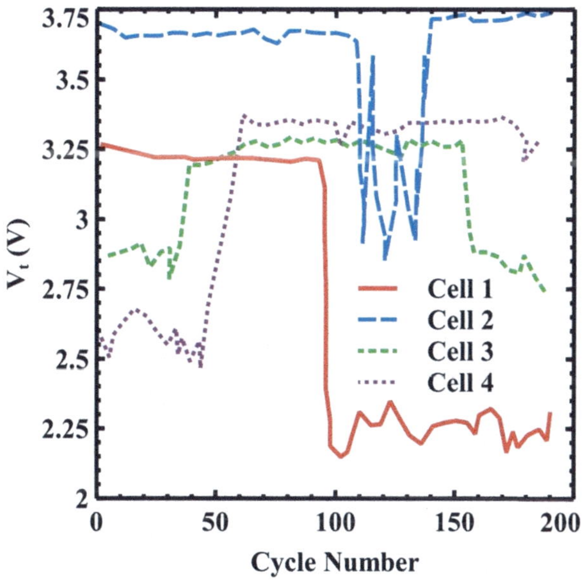

Fig. 8.17 Examples of erratic programming behavior through cycling in four different flash memory cells

can be occupied or unoccupied by a charge carrier, and the occupation can change through time, voltage stress, or temperature. The occupation of the trap or lack of occupation modulates the tunneling current, resulting in the characteristic quantum jumps in the threshold voltage shown in Fig. 8.17. In addition, process damage can create traps in the dielectrics, increasing the propensity for erratic behavior.

8.2.7 Electrical Disturbance Mechanisms (Read Disturb, Erase Disturb, Program Disturb)

NVM cells are operated in a memory array consisting of cells connected in rows and columns. Operating voltages are applied to the rows and columns to make the array function. Unfortunately, the operating voltages are also applied to cells that are not intended to be read, programmed, or erased along the same row or column as the modified cell, creating disturb conditions. These unintended (unselected) cells may receive a lower voltage than those being modified (selected). As a result, process damage can increase the probability of defective cells that have disturb issues.

The unselected cells must be able to withstand many disturb cycles consistent with the datasheet. Typically, disturb measurements are performed by stressing cells for different amounts of time with different voltages and temperatures to develop a mathematical model of the disturb kinetics.

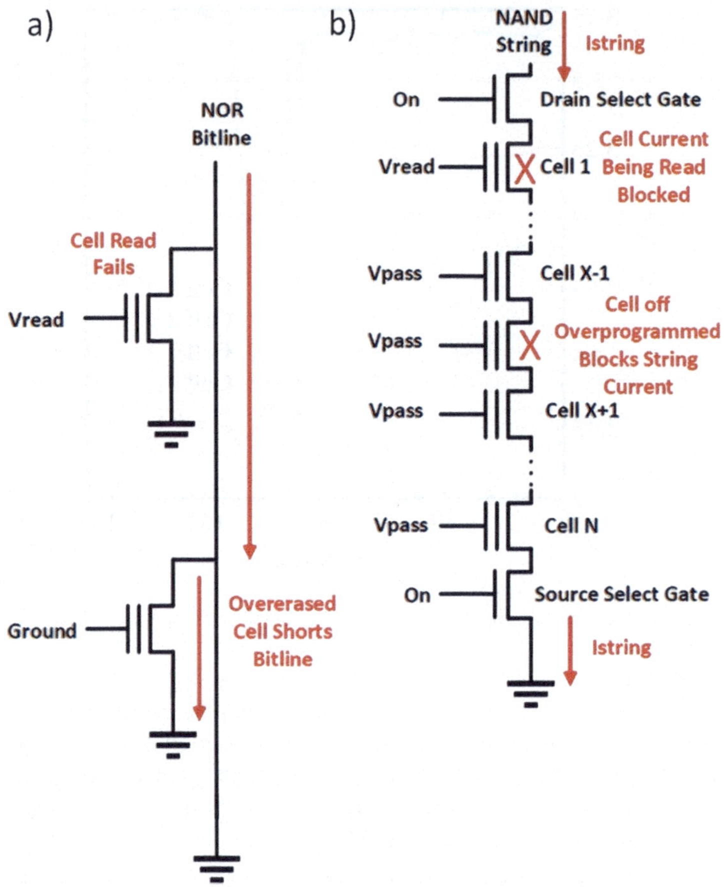

Fig. 8.18 Sketches of NOR and NAND flash memory bit lines. (**a**) NOR bitline showing overerase where the bitline column is shorted to ground, causing failure of the entire bitline by overwhelming the cell current of cells being read. (**b**) NAND array showing a passing wordline cell that is over programmed, causing read failure of a different cell by blocking current in the string

The memory chip is designed to be operated without disturbing or modifying the state of the unselected cells. The unselected cells must withstand the repeated operation of the other cells on the same row or column consistent with the product datasheet. The array must survive disturb conditions for the worst-case conditions, where all other cells on the same wordline or bitline are programmed and erased. The stress of the unselected cells is called disturb and is typically named by which array operation occurs on the selected cells.

Figure 8.18 shows some examples of disturb in a flash memory array (Chimenton et al., 2011). Figure 8.18a shows a NOR NVM architecture bitline schematic with two connected cells. The NOR architecture connects the drain of every cell on the bitline, forming a common node with all the transistors on the bitline connected in parallel. The selected cell on the bitline is read by applying a voltage (V_{read}) to the gate (wordline) of the selected cell and sensing the current on the bitline. The gates (wordlines) of unselected cells on the bitline are biased to ensure that the unselected transistors are off, preventing any leakage through the unselected transistors. If V_t of the unselected transistor is sufficiently negative, current leaks through the transistor, causing the bitline to appear shorted to ground, resulting in the failure of the entire column. Over-erase or erase-disturb can cause V_t to be negative. Process damage can increase the probability of defective cells that cause erase-disturb.

A similar effect can occur if the NVM cell is over-programmed in a NAND array. If V_t of the programmed cell in Fig. 8.18b is too high, V_{read} is not sufficiently high to turn on a programmed cell, and no current flows in the NAND string. As a result, the sensing circuit cannot detect any NAND cell in the serially connected string, causing a bit-level chip failure. This problem can degrade multilevel chip performance in particular.

8.2.8 NOR Read Disturbance

Figure 8.19 shows a plot of the read disturb for a NOR transistor as a function of the drain voltage and time for various drain voltages (Lee et al., 2008). As the drain voltage increases, an infinitesimal fraction of electrons in the channel becomes more energetic, gaining enough energy to overcome the quantum barrier formed with the tunnel oxide. As a result, electrons accumulate on the floating gate or trapping layer, shifting the threshold voltage higher and disturbing the transistor V_t to an unintended value. The existence of read disturbance requires that NOR cells be sensed with a very low drain voltage (~100 mV) to avoid slowly programming the cell, causing read disturb.

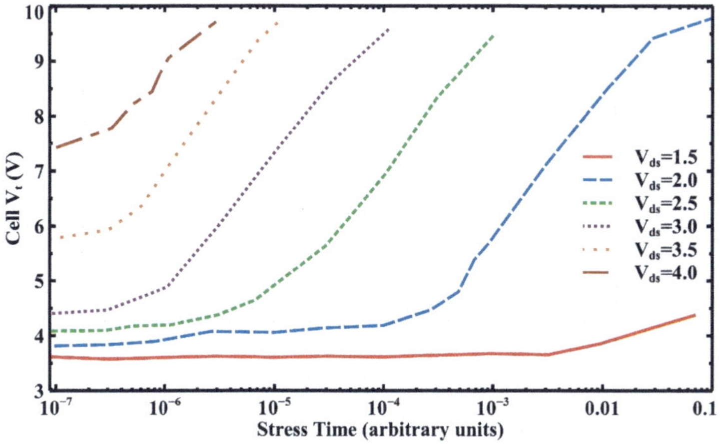

Fig. 8.19 NOR read disturb plotted against time (arbitrary units) for different drain voltages

8.2.9 NAND Read Disturb

A NAND flash array is configured with multiple cell transistors connected in series, as shown in Fig. 8.18b. All transistors in the NAND string have to be biased in order to read any cell. Read disturb can affect either the one cell being read or the rest of the cells biased to enable reading the sensed cell. The cell being read slowly programs through many read cycles, eventually failing by being read as an incorrect state.

A single cell in the NAND transistor string is read at a time by applying a read voltage to the wordline while applying a higher pass voltage to all the other wordlines to ensure that the current flows through the string. The pass voltages supply sufficient voltage to enable a sensing current to read the flash cell. However, the passing transistors undergo a gate disturb stress while other cells are read on the NAND string. As a result, the passing transistors slowly gain charge during many reads, causing the cell's threshold voltage to increase. Eventually, the V_t increase causes the disturbed cell to fail to pass the current necessary for reading the string, causing the entire cell string to fail by blocking the current flow. Figure 8.20 shows an example of read disturb in a 3D NAND string (Du et al., 2019). The threshold voltage of the disturbed cells increases as a function of time, as indicated by the number of reads. An erased cell will slowly program until it reaches the threshold

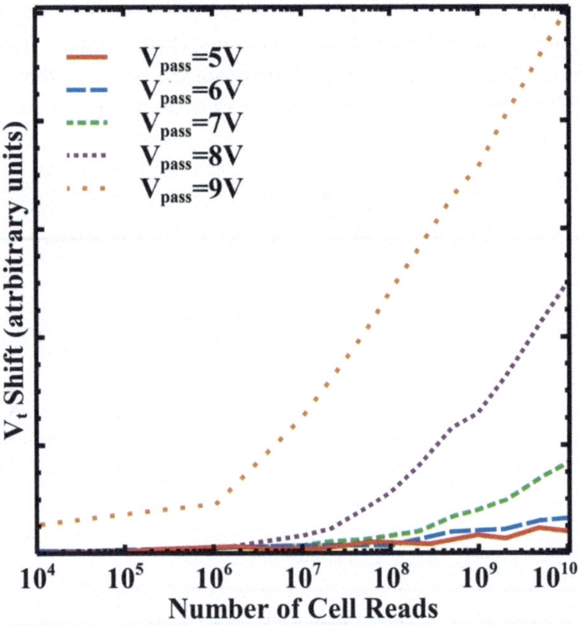

Fig. 8.20 V_t shift plotted against the number of read disturb stress cycles for a 3D NAND array for various V_{pass} voltages

voltage of a programmed cell, causing the bit to fail. A programmed cell will slowly program further until the pass voltage fails to turn on the cell to read the string current, causing the entire string to fail to read. The V_t shift increases as the gate voltage increases, resulting in a reduced time to failure. Tunnel dielectric defects induced by process damage can degrade NAND read disturb performance, causing reduced disturb immunity. This problem sets a limit on the maximum number of read cycles without rewriting the data.

8.2.10 Giant Random Telegraph Signal

Nonvolatile memories are susceptible to giant RTS, as described in Chap. 7. In addition, multilevel cells have minimal state separation between the logical bits, and noise can cause the cell to erroneously change states during cell operations. Finally, NVM cells generally have a small area with minimum transistor length and width that is more susceptible to a small number of traps causing filamentary current conduction behavior (see Chap. 7) in the channel as the current avoids the traps, dopant atoms, etc. Figure 8.21 shows an example of giant RTS in a floating gate flash memory cell (Fantini et al., 2007). The cell drain current oscillates between two metastable states, varying by a factor of 2 due to giant RTS.

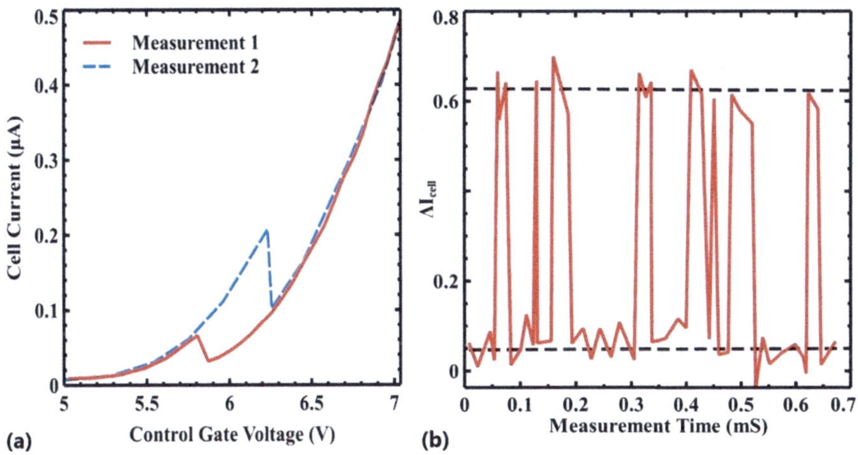

Fig. 8.21 Giant RTS impact on the cell current of an NVM cell. (**a**) Cell current transition as a defect captures an electron in a flash memory cell observable as two different curves in the drain current versus gate voltage. (**b**) Static measurement showing a cell current transitioning between two levels as a function of time

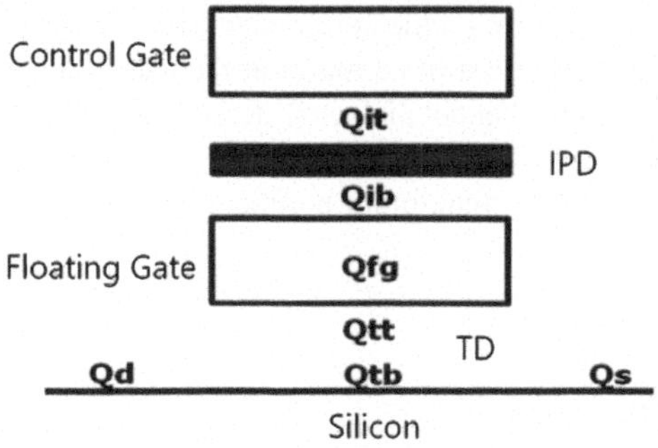

	50nm	35nm	25nm
Qtb	4	2	1
Qtt	9	7	4
Qib	22	17	9
Qit	149	103	100
Qs	33	9	5
Qd	61	16	10
Qfg	18	12	10

Fig. 8.22 Sketch of a NAND floating gate cell with parasitic trapping locations labeled in the interpoly dielectric, tunnel dielectric, and over the source and drain. The table shows the number of electrons trapped in the parasitic labeled locations based on technology computer aided design simulations necessary to shift the cell threshold voltage by 100 mV

8.2.11 Parasitic Charge Trapping

Planar nonvolatile memories were aggressively scaled until scaling was no longer possible. The amount of charge stored on the NVM cell continuously decreased until the number of electrons stored in the trapping region per state decreased to less than 100 (Prall, 2007). Planar scaling of NAND stopped, and NAND scaling transitioned to three dimensions, increasing the number of electrons stored per cell (Micheloni, 2016). Cells with just tens of electrons per state are extremely sensitive to charge trapped in unintended locations outside the trapping layer. A small amount of charge trapped in unintended locations can significantly affect the state (V_t) of the cell and the data stored in it. Figure 8.22 shows an example of sensitivity to trapped parasitic charge for advanced planar NAND flash memory cells (Prall & Parat, 2010). Technology computer-aided simulations are performed for three different technology nodes for trapped charge in parasitic locations. The number of trapped electrons necessary to shift the threshold voltage by 100 mV is shown. The number of electrons needed to shift the threshold voltage decreases as the cell is scaled, showing an increased sensitivity. Process damage can easily result in a few trapped electrons in the flash cell. Other advanced technologies can be similarly sensitive to the trapped charge. For instance, this effect can occur in a small finfet or SOI device.

8.3 Process Damage in CMOS Imagers and CCDS

8.3.1 Introduction

CMOS imagers are extremely sensitive to process damage because they are designed to be very sensitive to light for low light detection reasons. This sensitivity makes low-level process damage detectable in imagers beyond what is detectable in CMOS

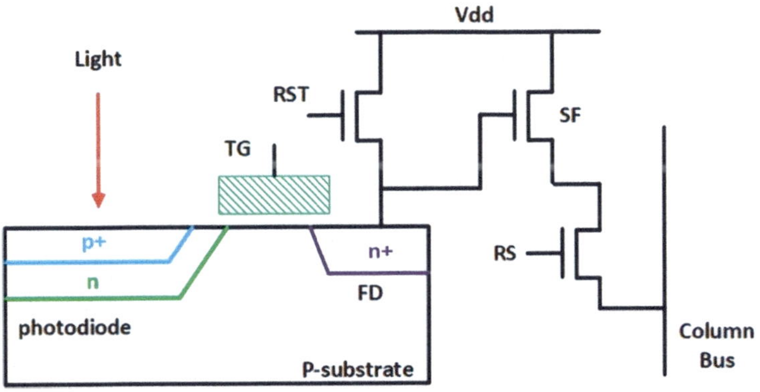

Fig. 8.23 Sketch of a CMOS imager pixel cell diagram and schematic

logic. The pixel sensitivity can reach single-photon levels of detection (Janesick, 2001). A similar older technology, charge-coupled device (CCD), is also susceptible to process damage, with nearly identical results to the degradation in imagers. CMOS imagers consist of a unit cell called a pixel arranged in an orthogonal array. The pixel consists of a light-detecting element, such as a photodiode, connected to a circuit that amplifies and stores the light signal. The signal is read from the array, transferring the image to the external world. The light from an optical image strikes a diode detector in a CMOS imager cell, creating a pocket of charge that converts the light to an electronic charge stored in the diode. An atomic-level defect in the imager cell can drain the diode's charge, causing a loss of the image data, resulting in reduced sensitivity or a defective light or dark pixel.

Figure 8.23 shows an example sketch of a CMOS imager circuit (Wang, 2008). The light is collected by a photodiode, creating a charge in the pn junction. The transfer gate TG transfers the charge to the source follower transistor SF, amplifying the charge. The amplified signal is transferred to the column outside the pixel by the RS transistor. The RST is used to reset the pixel before the light is collected.

CMOS imagers consist of two primary types of imagers based on the location of where the imager collects the light, as shown in Fig. 8.24 (Vici et al., 2020). The two types are frontside and backside imagers. For frontside imagers, the light enters the wafer frontside and passes through the metal, intermetal, and passivation layers before reaching the wafer surface, where a photodiode collects the light. Unfortunately, the complicated stack of metal and dielectric layers degrades the amount of light reaching the wafer surface, reducing the scalability of the frontside imager pixel. Backside imagers are formed with the photodiodes on the backside of the wafer, eliminating the metallization layers in the light path. As a result, backside imagers have the technical advantages of better scaling and performance. However, the process requirements for backside imagers are more costly and complex, requiring wafer-to-wafer bonding, wafer thinning, etc. (Durini, 2014).

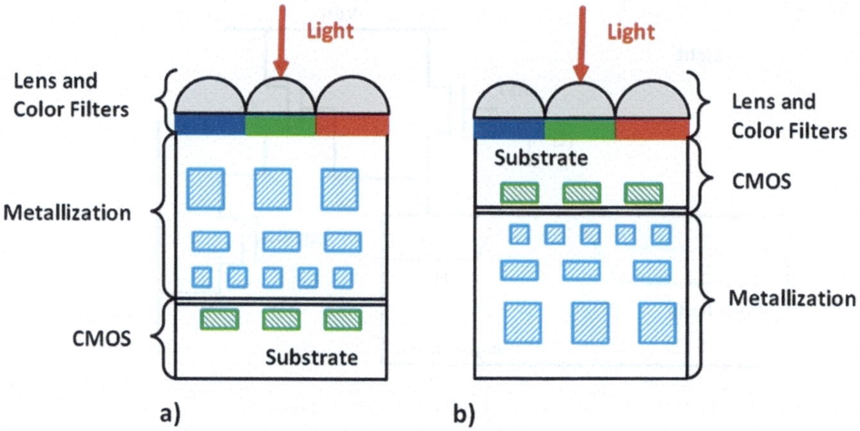

Fig. 8.24 Process cross-section showing different imager configurations. (**a**) Frontside imager, (**b**) backside imager

8.3.2 Atomic-Level Defects in CMOS Imagers

CMOS imagers are extremely sensitive to atomic-level defects, making them susceptible to process damage. Imagers are designed to generate as many electron-hole pairs as possible per photon. As a result, the multicomponent pixel can have different electrical signatures based on where the atomic-level defect occurs. For example, Fig. 8.25 shows a pixel performance metric called temporal read noise (TRN) measured by repeated pixel reads (Boukhayma, 2018). In this example, this metric is exceptionally well controlled, reaching a noise level of less than one electron.

CMOS imagers suffer from random telegraph signal (RTS) noise, as described in Chap. 7, which can severely affect the cell operation similar to the noise problems observed in nonvolatile memories described in Sect. 8.2.10. RTS noise results in a variable noise floor in the pixels as a single-atomic-level defect traps or discharges a charge carrier, resulting in varying results with time. As a result, pixels may flicker between light and dark, as the RTS leakage transitions between low and high values, in addition to a high noise floor. The operation of the CMOS imager can be optimized algorithmically to improve the image results. For example, the signal from a pixel may be averaged by reading the pixel multiple times and processing the result. A correlated double-sampling technique (CDS) reads the pixel twice to average out the noise, subtracting two consecutive readings to eliminate the noise (Wang, 2008). This technique is ineffective for a pixel with RTS noise because the random transitions are not averaged. Figure 8.26 shows an example result of a CDS algorithm on a pixel with RTS (Wang, 2008). The resultant pixel signal is trimodal based on the time correlation between the RTS transitions and the sampling.

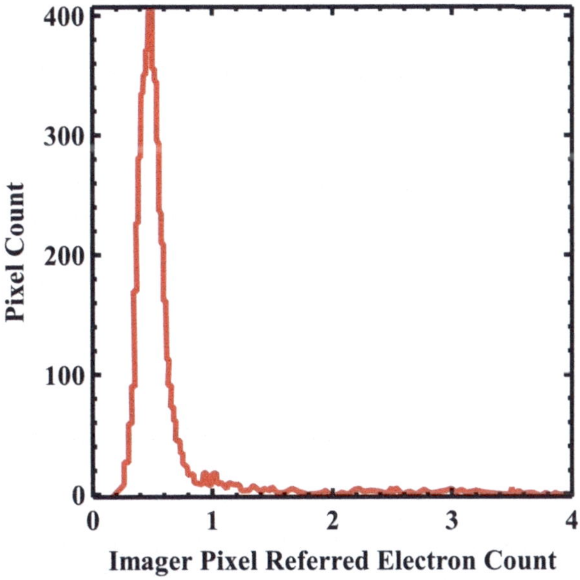

Fig. 8.25 An example of imager sensitivity to process damage showing the average for a noise metric averages about half of an electron

8.3.3 Process Damage in CMOS Imagers

CMOS imagers and CCDs are sensitive to all of the electrical defects described in Chap. 5. All of the process damage phenomena described previously can occur in a CMOS imager or CCD. Process damage in imagers can cause multiple problems, including gate leakage, junction leakage, STI leakage, and various types of noise (Wang, 2008). Defects at a silicon-dielectric interface at the wafer surface or shallow trench isolation interface can be electrically active, degrading the performance of the pixel. Defects in the bulk of the silicon can reduce the efficiency of the pixels by generating carriers in the junction depletion region or diffusing from outside the junction.

Imagers and CCDs are extremely sensitive to process damage. Common process technologies can degrade the imager performance. The steps include gate etching, contact etching, metal etching, sputtering, implantation, etc. (Janesick, 2001). All of the sources of process damage described in Chap. 9 can be expected to affect CMOS imagers. Additionally, CMOS imager fabrication has unique process steps and fabrication methods that can also be expected to cause process damage. These steps include cavity etching, backside films, etc. The architecture of the imager pixel, design, and process are heavily optimized to reduce the effects of process damage.

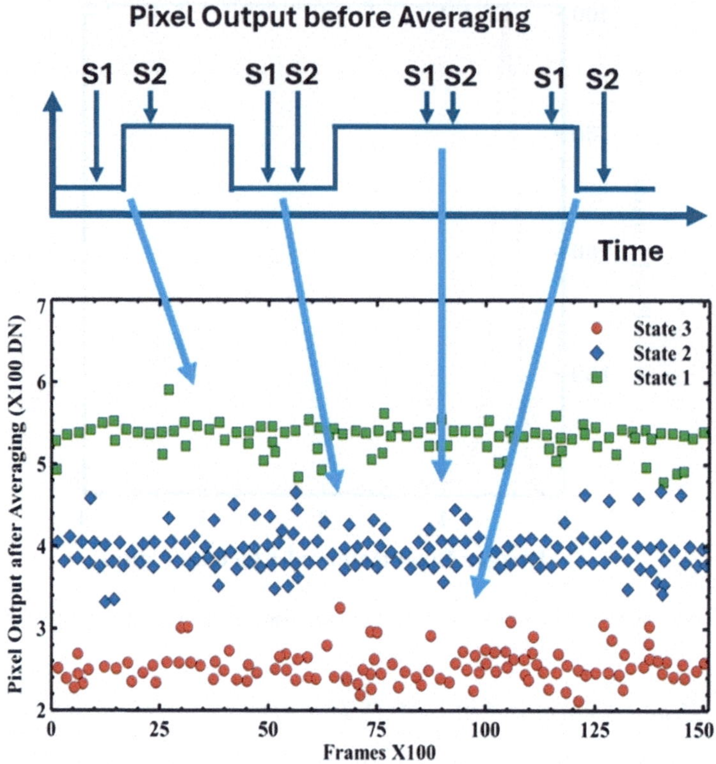

Fig. 8.26 Example of random telegraph noise in a CMOS imager operated in a double sampling mode (CDS). The pixel output is expressed as a digital number from an analog to digital converter. S1 and S2 are the locations in time where the pixel output is read, creating a trimodal noise signal based on when the pixel is sampled

8.3.4 Dark Current

The current generated when a pixel is in the dark is called dark current. Dark current is a critical metric for imagers. The dark current sets the noise floor for the imager. Advanced imagers can obtain noise floors at the single-digit electrons/s level (Kim et al., 2006). Therefore, reducing the dark current results in improved sensitivity and performance. A few examples of process damage degrading dark current in imagers follow.

A single-atomic-level defect in the dielectric-silicon interface can cause leakage through an interface trap under the gate oxide or the shallow trench isolation, draining the photodiode's collected charge. These defects occur at the semiconductor-dielectric interface or in the dielectric bulk. Figure 8.27 shows a CMOS imager cell sketch, with the surface generation regions in the dielectrics marked with a red X.

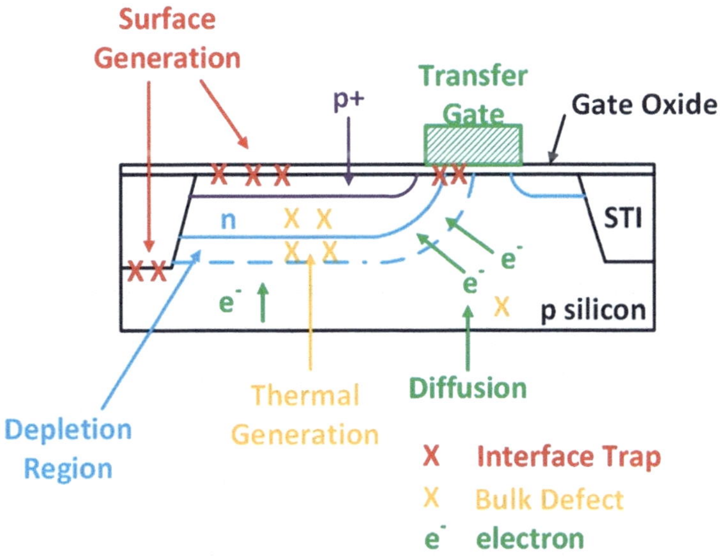

Fig. 8.27 Sketch of a CMOS imager cell showing potential locations for defects caused by process damage

A single-metal atom forming a bulk defect can damage the junction integrity by draining the charge from the diode, causing data loss (see Chap. 15). This type of leakage mechanism is created when a defect exists in the junction's depletion region, creating an additional leakage path beyond the intrinsic generation and diffusion mechanisms, as shown in Fig. 8.27 (Tan, 2013). Defects at a silicon-dielectric interface at the wafer surface or shallow trench isolation interface can be electrically active, degrading the performance of the pixel. Defects in the bulk of the silicon can reduce the efficiency of the pixels by generating carriers in the junction depletion region or diffusing from outside the junction.

CMOS imager pixels are optimized to mitigate the electrical effects of the surface defects shown in the figure. For example, a surface implant is used to optimize the photodiode, minimizing the surface electrical problems. This structure is called a pinned photodiode (Boukhayma, 2018).

Dark current distributions typically consist of intrinsic and extrinsic distributions, with the extrinsic distribution (kurtosis) consisting of defective pixels and intrinsic pixels without defects. Figure 8.28 shows a sketch of an example of a dark current distribution (Carrere et al., 2014). As described in Chap. 4, process damage can affect both distributions by creating defective pixels or shifting the intrinsic distribution.

Figure 8.29 shows a CMOS imager experiment (Sacchettini et al., 2019). A cavity is etched in a dielectric over the photodiode for optical reasons. The dielectric etch process increases the dark current, adding noise and reducing the pixel sensitivity. A metal plate is placed atop a nonfunctional pixel as an experiment. The metal

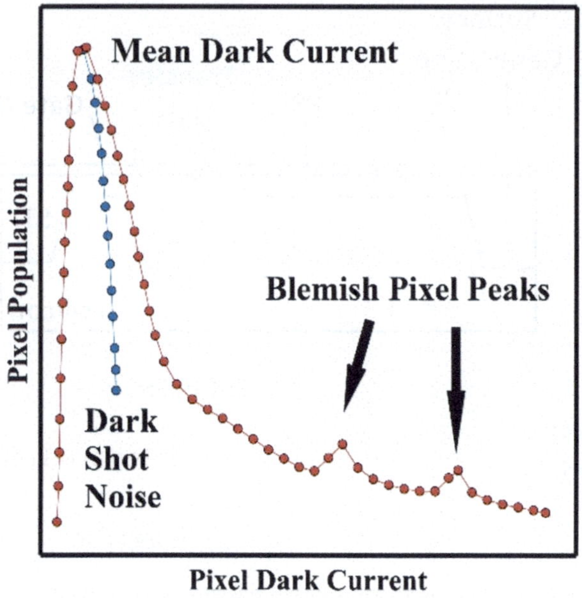

Fig. 8.28 Sketch showing an example distribution of a CMOS imager dark current measured by the pixels. The small peaks labeled blemish pixel peaks are attributed to molybdenum (Mo) contamination

plate pixel has no increase in dark current due to the cavity etch, proving that the process damage is likely UV light emission from the plasma.

Figure 8.30 shows an experiment where the resist ashing process was performed in two different chambers (Sacchettini et al., 2019). A chamber where the plasma is created in the same chamber as the wafer shows a degraded dark current. A remote plasma chamber where the plasma is created in a separate chamber from the wafer has a much reduced dark current because the wafer is not exposed to UV light and the plasma.

The dark current is extremely sensitive to the dielectric film on the silicon surface over the photodiode. Traps and interface states at the silicon-dielectric interface can degrade the dark current. Process or plasma damage in this film can degrade the dark current. Figure 8.31 compares different antireflective coatings on the silicon surface of a backside CMOS imager (Sacchettini, 2019). The ONO film is susceptible to process damage, as charge can be trapped in the nitride layer, forming a nonvolatile memory-like structure, as described in Sect. 8.2, with the surface properties varying with the charge trapped in the nitride layer.

As an example of the importance of trapping in a nitride layer, decreasing the electrical resistance of the nitride layer can reduce the propensity for trapping and reduce the dark current impact from trapping in the nitride. Lower resistance films over the pixel have been shown to reduce the dark current.

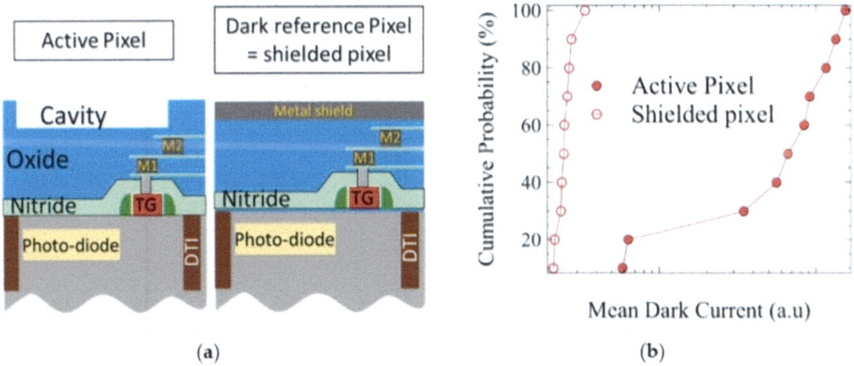

Fig. 8.29 Example of pixel degradation of dark current (no light) by process damage by a dielectric etch creating an optical cavity. The cause of the process damage is UV light degradation from the plasma process. (**a**) An experimental metal shield on a non-optical pixel protects the cell from UV damage. (**b**) Comparison of the dark current for an unshielded pixel and a pixel shielded with a metal layer showing reduced leakage for the protected pixel

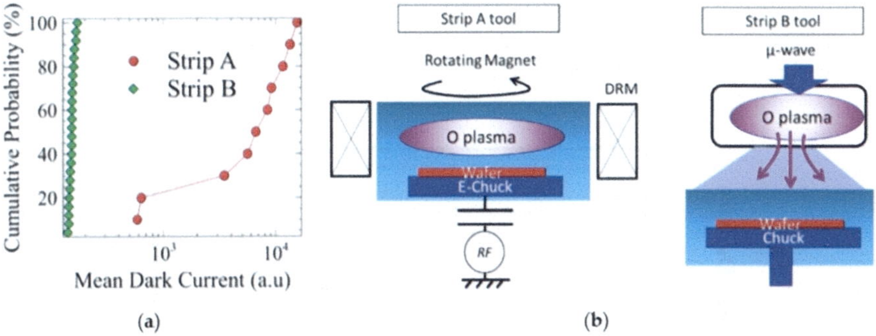

Fig. 8.30 Degradation of a CMOS imager pixel dark current due to a resist strip process showing two different resist strip tools. (**a**) Plot comparing dark current for two different strip tools. Process A is performed in a tool with the plasma generated in situ in the same chamber as the wafer, exposing the wafer directly to the plasma and UV light. Process B uses a remote chamber to separate the plasma generation physically from the wafer (**b**) Sketch of the two tools

8.3.5 High-Energy Light Damage in CMOS Imagers

CMOS imagers are sensitive to high-energy photons that can damage films and devices, as described in Chap. 9. Dielectric films adjacent to the photodiode are susceptible to photon damage, as shown in the following examples.

An experiment was performed in Fig. 8.32, where a CCD pixel is exposed to UV laser, degrading the dark current as a function of the laser energy, showing increasing dark current with increasing energy (Li et al., 2004). However, this experiment is not entirely realistic, as the wafer is exposed directly to light rather than through a scanner with a resist on the wafer. Virtually all photons should be trapped in the resist layers, making lithography less likely to be a source of process damage.

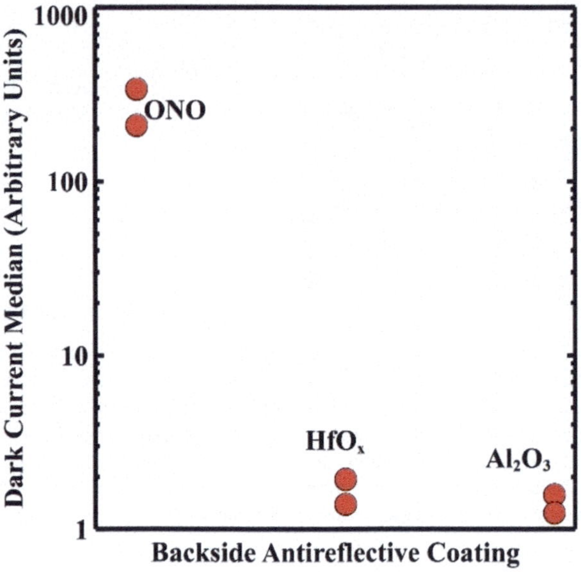

Fig. 8.31 Dark current dependence on the backside anti-reflective coating applied to the backside surface of a backside CMOS imager

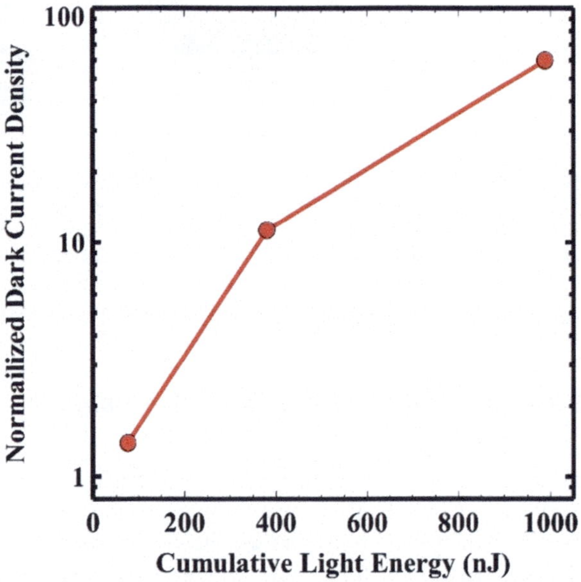

Fig. 8.32 Plot of charge couple device (CCD) dark current as a function of exposure to a 157 nm UV laser simulating UV lithography showing increasing dark current with increasing laser energy. This degradation is a worst case condition as the wafer is not covered with resist or other layers which will partially or fully block the UV light. Note that this wavelength has not been used for production lithography

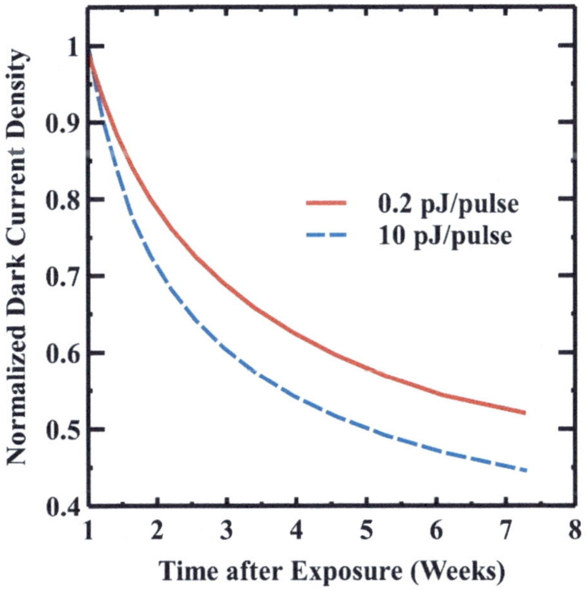

Fig. 8.33 Dark current recovery with time at room temperature after exposure to a 157 nm light

Photon damage from plasma processes, as described in Chap. 9, can degrade many CMOS devices.

The dark current can recover with time as the broken bonds from the damage can rebond, decreasing the dark current. Figure 8.33 shows an example of dark current recovery after exposure to a 157 nm laser at room temperature (Li et al., 2004). More than half of the damage can be recovered after waiting a few months at room temperature, requiring careful record-keeping during experiments and analysis to avoid incorrect conclusions.

8.3.6 White Pixel Process Damage

White pixel defects in CMOS imagers are caused by excessive leakage that electrically behaves as a pixel exposed to intense light with or without light exposure. White pixels are caused by common CMOS defects such as particles, metal contamination, etc. Additionally, plasma damage can also cause white pixel defects through the mechanisms described in this book. Figure 8.34 shows an example of white pixel defects caused by damage due to a nitride spacer etch process during the formation of the transfer gate transistor (Tokashiki et al., 2007). This type of damage is described in Chap. 9. The white pixel damage defect density increases when a magnetic field is applied during the spacer etch (see Chap. 9 for a magnetic field discussion).

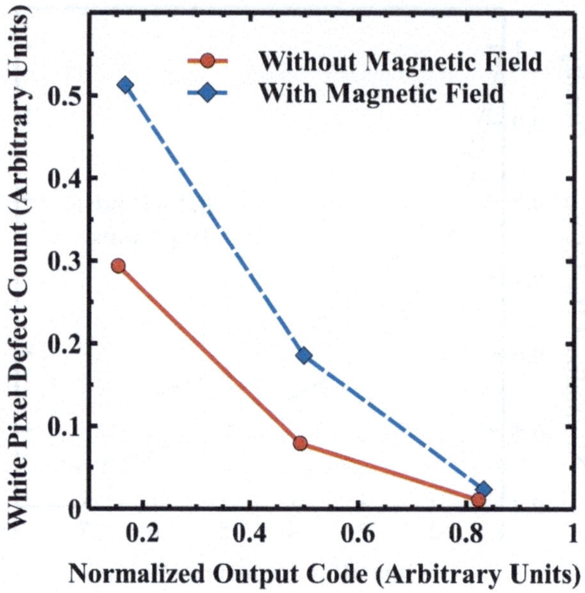

Fig. 8.34 Example of a white pixel defect caused by a nitride spacer etch process with and without a magnetic field. The plot shows the defective white pixel count as a function of the analog to digital converter output code

8.4 Process Damage in DRAM Memories

Dynamic random access memory (DRAM) cells consist of a single transistor and capacitor. A DRAM cell is also a device with high sensitivity to atomic-level defects. The binary data is stored as a charge in a DRAM cell temporarily retained on the cell capacitor and the junction. The data is volatile and is lost in milliseconds to seconds unless the data is rewritten, refreshing the data. Any leakage mechanism causing a loss of charge through the dielectric, isolation, junction, or transistor can cause the data to be lost or cause performance degradation. Figure 8.35 shows a cross section of a DRAM cell with common sources of leakage labeled (Lee et al., 2012). Most of the mechanisms shown have been previously described as being susceptible to process damage (see Chap. 5). For example, charge can leak through the DRAM capacitor dielectric, degrading retention through a stress-induced leakage mechanism caused by process stress. The junction can leak excessively if damaged by process damage, causing data loss. A DRAM cell transistor subject to process damage can leak through multiple mechanisms, including subthreshold conduction, gate-induced diode leakage, shifted V_t, etc. In addition, parametric

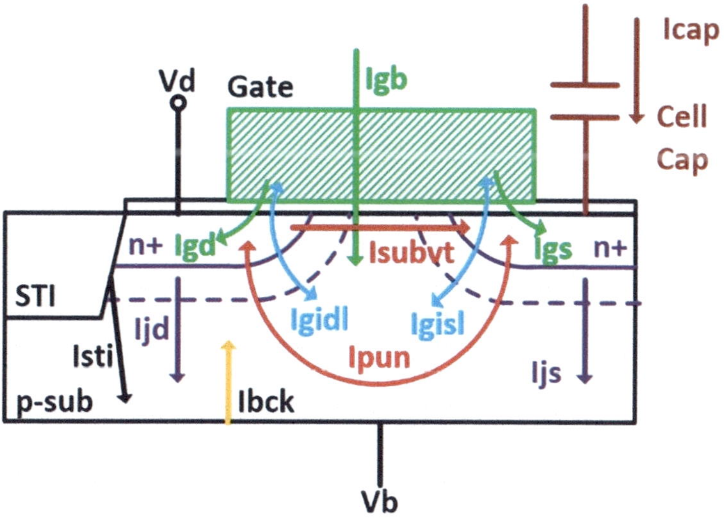

Fig. 8.35 Cross-section of a DRAM cell showing leakage sources

shifts in the cell transistor can change the dynamic performance of the cell, affecting the access speed.

8.4.1 DRAM Refresh

The data retention of a DRAM is determined by the loss of charge from the capacitor through all leakage mechanisms that drain the charge, destroying the binary data. Most DRAM leakage mechanisms can be degraded by plasma and process damage, reducing the refresh time. The mechanisms are described in this book, mostly in Chap. 5. The access transistor can leak through subthreshold leakage at the surface or in the bulk silicon or gate-induced diode leakage. The isolation can leak, causing cell-to-cell leakage. The diode junction can leak through reverse leakage current, with a minority current diffusing from the wafer bulk or a minority current generated from the backside surface of the wafer. The gate dielectric can leak (Chap. 5). Also, the cell capacitor dielectric can leak by tunneling, similar to the MIM capacitor (Sect. 8.5).

Each refresh degradation mechanism can have both an intrinsic component limited by primary device physics and an extrinsic component limited by fundamental physics combined with a defect mechanism that can be degraded by plasma and process damage. For example, the gate-induced diode leakage mechanism shown in Fig. 8.35 can be degraded by plasma damage inducing SILC and degrading the retention (Kim, 2015).

8.4.2 Variable Retention Time in DRAMs

A DRAM memory chip only retains its data for a short amount of time. The DRAM manufacturer must guarantee that the DRAM chip can meet this minimum time requirement called the refresh time. The manufacturer tests the DRAM chip for extended times and temperatures to ensure that every bit in the array can meet the refresh time requirement.

A DRAM cell's refresh or retention time can be highly variable, ranging from the bare minimum specification of around 64 ms to several seconds, depending on the quality of the cell. In addition, the refresh time is strongly temperature-dependent, with higher temperatures being worse due to increased leakage from most mechanisms. A near-perfect DRAM cell can retain data for a few seconds. A less-than-perfect DRAM cell has a lower retention time depending on the severity of the defects in the cell. DRAM manufacturers replace low refresh cells with redundant cells, eliminating them from the chip operation, as observed by the customer. Figure 8.36 shows an example of a retention probability distribution plot of a DRAM array (Hamamoto et al., 1998). The nearly perfect bits form the intrinsic part of the distribution, and the less-than-perfect bits form an extrinsic low retention tail. Process damage can degrade the retention characteristics of a DRAM chip,

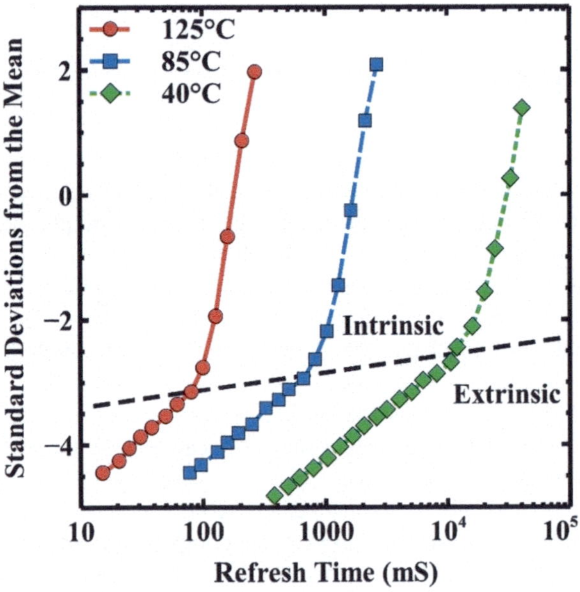

Fig. 8.36 Probability plot showing an example of DRAM retention time for three different temperatures. The black dashed line (around -3σ from the mean at around 0.13% of the bits are defective) shows the boundary between the intrinsic distribution above the dashed line and the defective extrinsic distribution below the dashed line. This distribution is worse than typical production quality DRAMs

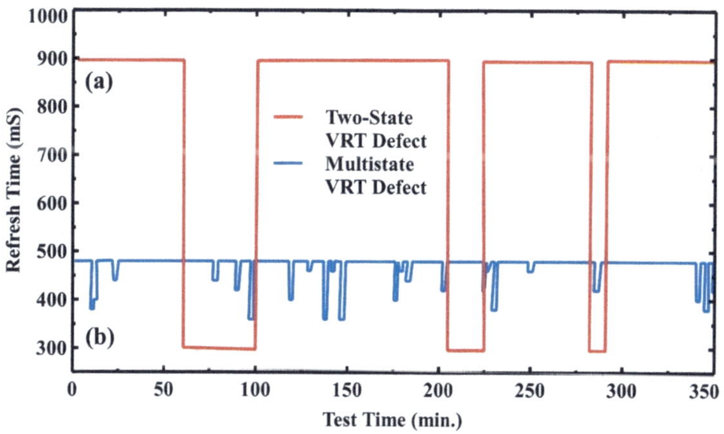

Fig. 8.37 Example of DRAM variable retention time (VRT). (**a**) Two-state VRT caused by a single atomic level defect. (**b**) Multi-state VRT showing six states likely caused by three atomic level defects

increasing the extrinsic part of the distribution or causing cells to fail the retention specifications. Single or a few atomic-level defects in a DRAM cell are sufficient to cause refresh failures.

In addition, single-atom defects can cause a cell to have an erratic refresh time. This phenomenon is called variable retention time (VRT). Figure 8.37 shows two example plots of the variable retention time in a DRAM memory array (Bacchini et al., 2014). Figure 8.37a shows a refresh time versus elapsed time plot of VRT caused by a single-atomic-level defect, which causes the retention time to vary between two different values depending on the state of the defect. The two states are caused by the defect trapping or emitting charge. Figure 8.37b shows a refresh time versus time plot for a DRAM cell with three defects. Each of the three defects can trap or emit charge, resulting in six different retention or refresh times.

Variable retention time is strongly temperature-dependent with increased leakage at a higher temperature. Typically, the worst-case refresh conditions are at elevated temperatures. The VRT defects trap and detrap charge carriers more rapidly, increasing the oscillation frequency at high temperatures. Figure 8.38 shows an example of variable retention time at three different temperatures (Ohyu et al., 2006). The refresh time was monitored through time at three different temperatures. The two-state oscillations can be seen to increase with temperature. The degradation in the refresh time with temperature is evident. VRT problems create testing difficulties for the DRAM manufacturer, as observed in the plot. The long time constant for the refresh time transition on the order of hundreds of minutes makes detecting variable retention time during chip testing difficult and expensive. A variable retention time defective bit can alternatively pass and fail the refresh specification through time, creating a propensity for a defective chip to escape the functional test without repair. A defective VRT bit can escape testing by remaining in a good

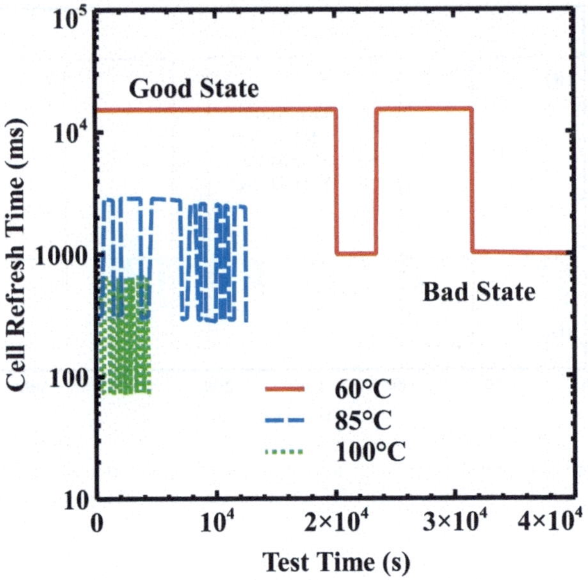

Fig. 8.38 DRAM data retention time vs. elapsed time for a defective VRT DRAM cell with VRT for three different temperatures

state during chip testing, only failing in the customer's possession, posing a severe DRAM quality and reliability problem.

The behavior of defective VRT bits can vary with time with changing refresh time and oscillation frequency. A DRAM cell that is good at time zero during testing can become defective later in time, creating an untestable defect. Also, a defective DRAM cell can become good over time, with the varying refresh time and bi-stable states disappearing and reappearing. VRT is essentially an unstable phenomenon.

After DRAM chips have completed testing, they are mounted on a board such as a dual inline memory module (DIMM) for insertion into a computer. The DRAM is attached to the board with a conducting solder that connects the DRAM chip's electrical pins to traces on the board. The solder must be melted to form this connection. An infrared thermal heat step of about 250 °C for a few minutes melts the solder physically and electrically bonds the chip to the board. This short, low-temperature heat step provides enough thermal energy to disturb or create defective VRT bits in a DRAM chip. Figure 8.39 shows a plot of the change in the refresh time of the weakest bit in a DRAM chip for 10,000 packaged chips (Chang et al., 2005). The chips are exposed to IR-mounting thermal process three times to simulate a worst-case condition. The refresh time of the worst bit can be observed to shift in a positive or negative direction for a portion of the chips tested, indicating that the weak VRT bits are unstable. This result demonstrates that the board-mounting process can shift or create VRT bits since these chips passed the refresh time specification during testing. DRAMs commonly have post-package repair capability, allowing

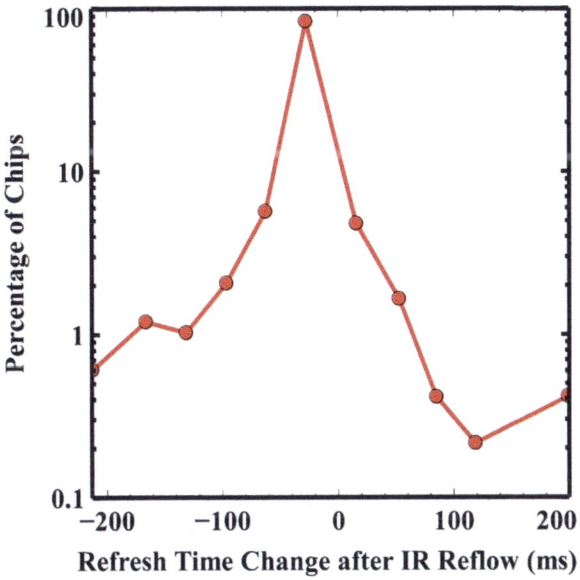

Fig. 8.39 Plot showing the shift in the weakest bit refresh time for 10,000 (256 Mb) DRAM chips to 3 cycles of infrared (IR) reflow

the replacement of defective cells after a package has been mounted to a board (Takahasi et al., 2001).

The single-atom defect mechanism causing the variable retention time problem has been heavily researched over the years. Understanding the mechanism is complex as the DRAM cell itself is needed to investigate VRT. Parametric structures cannot be directly used because they are not sensitive to the necessary leakage levels. However, parametric structures can indirectly be used to correlate product VRT problems with related parametric degradation. The underlying physics has been partially understood, yielding insights into solutions to mitigate the problem. Variable retention time in DRAMs has very similar characteristics to a common noise problem in CMOS devices called random telegraph signal (RTS). RTS typically has time-variant two-state behavior like VRT behavior. Figure 8.40 shows an example of RTS in the drain current of a MOS device with similar behavior as DRAM RTS (Chen, 2010).

Several possible atomic-level defects can potentially cause RTS and VRT, as shown in Fig. 8.41. A bulk VRT defect (C) in the depletion region of the DRAM storage node junction can cause intermittent variable junction leakage from the storage node to the substrate, killing the refresh time of the cell. This bulk defect may be a metal ion, lattice defect, etc.

An interface trap at the isolation edge (B) or the gate-to-source/drain overlap region (D) can create a bistable leakage location (Chang et al., 2005). The type of defect needs to exist in an unfavorable location instead of anywhere in the device. The channel leakage can also be modulated when a bistable defect exists in the gate

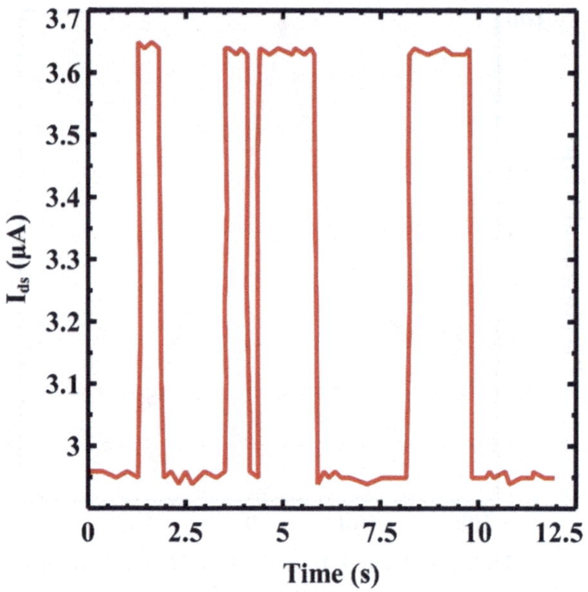

Fig. 8.40 Example of random telegraph signal noise in a mos device

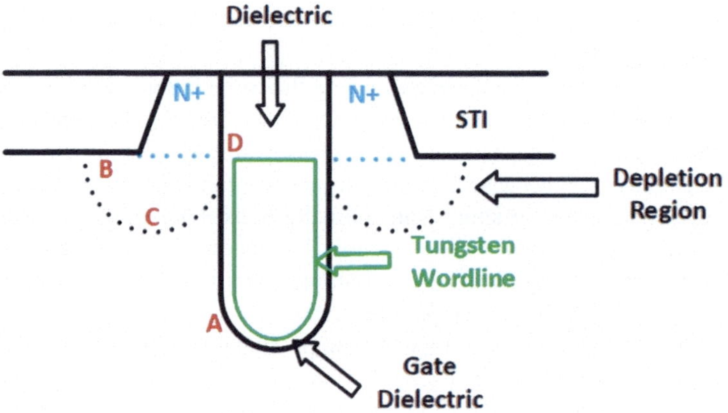

Fig. 8.41 Cross-section sketch of CMOS DRAM access transistor showing potential causes of VRT in DRAMs and RTS in MOS transistors. Interface defects are located at the isolation edge and the gate drain overlap region. Bulk defects are located in the depletion regions of the cell node junction

dielectric or the surface (A). The electric field from a charge carrier trapped in the defect can interact with the carriers flowing in the channel, modifying the current depending on the occupation of the trap (Chen, 2010).

This assortment of potential VRT defects can be caused by process damage at minute levels, far below levels visible through other types of damage (Goiffon et al., 2020). Additionally, hydrogen can play a role in VRT, as described in Chap. 14.

8.4.3 DRAM Row Hammer Problem

The DRAM row hammer problem appeared in the last few decades as DRAMs aggressively scaled to ever-smaller dimensions (Kim, 2017). As a result, the DRAM storage node diode junction is closely adjacent to an unrelated passing wordline used to access an unrelated cell. The unrelated passing wordline can toggle millions of times during the data retention period depending on the address data pattern used to select wordlines.

Rarely, a worst-case condition can occur during operation when a single rowline (wordline) is toggled continuously during regular operation. The DRAM datasheet allows continuous access to one row during a refresh cycle. Practically, there is seldom a good reason for continuous row access. However, row hammer is a security risk that can be used to attack a computer. Row hammer can cause a change in the data state of bits adjacent to the hammering row that can be used to access unauthorized regions of the memory space for nefarious reasons (Gruss et al., 2018). Figure 8.42 shows an example of data on a row hammer test from three different DRAM manufacturers (Park et al., 2014). The percentage of failing bits and rows is plotted against the cumulative number of hammers for an entire 4 Gbit chip. The

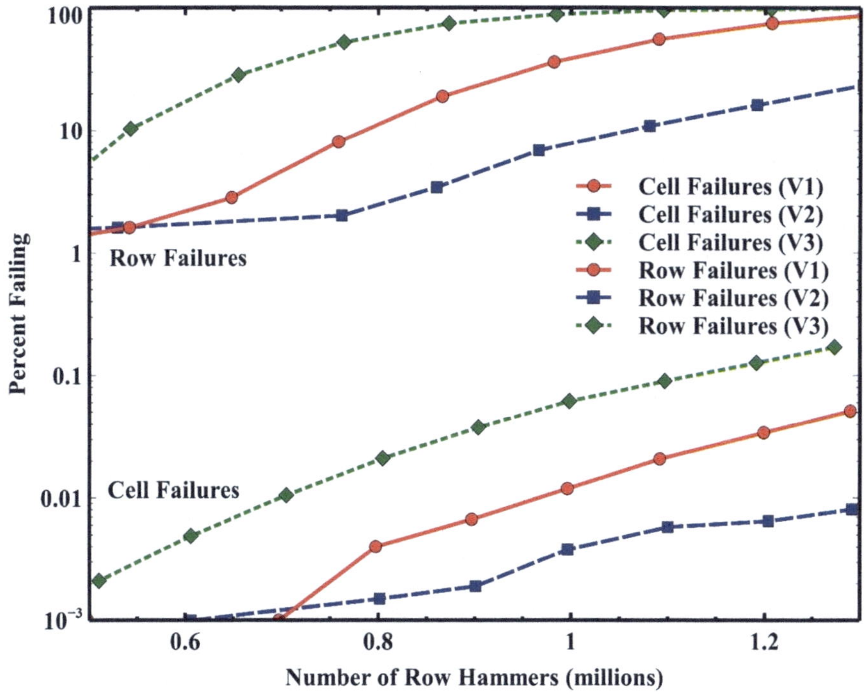

Fig. 8.42 Example of the percentage of failing cells and rows for millions of row hammer events in 3Xnm 4Gb DRAM chips for three different DRAM manufacturers (V1, V2, V3)

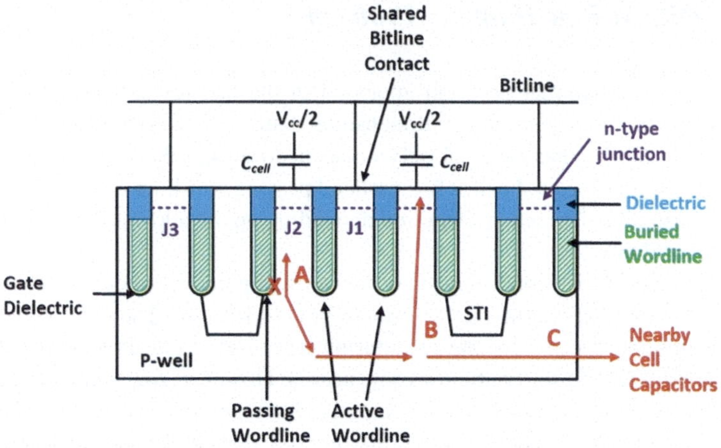

Fig. 8.43 Sketch of a DRAM array showing the location of a single atomic defect trap X in the passing wordline. The ejected current flows from trap X near the edge of the deeper passing wordline towards adjacent storage node junctions (J1 and J2). The ejected carrier is collected by a nearby storage node junction, causing a refresh failure. The ejected carrier can also travel across multiple cells to be collected by nearby nonadjacent storage nodes

failure percentage increases as the number of hammers increases. The number of possible hammers in a typical DRAM for a refresh cycle can reach a few million, as specified in the DRAM datasheet. The data shown here is for chips with no row hammer mitigation applied.

The underlying physics of row hammer has been heavily researched and is partially understood. The passing wordline adjacent to the storage node can electrically impact the surrounding region, essentially hammering this region. Figure 8.43 shows a cross-section sketch of a DRAM cell. Two DRAM cells share a single bitline contact, as labeled in the sketch. Two DRAM wordlines are positioned on either side of the contact, forming the recessed cell transistor's gate. The cell capacitor is connected to the opposite side of the cell transistor gate (labeled C_{cell}). There are n-type diode junctions underneath the bitline and cell capacitor contacts. The passing wordlines are adjacent to the cell capacitor contact partially embedded in the shallow trench isolation.

As the passing wordline toggles high and low, the silicon region adjacent to the passing wordline transitions between accumulation, depletion, and inversion. If a trap exists along the silicon surface of the passing WL, such as in the location labeled X, the trap can fill with electrons and eject electrons as the passing wordline is cycled. An electron can fill the trap when the passing wordline goes positive, attracted by the positive electric field. Additionally, the electron can be ejected when the passing wordline goes negative, repelled by the negative electric field.

The ejected electron must travel somewhere driven by the electric field. It may travel deep into the substrate, recombining and disappearing in a recombination center with no impact on the electrical state of the DRAM. The electron ejected from point X can also follow the two paths labeled A and B in the sketch, attracted by the positive electric field of the junctions when a positive voltage is stored in the capacitor. The carrier may travel path A toward the silicon surface, where it can compensate the charge on the capacitor, degrading the retention time. The electron can also travel path B, repelled by a negative voltage and attracted to the positive potential in the cell capacitor junction. Ejected electrons that collect in the path B junction compensate for the positive charge stored in this junction, degrading the refresh, potentially causing a bit failure in the DRAM chip.

The ejected electrons can also travel to junctions further away from the hammering row in either direction, where they are also collected, degrading refresh. Path C in the figure shows an example of a junction farther away from the passing wordline collecting the injection charge.

About 10,000 electrons are stored in a DRAM capacitor for a modern DRAM, enabling digital data storage. This value can be calculated from $Q = CV$ with a 7 fF capacitor storing a 0.5 V signal. However, if a single electron is ejected for each of a million hammers, the 10,000 stored electrons are easily compensated, causing loss of the stored data. In this example, only one ejected electron in a thousand hammers collected by the junction is necessary to cause data loss. This rare event is an example of a single-atom defect causing chip failure.

Customers and users in the field found the DRAM row hammer problem. Unfortunately, DRAM manufacturers missed the problem, creating a reliability and security risk. Many solutions to the DRAM row hammer problem have been developed and implemented. As a result, the row hammer problem is currently not a significant issue, as solutions have mitigated the problem. Some of the solutions were brute force solutions, allowing fixes in the field.

One example is increasing the refresh frequency, which reduces the time available for data loss, with performance and power penalties (Kim et al., 2020). Other solutions include adding intelligence to the DRAM chip to detect hammer situations and mitigate the problem by refreshing the victim rows. Plasma and process damage can potentially degrade the DRAM row hammer performance by creating single-atom defects that are the root cause of the problem. The process modules with an elevated risk of row hammer degradation include steps associated with the DRAM cell construction such as the wordline, contacts, capacitor, cell plate, etc.

8.5 Metal-Insulator-Metal Capacitor Damage

8.5.1 Introduction

Metal-insulator-metal (MIM) capacitors (also called metal oxide metal (MOM) capacitors) are commonly added to a CMOS process to increase circuit performance and capability by enabling high-quality, high-capacitance capacitors. MIM dielectrics can suffer from process damage similar to that of gate dielectrics (Wang,

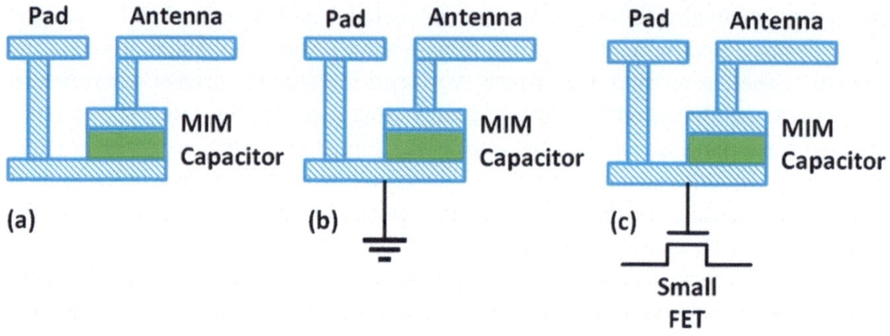

Fig. 8.44 Diagram of example structures for metal-insulator-metal related plasma damage. (**a**) Floating MIM capacitor connected in the metallization layers. (**b**) MIM capacitor connected to an antenna and ground. (**c**) MIM capacitor connected to an antenna and the gate of a transistor

2004). Damage mechanisms, including stress-induced leakage, dielectric breakdown, degradation in dielectric breakdown, etc., can occur in MIM capacitors. MIM capacitors commonly use high-k dielectrics to provide high capacitance, which can be more susceptible to process damage than silicon dioxide. The presence of MIM capacitors can also cause damage by capacitively coupling charge and voltage into sensitive transistors electrically connected to the capacitor. Figure 8.44 shows sketches of common MIM configurations related to plasma damage (Weng et al., 2009). Figure 8.44a, b shows configurations where the MIM capacitor itself can be damaged. Figure 8.44c shows a sketch where the MIM capacitor can damage underlying CMOS transistors by coupling the antenna energy from a metal line connected to the capacitor into the transistor.

8.5.2 *Process Damage to MIM Capacitors*

An example study of plasma damage and MIM capacitors is described in the following paragraphs (Weng et al., 2009). This work studied plasma damage caused by high-density plasma dielectric deposition (HDP) on top of a metal antenna. The metal antenna area is varied with the MIM capacitor area held constant at 5 μm^2 with a 30 nm SiO_2 dielectric thickness. Modern high-k MIM capacitors behave differently, but the lessons learned here apply. Figure 8.45 shows a probability plot of MIM leakage with three different top metal antenna-to-MIM ratios for a MIM capacitor connected directly to the ground. The antenna ratio used here is the ratio of the metal antenna area to the MIM area. MIM leakage at ~1 nA indicates that dielectric breakdown has occurred (right side of the plot). A small 100 antenna ratio (AR) has no dielectric breakdown, while a 1000 antenna ratio has a failure rate of 30% of the MIM capacitors with degraded leakage or breakdown.

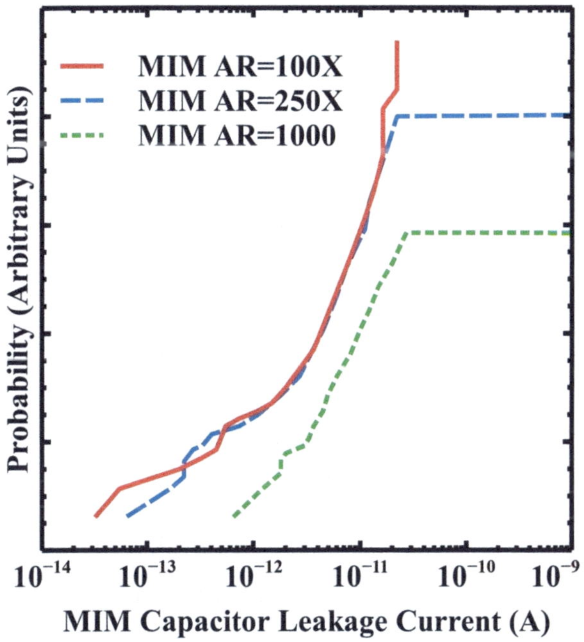

Fig. 8.45 Probability plot of MIM capacitor leakage as a function of antenna ratio of upper metal on top of the MIM capacitor during dielectric deposition on top of the capacitor

8.5.3 Process Damage Induced by MIM Capacitors

In addition to the direct MIM damage described in this study, when one terminal is grounded, MIM and other high-capacitance structures in a CMOS process can interact with damage-sensitive structures or act as a conduit to damage other sensitive structures. For example, charge and voltage collected by antennas above or below a MIM capacitor can couple voltage through the MIM capacitor, increasing MIM damage or damaging underlying structures. Figure 8.46 shows a sketch illustrating this problem structure (Weng et al., 2009). Plasma current collected by the metal antenna above the MIM capacitor charges the MIM upper plate. As a result, the upper plate electrically couples to the lower plate, inadvertently raising the voltage on the CMOS transistor gate. The voltage stress can damage or destroy the transistors through the degradation mechanisms described in this book.

Damage in this type of structure can occur in one of the two devices (MIM or MOS) or both, depending on the resistance to damage of the dielectric and many other factors. The antenna ratio calculation for this problem is complicated, and several different methods can be used to estimate the antenna. One factor is the capacitive coupling ratio between the two capacitors. The voltage stress applied to the MIM capacitor is described by Eq. (8.4), where V_p is the stress voltage on the top electrode of the MIM capacitor (Weng et al., 2009). Increasing the gate capacitance

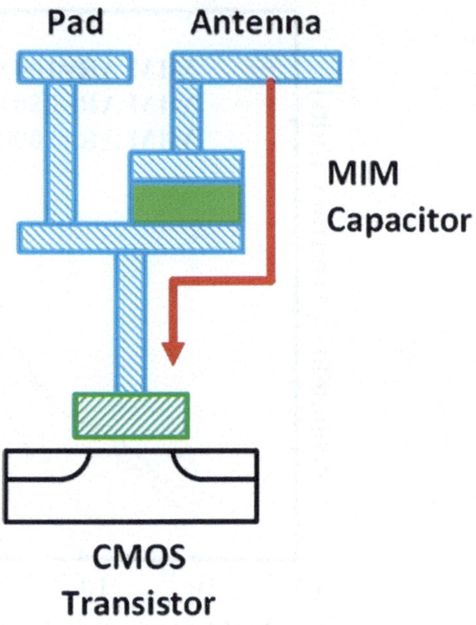

Fig. 8.46 Sketch showing plasma damage caused by a metal antenna on top of a MIM capacitor, coupling voltage to an underlying gate

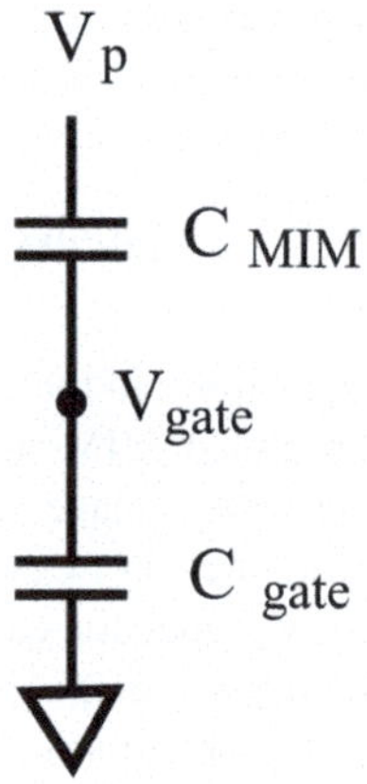

Fig. 8.47 Simple equivalent circuit for a MIM capacitor coupling to a transistor gate capacitor

increases the stress across the MIM dielectric. More analysis is needed in addition to using a simple antenna ratio of the metal area divided by the gate area, as described in Chap. 2.

The antenna ratio based on the coupling capacitance ratios of the MIM capacitor to the gate capacitor can be used to estimate the stress voltage across the gate dielectric. The capacitance ratio depends on the capacitor area, dielectric constants, and dielectric thicknesses for simple two capacitors in a series circuit, as shown in Fig. 8.47.

Similarly, the stress applied to the gate dielectric is the plasma potential V_p multiplied by the ratio of the MIM capacitance to the total capacitance consisting of the MIM capacitance and the gate capacitance, as given by Eq. (8.5). Increasing the MIM capacitance increases the stress on the gate dielectric. High values of V_{gate} can degrade or destroy the gate dielectric.

$$V_{MIM} = V_p \frac{C_{gate}}{C_{MIM} + C_{gate}} \tag{8.4}$$

$$V_{gate} = V_p \frac{C_{MIM}}{C_{MIM} + C_{gate}} \tag{8.5}$$

The damage to the MIM dielectrics in this study also occurred when the MIM capacitor was connected to a MOS transistor. The MIM breakdown correlates with the metal antenna-to-gate area ratio. The results are very similar to the condition where a terminal of the MIM capacitor is connected to ground. This correlation occurs for both NMOS and PMOS transistors.

Additionally, the transistors were damaged when the antenna above the MIM capacitor was exposed to a HDP dielectric deposition on top of the MIM. Figure 8.48 shows the gate dielectric breakdown degradation due to this damage (Weng et al., 2009). The breakdown distribution of the thicker 3.5 nm dielectric shows a fraction of the distribution with a reduced breakdown voltage and an increasing defect density for increasing antenna ratios. The plot shows the results for a thinner 1.5 nm dielectric. The 1.5 nm dielectric shows no evidence of increasing breakdown with an increasing antenna ratio. The significant difference in the results as a function of

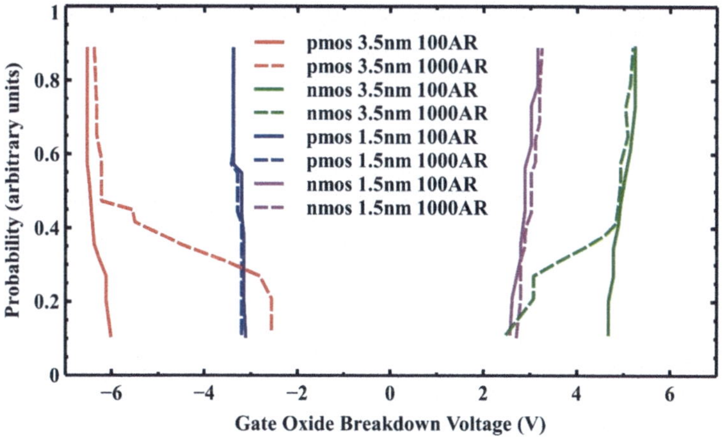

Fig. 8.48 Gate dielectric breakdown due to MIM related plasma damage for two different transistor gate dielectric thicknesses for both nmos and pmos transistors. The transistors are connected to a metal antenna with different areas through a MIM capacitor, creating different antenna ratios. The breakdown of the thicker 3.5 nm dielectric is degraded with increasing antenna ratio, while the thinner 1.5 nm gate dielectric is not degraded

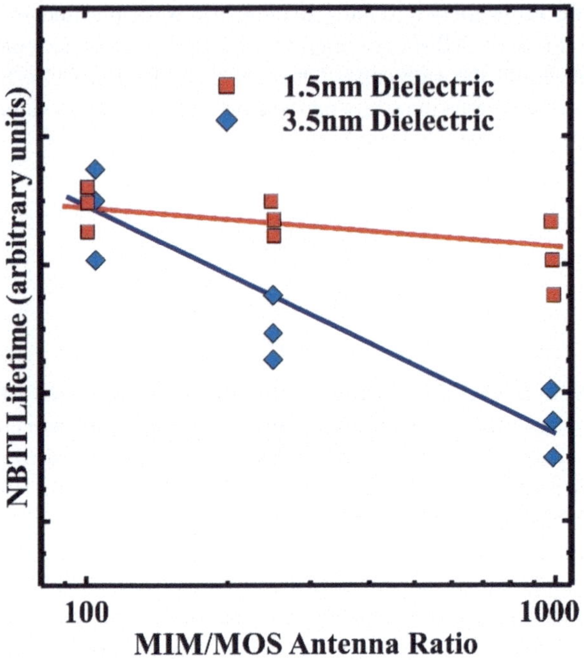

Fig. 8.49 Plot of NBTI lifetime (arbitrary lifetime) as a function of MIM to MOS area antenna ratio for two different dielectric thicknesses. The lifetime in the 3.5 nm group is a strong function of the antenna ratio, while the 1.5 nm group is a weak function of the antenna ratio

the dielectric thickness is due to several reasons, including differences in the propensity for damage as a function of the gate dielectric thickness, differences in coupling, and differences in intrinsic leakage (Weng et al., 2009).

The transistors surviving without dielectric breakdown in this study have no time-zero degradation but have degraded reliability, creating a risk of product failure. Figure 8.49 shows a plot of the PMOS negative-bias temperature instability (NBTI) as a function of the MIM/MOS antenna ratio for two different dielectric thicknesses. The thicker 3.5 nm dielectric data shows that the NBTI lifetime is degraded with the antenna ratio, indicating that the plasma damage also creates latent damage. The 1.5 nm group may show a slight degradation with an increasing antenna ratio (Weng et al., 2009).

CMOS processes commonly have multiple types of transistors with multiple dielectric thicknesses to support the different voltages needed in a modern chip. Design verification practices must ensure that the correct voltages are applied to the CMOS circuits. These verification techniques should also include preventative measures to minimize process damage. Plasma design rules controlling the configuration of the MIM circuits can minimize the propensity for damage by eliminating the worst possible process damage conditions.

8.6 Process Damage in Low-k Dielectric Films

8.6.1 Introduction

Low-dielectric constant (low-k) films are commonly used in the CMOS interconnect metallization layers to reduce the metal capacitance, improving speed and reducing power dissipation (Baklanov et al., 2007). Low-k films achieve a low dielectric constant by choosing specific materials, including silica-based compounds, silsesquioxane (SSQ), polymers, amorphous carbon, porous films, etc. Low-k dielectrics tend to be of low density, partially replacing the material with tiny pockets of air to reduce the dielectric constant. Low-density films such as these tend to be unstable and easily degraded by process damage. Low-k films can be degraded or compromised by process damage, causing multiple changes to the film. Process damage may cause a change in density, dielectric thickness, dielectric constant, wet and dry etch rates, electrical leakage, feature angle, reduction in the time-dependent dielectric breakdown, etc. Ion bombardment from plasmas can modify the dielectric near the surface, densify it, or change the chemical composition. Chemical and radical reaction mechanisms can modify the film near the surface, changing its molecular composition. High-energy photons (UV, etc.) can break atomic bonds, modifying the material.

Porous low-k dielectrics achieve low permittivity by achieving a high porosity level in the films. The porosity achieves a low dielectric constant because the dielectric constant of the air contained in the pore is the minimum possible at a value of 1, significantly reducing the overall dielectric constant of the film by compensating for the higher dielectric constant of the rest of the film. The porous films have a large surface area exposed to the chemicals and plasmas during processing. The film may shrink due to pore collapse during processing. In addition, the dielectric constant may change as a function of the dimension, complicating the modeling of electrical performance.

Many crucial factors are necessary for successfully implementing a low dielectric constant dielectric technology, as shown in Table 8.2 (Ho et al., 2003; Murarka et al., 2003). Low-k materials are among the most sensitive materials used in

Table 8.2 Important properties of low-k dielectrics

Dielectric constant, dielectric constant anisotropy, refractive index, dissipation factor
Dielectric leakage, time-dependent dielectric breakdown
Density, porosity
Tensile strength, thermal expansion, glass-transition temperature, Young's modulus
Moisture uptake, diffusion resistance, gas permeability, shrinkage, weight loss
Adhesion, crack resistance, structural collapse, chemical resistance
Chemical structure, phase transition, glass-transition temperature, thermal stability, change with thermal history
Etch profile control

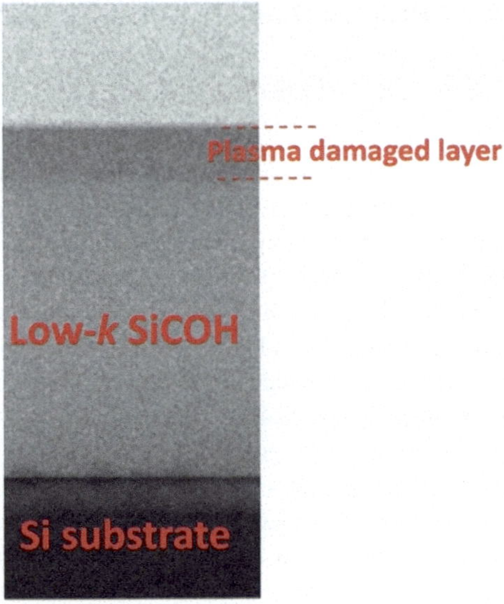

Fig. 8.50 Transmission electron microscope image of a low-k dielectric layer damaged by exposure to an O_2 plasma

semiconductor processing. Therefore, most processes must be optimized for minimum damage due to the sensitivity of the material because of its density and porosity.

Plasma damage and other process damage can modify an exposed low-k surface, modifying the properties described in Table 8.2. Figure 8.50 shows a TEM cross-section image of a low-k dielectric film exposed to an O_2 plasma that modified the surface properties observable as a dark layer on the film's surface (Cheng et al., 2019).

8.6.2 *Low-k Dielectric Damage Examples*

A few examples of process damage follow. Figure 8.51 shows the dielectric leakage and degradation in the breakdown voltage for a low-k dielectric with and without an O_2 plasma treatment commonly used in resist ashing (Cheng et al., 2019).

The dielectric constant of low-k materials can be degraded by process damage, reducing the resistance-capacitance (RC) speed advantage of low-k materials. Figure 8.52 shows a plot demonstrating the increase in the dielectric constant of a low-k dielectric by exposure to a plasma containing N_2, NH_3, and O_2 at various ratios (Cheng et al., 2019). The *y*-axis shows the absolute value of the increase in the dielectric constant. Low-k materials typically have a dielectric constant of 2–3. The plot shows an increase of ~25%, which is a substantial shift. The degradation in the dielectric constant may be anisotropic, favoring a specific direction in the material. The directionality may be caused by the film being degraded in the same direction as the plasma exposure.

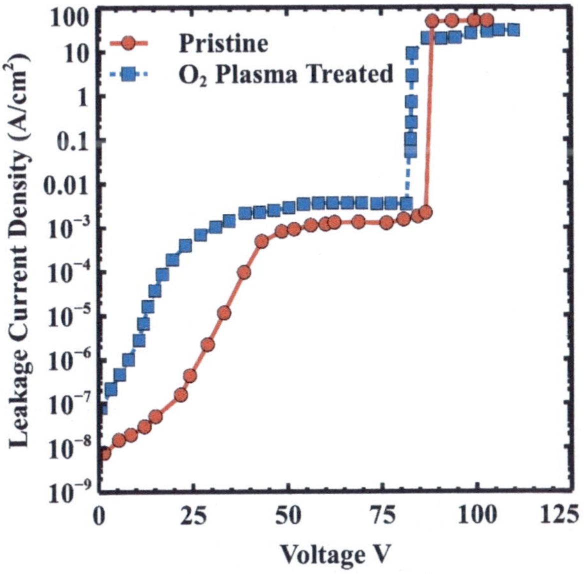

Fig. 8.51 Low-k dielectric leakage with and without exposure to an oxygen plasma using in resist ashing

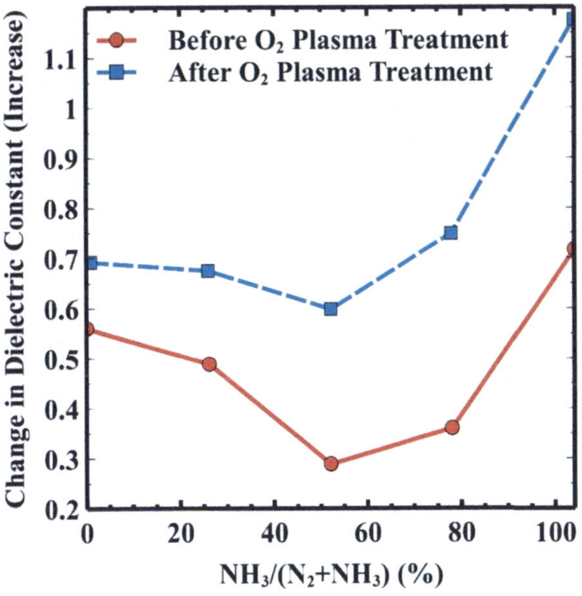

Fig. 8.52 Change in the dielectric constant of a low-k dielectric film exposed to N_2 and NH_3 containing plasmas with different gas ratios, followed by an oxygen plasma treatment. The change in dielectric constant with and without an O_2 plasma treatment is shown

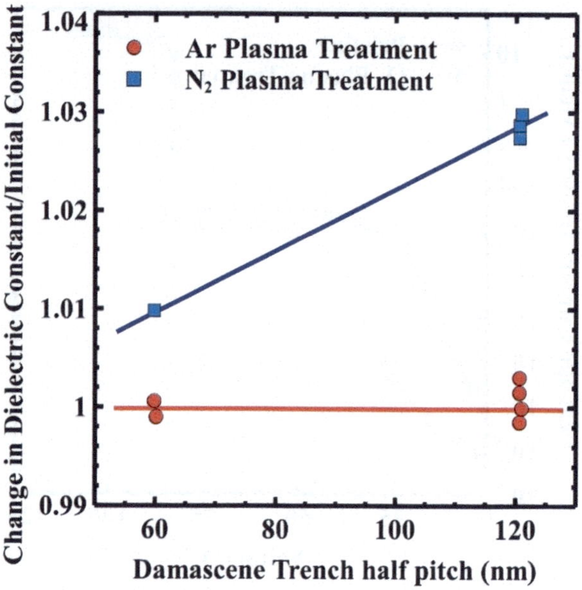

Fig. 8.53 Change in low-k dielectric constant with damascene trench dimension for two different 10 min plasma treatments causing process damage due to light exposure

The dielectric degradation may also be a function of design rules such as the metal pitch. An increasing surface area exposed to the plasma causes rule-dependent degradation as the pitch is reduced. Figure 8.53 shows an example of the dielectric constant changing as a function of the pitch for two different plasma treatments, expressed as a change from the initial value (Ikeda et al., 2015). This phenomenon must be carefully understood and characterized to enable accurate design models. Typically, surface degradation in damascene channels causes a space-dependent dielectric constant shift, resulting in a change in the dielectric constant with space.

Individual recipe parameters in a single step can cause a drastic shift in low-k dielectric parameters. For example, Fig. 8.54 shows the change in the dielectric constant for a porous low-k dielectric exposed to an oxygen plasma (Cheng et al., 2019). Increasing time or power increases the damage depth, resulting in an increased or decreased dielectric constant.

Plasma degradation of low-k dielectric films occurs for most common gases in plasma processing. Figure 8.55 shows the density change for a low-k dielectric exposed to different typical process gases in a plasma (Bao et al., 2007). All process gases in this experiment densified the film, with a likely degradation in electrical properties and film thickness shrinkage.

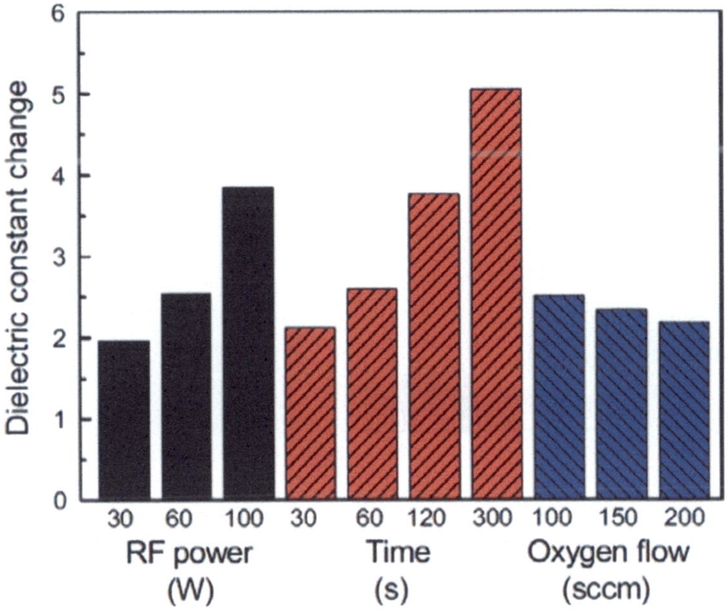

Fig. 8.54 Porous low-k dielectric constant change (absolute) as a function of recipe parameters for an O_2 plasma exposure. The untreated (control) dielectric constant was 2.56

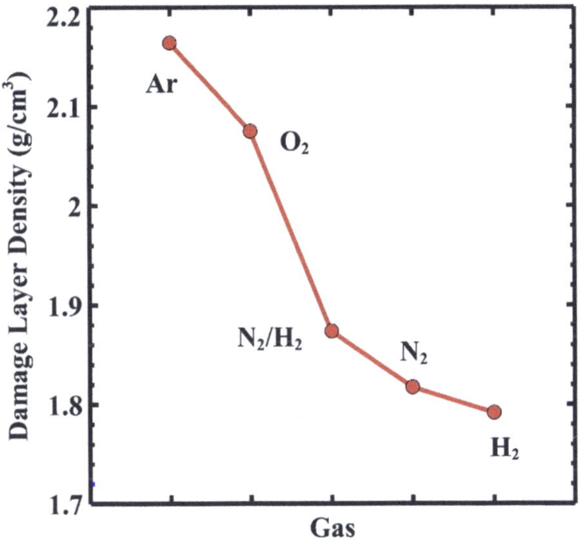

Fig. 8.55 Plot showing the density of a surface damaged layer in a low-k film with plasma exposure to various elements

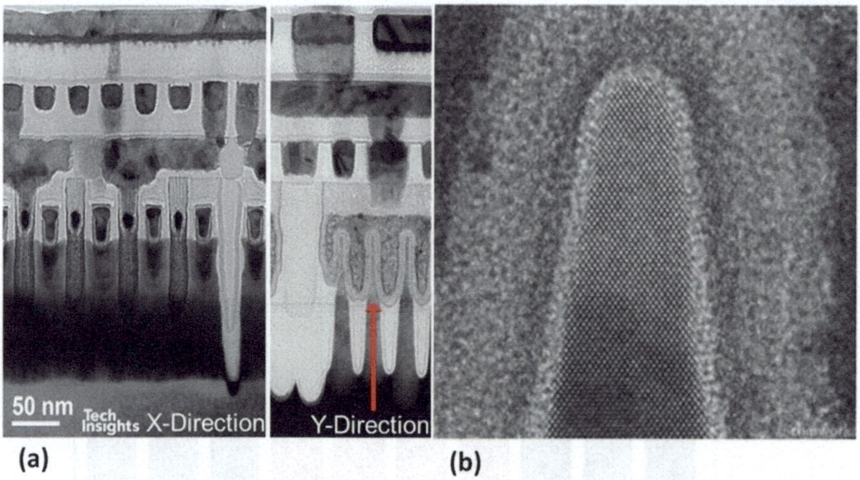

Fig. 8.56 TEM Cross-section of a finfet (**a**) an Intel 10n Finfet. The red arrow points to the small feature fin. (**b**) High magnification of the tip of an Intel 22 nm fin. Both pictures courtesy of Semiconductor Insights

8.7 Damage in Finfet Technology

Finfet CMOS transistors are natural successors to planar CMOS technology. The transistor has a common gate on three sides of a silicon pillar, significantly improving the gate density due to its three-dimensional structure. In addition, the three-sided gate provides increased control of the silicon channel in the pillar, improving the transistor's electrical characteristics. Finfets have been relentlessly scaled since their introduction (Park et al., 2006). Finfet dimensions are approaching atomic dimensions, with fin widths of the order of tens of atoms, resulting in increased sensitivity to process damage. Scaling causes finfets to enter the realm where additional erratic behaviors occur, similar to those observed in nonvolatile memories described in this chapter. The minute channel volume in a scaled finfet reduces the ability to dissipate energy, making the transistor more susceptible to plasma damage due to the minute fin width. Figure 8.56 shows an example TEM of a finfet transistor showing a small-dimension fin (Gelsinger, 2022; James & Current, 2018).

The CMOS process necessary to fabricate a finfet exposes the channel of the three-sided fin to damage from multiple process steps, including the silicon fin etch, isolation formation, isolation recess to form the fin, implantation, contact formation, etc. Figure 8.57 shows a sketch highlighting etch damage during the fin formation step (Eriguchi et al., 2014). The ions in the etch impact the sidewalls, where they can knock atoms into the silicon. In addition, ions can be sputtered off the horizontal silicon surface and knocked into the fin sidewalls. The surface of the etch front breaks bonds, creating a damaged surface. Careful optimization of the finfet processing conditions is required to minimize process damage. Post-etch processes to remove the damaged surfaces are used to mitigate the damage from the etch.

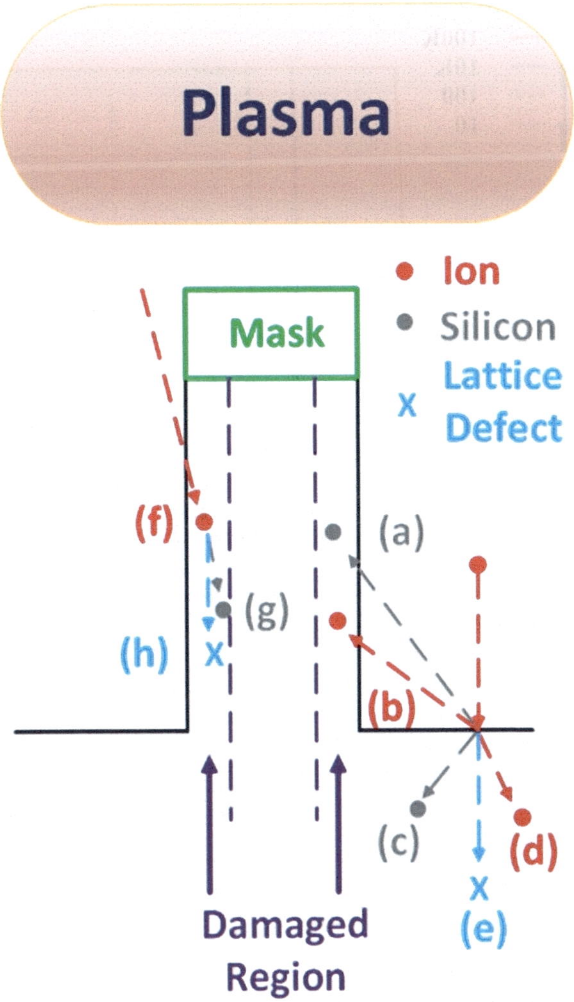

Fig. 8.57 Cross-section sketch showing etch damage where lateral damage on the sidewall is a significant fraction of the channel. (a) Backsputter of silicon atoms. (b) Backsputter of etchant ions. (c) Knock on of silicon atoms forming an interstitial. (d) Knock on of etchant ions forming an interstitial. (e) Lattice damage formed by broken bonds. (f) Vertical sidewall knock on of etchant ions forming an interstatial. (g) Vertical sidewall knock on of silicon atoms forming an interstitial. (h) Vertical sidewall lattice damage formed by broken bonds

The industry is transitioning from finfets to more advanced CMOS structures, such as gate all around transistors (GAA), as finfet scaling has approached its limits. These devices are also susceptible to process and plasma damage, requiring extensive optimization.

Finfets are susceptible to all the problems described in this book and additional problems like the fin etch described above. The minute dimensions of finfet

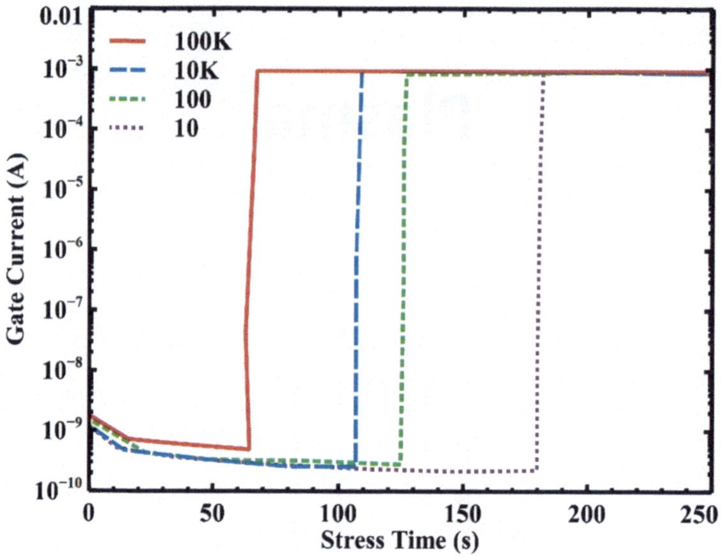

Fig. 8.58 Plot showing 7 nm finfet TDDB after plasma damage as a function of the antenna ratio for back end of line damage

transistors limit the dissipation of heat and energy, reducing the overall energy required to damage the transistor. As an example, Fig. 8.58 shows a plot highlighting time-dependent dielectric breakdown due to plasma damage in a 7 nm finfet technology (Tsai et al., 2018). The TDDB lifetime decreases with increasing back-end antenna ratio.

8.8 Silicon-on-Insulator Technology

8.8.1 Introduction

Silicon-on-insulator (SOI) technology can be very sensitive to plasma and process damage. For an SOI wafer, the semiconductor is formed on top of a wafer with an insulating dielectric, forming a silicon region on an insulator layer (Kononchuk & Nguyen, 2014). This technology has many technical benefits, but it also has the disadvantage of being an expensive wafer. However, an SOI transistor can behave differently due to the structural and electrical differences compared with bulk silicon technology. Plasma and process damage behavior is different for SOI technology, requiring consideration of the following items: more severe damage from the source and drains, damage from differential antennas, difficulty in diode protection, and channel conduction during plasma damage.

Dielectrics on all sides surround the SOI transistors, and as a result, the small silicon region lacks the thermal conduction capability of bulk silicon technology, where the silicon wafer acts as a heat sink (Martin, 2021). The thermal conductivity of silicon is about 100 times higher than that of silicon dioxide. As a result, heat tends to be trapped in the SOI silicon region, resulting in channel temperature increases of more than 100 °C during operation. As a result, the transistor behavior can change as a function of operating parameters such as frequency and voltage (Tenbroek et al., 1996). Process damage may be worse on SOI structures due to their higher sensitivity.

8.8.2 *Effects of Buried Oxide*

The underlying dielectric layer beneath the transistor, called a buried oxide (BOX) layer, can be degraded by process damage, trapping charge, and forming interface states that can cause leakage by forming a parasitic channel underneath the normal silicon channel. This trapping region does not exist in bulk silicon, where charge is typically trapped in the gate dielectric and isolation edges. Figure 8.59 shows a sketch showing the location of the parasitic channel in the BOX layer. Damage to the BOX layer can also cause more subtle transistor degradation, such as threshold voltage variation. The impact of BOX damage is strongly dependent on the SOI silicon thickness.

The floating body prevents the formation of a protective diode connection from the gate to the substrate. Therefore, diode protection in SOI technology requires

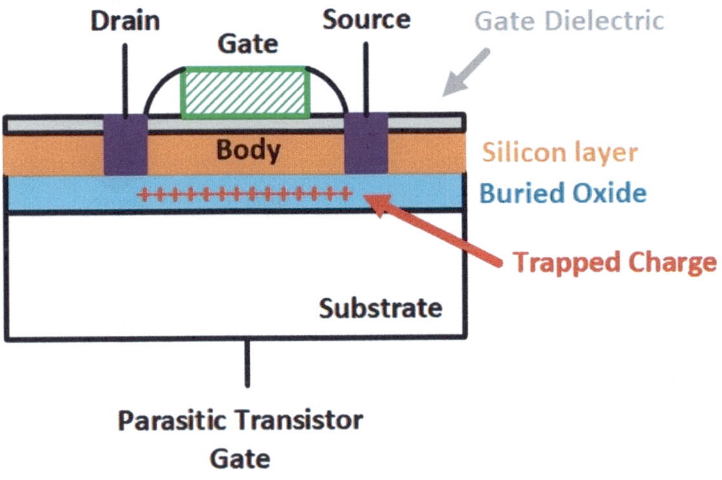

Fig. 8.59 Sketch showing a cross-section of an SOI transistor with the buried oxide (BOX) layer under it showing the location of trapped charge. The parasitic transistor with a backside gate is shown

either a two-terminal lateral surface diode or a special connection to the substrate through the buried oxide layer (Martin, 2021). More details are covered in the SOI diode section (see Chap. 3).

Unlike bulk silicon, the SOI body silicon is isolated from the underlying substrate, creating a floating body that is not electrically connected. The floating body can be capacitively coupled and electrically biased by voltages applied to the gate, source, and drain, changing the electrical behavior of the transistor by charging until equilibrium is reached. This phenomenon is called the floating body effect in SOI technology. From the plasma damage perspective, the floating body changes the electrical degradation. The floating body charges up toward the gate potential during plasma damage from the gate, reducing the dielectric current. The charge flowing through the gate dielectric collects in the isolated silicon region and is unable to flow into the bulk. The damage from the gate current is reduced as the bottom oxide (BOX) insulator inhibits the gate current relative to a bulk wafer. Therefore, the charged body can potentially reduce the amount of damage to the dielectric.

Plasma damage from the drain or source can charge up the floating body, increasing the dielectric current and potentially worsening plasma damage. The charge flowing into the source/drain region also collects in the isolated silicon region. An SOI device is more sensitive to source/drain charge as the charge cannot dissipate into the wafer's bulk.

Figure 8.60 shows a cross-section sketch showing damage locations for CMOS formed in SOI wafers compared to a bulk silicon wafer (Yamamoto et al., 2019). Figure 8.61 shows a plot comparing devices with source/drain antennas and gate antennas, resulting in nearly identical shifts since both charge sources accumulate in the isolated semiconductor region (Yamamoto et al., 2019). The figure shows that the threshold voltage shift is worse for a 15 nm silicon layer than for a 10 nm layer. Experiments varying the gate dielectric and BOX thicknesses can determine the roles of damage to the gate dielectric or buried oxide in transistor degradation.

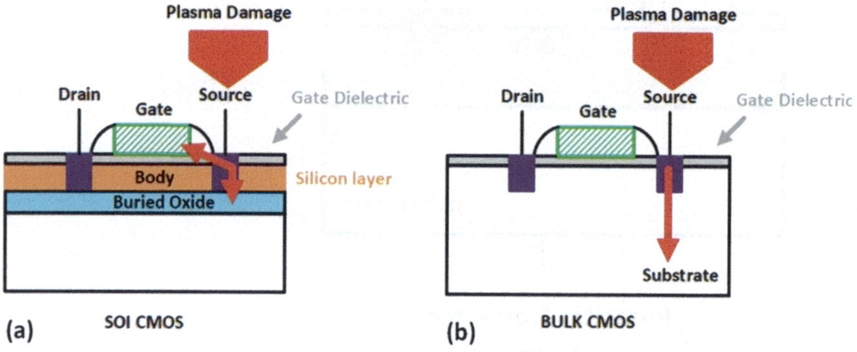

Fig. 8.60 Cartoon comparing process damage between a bulk wafer and an SOI wafer. (**a**) Drain damage discharges to gate and bottom oxide in SOI CMOS. (**b**) Drain damage discharges to the substrate for bulk CMOS

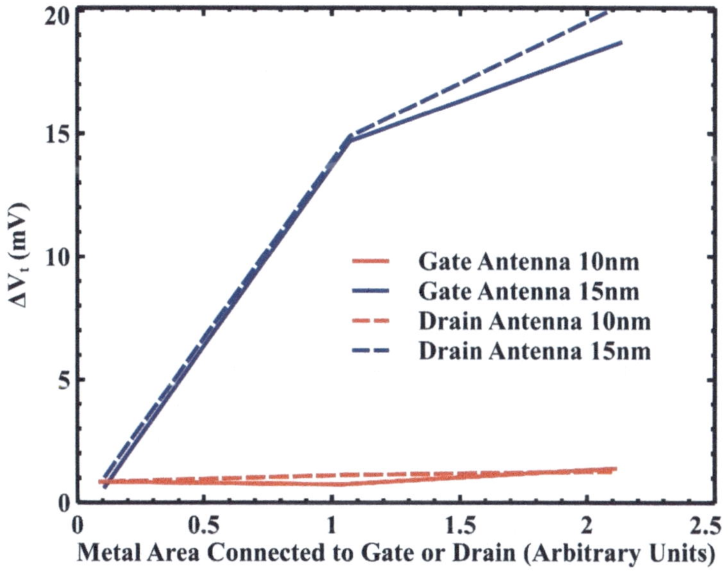

Fig. 8.61 V_t shift plotted against metal area (arbitrary units) for two different bottom oxide dielectric thicknesses (BOX). The antenna are stacked multiple copper layers

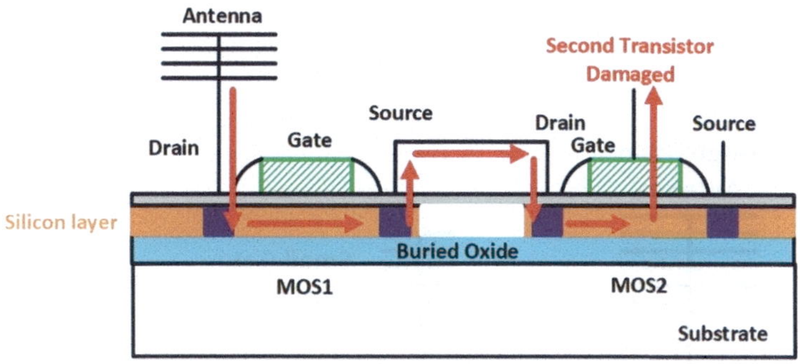

Fig. 8.62 Sketch showing charge collected on a drain antenna propagating through the first transistor channel (MOS 1) and damaging a second transistor (MOS 2) in a different buried oxide region (floating body)

8.8.3 *Propagation Between Floating Bodies*

Floating body effects can cause plasma damage to propagate between different transistors, hopping from floating body to floating body, and extending the damage to connected transistors as multiple floating bodies charge up. For example, Fig. 8.62 shows a sketch of two transistors illustrating transistor-to-transistor damage (Martin, 2021). A metal antenna is connected to the first transistor drain. The source of the first transistor is connected to the drain of a second transistor in a separate buried layer. The charge flowing from the antenna into the first floating body flows through the channel and into the second transistor drain, charging the body of the second transistor and damaging the gate dielectric of the second transistor. In SOI technology, damage can propagate from transistor to transistor, complicating process damage and reliability.

8.8.4 SOI Antenna Ratios

Silicon-on-insulator complicates the calculation of an antenna ratio used to minimize plasma damage. Bulk silicon ratios are typically calculated based on the collecting antenna area (or edge) ratios to the gate or well area. SOI calculations should additionally include the ratio of the charging antenna area (or edge) to the floating body area, downstream connected floating body areas, etc.

Additionally, differential antenna ratio calculations are more critical for SOI technology relative to bulk technology. Figure 8.63 shows a sketch of differential antenna ratios connected to an SOI transistor (Martin, 2021). A differential calculation is needed since the floating body can be charged from both the source or drain and gate in this case. The areas (A1 and A2) of the two antennas connected to the source and drain are added or subtracted and divided by the area of the floating body region or the gate area (Yamamoto et al., 2019). Charging from both the gate and drain antennas can increase the stress on the BOX layer, damaging the BOX

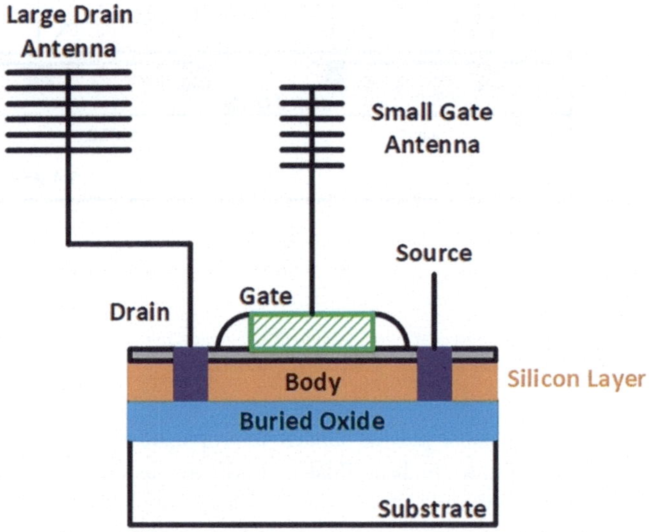

Fig. 8.63 Sketch showing differential antennas with a larger antenna on the drain and a smaller antenna on the gate

dielectric for the additive case shown in Chap. 5. The worst-case conditions occur when the polarities are opposite, creating higher stress than similar polarity charging (Martin, 2021):

$$AR_{DIFF} = \frac{A_1 \pm A_2}{A_{BOX}} \tag{8.6}$$

References

Alba, S., et al. (1997). Study of plasma charging and photoresist coating effects using UV-erased FLASH cell as charging sensor. In *2nd International Symposium on Plasma Process-Induced Damage*. https://doi.org/10.1109/PPID.1997.596759

Arai, F., et al. (1998). Extended data retention process technology for highly reliable flash EEPROMs of 1E6 to 1E7 W/E cycles. In *IEEE IRPS*. https://doi.org/10.1109/RELPHY.1998.670672

Bacchini, A., et al. (2014). Characterization of data retention faults in DRAM devices. In *2014 IEEE International Symposium on Defect and Fault Tolerance in VLSI and Nanotechnology Systems (DFT)*. https://doi.org/10.1109/DFT.2014.6962069

Baklanov, M., Green, M., & Maex, K. (2007). *Dielectric films for advanced microelectronics*. John Wiley and Sons. https://doi.org/10.1002/9780470017944

Bao, J., Shi, H., Liu, J., Huang, H., Ho, P. S., Goodner, M. D., Moinpour, M., Kloster, G. M., Ogawa, S., Ho, P. S., & Zschech, E. (2007). Mechanistic study of plasma damage of low k dielectric surfaces. *AIP Conference Proceedings, 10*(1063/1), 2815773.

Beug, M. F., Melde, T., Paul, J., & Knoefler, R. (2011). TaN and AL2O3 sidewall gate-etch damage influence on program, erase, and retention of sub 50-nm TANOS NAND flash memory cells. *IEEE Transactions on Electron Devices, 58*(6), 1728–1734. https://doi.org/10.1109/ted.2011.2121070

Boukhayma, A. (2018). *Ultra low noise CMOS image sensors*. Springer. https://doi.org/10.1007/978-3-319-68774-2

Brewer, J. E., & Gill, M. (2008). *Nonvolatile memory technologies with emphasis on flash*. IEEE Press.

Cai, T., et al. (2015). Data retention in MLC NAND flash memory: Characterization, optimization, and recovery. In *IEEE 21st International Symposium on High Performance Computer Architecture (HPCA)*. https://doi.org/10.1109/HPCA.2015.7056062

Carrere, J. P., et al. (2014). CMOS image sensor sensor: Process impact of dark current. In *IEEE International Reliability Physics Symposium*. https://doi.org/10.1109/IRPS.2014.6860620

Cha, C. L., et al. (1998). Threshold voltage instabilities of fresh flash memory devices caused by plasma charging. In *3rd International Symposium on Plasma Process-Induced Damage*. https://doi.org/10.1109/PPID.1998.725589

Chang, M., et al. (2005). Si–H bond breaking induced retention degradation during packaging process of 256 Mbit DRAMs with negative wordline bias. *IEEE Transactions on Electron Devices, 52*(4), 484–491. https://doi.org/10.1109/TED.2005.844743

Chen, C.-Y. (2010). *Low frequency noise in advanced CMOS technology*. Stanford University.

Cheng, Y.-L., Lee, C.-Y., & Haung, C.-W. (2019). Plasma damage on low-k dielectric materials. In *Plasma science and technology—Basic fundamentals and modern applications*. IntechOpen. https://doi.org/10.5772/intechopen.79494

Cheung, K. P. (1999). *Plasma charging damage*. Springer. https://doi.org/10.1007/978-1-4471-0247-2

Chimenton, A., et al. (2011). A statistical model of erratic behaviors in flash memory arrays. *IEEE Transactions on Electron Devices, 58*(11), 3707–3711. https://doi.org/10.1109/TED.2011.2165722

Chiu, Y.-Y., et al. (2016). Impact of string pattern on the threshold-voltage spread of program-inhibited cell in NAND flash. *IEEE Journal of Electron Devices Society, 4*(4), 1. https://doi.org/10.1109/JEDS.2016.2565820

Du, P.-Y., et al. (2019). Read disturb evaluations of 3D NAND flash for highly read-intensive edge-computing inference device for artifical intelligence applications. In *IEEE 11th International Memory Workshop (IMW)*. https://doi.org/10.1109/IMW.2019.8739755

Durini, D. (2014). *High performance silicon imaging—Fundamentals and applications of CMOS and CCD sensors*. Elsevier-Woodhead Publishing. https://doi.org/10.1016/C2017-0-01564-1

Eriguchi, K., et al. (2014). A new aspect of plasma-induced physical damage in three-dimensional scaled structures sidewall damage by stochastic straggling and sputtering. In *International Conference on IC Design and Technology*. https://doi.org/10.1109/ICICDT.2014.6838597

Fan, Y., et al. (2022). Investigation of retention characteristics in a triple level charge trap 3D NAND flash memory. In *IEEE International Reliability Physics Symposium*. https://doi.org/10.1109/IRPS48227.2022.9764506

Fantini, P., et al. (2007). Giant random telegraph signals in nanoscale floating-gate devices. *IEEE Electron Device Letters, 28*(12), 1114–1116. https://doi.org/10.1109/LED.2007.909835

Gelsinger, P. (2022). *Pat Gelsinger takes us on a trip down memory lane—And a look ahead*. Retrieved January 24, 2023, from Techinsights.com

Gilmer, D. C., et al. (2010). Impact of plasma gate reoxidation on the non-volatile charge trap memory device. In *IEEE International on Memory Workshop*. https://doi.org/10.1109/IMW.2010.5488408

Goiffon, V., et al. (2020). Radiation-induced variable retention time in dynamic random access memories. *IEEE Transactions on Nuclear Science, 67*(1), 234–244. https://doi.org/10.1109/TNS.2019.2956293

Gruss, D., et al. (2018). Another flip in the wall of rowhammer defenses. In *IEEE Symposium on Security and Privacy*. https://doi.org/10.1109/SP.2018.00031

Hamamoto, T., et al. (1998). On the retention time distribution of dynamic random access memory (DRAM). *IEEE Transactions on Electron Devices, 45*(6), 1300–1309. https://doi.org/10.1109/16.678551

Ho, P. S., Leu, J., & Lee, W. W. (2003). *Low dielectric constant materials for IC applications*. Springer. https://doi.org/10.1007/978-3-642-55908-2

Ielmini, D., et al. (2002). Localization of SILC in flash memories after program/erase cycling. In *IEEE International Reliability Physics Symposium*. https://doi.org/10.1109/RELPHY.2002.996603

Ikeda, T., et al. (2015). Plasma induced photo irradiation damage on low-K dielectrics enhanced by Cu line layout. In *International Conference on IC Design and Technology*. https://doi.org/10.1109/ICICDT.2015.7165901.

James, D., & Current, M. I. (2018). Analytical techniques for ion implantation. In *Unknown book*.

Janesick, J. R. (2001). *Scientific charge-coupled devices*. The International Society for Optical Engineering.

Kameyama, H., et al. (2000). A new data retention mechanism after endurance stress on flash memory. In *IEEE IRPS*. https://doi.org/10.1109/RELPHY.2000.843914

Kim, D.-H. (2015). AVERT: An elaborate model for simulating variable retention time in DRAMs. *Microelectronic Engineering, 55*, 1313. https://doi.org/10.1016/j.microrel.2015.06.064

Kim, D. H. (2017). *Design methodologies for scalable and reliable memory system*. Georgia Institute of Technology.

Kim, J., et al. (2006). 1/2.5″ 8 mega-pixel CMOS image sensor with enhanced image quality for DSC application. In *IEEE IEDM*. https://doi.org/10.1109/IEDM.2006.346975

Kim, J. S., et al. (2020). Revisiting rowhammer: An experimental analysis of modern DRAM devices and mitigation techniques. In *ACM/IEEE 47th Annual International Symposium on Computer Architecture (ISCA)*. https://doi.org/10.1109/ISCA45697.2020.00059

Kononchuk, O., & Nguyen, B.-Y. (2014). *Silicon-on-insulator (SOI) technology—Manufacture and applications*. Elsevier. https://doi.org/10.1016/C2013-0-16306-4

Lee, D.-K., et al. (2002a). Influence of plasma edge damage on erase characteristics of NOR flash EEPROM using channel erase method. In *IEEE IRPS*. https://doi.org/10.1109/RELPHY.2002.996659

Lee, W. H., et al. (2002b). Impact of floating gate dry etching on erase characteristics in NOR flash memory. *IEEE Electron Device Letters, 23*(8), 476–478. https://doi.org/10.1109/LED.2002.801305

Lee, Y.-H., et al. (2008). Drain read disturb assessment of NOR flash memory. In *IEEE International Symposium on VLSI Technology, Systems, and Applications (VLSI-TSA)*. https://doi.org/10.1109/VTSA.2008.4530809

Lee, C.-Y., Lai, C.-S., Yang, C.-M., & Wang, D. H. L. (2012). DRAM data retention and cell transistor threshold voltage reliability improved by passivation annealing prior to the deposition of plasma nitride layer. *IEEE Transactions on Device and Materials Reliability, 12*(2), 406–412. https://doi.org/10.1109/TDMR.2012.2188895

Li, F. M., Nixon, O., & Nathan, A. (2004). Degradation behavior and damage mechanisms of CCD image sensor with deep-UV laser radiation. *IEEE Transactions on Electron Devices, 51*(12), 2229–2236. https://doi.org/10.1109/ted.2004.839758

Martin, A. (2021). Reliability characterization of plasma induced charging damage (PID) for MOS transistors of a silicon on insulator (SOI) process using product relevant test structures and fWLR methods. In *IEEE International Integrated Reliability Workshop*. https://doi.org/10.1109/IIRW53245.2021.9635611

Micheloni, R. (2016). *3D flash memories*. Springer. https://doi.org/10.1007/978-94-017-7512-0

Murarka, S. P., Eizenberg, M., & Sinha, A. K. (2003). *Interlayer dielectrics for semiconductor technologies*. Academic Press.

Nicosia, G., et al. (2015). A single-electron analysis of NAND flash memory programming. In *IEEE IEDM*. https://doi.org/10.1109/IEDM.2015.7409700

Oh, D., et al. (2008). Program disturb phenomenon by DIBL in MLC NAND flash device. In *Joint Non-volatile Semiconductor Memory Workshop and International Conference on Memory Technology*. https://doi.org/10.1109/NVSMW.2008.7

Ohyu, K., et al. (2006). Quantitative identification for the physical origin of variable retention time: A vacancy-oxygen complex defect model. In *IEEE IEDM*. https://doi.org/10.1109/IEDM.2006.346792

Okuyama, Y., et al. (1998). Monte carlo simulation of stress-induced leakage current by hopping conduction via multi-traps in oxide. In *IEEE IEDM*. https://doi.org/10.1109/IEDM.1998.746501

Park, K. T., et al. (1995). Degradation of the tunnel oxide and the endurance characteristics of the NAND flash memory due to plasma CVD charging-up damage. In *IEEE ESSDERC*.

Park, T.-S., et al. (2006). Characteristics of the full CMOS SRAM cell using body-tied TG MOSFETs (bulk FinFETs). *IEEE Transactions on Electron Devices, 53*(3), 481. https://doi.org/10.1109/TED.2005.864392

Park, K., et al. (2014). Active-precharge hammering on a row induced failure in DDR3 SDRAMs under 3x nm technology. In *IEEE International Integrated Reliability Workshop*. https://doi.org/10.1109/IIRW.2014.7049516

Prall, K. (2007). Scaling non-volatile memory below 30nm. In *IEEE Non-Volatile Semiconductor Memory Workshop*. https://doi.org/10.1109/NVSMW.2007.4290561

Prall, K., & Parat, K. (2010). 25nm 64Gb MLC NAND technology and scaling challenges. In *IEEE IEDM*. https://doi.org/10.1109/IEDM.2010.5703300

Ranuárez, J. C., Deen, M. J., & Chen, C.-H. (2006). A review of gate tunneling current in MOS devices. *Microelectronics Reliability, 46*(12), 1939–1956. https://doi.org/10.1016/j.microrel.2005.12.006

Roizin, Y., et al. (2003). Plasma-induced charging in two bit per cell SONOS memories. In *8th International Symposium on Plasma and Process-Induced Damage*. https://doi.org/10.1109/PPID.2003.1200914

Sacchettini, Y. (2019). A highly reliable back side illuminated pixel against plasma induced damage. In *IEEE IEDM*. https://doi.org/10.1109/IEDM19573.2019.8993561

Sacchettini, Y., Carrere, J. P., Duru, R., Oddou, J. P., Goiffon, V., & Magnan, P. (2019). CMOS image sensors and plasma processes: How PMD nitride charging acts on the dark current. *Sensors (Basel), 19*(24), 5534. https://doi.org/10.3390/s19245534

Takahasi, T., et al. (2001). A multigigabit DRAM technology with 6F2 open-bitline cell, distributed overdriven sensing, and stacked-flash fuse. *IEEE Journal of Solid State Circuits, 36*(11), 1721–1727. https://doi.org/10.1109/4.962294

Takebuchi, M., et al. (1996). New interconnect plasma induced damage analyzed by flash memory cell array. In *IEEE IEDM*. https://doi.org/10.1109/IEDM.1996.553150

Takebuchi, M., et al. (1998). Tunnel oxide degradation mechanism in flash devices for via etching process. In *3rd International Symposium on Plasma Process Induced Damage*. https://doi.org/10.1109/PPID.1998.725588

Tan, J. (2013). *4T CMOS active pixel sensors under ionizing radiation* [PhD, TU Delft].

Tenbroek, B. M., et al. (1996). Self-heating effects in SOI MOSFET's and their measurement by small signal conductance. *IEEE Transactions on Electron Devices, 43*(12), 2240–2248. https://doi.org/10.1109/16.544417

Tokashiki, K., Bai, K., Baek, K., Kim, Y., Min, G., Kang, C., Cho, H., & Moon, J. (2007). Study of plasma charging-induced white pixel defect increase in CMOS active pixel sensor. *Thin Solid Films, 515*(12), 4864–4868. https://doi.org/10.1016/j.tsf.2006.10.065

Tsai, Y. P., et al. (2018). 7nm FinFET plasma charge recording device. In *IEEE IRPS*. https://doi.org/10.1109/IEDM.2018.8614520

Tsui, B. Y., Yen, C. C., Li, P. H., & Lai, J. Y. (2011). Effects of EUV irradiation on poly-Si SONOS NVM devices. *IEEE Electron Device Letters, 32*(5), 614–616. https://doi.org/10.1109/led.2011.2121052

Vici, A., Russo, F., Lovisi, N., & Irrera, F. (2020). On border traps in back-side-illuminated CMOS image sensor oxides. *IEEE Transactions on Electron Devices, 67*(5), 2022–2027. https://doi.org/10.1109/ted.2020.2983039

Wang, Z. (2004). *Detection of and protection against plasma charging damage in modern IC technology*. Twente University.

Wang, X. (2008). *Noise in sub-micron CMOS image sensors* [PhD, University of Southampton, UK].

Weng, W.-T., Lee, Y.-J., Lin, H.-C., & Huang, T.-Y. (2009). Effects of plasma damage on metal–insulator–metal capacitors and transistors for advanced mixed-signal/radio-frequency metal–oxide–semiconductor field-effect transistor technology. *Japanese Journal of Applied Physics, 48*(8). https://doi.org/10.1143/jjap.48.086001

Wu, C. H., et al. (2015). Mapping of wafer-level plasma induced charge contour by novel on-chip in-situ recorders in advanced FinFET technologies. In *IEEE IEDM*. https://doi.org/10.1109/IEDM.2015.7409644

Yamamoto, Y., Segi, K., Tsuda, S., Makiyama, H., Hasegawa, T., Maekawa, K., Shinkawata, H., & Yamashita, T. (2019). The study of plasma induced damage on 65-nm silicon on thin BOX transistor. *IEEE Journal of the Electron Devices Society, 7*, 825–828. https://doi.org/10.1109/jeds.2019.2917893

Zambelli, C., et al. (2020). Points-over-threshold statistics for post-retention read disturb reliability in 3D NAND flash. In *IEEE International Integrated Reliability Workshop*. https://doi.org/10.1109/IIRW49815.2020.9312860

Chapter 9
Common Sources of Process Damage

9.1 Introduction

Many process steps have the propensity to create damage by different means. Plasma damage through charge accumulation on the wafer surface is one of the most common problems. However, there are many other cases where damage occurs through various means other than plasma damage. For example, damage can occur through triboelectric effects caused by friction through simple processes such as deionized water rinses (Sect. 9.6). Processing may contaminate the wafer with unintended materials that degrade and destroy the electrical function. Structural damage can occur by processing damage physically, destroying the intended structures on the wafer surface. This chapter reviews the most commonly known sources of process damage and the mechanisms involved.

9.2 Accelerated Degradation Under Processing Conditions

CMOS wafers are subject to extreme conditions during processing in semiconductor manufacturing equipment. These conditions are much more extreme than the chip's operating conditions during its lifetime in the field. Processing conditions can expose a wafer to temperatures of hundreds of degrees centigrade, while chip operating temperatures are usually limited to a maximum temperature of 125 °C for high-temperature applications such as automotive. Many processing steps have a maximum temperature of around 400 °C to prevent excessive heat from damaging the metallization after the transistor is formed, which can degrade the chip's electrical characteristics. Exposure to plasma damage at elevated temperatures can degrade or destroy dielectrics rapidly, far faster than under normal operating conditions. The acceleration of degradation under process conditions can occur a billion times faster

K. Prall, *CMOS Plasma and Process Damage*,
https://doi.org/10.1007/978-3-031-89029-1_9

than during normal chip operation (Cheung, 2001). As a result, materials that could survive for the chips' required lifetime can be degraded to the point of failure during a few minutes of processing.

Figure 9.1 shows the gate dielectric breakdown failure acceleration for a thick dielectric with temperature and voltage acceleration. The lifetime of the dielectric at a processing condition of 400 °C is several orders of magnitude shorter than at a maximum operating temperature of 125 °C (Strong et al., 2009, pg. 272). The lifetime at 400 °C can be very short in the range of seconds, which is a similar time as many process recipe conditions. The charging conditions determine the acceleration due to the voltage stress.

As an example, Fig. 9.2 shows the charge to breakdown (QBD) degradation as a function of temperature for a plasma ashing process (Ma et al., 1995). The QBD at 340 °C is severely degraded for an antenna ratio of 10^4, with the capacitors breaking down at zero charge, indicating dielectric breakdown during plasma damage before the QBD test. The lower temperature ashing process is significantly better. Acceleration of damage by temperature, voltage, time, and chemical action occurs in many CMOS manufacturing processes.

High-k dielectric degradation is similarly accelerated with temperature. High-k dielectrics are typically easier to damage compared to SiO_2 because the bond strength in the material is weaker (McPherson et al., 2002). Figure 9.3 shows a plot of charge to breakdown (QBD) for a high-K dielectric with a SiO_2 interfacial layer. The QBD is degraded by about 6 orders of magnitude between 125 °C and 290 °C, indicating that damage during processing may be a severe risk (Wu & Sune, 2013).

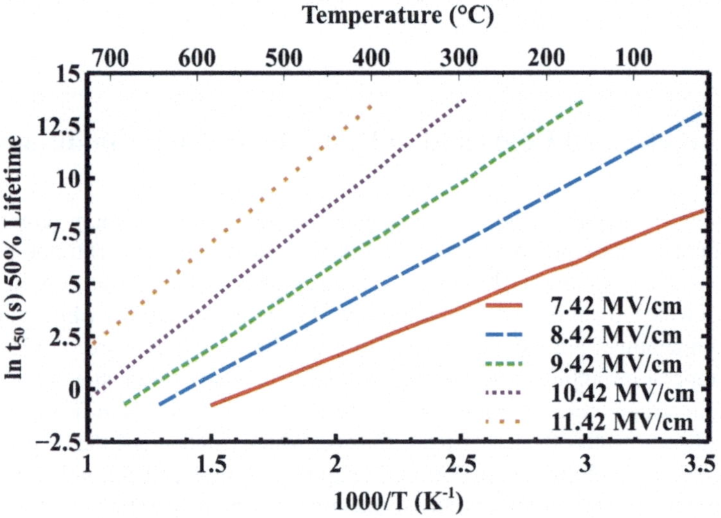

Fig. 9.1 Time-to-breakdown at 50% failure as a function of temperature for 15 nm gate oxide at different electrical field strengths

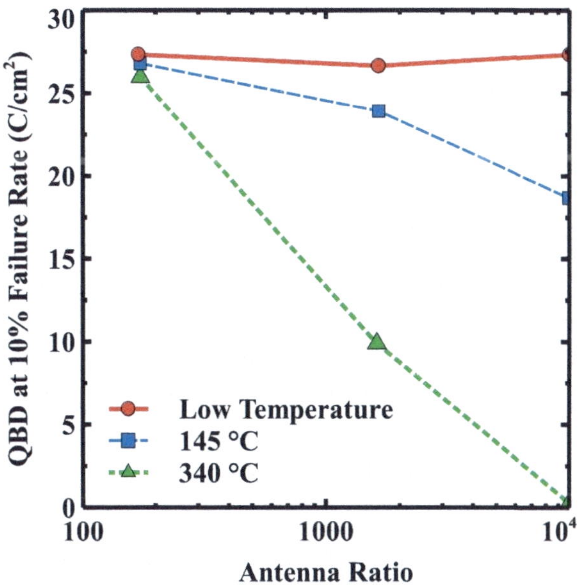

Fig. 9.2 Plot of QBD at 10% failure rate for a 10 nm gate dielectric during plasma ashing at different temperatures for different gate antenna ratios. The control is not exposed to the plasma. The plasma ash exposure consisted of 15 1 min. steps

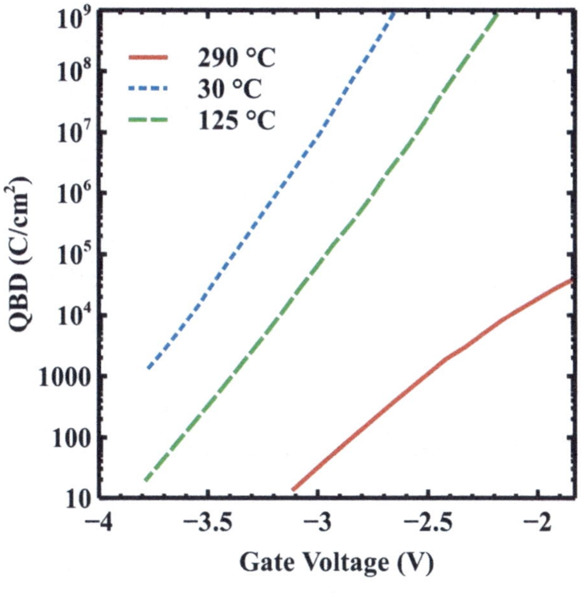

Fig. 9.3 Plot of QBD versus gate voltage for different temperatures for a high-k/SiO_2 dielectric

Table 9.1 Common processes causing process damage

Ion implantation
Plasma etching
Electron and ion beam exposure, including inspection
Plasma film deposition
Bond pad and passivation processing
Photoresist and masking materials processing, ashing
Deionized water processing
Triboelectric effects in wafer spin processes, scrubbing, chemical mechanical polishing, etc.
Exposure to high-energy light sources
Sputter etch
Physical vapor deposition (sputter deposition)
Through-silicon vias and advanced packaging processing

9.3 Common Sources of Process Damage

Many commonly used semiconductor processes can cause process damage in a CMOS manufacturing process. This section reviews some of these sources of damage. Almost all processes have some potential for damage. A list of common process damage sources covered here is shown in Table 9.1.

9.4 Ion Implant Damage

Ion implantation is used multiple times in a CMOS manufacturing process to provide a broad spectrum of dopant atoms at different energies and doses into the wafer to tailor the CMOS device properties (Rimini, 1995). Ion implantation exposes the wafers to a beam of positively charged ions that collect on the wafer, creating a situation where damage to devices and structures can occur as charge accumulates in the wafer. The ion beam is scanned across the wafer, providing a uniform dose across the wafer. The structures on the wafer surface are intermittently exposed to the ion beam, charging the structures. Ion implantation is a good case study for process damage due to the simplicity of the physics involved.

9.4.1 Ion Implantation Charging Damage

Charge collects on structures on the wafer surface, such as transistor gates, as the wafer is exposed to an ion beam, as shown in Fig. 9.4. The ion current from the implanter beam strikes the wafer surface, and the ion charge collects on conducting and nonconducting structures on the wafer surface. The implanter ion beam current projects a current I_i (usually mA) onto a region of area A_b determined by the beam

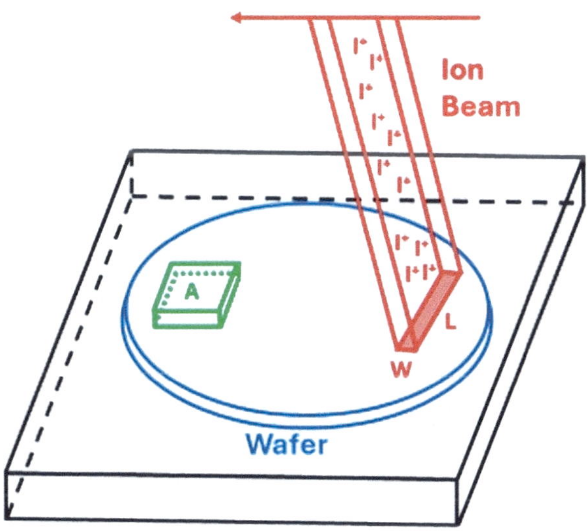

Fig. 9.4 Sketch showing an ion implant beam scanning across a wafer

dimensions as it scans across the wafer. A structure on the wafer with area A that receives charge Q during exposure time t is given by Eq. (9.1) (Mack et al., 1985).

$$Q = A\,I_i t \,/\, A_b \tag{9.1}$$

The total charge the structure receives is proportional to its area and the total dose D (ions/cm^2), as shown in Eq. (9.2).

$$Q = DA \tag{9.2}$$

If the structure is a capacitor, the capacitor is formed by the gate metallization over the isolation oxide and the silicon substrate. The voltage V the structure reaches is given by Eq. (9.3), where q is the charge on an electron (Coulombs):

$$V = \frac{qD}{C} \tag{9.3}$$

As an example of a worse-case condition, a gate metal structure on top of an isolation (STI) oxide 200 nm thick can reach a voltage of nearly 100,000 volts for a high implant dose of 10^{16} ions/cm^2. A 1000 V across a 200 nm dielectric can create a 50 MV/cm electric field, easily destroying the capacitor. This worse-case condition assumes that the capacitor collects all the implant charge and that no charge leaks away by secondary emission or surface leakage.

Charging damage can be significantly worse for structures that collect substantial amounts of ion beam charge and discharge it into sensitive structures. Figure 9.5 shows a sketch of an example sensitive structure. This type of structure is called an antenna structure. A large-area metal gate structure on top of the STI isolation is

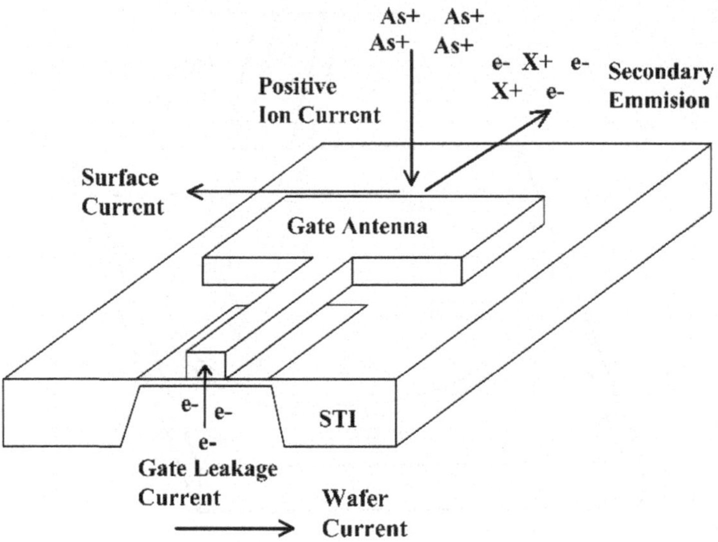

Fig. 9.5 Sketch showing a gate antenna on top of shallow trench isolation exposed to an ion implantation beam

connected to a small gate metal structure on active active with a gate dielectric underneath. The gate metal antenna structure on the dielectric isolation collects ions in proportion to its area from the implanter beam, building up charge with increasing voltage. The accumulated charge on the metal antenna attempts to find a leakage path to exit the antenna as the ion charges repel each other. The charge leaks from the antenna through the lowest resistance path. The gate dielectric is typically the lowest resistance path, conducting current by quantum mechanical tunneling. The gate dielectric leakage increases as the voltage on the gate metal increases through Fowler–Nordheim or direct tunneling. The antenna's voltage increases until the gate leakage from the antenna matches the ion current onto the antenna, reaching a quasi-steady-state condition where the currents are matched. The gate dielectric undergoes damaging stress under these conditions, causing degradation in electrical characteristics.

The worst-case example described previously is extremely rare under real-world conditions. The charge building up on the antenna may leak across the surface of the wafer in addition to the gate dielectric leakage, reducing the maximum voltage level. In addition, secondary emission of ions and electrons may be ejected from the antenna due to the impacting ion beam. The secondary emission changes the voltage and amount of charge on the antenna. The sketch in Fig. 9.5 shows these current paths. The beam current accumulates on the wafer surface and leaks across the surface.

9.4.2 Flood Gun

Ion implantation has a high propensity for process damage as it injects a beam of charged ions onto the structures on the wafer's surface. The fundamental mechanism for ion implant operation results in a likely occurrence of process damage. The ion beam current is a parameter that can be easily adjusted to modulate the charge flowing onto the wafer. Therefore, there is typically a direct correlation between process damage and beam current. This correlation can be used to prove that the cause of a problem is process damage. Due to the intrinsic process damage nature of an ion implanter, implanters are equipped with an electron source called a flood gun.

The flood gun injects negatively charged electrons into the positively charged ion beam in the implanter to neutralize the ion charge. The system is optimized to neutralize the beam charge by compensating the ions with an equal number of electrons, mitigating the plasma damage. Figure 9.6 shows a plot of the voltage induced on a MOS capacitor with varying flood gun current (Wu & Kolondra, 1991). The voltage across the capacitor decreases until the electron charge matches the ion beam and neutralizes it, compensating for the implant beam current. A negative voltage is induced on the capacitor as the flood current increases beyond the neutral point where the ions are precisely compensated. An excessive flood gun current stresses the capacitor negatively, causing process damage. The implant and flood gun conditions must be carefully optimized to mitigate plasma damage.

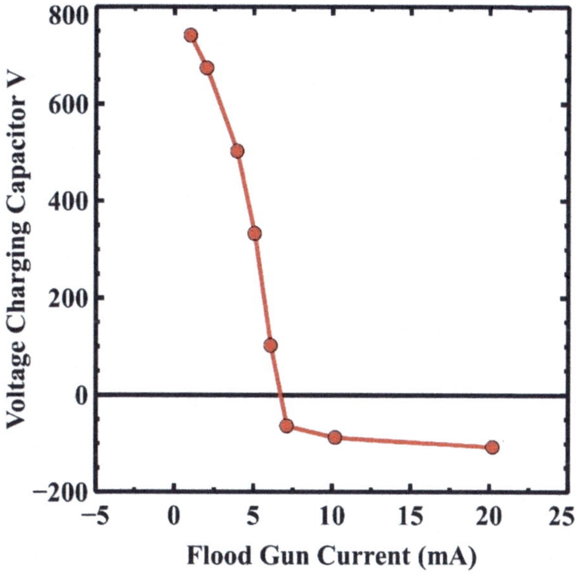

Fig. 9.6 Plot showing the charging voltage on a capacitor as a function of the flood gun current for a 100 keV Ar ion implant with a 1 mA beam current

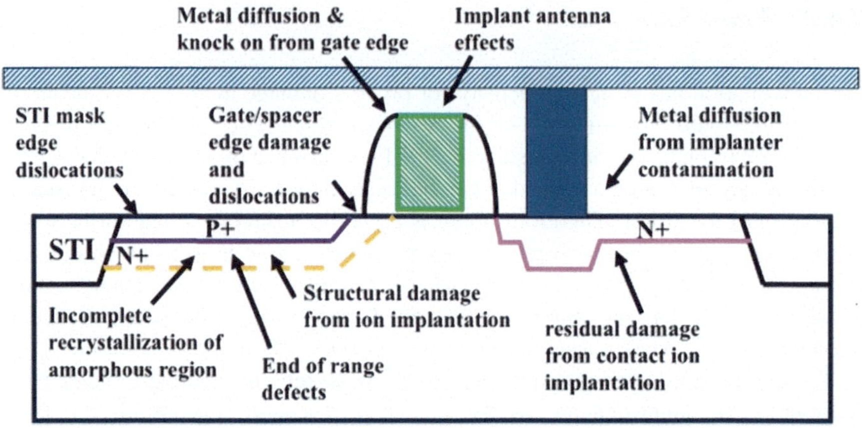

Fig. 9.7 Cross-section sketch showing implant related process damage mechanisms that can affect a CMOS imager cell. The sketch also applies generally to CMOS. The text highlights implant related mechanisms

9.4.3 Implant Damage Excluding Charging

In addition to ion charging damage, ion implantation can degrade CMOS chips through several other mechanisms (Romano et al., 2015). Figure 9.7 shows a CMOS wafer cross-section sketch showing potential mechanisms of implant-related damage (Ito, 2012).

Ion implanters can knock-on unintended atoms into the substrate, causing junction and transistor leakage. These knocked-on atoms can be from the films, photoresist, etc., on the wafer surface, or from contamination such as metal atoms on the wafer surface. In addition, the knocked-on atoms can become electrically active. For example, a Shockley–Read–Hall (SRH) site can cause leakage, resulting in single-atom defects.

9.4.4 Implant Structural Damage

Ion implantation imparts a large amount of kinetic energy into the implanted wafer, which can damage the crystalline semiconductor lattice and materials deposited on the wafer surface. The damage is dependent on multiple parameters in the implant recipe. Figure 9.8 shows a diagram of ion implant recipe parameters that can impact the severity of lattice damage (Rao, 2014). Implantation of heavy ions causes more damage than implanting light ions. Higher doses and higher energy cause more damage. Additional parameters are shown in the figure.

Ion implantation can cause structural damage to the semiconductor substrate for higher doses, creating electrical leakage. High doses greater than about

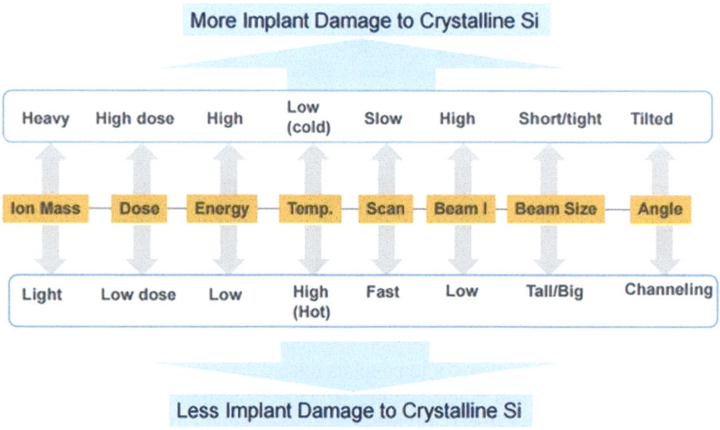

Fig. 9.8 Diagram showing important ion implant parameters related to crystalline damage

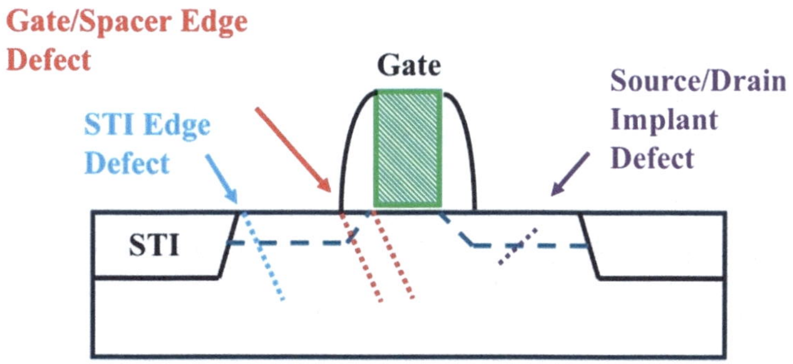

Fig. 9.9 Cross-section sketch showing location in an MOS transistor where implant related defects may occur

10^{14} ions/cm^2 cause enough damage to the semiconductor lattice to destroy the crystal structure, amorphizing it completely. The crystal structure is regrown after implantation with thermal annealing in a process called solid-phase epitaxial regrowth (SPEG). The silicon crystal regrows slowly along the crystal planes of the wafer substrate until the original crystal structure of the implant region is regrown. Defects may form during this regrowth process, specifically in regions where the regrowth regions contact a feature on the surface of the wafer, such as a gate edge or isolation (STI) field edge. A dislocation or stacking fault may form where the regrowth meets the surface feature, causing leakage or device failure. Careful optimization of the processing conditions during implantation and annealing may be needed to eliminate the dislocations. Figure 9.9 shows a cartoon of locations in a MOS transistor where dislocations tend to form, including the gate edge, spacer edge, and STI edge (Chang, 2011). The dislocations tend to form at nearly a 45°

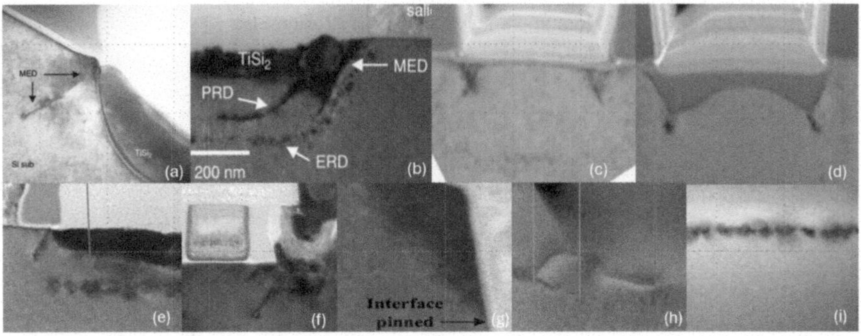

Fig. 9.10 Example TEM photographs of ion implant related defects. (**a**) Mask edge defect (MED) at the edge of a nitride spacer and salicide region. (**b**) Peak range defect (PRD), mask edge defect (MED), and end of range defect (ERD). (**c**) MED at a nitride edge. (**d**) MED at a nitride edge. (**e**) MED at the edge of a nitride spacer and salicide region with end of range defects below the salicide region. (**f**) MED at a nitride spacer, salicide region, and contact. (**g**) Pinned amorphous region at a STI edge. (**h**) End of range defects. (**i**) End of range defects

angle (for <100> substrates) caused by epitaxial growth from both the side and bottom of the amorphous region.

Figure 9.10 shows TEM examples of implant-related defects (Current, 2014, 2019; Lee, 2020; Rudawski et al., 2008; Rudawski et al., 2009). Figure 9.10a shows a mask edge defect (MED) TEM of a dislocation at the edge of the spacer in a CMOS transistor (Tung et al., 2003, pg. 165). Excessive stress at the bottom spacer corner results in a fracture of the silicon lattice or the SPEG regrowth from the bottom and side of the junction. This type of defect is commonly caused by ion implantation or the salicide metallization process. Figure 9.10c and d shows mask edge defects formed at the edge of two implanted nitride edges. Figure 9.10e shows another mask edge defect at the edge of a nitride spacer and salicide region. Figure 9.10g shows an amorphous region at the STI edge that is not recrystallized as the SPEG is pinned at the STI edge.

In addition to implant defects formed at feature edges, implant-related defects can form in regions unrelated to the surface features. For example, Fig. 9.10b shows three different defects, including a mask edge defect. There are two possible defects that can be formed by ion implantation without a feature edge. One defect is the end of range defects (EOR) formed at the maximum depth of the amorphous region where large supersaturations of silicon interstitial atoms exist that increase in size during annealing through an Ostwald ripening mechanism (Koffel et al., 2009). The third defect is the peak range defect (PRD) located at the peak dopant concentration in the implant doping profile, where the heavy concentration of dopants interferes with recrystallization. Figure 9.10h and i shows examples of end-of-range defects.

Ion implant defects are a complicated function of the wafer structures, implant conditions, and post-implant processing conditions. Process steps such as contact formation, salicide formation, and stressor formation interact with the implant and

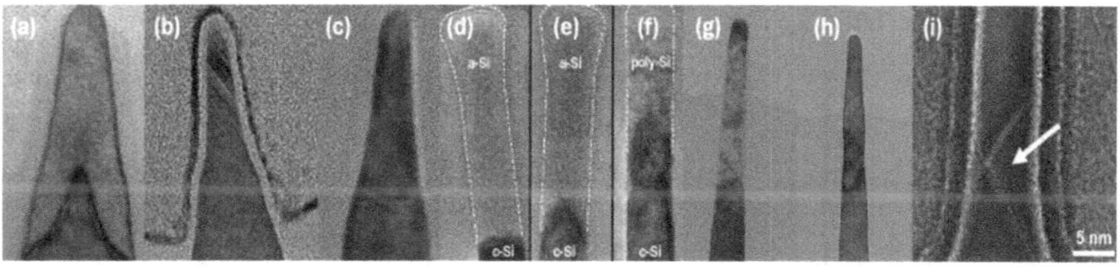

Fig. 9.11 Example TEM cross-sections of finfet structures after implant and annealing. (**a**) Implant performed at room temperature showing the amorphous region (As 1E15 ions/cm^2 at ~5 keV). (**b**) Room temperature implant followed by a 1000 °C 1 sec spike anneal showing dislocations. (**c**) High-temperature implant without anneal showing good crystallization and a defect. (**d**) As implanted fin. (**e**) After 650 °C show slight recrystallization. (**f**) After 1050 °C rapid thermal process showing partial annealing. (**g**) After low dose as ion implantation showing an amorphous region. (**h**) After spike anneal showing recrystallization. (**i**) Dislocations across the entire fin channel channel the transistor drain and source

play critical roles in defect formation. The annealing processes are critical for the elimination of implantation related defects.

The amorphous region created by a heavy-dose ion implant can be challenging to recrystallize in small amorphous features, such as finfets, as the amorphous–crystal boundary is very small and quasi-one-dimensional. Figure 9.11 shows example transmission electron microscope pictures of a finfet channel after implantation and different annealing steps (Fletch, 2014; Pipes et al., 2014; Wood et al., 2013; Ye et al., 2021). Figure 9.11a shows the fin as implanted where the arsenic implant has amorphized the fin's top and side. Figure 9.11b shows an attempt to recrystallize the fin with a rapid thermal process spike anneal. However, dislocations and stacking faults remain in the center of the fin. Figure 9.11c shows that hot ion implantation results eliminate the dislocations. Figure 9.11d shows an amorphous region formed by an implant. Figure 9.11e shows slight recrystallization after a 650 °C anneal, and Fig. 9.11f shows partial recrystallization after a 1050 °C rapid thermal processing (RTP). Figure 9.11g shows an amorphous region, and Fig. 9.11h shows recrystallization after a spike anneal. Figure 9.11i shows dislocations across the entire fin, shorting the drain and source of the transistor.

9.4.5 Ion Implantation Through the Gate

Ion implantation through the gate dielectric and metallization is sometimes performed to adjust the CMOS electrical parameters. The ions can break bonds in the dielectric and semiconductor, creating traps and interface states and degrading the electrical characteristics. In addition, the ions impacting the gate metallization or gate dielectric can knock on ions into the gate dielectric and substrate, degrading the electrical characteristics. Ion implantation at higher energies has a higher propensity for dielectric damage. Figure 9.12 shows a plot of dielectric breakdown yield as

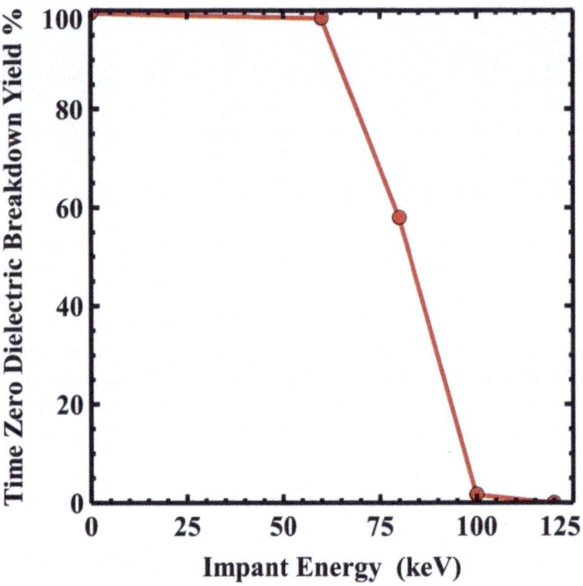

Fig. 9.12 Plot showing time-zero dielectric breakdown yield (TZDB) as a function of As ion implant energy at a dose of 1.5×10^{12} ions through the gate. The yield criteria is defined as greater than 8 MV/cm

a function of ion implant energy through the gate for a 7 nm dielectric (Shiraki, 2008). The criterion for failure is a breakdown voltage of less than 8 MV/cm.

Degradation due to ion implantation through the gate dielectric is also a function of the implant dose. Figure 9.13 shows an example plot of the charge to breakdown (QBD) as a function of the implant dose for different gate dielectric thicknesses (Jank et al., 2000). Thinner oxide layers are less likely to be damaged because the ions have a higher probability of passing through the dielectric without interacting with the fewer number of atoms in the dielectric.

9.5 Triboelectric Charging in Wet Processes

Some wet chemical processes can create process damage through electrostatic damage by triboelectric charging, where friction generates a charge on the wafer surface (Mack et al., 1985). Wafer scrubbing with a rotating noncontacting brush is commonly used in CMOS wafer processing for enhanced wafer cleaning to remove particles from chemical mechanical processing and other processes. Spin dryers are used to dry wafers by centrifugal force after wet cleaning steps. Process steps that spin the wafers, such as single wafer cleans, resist coatings, etc., can generate a triboelectric charge (Alba et al., 1997). Marangoni drying extracts moisture from the wafer surface with alcohol vapor. Processes like this can generate a triboelectric

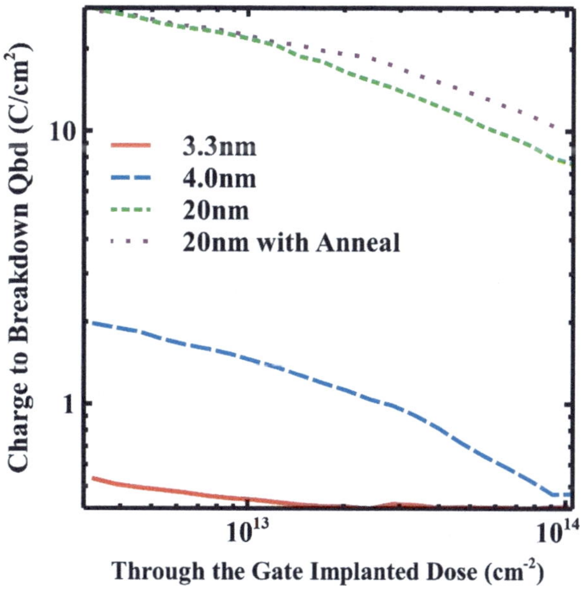

Fig. 9.13 Charge to breakdown for various oxide thicknesses for different implant (100 KeV boron) doses through gate dielectrics. The left most data points are unimplanted samples. One group added a furnace anneal of 950 °C for 30 min in an inert ambient after the implant

charge on the wafer surface, causing process damage as the charge is dissipated (Hagimoto et al., 2009). Additionally, visible electrostatic discharge defects can be created on the wafer, like the defects shown in Chap. 5.

Additionally, nonvisible damage discussed in the previous sections may occur but cannot be detected inline during processing. High-resistance deionized (DI) water is used in CMOS processing to avoid contamination. Water may not create a sufficient leakage path for charge dissipation from the wafer, leaving a charge on the wafer. In addition to process damage, triboelectric charge can attract particles onto a wafer, creating yield problems.

9.6 Damage Due to Spin and Deionized Water Processes

CMOS wafers commonly use spin processes where the wafer is held by a vacuum chuck and rotated at high speed while deionized water and chemicals are dispensed on the wafer surface. Process damage and electrostatic discharge can damage sensitive CMOS devices due to the friction of the high-resistance water across the wafer surface.

Figure 9.14a shows a cross-section sketch of an oxidized wafer on a spin chuck with deionized (DI) water flowing across the surface covered with SiO_2 (Istsumi et al., 2002, pg. 245,261). High-speed rotation of the wafer of thousands of

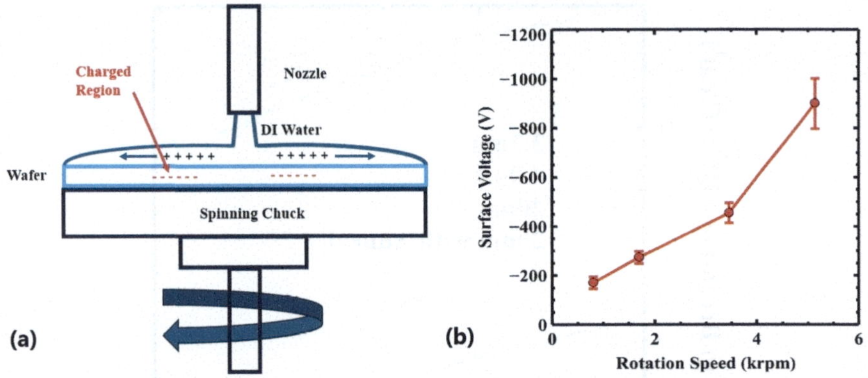

Fig. 9.14 (**a**) Cross-section sketch of a spin chuck with deionized water flowing on the surface of a wafer. (**b**) Showing the voltage induced on the surface of the dielectric versus the rotation speed during a deionized water spin rinse. The voltage was measured by an electrostatic meter

revolutions per minute causes friction on the wafer surface. The friction across the wafer surface generates heat and electrostatic charge that can break apart the water atoms into H+ and OH- ions. The difference in electronegativity causes the OH- atoms to become trapped in the silicon dioxide and the H+ to remain in the deionized water. The rotating ions generate a magnetic flux perpendicular to the wafer surface. The flux causes a nonuniform charge distribution across the wafer surface, pushing H+ ions to the wafer center where they can diffuse through the silicon dioxide layer, reaching the silicon–oxide interface and passivating defects like interface states in a similar manner as a hydrogen alloy.

Like a thermal hydrogen alloy, the passivating process can anneal plasma damage from previous process steps. Additionally, large voltages can be generated on the wafer surface with a DI water spin, as shown in Fig. 9.14b, where higher spin speeds generate higher surface voltages.

The charge induced in the dielectric has been shown to improve plasma damage problems through hydrogen alloying of defects and possibly compensation of charge remaining from damage at a previous step (Istsumi et al., 2002, pg. 253,257). Also, the induced voltage can damage the dielectric.

The magnetic field induced by the spinning ions results in nonuniform damage across the wafer, as shown in Fig. 9.15 as shown by measurements of minority carrier lifetime and flatband voltage shifts in the C–V characteristics.

9.7 Resist and Masking-Related Damage

Resists and other masking materials enable semiconductor processes to modify one region of the wafer while protecting other wafer regions from modification. Masking materials are generally insulators and can absorb charge on their surface. The

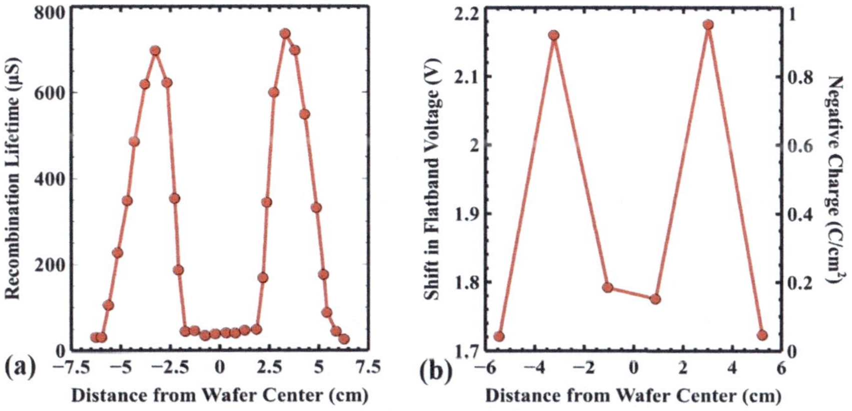

Fig. 9.15 Plot showing across-wafer nonuniformity caused by a DI spin rinse. (**a**) Across-wafer variation in minority carrier lifetime. (**b**) Across-wafer variation in C-V flatband voltage and corresponding negative charge trapped on the wafer surface (Sze & Lee, 2010, pg. 253,257)

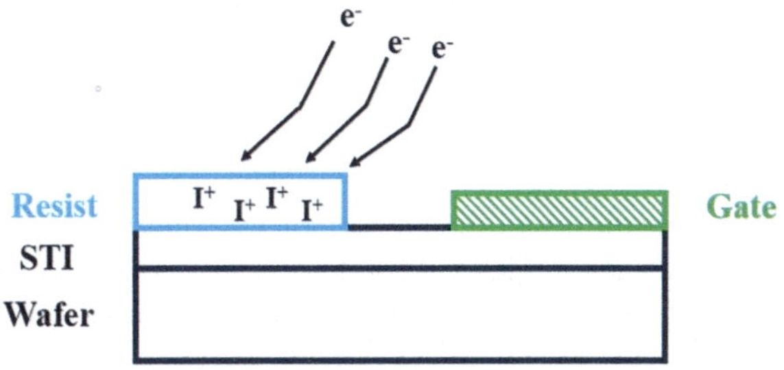

Fig. 9.16 Sketch showing resist charging modifying the plasma charge behavior near an antenna

collected charge can affect the processing conditions on the wafer surface. The collected charge can flow or dissipate across the wafer surface due to the mask pattern and resistivity. The masking materials can be impacted by the processing and change as the recipe progresses. High temperatures can cook the mask material, modifying it with chemical changes such as outgassing. In addition, the chemicals in the processing chamber may react with the masking material, changing its composition.

The collected charge may build up in the masking material and gain the capability to deflect ions and electrons, modifying the processing conditions at the wafer surface and deviating from the intended process conditions (Dixon et al., 1997). Figure 9.16 is a sketch showing a resist pattern near an antenna collecting charge, modifying the electric field at the antenna edge, changing the local process conditions, and process damage. Additionally, the charge on the wafer surface may bend the electron or ion path, changing the impact angle with the wafer and affecting local process conditions. This effect may occur in multiple types of CMOS processing, including ion implantation, plasma etching, sputtering, etc.

Fig. 9.17 SEM photograph of an example electrostatic discharge arc showing tendrils

The collected charge on the masking layer can build up to the point where a catastrophic electrical discharge can occur, damaging the wafer surface. Figure 9.17 shows a microscope photograph of a defect caused by an electrical arc (Wavelength Electronics, 2015). Note the distinctive arc pattern with tendrils of material migrated from one conductor to another. This kind of defect can correlate with specific resist patterns, design rules, and wafer structures that collect the most charge.

9.7.1 Resist and Masking Overlap-Related Damage

A resist or masking pattern overlapping a conducting antenna may collect and conduct charge, increasing the effective antenna area and forming a combined antenna consisting of the antenna and the additional antenna formed by the conducting resist (Dirnecker et al., 2002). A masking layer can also have the opposite effect, protecting an antenna by insulating it, providing that the masking layer is nonconductive. Figure 9.18 shows a plot of antenna yield for different values of resist overlap on a gate antenna during an ion implant. The figure also shows the effects of different resist patterns. This plot shows that increasing resist coverage reduces process damage and increases antenna yield. The worst-case conditions in this example are for structures with an extensive overlapping resist feature or a nearby resist pattern. The best-case conditions are for a structure where the resist completely covers the feature. The damage in this example is caused by an ion implantation step with the flood gun turned off as a worst-case condition.

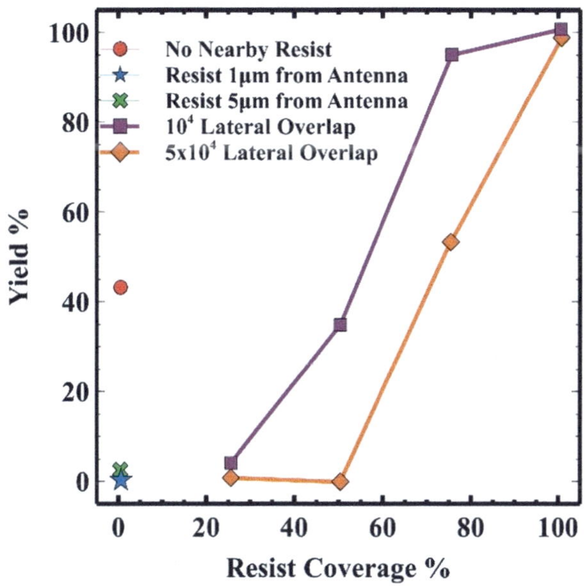

Fig. 9.18 Plot showing the yield of a polysilicon gate antenna as a function of the resist coverage for an ion implant into a gate antenna. Data is shown for no nearby resist and a resist pattern 1 μm and 5 μm away from the antenna, along with the resist overlapping the gate

9.7.2 Plasma Ashing

Photoresists and numerous types of masking materials block unwanted processing from selected regions of the wafer to form patterns essential to the formation of CMOS chips. Common process elements, such as etching, implantation, etc., use masking materials to achieve the required finished wafer structures. The masking materials become damaged through the bombardment of ions, chemical reactions, densification, etc., making them difficult to remove from the wafer and requiring complex stripping processes to remove the materials from the wafer. In addition, the wafer surface must be thoroughly cleaned to eliminate the possibility of contamination. Masking material removal by plasma processing is commonly used because of its aggressive ability to remove the difficult-to-remove damaged materials. Plasma stripping processes have the propensity to cause process damage.

Figure 9.19 shows a probability plot demonstrating a wide threshold voltage distribution caused by a plasma resist strip process for a 90 nm 50 Å CMOS imager process (Sacchettini et al., 2019). The structures evaluated only have a bond pad serving as an antenna. The group receiving a wafer strip process has a wide distribution with a defect tail. A 400 °C anneal after the plasma strip nearly recovers the distribution to its previous state.

Figure 9.20 shows an increased extrinsic defect density at low capacitor breakdown voltages for metal antennas during a resist strip process (Shin & Hu, 1993).

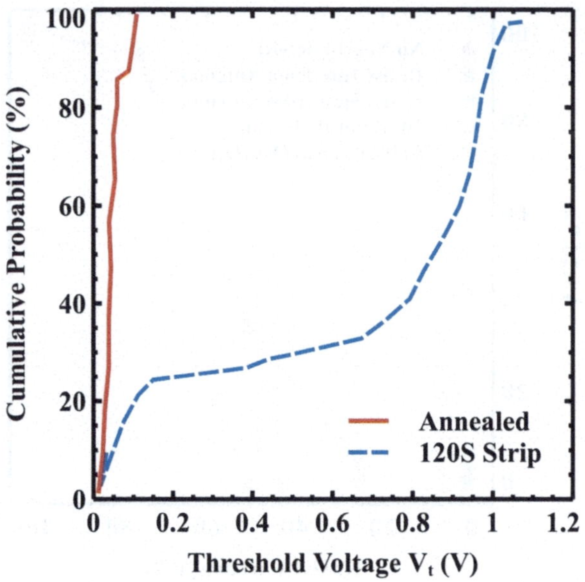

Fig. 9.19 Probability plot of threshold voltage shift of an nmos transistor connected to a bond pad antenna subject to an O_2 plasma resist strip. The wide V_t distribution is the same for different strip times. A 400 °C anneal partially heals the wide distribution (left side of the plot)

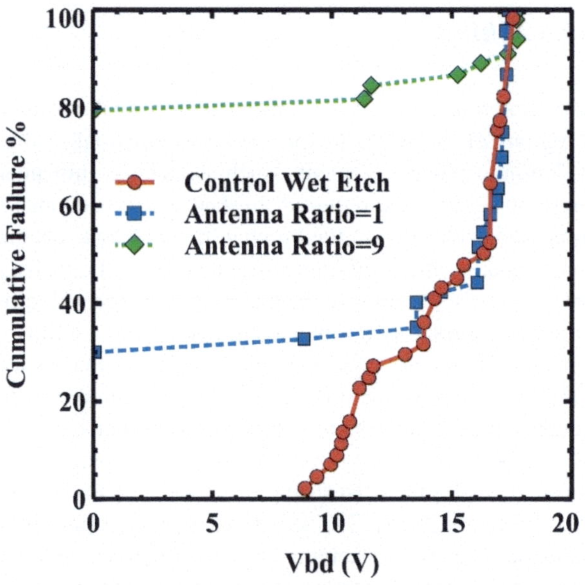

Fig. 9.20 Probability plot showing dielectric breakdown caused by resist strip after a metal etch process

The control group also has an extrinsic defect tail with different characteristics at higher breakdown voltages.

9.8 Electron and Ion Beam Damage

CMOS wafers are sometimes exposed to other types of electron or ion beams in addition to the most commonly used ion implants. Wafers are often inspected with electron beams used in critical dimension measurements, alignment measurements, or defect inspections. The entire wafer or small wafer regions are measured to collect data for process monitoring and control at multiple points in the process flow. An electron beam can cause process damage by several mechanisms, including charging the wafer surface, high-energy photon exposure, sputtering, wafer heating, or electrostatic discharge (Egerton et al., 2004). Electron beams are sometimes used for wafer curing, also creating the potential for damage (Pyo & Kim, 2013). The failure pattern on the wafer usually matches the electron beam pattern across the wafer, giving a specific signature to isolate the cause of the failure. Table 9.2 shows a chart depicting the different types of beam damage from electron and ion beams, including charging, electrostatic discharge, heating, structural damage, etc. (Egerton et al., 2004).

Ion beams are used in CMOS failure analysis to cut holes in the wafer and expose underlying features for examination (Fleetwood, 2000). The beam can cause process damage to structures and devices near the location where the beam touches the wafer. Figure 9.21 shows a plot of an NMOS transistor damaged by exposure to an ion beam (Cambell et al., 1997). The subthreshold current is shifted to the left, showing evidence of the creation of a large number of fixed oxide charges by the damage. Accidental transistor damage during failure analysis can confound failure analysis, leading to false conclusions.

Ion and electron beam processing are sometimes used in CMOS processing for purposes other than ion implantation, such as curing films (Pyo & Kim, 2013). Figure 9.22 shows an example of electron beam degradation causing a threshold voltage shift, with little or no dependence on the antenna ratio, indicating a mechanism other than charging.

Table 9.2 Damage effects from electron and ion beams

Electron irradiation damage	
	Electrostatic charging
Elastic scattering (electron–nucleus)	Inelastic scattering (electron–electron)
Atomic displacement (knock on)	Impact heating
Sputtering	Structural damage (knock-on, amorphization)
	Mass loss
	Contaminant deposition

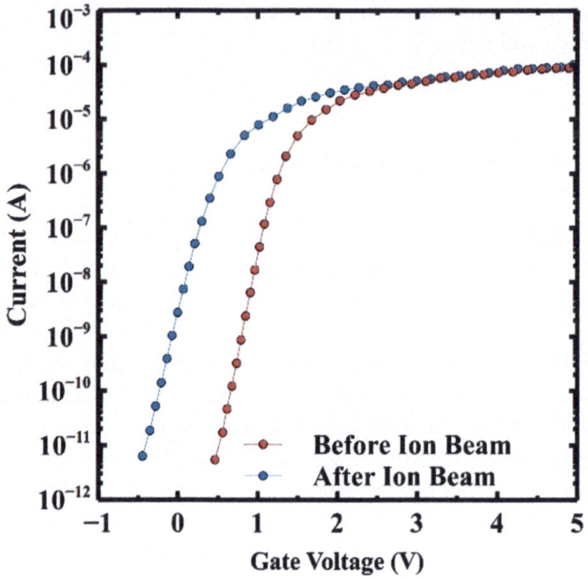

Fig. 9.21 Plot of subthreshold conduction at low and high biases in an MOS transistor showing process damage from exposure to an ion beam. The threshold voltage is shifted to the left after exposure to an ion beam

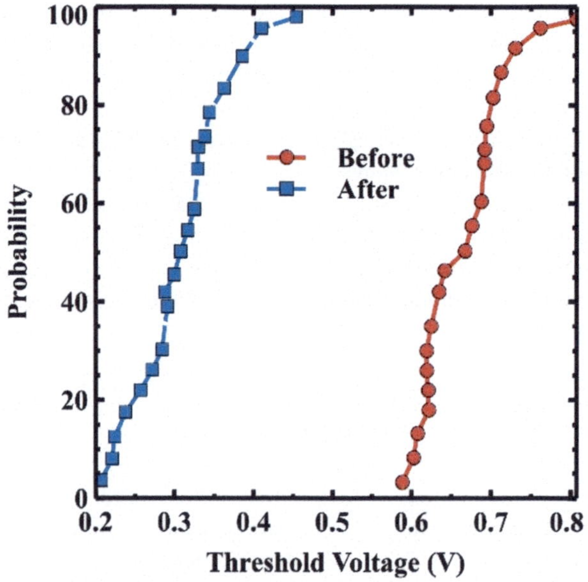

Fig. 9.22 An nmos threshold voltage shift caused by exposure to electron beam used to cure a spin on glass dielectric. The antenna ratio is 1000

9.9 Bond Pad Damage

Bond pads naturally form an antenna structure due to the large area required to connect the chip to the outside world. Therefore, the bond pad must be a relatively large minimum size as required by assembly processes and probe card requirements. A typical bond pad size is around 50–100 μm in length and width. Devices and structures connected to a bond pad at the chip to the external world interface are more likely to be damaged than regions internal to the chip. A bond pad on a product chip is connected to electrostatic discharge (ESD) protection circuits that protect the chip from damage originating from outside of the chip. The ESD protection circuits also tend to partially protect against process damage after the circuit is completed during processing, making product bond pads more resistant to process damage.

Bond pads also connect to parametric test structures placed in the scribe line or other wafer regions. However, ESD protection circuits are not generally added to parametric bond pads because they interfere with some parametric measurements. In addition, the lack of ESD protection circuits makes parametric structures more likely to be damaged. As a result, parametric structures can show severe electrical shifts or variations while the product chip is unaffected, creating a potential misdiagnosis of chip failure mechanisms (Lee et al., 2009).

As a result of this requirement, processes performed during or after bond pad formation must be carefully optimized to minimize process damage. The process steps related to the bond pad formation that are likely to create process damage include bond pad metal etch, passivation deposition, passivation removal from the pad, etc.

Additionally, pads much smaller than bond pads with dimensions of a few microns are used for microprobing to characterize the electrical behavior of the product and parametric devices. These pads are less likely to cause process damage. Comparison of identical structures with microprobe pads to regular bond pads can provide insights into pad-related process damage.

9.10 General Considerations for Plasma Processing

Plasma processes are used ubiquitously throughout a CMOS process flow for many different purposes, including film deposition, etching, cleaning, etc. The processing tools used for different purposes have some common elements and common behaviors affecting process damage. This section explores several common elements of plasma processing that affect process damage.

9.10.1 Equivalent Circuit for a Plasma System

A wafer placed in a plasma is electrically coupled to the plasma, creating an electrical interaction between the chamber, plasma, wafer, and structures on the surface of the wafer. Figure 9.23 shows a simple circuit schematic for the convolution of these elements (Cheung, 1999, pg. 63). The plasma in the chamber is an electrical conductor due to the large concentration of electrons, ions, and radicals. The plasma is surrounded by a sheath or dark space where a high electric field exists containing few charge carriers. The sheath electrically behaves as a capacitor labeled C_{sheath} on either side of the plasma. There is a circuit element consisting of a blocking capacitor connected to the radio frequency (rf) supply, used to ensure that high voltages are dropped across the sheath. This element is labeled C_{block} in the figure. The CMOS element subject to process damage forms a dielectric labeled C_{ox} in the figure. A direct current (DC) voltage bias is commonly applied across the plasma in the system. The voltage can be high, sometimes reaching approximately a thousand volts. This voltage level can quickly destroy a CMOS wafer. However, most of the voltage drops across the plasma sheath, and only a small voltage reaches the wafer, rarely causing damage. Process tools are optimized to couple appropriate voltages to the wafer backside. The backside condition can affect the equivalent circuit by adding a backside capacitor if there is a dielectric on the backside.

Plasma damage is caused by a combination of the plasma processing system and the CMOS wafer. The complexity of the phenomenon is partly due to the interaction of the wafer controlled by the CMOS manufacturing company and the processing tool controlled by the tool vendor. Combining the two factors in modeling or simulation is difficult and not often achieved. C_{ox} represents the gate dielectric in Fig. 9.23, as the CMOS manufacturer controls this component, and the equipment manufacturers and the recipe control the rest of the capacitors in the figure.

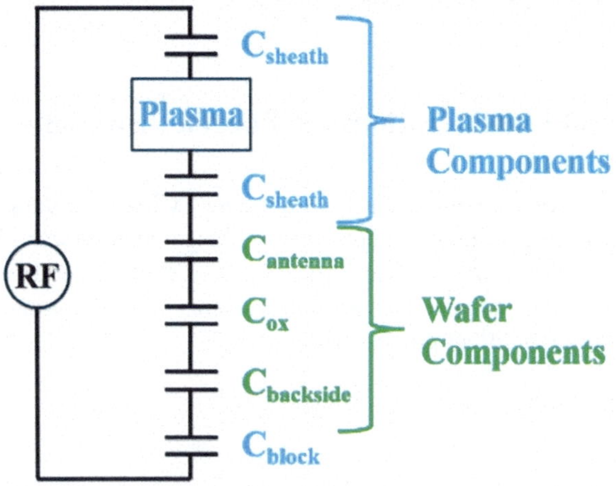

Fig. 9.23 Equivalent circuit for a wafer in a plasma system

9.10.2 Alternating Current Bias

A plasma system commonly uses a radio frequency (RF) signal that couples to the floating wafer. For example, a frequency of 13.56 MHz is commonly used. The bias across the wafer varies positively and negatively following the RF signal. The stress and process damage to the CMOS structures vary depending on the AC cycle. Figure 9.24 shows a sketch of AC waveforms creating process damage following the applied RF signal (Cheung, 1999, pg. 62). The floating wafer capacitively coupled to the RF signal follows the signal. Electron current spikes occur over a portion of the RF cycling, creating dynamic time-varying process damage. The combination of direct current (DC) bias and the RF signal must be carefully optimized to minimize process damage.

9.10.3 Across-Wafer Uniformity

Modern processing tools are highly optimized to achieve a uniform plasma across the wafer. Variations in the plasma across the wafer can cause a wide variation in process damage. Plasma variation can cause charge to flow into the wafer in some regions and out of the wafer in other regions. This nonuniformity can cause considerable variation in the electrical damage and symptoms across the wafer. Current flows in the bulk of the silicon wafer as charge flows in and out of the wafer in different regions. Figure 9.25 shows a circuit schematic showing the current flow under nonuniform plasma conditions (Cheung, 1999). Due to plasma damage, the threshold voltage may increase or decrease on the wafer edge due to electron

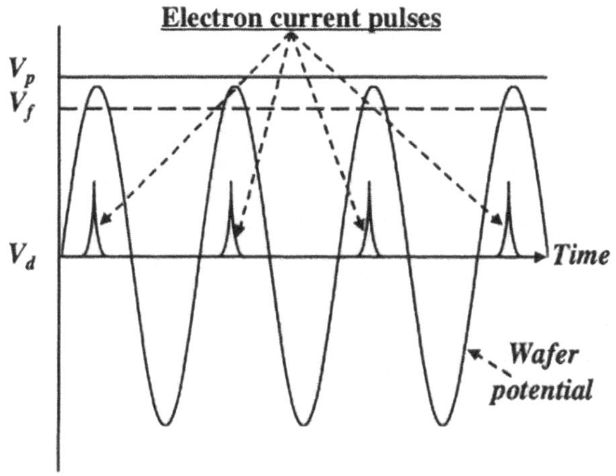

Fig. 9.24 Sketch showing AC electron current process damage following the RF waveform offset by the DC bias. V_p is the plasma potential, V_d is the DC bias, and V_f is the floating potential

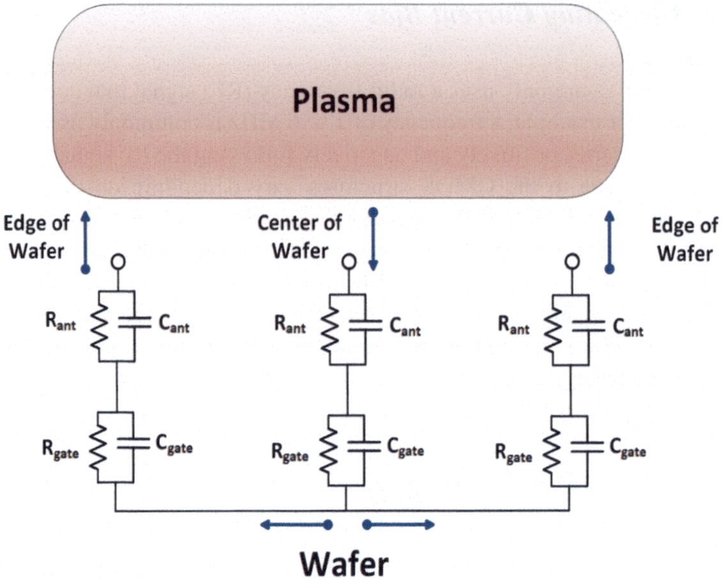

Fig. 9.25 Circuit model for current flow in a wafer exposed to a nonuniform plasma showing current going into the wafer at the center and out of the wafer at the wafer edge. C_{ant} represents the antenna capacitance, and C_{gate} represents the gate dielectric capacitance. Arrows show the direction of current flow in this example

injection into the gate dielectric, while the center of the wafer shifts in the opposite direction.

The across-wafer signature in electrical parameters for a nonuniform plasma typically varies widely. Figure 9.26 shows a probability plot of the V_t shift across the wafer for a floating gate plasma damage finfet test structure (Wu et al., 2015). V_t shifts in both positive and negative directions across the wafer due to the change in the direction of the charging current.

Modern semiconductor fabrication tools have been heavily optimized to expose the wafer to a uniform plasma over the entire surface to mitigate the cross-wafer process damage variation. The plasma must be uniform across the wafer for all process conditions, including starting and stopping the plasma. Starting or stopping the plasma can create a transient nonuniformity, causing process damage at the beginning or end of the plasma recipe. However, variation in process damage across the wafer is quite common, as shown in Fig. 9.27. The V_t variation across the wafer is recorded by a finfet based floating gate structure with connected metal antennas (Tsai et al., 2018). There is a difference in the across-wafer signatures depending on the metal layer.

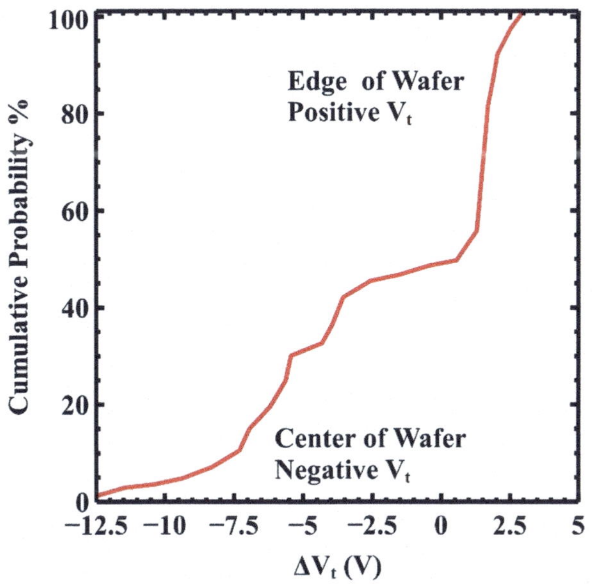

Fig. 9.26 Plot showing the nonuniform back end of line plasma damage induced V_t shift for a floating gate nonvolatile memory cell across the wafer in a finfet technology

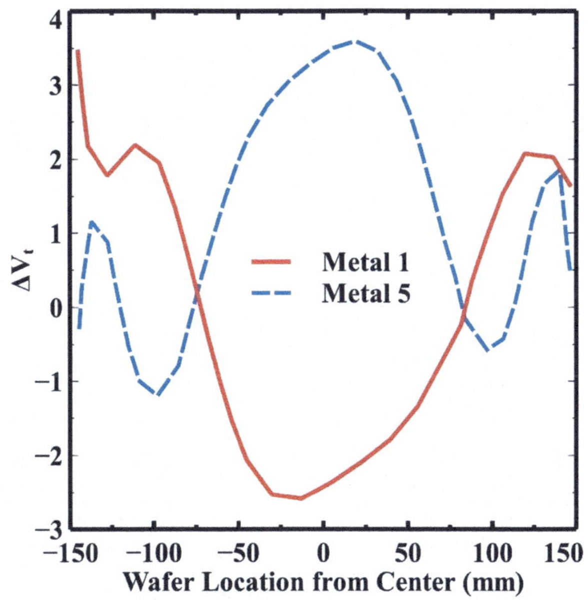

Fig. 9.27 Plot comparing plasma damage V_t variation across a wafer recorded by a nonvolatile memory detection test structure for two different layers of metal (M1 and M5)

9.10.4 Electrostatic Chuck

Semiconductor fabrication tools must hold the wafer securely inside the tool for processing. Electrostatic chucks hold the wafer in place with electrostatic force. This technique is superior to other methods, such as clamping the wafer for several reasons, including reduced particle generation, reduced wafer breakage, improved edge of wafer processing, backside cooling, etc. Electric charge is injected into the wafer chuck to clamp the wafer firmly. The operation of the electrostatic chuck creates the potential for process damage in addition to the plasma used in the processing. Lift pins raise the wafer from the chuck and can cause process damage by changing wafer potentials.

Figure 9.28 shows sketches of an electrostatic chuck operation (Nojiri, 2012). Figure 9.28a shows a unipolar electrostatic chuck. A voltage is applied to the chuck, inducing a sheet of charge on the back of the wafer that clamps the wafer to the chuck. Figure 9.28b shows a sketch of a bipolar electrostatic chuck. The bipolar chuck operates by applying a voltage potential difference between two wafer regions, inducing a positive charge in one region of the wafer and a negative charge in another region, and clamping the wafer to the chuck. Figure 9.28b shows the charge induced by a bipolar clamp with a positive charge on one portion of the wafer and a negative charge on another. The bipolar clamp can create a voltage potential variation across the wafer, affecting the electrical stress on process damage-sensitive structures.

High voltages operate an electrostatic chuck by inducing a charge in the wafer, creating an electrostatic force that holds the wafer on the chuck. For a bipolar chuck, opposite polarities are applied to the two electrodes, inducing an opposite polarity charge across the wafer. Figure 9.29 shows an example waveform (Nojiri, 2012). At the end of the process, the chuck polarities are reversed to flip the electrostatic force to release the wafer. Process damage can occur if the high voltages in the electrostatic chuck affect devices on the wafer surface.

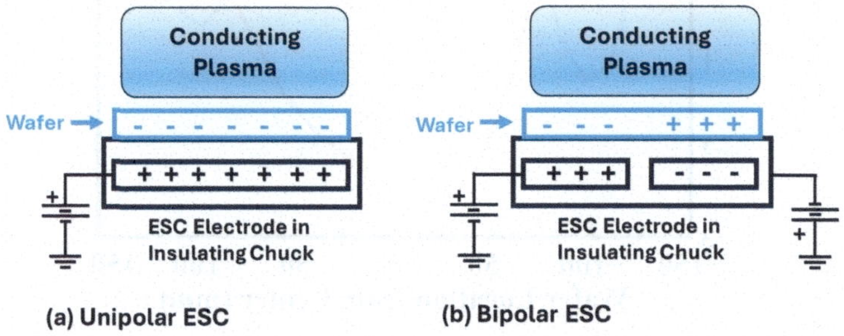

Fig. 9.28 Sketch of the operation of an electrostatic chuck. (**a**) Sketch and schematic of a unipolar electrostatic chuck. (**b**) Sketch and schematic of a bipolar electrostatic chuck

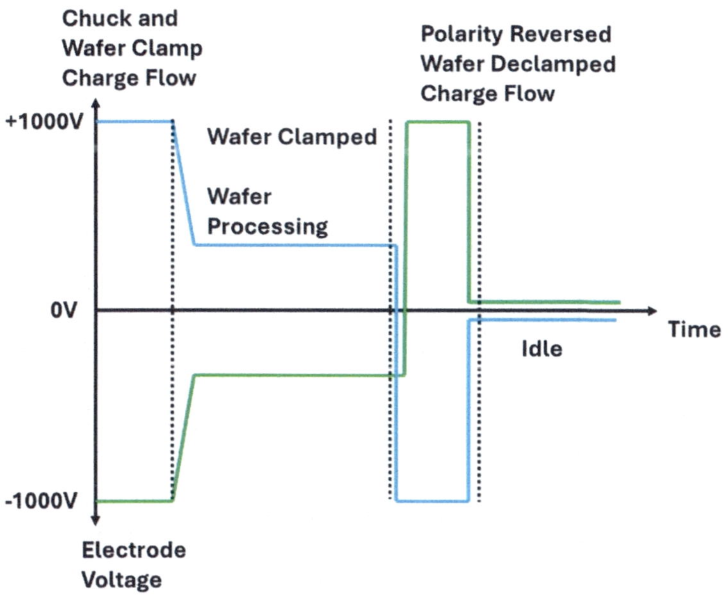

Fig. 9.29 Electrostatic chuck voltage waveform for an etch process using a bipolar chuck. The blue line is the voltage applied to the positive electrode, and the green line is the waveform applied to the negative electrode

The wafer can experience a voltage and current spike when electrostatically clamped or declamped, creating a process damage event. Therefore, the recipe must be optimized to prevent process damage. The waveform typically requires a slow, gentle voltage ramp to prevent process damage and optimization of the plasma start conditions. Figure 9.30 shows a histogram of voltages measured with a special dedicated nonvolatile memory cell test wafer (Siu et al., 2002). A nonoptimized recipe has a significant threshold voltage of around 15 volts, indicating severe plasma damage and an optimized recipe has a threshold voltage of around 2 volts, indicating significantly reduced process damage. The 2 V shift of the NVM structure cannot predict the actual threshold shift on a CMOS transistor. The plasma start conditions played a critical role in the damage, in this case, causing yield loss. The lift pins inadvertently touched the wafer during processing, worsening the problem.

The chucking process may change the wafer map of plasma damage failures by creating variation in potential across the back of the wafer and inducing a current in the wafer. Figure 9.31 shows a wafer map of a unipolar chuck recipe tunnel dielectric breakdown (Lee et al., 2012). The breakdown voltages are degraded on the edge of the wafers in the unipolar chuck group, while a bipolar chuck recipe had no edge of wafer breakdown problems. However, in this example, the chuck configuration is confounded with other process parameters changed by the chucking method, such as power.

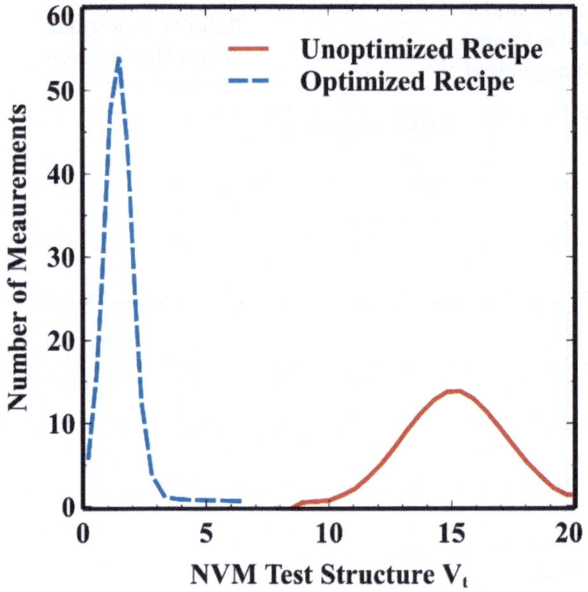

Fig. 9.30 Histogram showing the V_t of a nonvolatile test structure measured on the front of a wafer due to the chucking process for a nonoptimized recipe showing a high V_t shift and for optimized chucking recipe showing a low V_t shift

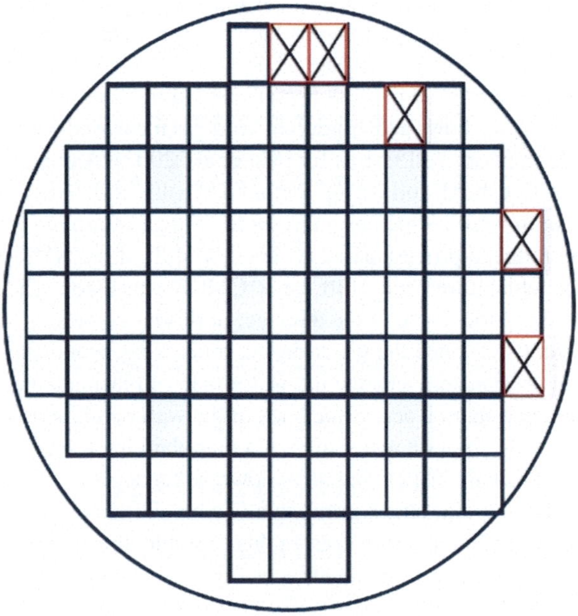

Fig. 9.31 Wafer map of a 6 nm low tunnel oxide breakdown in a 30 nm NAND flash process during a field oxide recess process in a self-aligned floating gate process showing a 3000 W process with a unipolar electrostatic chuck (ESC). A bipolar process didn't show an edge of wafer problem

9.10.5 Plasma Ignition and Extinction

Starting and stopping the plasma in a process tool is a critical step that can cause process damage. The ignition and extinction of the plasma are problematic steps in the recipe that control the plasma uniformity and avoid process damage. This transient step can cause process damage. Therefore, optimization of the recipe is required to minimize process damage. Figure 9.32 shows examples of voltage waveforms measured in situ for various plasma start-up conditions (Cheung, 1999, pg. 218).

9.10.6 Recipe Transients

Plasma transients can occur between recipe steps, in addition to the plasma ignition and termination described in Sect. 9.10.5. The plasma conditions, gas flow, etc., may change as the etch recipe transitions between recipe steps, creating a spike in the current and voltage across a stressed structure such as the gate dielectric. Additionally, the plasma conditions may undergo an abrupt transition when the etched surface conditions change. For example, near the endpoint, the etched pattern may transition from a solid conductor to a patterned line and space pattern where the only exposed conductor is the line edges. The pattern transition may induce a transient in the plasma conditions, inducing a spike in the current and voltage across the antenna. Figure 9.33 shows a simulation of the current flowing through a gate dielectric as the wafer transitions from a solid pattern to an etched

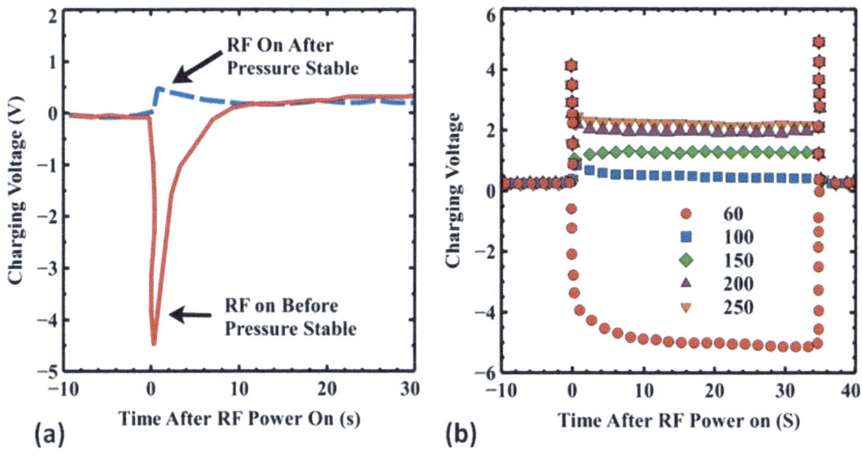

Fig. 9.32 Examples of waveforms for various plasma ignition and extinction conditions. (**a**) Differences in the ignition waveform with unstable chamber pressure and stable chamber pressure. (**b**) Differences in the plasma extinction waveform with different chamber pressures (mTorr) showing a voltage spike with decreased pressure

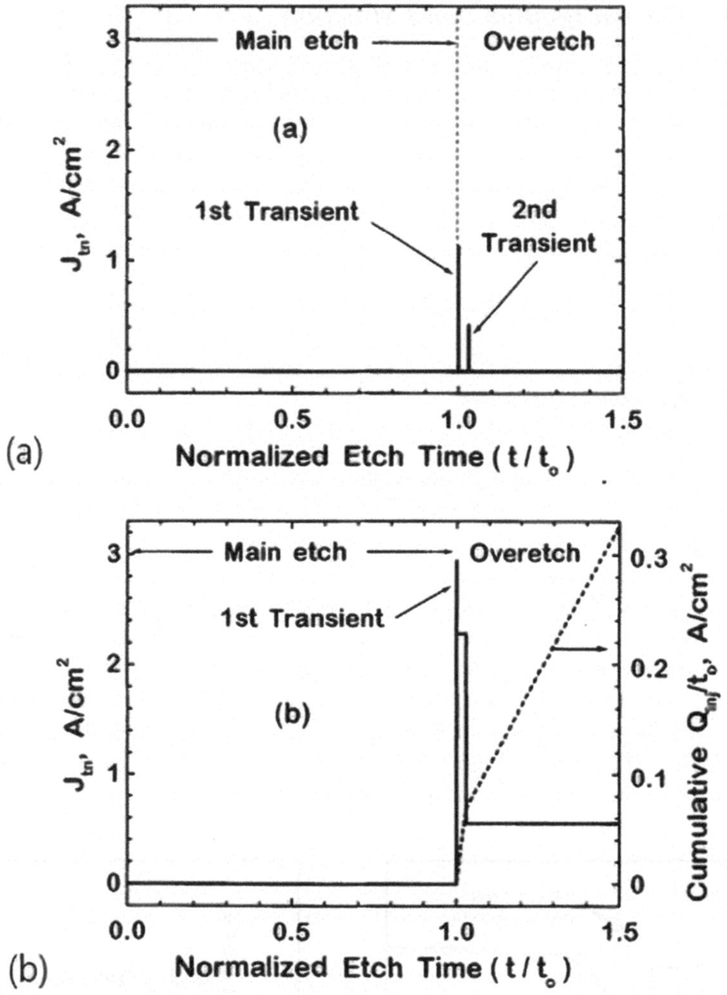

Fig. 9.33 Simulated current (J_{tn}) through a gate dielectric shows a transient spike at the point where the pattern transitions from a solid conductor to a patterned conductor. (**a**) With a floating substrate. (**b**) With a grounded substrate

pattern as the endpoint occurs (Shul & Pearton, 2000, pg. 294). Minimal current flows through the dielectric when there is a solid conductor before the endpoint. After the endpoint, the current can only flow through the pattern edges, creating a current spike. As shown in the figure, the current spike is much worse with a grounded substrate than with a floating substrate.

9.10.7 Magnetic Fields

Magnetic fields are often used in CMOS processing tools to increase the plasma density in some types of semiconductor fabrication equipment. The motion of individual particles in a plasma is controlled by the Lorentz force consisting of a magnetic field and an electric field (Shul & Pearton, 2000). The low-mass electron follows the magnetic field lines in a helical path, as shown in Fig. 9.34a (Nojiri, 2012). Some semiconductor tools contain a permanent magnet or temporary electrically controlled magnetic field. Figure 9.34b shows a sketch of an etch chamber with a permanent magnet as an example. The magnetic field causes the electrons to flow across the wafer, as shown in Fig. 9.34c. A magnetic field-driven chamber configuration can sometimes cause plasma density nonuniformity, playing a role in plasma damage.

Figure 9.35 shows an example of plasma damage caused by a magnetic field in a magnetron-reactive ion etch system (Nojiri, 2012). There is a considerable improvement in the flatband voltage shift and gate dielectric breakdown voltage without a

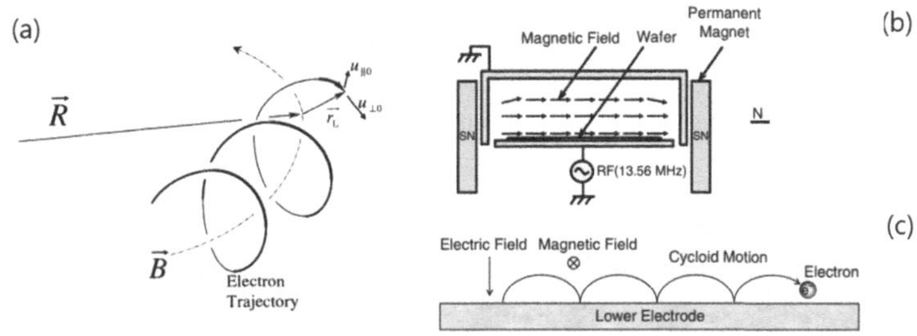

Fig. 9.34 Electron trajectory in a magnetic field. (**a**) Helical electron path in a magnetic field. (**b**) Sketch of a Tokyo Electron (TEL) dipole ring magnet (DRM) etch chamber showing the magnetic field lines across a wafer. (**c**) Electron path across the wafer surface in the TEL DRM

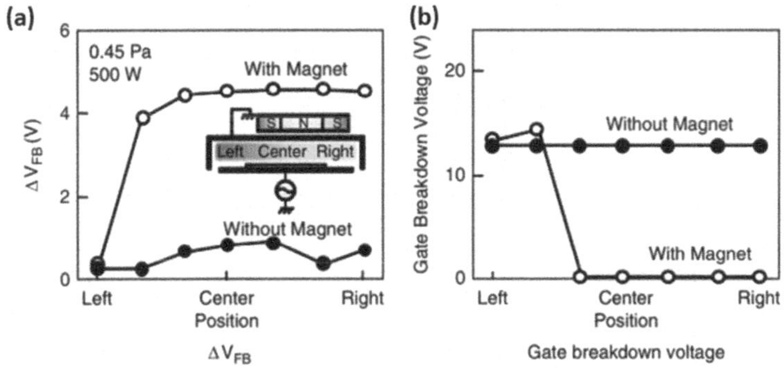

Fig. 9.35 Example of magnetic field-dependent plasma damage in a magnetron reactive ion etching system. (**a**) Chamber sketch and a plot of flat band voltage shift with and without a magnetic field as a position across the wafer. (**b**) Gate electric breakdown with and without a magnetic field as a function of wafer position

magnetic field. This type of chamber configuration is mostly obsolete partially due to plasma damage problems.

9.10.8 Plasma Damage in Exposed Silicon Regions

The layout pattern can also cause variation across the die and wafer. For example, an open region where the silicon surface is exposed can provide a direct path between the plasma and substrate (Shul & Pearton, 2000, pg 259). An example is shown in Fig. 9.36, where current flowing through the gate dielectric can follow a path through the substrate to an exposed silicon region. This local conduction can cause a process damage pattern that can follow the layout or distance to the exposed silicon regions. A nonuniform plasma between the gate and the open substrate can create conditions where the ion and electron fluxes between the two locations are different, aggravating the plasma damage.

Scribe line structures with exposed silicon placed between the chips on the wafer can result in processing differences affecting process damage. These types of structures are commonly used in scribe lines for alignment marks, metrology marks, etc. Sometimes, the underlying silicon can be directly exposed on the surface of the wafer, which is different from the chip patterning covering the wafer.

9.10.9 Wafer Bevel and Backside Substrate Effects

The wafer substrate can play a significant role in process damage (Carrere et al., 2003). The wafer can serve as a current conduction path between features on the wafer surface or between the CMOS wafer and the processing chamber. Figure 9.37 shows a wafer sketch showing the current flowing through the wafer between different metal lines. Current flow can also occur between the structures being damaged

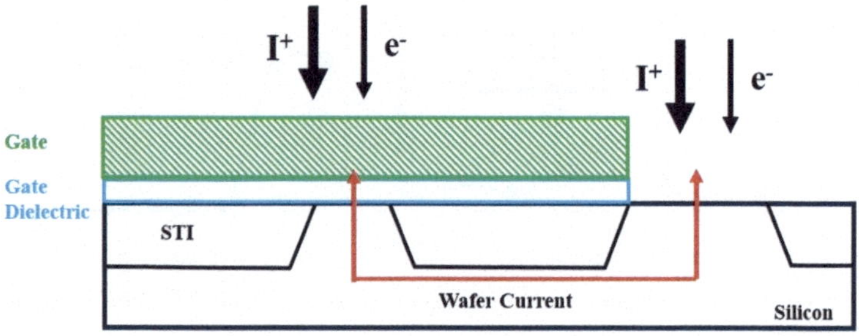

Fig. 9.36 Process cross-section sketch showing exposed silicon creating a conduction path from the gate antenna to an exposed substrate region. The plasma-induced wafer current flows from the gate dielectric to the exposed silicon region or in the reverse direction

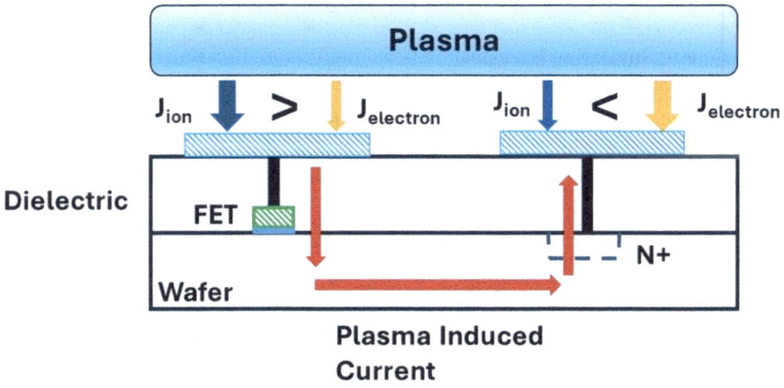

Fig. 9.37 Wafer sketch showing current flow through the substrate during plasma damage

and the processing chamber. The wafer's resistivity can be a crucial factor in process damage caused by across-wafer nonuniformity, with high-resistivity wafers having a significant voltage difference between the injection and collection regions. Low-resistivity substrates have the propensity for a large damaging current flow between locations.

The edge of the wafer and any exposed region where bare silicon is exposed can play a significant role in process damage. An exposed substrate region allows a direct path for the plasma to reach the wafer, allowing current to flow in the wafer that interacts with the CMOS structures, impacting process damage.

The edge of a wafer is manufactured with a bevel where the edge angles are optimized to avoid abrupt angles, reducing defects and fractures of the wafer edge. In addition, processing on the bevel is optimized to eliminate yield-killing defects peeling from this region of the wafer, where the wafer edge shape creates processing conditions much different from those of the flat region of the wafer. The difference in wafer processing can result in the wafer edge structures having exposed silicon, which affects process damage.

The backside of the wafer has different films deposited on it as the front side is processed. These films can electrically isolate the wafer backside from the wafer chuck holding it during processing. However, dielectric films on the backside of the wafer can create a capacitor between the wafer and the chuck. This inadvertent capacitor can modify the plasma processing equivalent circuit shown in Fig. 9.23, adding a capacitor in series to the circuit. The additional capacitor can adversely or positively affect process damage depending on the impact on the processing conditions. The presence of a dielectric on the backside of the wafer can change the plasma damage characteristics, improving or degrading the damage. The absence of films on the backside may increase the voltage and stress across the CMOS devices. Films on the backside may mitigate the damage by decreasing the current stress or increase the damage by shifting the plasma conditions. Figure 9.38 compares the damage caused by a via etch with different backside conditions by comparing the threshold voltage shift difference between an antenna structure and a reference

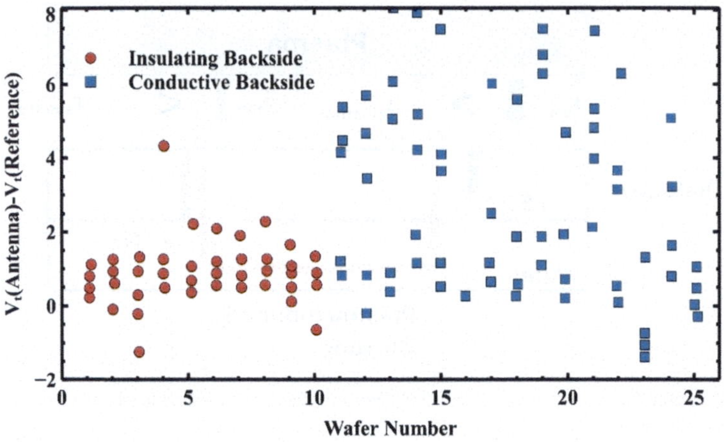

Fig. 9.38 Plot showing an increased V_t shift and variation for a conductive backside compared to an insulating backside during a via etch

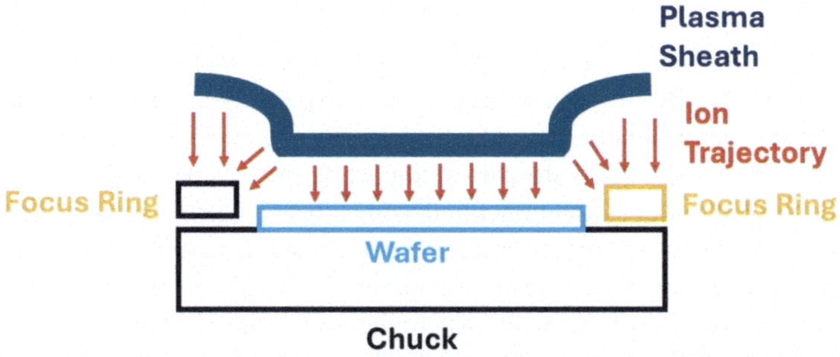

Fig. 9.39 Cross-section sketch showing a focus ring at the edge of the wafer used to modify the plasma for improved uniformity

structure (Cheung, 1999). The threshold voltage variation and shift may increase with a conductive backside compared to an insulating backside.

Modern processing tools are heavily optimized to enable a uniform process across the entire wafer. As a result, the wafer edge physically and electrically disrupts the plasma processes. Modifications to the processing chamber are performed to mitigate the wafer edge disruption. These modifications can improve process damage by making the plasma uniform but can cause process damage, specifically at the wafer edge. A chamber design may contain a focus ring or other techniques used to modify the plasma at the wafer edge to improve process uniformity. The difference in the plasma at the edge of the wafer can modify the plasma damage (Yang et al., 2017). Figure 9.39 shows a sketch of a plasma chamber with a focus ring at the edge of the wafer that improves uniformity by modifying the electric and magnetic behavior. The focus ring must minimize process damage while improving uniformity.

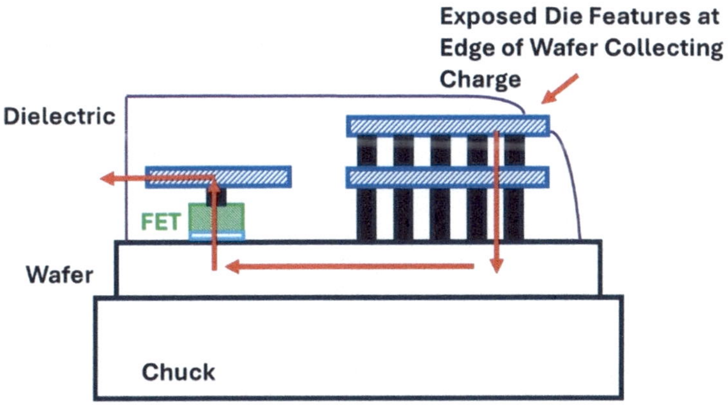

Fig. 9.40 Cross-section sketch of a wafer edge and bevel. The arrow points to the chip region attacked at the wafer edge

The wafer bevel at the edge of the wafer accumulates and loses films during processing in a somewhat erratic and uncontrolled manner as the process is optimized for processing the chip and not the wafer edge. As a result, the wafer bevel is sometimes electrically isolated from the plasma and tool when it is covered with dielectrics. Other times, the bevel is directly exposed to the plasma and chuck. As the wafer transitions between exposed and insulated edges, the wafer's electrical characteristics in the plasma are changed, potentially increasing plasma damage. In addition, electrical currents may intermittently flow through the wafer when this transition occurs. As a result, the bevel conditions during processing may significantly impact process damage and potentially create a worst-case arcing condition. Figure 9.40 shows a sketch of the bevel on the edge of a wafer (Ma, 2003). The metal structures on the right side of the sketch can be intermittently covered and uncovered, creating a transient damage condition where current flows through the wafer.

Lithography tools use a scanner or stepper job (program) to control the location on the wafer where the chips are placed. The stepper job controls the placement of each reticle field on the wafer. The job is typically optimized to avoid printing partial chips on the wafer and reduce the number of locations where a partial chip exists at the bevel. This optimization improves particle problems and reduces the occurrence of problem areas like the one shown in Fig. 9.40.

9.10.10 Chamber Cleaning

The chamber in a plasma processing system can inadvertently deposit materials and byproducts inside the chamber as many wafers are processed. The unintended deposition increases as each wafer is processed, increasing the cumulative build-up

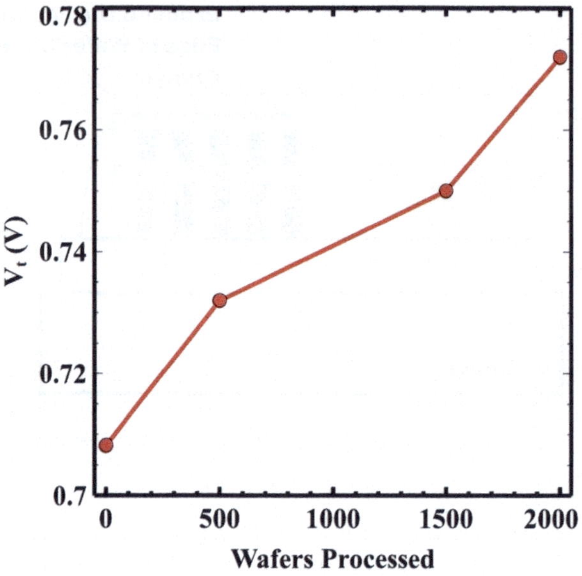

Fig. 9.41 Plot showing a pmos threshold voltage shift as a function of the number of wafers processed

inside the chamber. Control of the deposits in a chamber requires periodic maintenance to clean the processing chamber to remove the deposited materials. The clean can be performed manually by opening the chamber or in situ by cleaning the chamber with a plasma step. Depending on the cleaning requirements, the in situ clean can be performed with a product wafer, dummy wafer, or no wafer in the chamber. The processing conditions may change as material builds up in the chamber, affecting process damage. Figure 9.41 shows an example of processing damage increasing with the number of wafers processed between cleans in a metal etch chamber, showing a threshold voltage shift and a large increase in variability (Krishnan et al., 1996).

9.10.11 Topology Effects

Plasma damage interacts with structures on the wafer surface. Three-dimensional structures can have significantly different plasma damage from structures with no topology. Some CMOS technologies, such as finfets, DRAM, and three-dimensional NAND, extensively use three-dimensional structures that tend to aggravate plasma damage conditions. For example, Fig. 9.42 compares the plasma damage during a plasma nitridation process for a planar transistor and a three-dimensional recessed-gate transistor. The gate dielectric of the 3D transistor is heavily degraded by a low-pressure process, resulting in a low charge to breakdown (Cho et al., 2007).

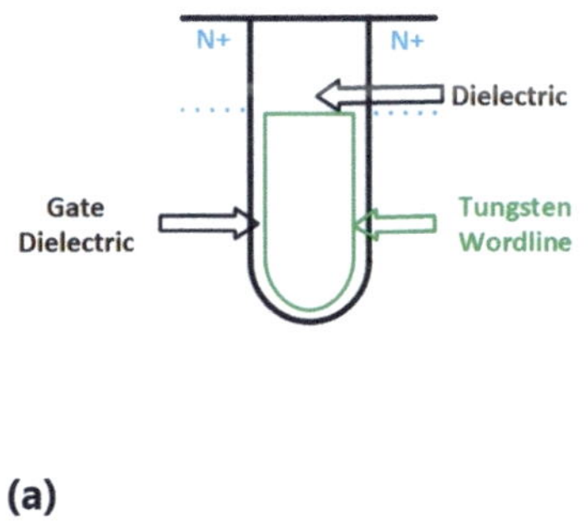

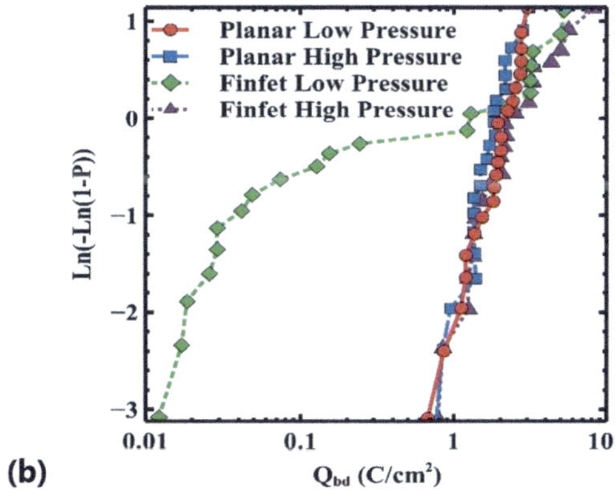

Fig. 9.42 Comparison of plasma damage between a planar structure and a three-dimensional recessed gate structures during the gate dielectric plasma nitridation (used in DRAM). (**a**) Sketch of a recessed gate structure. (**b**) Probability plot of charge to breakdown (QBD) comparing a planar and a 3D transistor for two different pressure plasma nitridation processes

Plasma processing in 3D structures can differ substantially from that of planar structures, with ions and electrons reflecting off surfaces, different plasma densities inside 3D features, etc. Processing finfets (see Fig. 9.58), DRAMs, 3D NAND, etc., can be challenging with the added propensity for damage because of the tall 3D structure (see Fig. 9.62 and Chap. 3).

9.11 Plasma Film Deposition

Plasma processing is commonly used for film deposition because of its high throughput and other capabilities. The surface of the wafer is exposed to the plasma during the deposition. Process damage can occur due to high-energy light or wafer charging (Lee et al., 2004). There is a significant difference in process damage between the deposition of conductors and that of insulators. Plasma film deposition is ubiquitous throughout a CMOS process, and is commonly used for the deposition of dielectrics and less frequently for conductors. Table 9.3 shows a partial list of films known to cause process damage when deposited with a plasma process.

9.11.1 Plasma Dielectric Deposition

For dielectrics, damage is more likely to occur when processing begins and structures on the wafer surface are exposed. After some of the insulating film is deposited, the film protects the structure from direct charge as the insulator prevents current flow onto the underlying structures. As the dielectric film deposition

Table 9.3 Partial list of common plasma-deposited films known to cause plasma damage

Silicon nitrides
Silicon oxynitrides
Tetraethylorthosilicate glass (TEOS)
Undoped silicate glass (USG)
Phosphorous silicate glass (PSG)
Boron phosphorus silicate glass (BPSG)
Carbon
Stressor films
Low-k dielectrics
Antireflective coatings
Polysilicon
Contact metallization

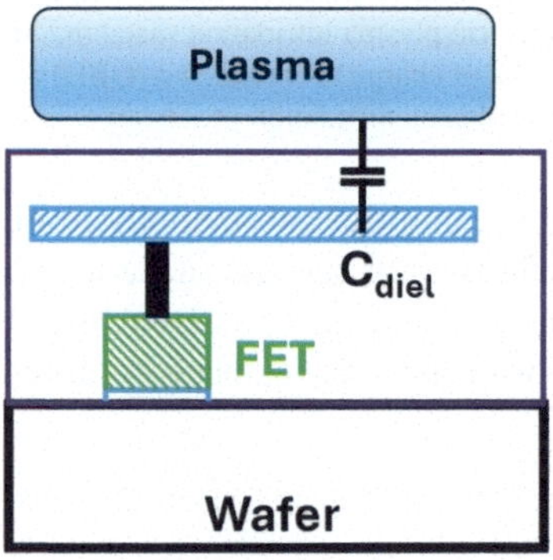

Fig. 9.43 Sketch of dielectric deposition on top of a metal line and equivalent circuit showing coupling through the dielectric layer into the metal line

progresses, the underlying conductors are covered with the increasing dielectric film, and the plasma no longer directly contacts the metal layer (Cheung, 1999). The plasma capacitively couples to the metal layer through the deposited dielectric (C_{diel}), creating the potential for plasma damage even after the conductor is covered, as charge is coupled into the metal layer and associated antenna, driving current into the gate dielectric. Figure 9.43 shows a sketch of a metal line with a dielectric deposited on it and an equivalent circuit with the additional capacitive coupling.

Coupling through the deposited dielectric can generate plasma damage throughout the deposition process, continuing to cause damage with increasing thickness. Figure 9.44 shows a plot of a hot carrier reliability test plotted against the deposited dielectric thickness. The reliability is degraded with continued deposition. Several potential causes for the degradation include capacitive coupling, high-energy photons, increased photoconduction, or barrier lowering (Cheung, 2001).

A large spectrum of plasma deposition process options are used in CMOS processing. Consequently, there can be a considerable variation in process damage across the potential processes for similar applications. Figure 9.45 shows an

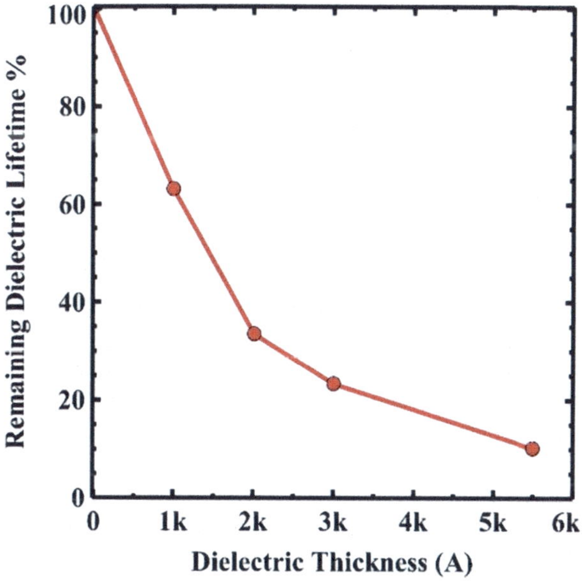

Fig. 9.44 Remaining hot carrier lifetime plotted against deposited plasma enhanced tetraethyl orthosilicate (TEOS) glass thickness. The plot shows dielectric damage during a dielectric deposition, even though the conductor is covered with a thick dielectric

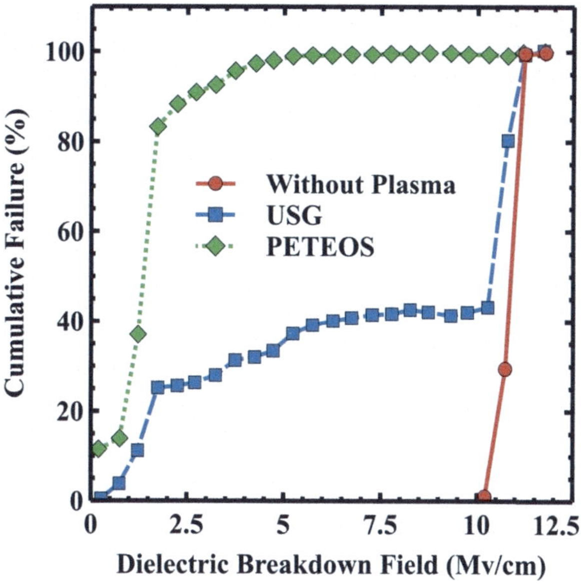

Fig. 9.45 Probability plot of gate dielectric wearout with different intermetal dielectric deposition processes. PE-TEOS is a plasma deposited tetraethylorthosilicate based process. USG is a plasma deposited undoped silicate glass. The rightmost curve is the control group without film deposition and plasma exposure

example of two different dielectric deposition processes with a significant difference in plasma damage results (Cheung, 1999, pg. 174). Optimization of the process, tool, etc., is commonly required to correct plasma damage during processing.

9.11.2 Plasma Conductor Deposition

There are very few publications on plasma damage during plasma conductor deposition. During deposition, a conducting film covers the wafer throughout the deposition process, starting with the initial few atomic layers. The conductors are in electrical contact with the underlying structure throughout the deposition process and can prevent charge accumulation, which can cause plasma damage. In addition, plasma processing typically involves less energy at the wafer surface than does physical vapor deposition (PVD) or sputtering. These factors make plasma damage less likely for plasma conductor deposition. Conductors typically block light, making high-energy light damage less likely for plasma conductor deposition.

9.11.3 High-Density Plasma Deposition

Some plasma film deposition processes are optimized for good film step coverage and gap-fill. For example, high-density plasma (HDP) film deposition is used in some process modules such as shallow trench isolation (STI) or subtractive metallization, where good gap-fill characteristics are required (Wolf, 2002). An HDP process used for gap-fill applications includes a sputtering component that facets the sharp corners of the dielectric to expand the opening, enabling better gap-fill. The HDP process typically oscillates between deposition and faceting, potentially exposing the metal corners many times. Additionally, coupling through the dielectric during processing can worsen the damage. Process damage for an HDP process can occur at the beginning of the deposition, where metallization is directly exposed to the plasma, or at any time during the process, where the corners of the feature are exposed, creating a path for the charge to flow into the corners of the conductor (Ackaert et al., 2000). Figure 9.46 shows a sketch of an HDP gap-filling progression where the corners of the metal may become exposed.

9.12 Sputter Deposition

Physical deposition techniques known as sputter or physical vapor deposition (PVD) are used in specific CMOS process applications like the deposition of metals. The sputtering process is a physical process using kinetic energy imparted into noble gas ions to knock ions off of a target and onto the wafer surface, depositing a

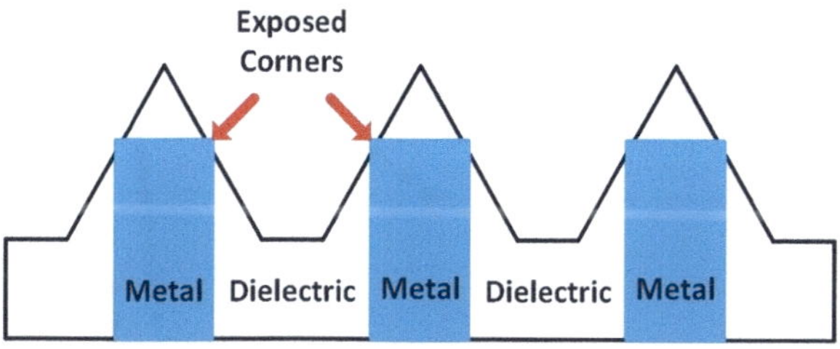

Fig. 9.46 Sketch showing the progression of dielectric deposition in a high-density plasma dielectric deposition using a sputtering component for gap fill between metal lines

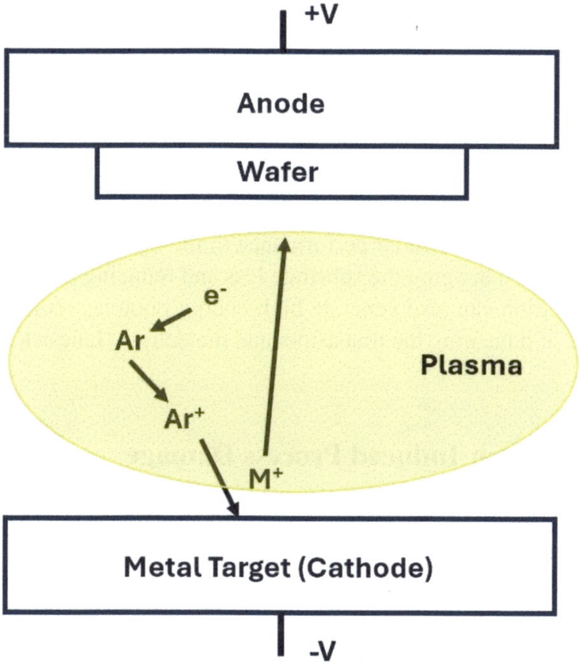

Fig. 9.47 Sketch showing the operation of a sputter deposition system

film on the wafer surface. Figure 9.47 shows a sketch of sputter deposition. The metal and noble gas ions impact the wafer surface, transferring charge to the wafer and may cause process damage.

Energetic ions impacting the wafer surface can create defects, including traps and interface states that degrade oxide quality. Figure 9.48 compares PVD gate metal deposition with a low-pressure chemical vapor deposition (LPCVD) polysilicon deposition (Ushiki et al., 1998). The LPCVD process uses a high-temperature thermal process to deposit films without plasma components. The charge to breakdown, which is a measure of dielectric quality, is degraded with PVD. Changing the

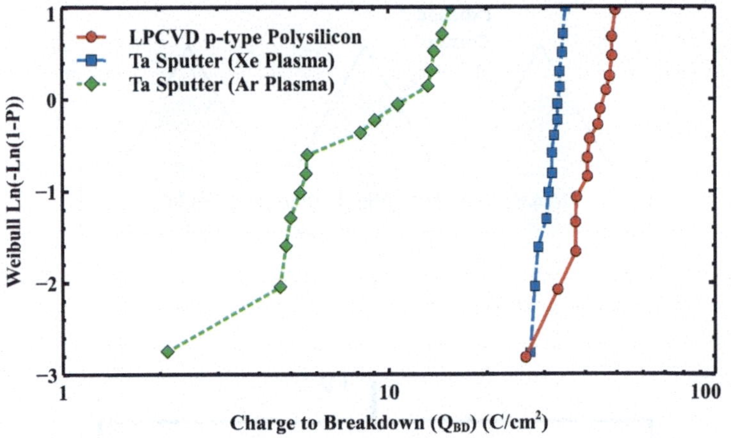

Fig. 9.48 Probability plot of gate dielectric charge to breakdown (QBD) showing a comparison of a sputtered gate metallization with a low-pressure chemical vapor deposition (LPCVD) polysilicon

sputtering ion from argon to xenon has improved QBD considerably. The heavier xenon atom can provide a similar performance to the lighter argon atom using less accelerating energy, damaging the substrate less and reducing the damage.

Sputter deposition can also generate high-energy photons, including secondary X-rays capable of damaging the transistors and dielectrics (Janesick, 2001).

9.13 Plasma Etch-Induced Process Damage

9.13.1 Introduction

Plasma etching is a widespread cause of process damage. The wafer is bombarded with ions, electrons, and radicals, which can charge structures on the wafer as described in Sect. 9.10. Other sources of process damage from plasma etching include high-energy photon damage, kinetic damage, and the knock-on of atoms into the wafer from the plasma or materials on the wafer surface. Figure 9.49 shows a sketch highlighting the different ways plasma etching can damage a wafer. Modern CMOS processes require high-energy plasmas to achieve straight profiles in the high-aspect ratio features etched into the wafer. High-energy processes can achieve high throughput, resulting in a lower wafer cost.

Plasma etching is performed many times throughout a CMOS manufacturing flow, increasing the likelihood of damage (Shul & Pearton, 2000). Many etch steps routinely cause plasma and process damage, including isolation-related etches, gate-related etches, contact and via etches, metal etches, bond pad etches, etc. Most electrical damage mechanisms described previously can be caused by plasma etching.

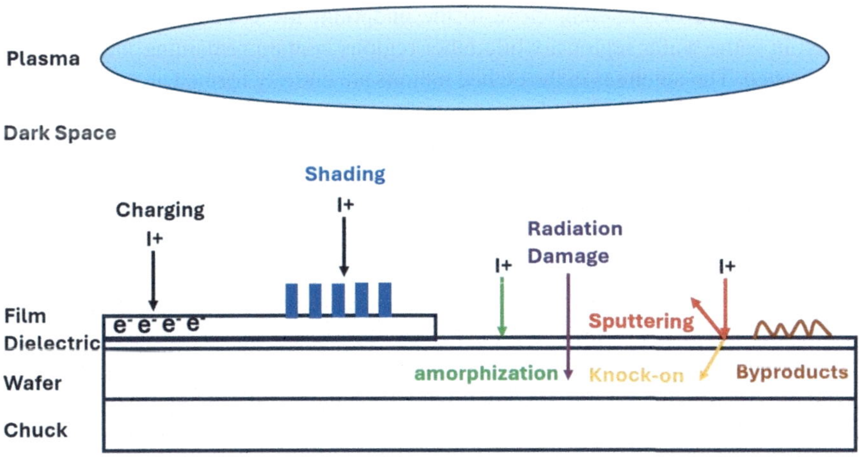

Fig. 9.49 Sketch showing possible sources of process and plasma damage during plasma etch

The damage mechanisms described in Fig. 9.49 include charging damage from ion and electron charging of the wafer surface and high-energy photon damage (Ma, 2004). Residuals from the masking materials and polymer deposition from the etch byproducts can contaminate the wafer surface. The masking materials may contain contaminants like mobile ions. The ion kinetic energy can sputter atoms from the surface of the wafer, contaminating nearby structures. Additionally, surface atoms can be knocked onto locations deeper into the substrate.

9.13.2 Overetch

Overetch is an additional etch time added to the etch recipe, which etches for an additional amount of time beyond the point where the etch nominally clears the material from the wafer (Wang, 2004). The additional time provides a margin to ensure that the complete pattern is etched into the entire wafer, accounting for variation across patterns, time, etchers, etc. Finally, etch recipes terminate by commonly using a fixed time or an endpoint that detects the changing conditions of the plasma as the wafer nears completion.

If a conductor is etched, the entire wafer is covered with a conductor at the start of the etch step. During the initial segment of the etching process, the entire wafer serves as a giant connected antenna, limiting damage. The charging current can flow through all dielectrics capable of conducting current under the charging conditions. The charge may flow through all conducting regions of the wafer in a distributed manner, or if a low resistance defective path exists, the current can flow through the defect, discharging the entire wafer charge energy locally with a high likelihood of damage. In addition, any defects with weak structures may discharge the entire wafer's charge, creating a locally damaged region.

Near the end of the etch, close to the endpoint, the conductor is completely etched in some wafer regions, while other regions contain remaining material still to be etched. The antennas in the etched regions are entirely formed and can amplify the plasma charging. The fully formed antennas can collect charge only on the edges as the top of the antenna is covered with nonconducting resist or other masking material preventing charging from the top. The edges may not be vertical before the etch completion, increasing the collection area of the conductor sidewall.

The antennas etched into a conductor vary dynamically with time, changing area, or perimeter. As a result, the dynamic antenna can differ from the final pattern. This variation can cause damage that differs from the straightforward antenna ratio calculation in nonobvious ways.

The process damage during the overetch can also be a function of the design rules, as wide-open areas tend to overetch faster, increasing the degree of process damage. Once the entire wafer is etched, all antennas are formed, and process damage from the patterned antenna occurs over the entire wafer on all the structures.

A simple pattern of repeating lines and spaces may have significant etch variation between lines due to these effects, as shown in Fig. 9.50 based on computer simulations (Shul & Pearton, 2000). The line and space patterns are shown with

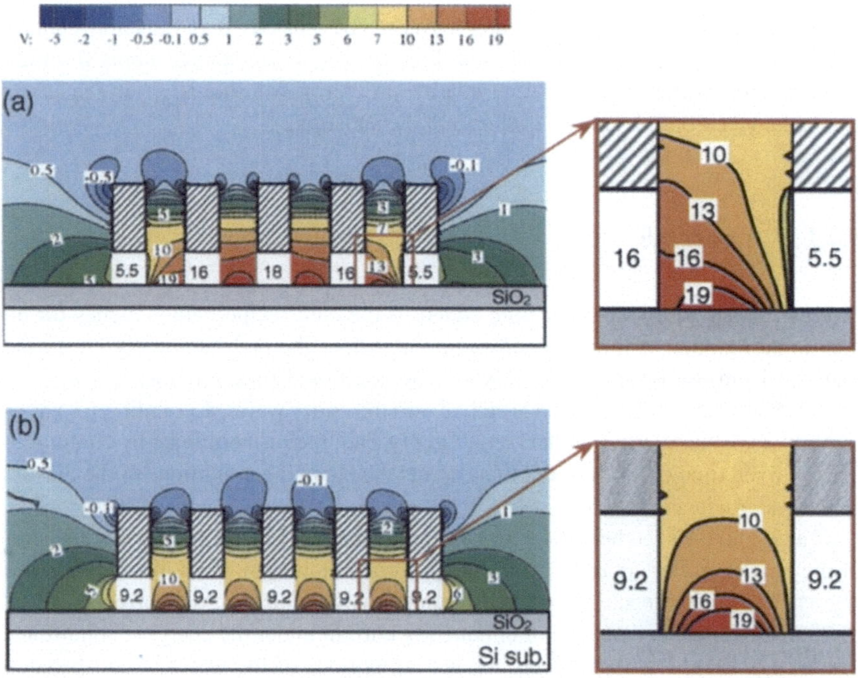

Fig. 9.50 Cross-section cartoon showing the results of a simulation of a plasma etch of lines and spaces. (**a**) Simulation with the lines electrically separated and isolated. (**b**) Simulation with the lines electrically connected. The equipotential contour lines and colors represent the variation in the voltage potential

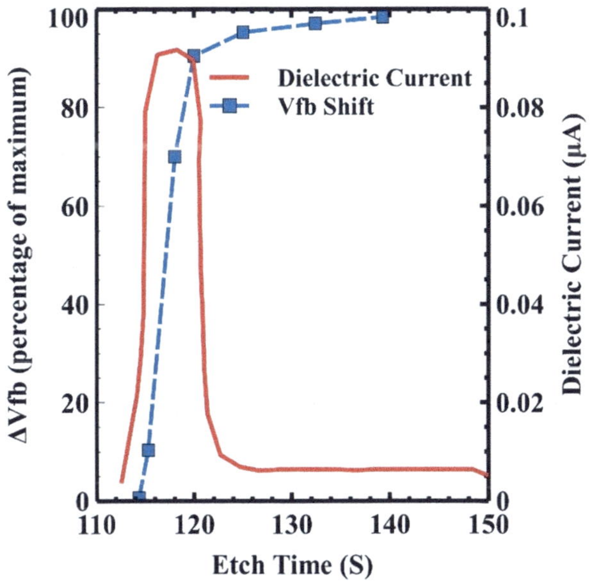

Fig. 9.51 Plasma damage measured as a function of capacitor flat band voltage shift plotted against the etch time (dashed line). Plasma damage begins near the end point, where the etch completes at around 115 seconds. The solid line shows a simulated charging dielectric current. The measurements are made through capacitance-voltage analysis of mos capacitors by exposing different wafers to different etch times

equipotential contour lines and colors. The first space from each edge has a very different asymmetric potential distribution compared to the center lines. Furthermore, the potential lines change when the metal lines are connected or unconnected, which can be observed by comparing Fig. 9.50a and b. These subtle local variations in the potential due to layout differences can cause process damage to varying layout.

Figure 9.51 shows in situ damage measurements as a function of the etch recipe step and time (Fang et al., 1992). The damage begins as the etched material is cleared, exposing the structure sidewall, and continues through the overetch. The Fowler–Nordheim current is shown as it spikes at the start of the overetch and decays to a steady state value.

9.13.3 *Layout and Topology Dependent Process Damage*

Plasma etch processes have a natural variation in the local etch rate which is dependent on the space between features and the pattern density. The difference in behavior between ions and electrons at the wafer surface contributes to this phenomenon. The electrons impact the wafer over an extensive range of angles due to its low mass, repulsion by the sheath, and negative charge trapped in the masking layer. The ions tend to be highly collimated, impacting the wafer over a narrow range of angles. The electron and ion difference causes a difference in the etch performance in narrow regions.

Figure 9.52 is a sketch comparing the impact angle difference between electrons and ions (Shul & Pearton, 2000). The low-mass, high-energy electrons with a large mean free path can impact the surface of the wafer at different angles. These etch differences cause a difference in the etch performance, and plasma damage creates a difference in process damage. Figure 9.52d compares the angular distributions of electrons and ions based on a simulation.

The ions impact the wafer over a narrow range, while electrons impact the wafer at all angles. The ions impact the wafer unidirectionally for an open region of the wafer, and the electrons impact the wafer isotropically, as shown in Fig. 9.52a. Figure 9.52b shows the ion and electron impact on the edge of a patterned feature. The electrons impact the etched regions over a 90-degree angle, while the ions impact vertically. Figure 9.52c shows the ion and electron impact in a small space between features, with electrons and ions both impacting the wafer vertically. The electrons tend to be trapped in the resist layer, while the ions become trapped on the wafer surface. These differences in impact angles create a difference in etch performance and damage. The ions and electrons trapped in the features bend the ion beam and cause reflections, modifying the charging and patterning. Figure 9.52d shows the angular distributions of electrons and ions in a plasma system.

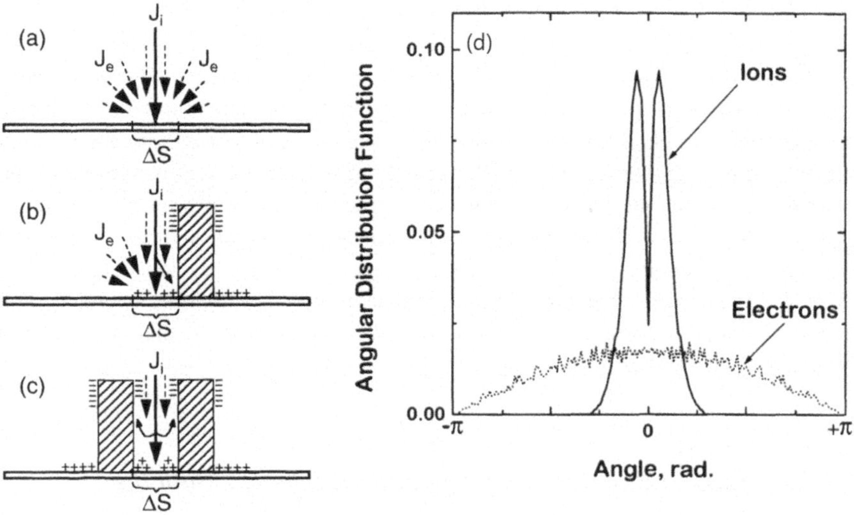

Fig. 9.52 Sketch showing the difference in electron (J_e) and ion angle (J_i). (**a**) Angles of ions and electrons impacting a flat wafer. (**b**) Angles of ions and electrons on the side of a feature. (**c**) Angles of ions and electrons in a small space. Dashed-line arrows show the electron flux and solid-line arrows depict the ion flux. The decreasing length of the dashed lines indicates a smaller electron flux. The symbols + and – depict where electrons and ions become trapped on the wafer surface and in the resist. (**d**) Plot showing the angle distribution and ions and electrons

This effect results in a difference in antenna effects with variation in layout. Figure 9.53 shows the etch time required to clear different spaces (Cheung, 1999). Smaller spaces require longer etch times as fewer charged species impact the wafer over a narrow range of angles. Conversely, wider spaces allow for an extensive range of angles, etching faster. As a result, the structures with larger spaces undergo longer over-etches and increased plasma damage. Figure 9.54 shows an example plot of the etch time required as a function of the space between lines, showing the longer etch needed for smaller spaces (Simon et al., 2000). Micro-trenching

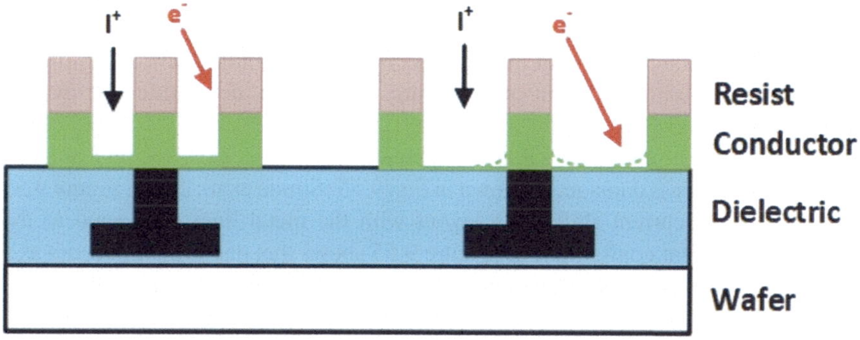

Fig. 9.53 Sketch showing etch differences between a large space and s small space due to electron and ion angle differences

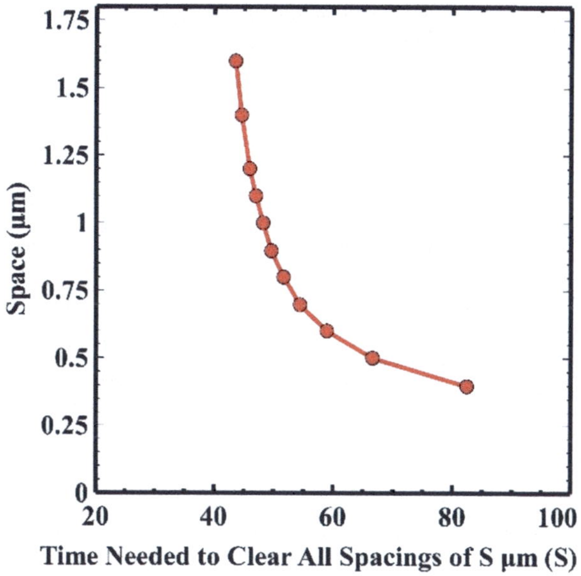

Fig. 9.54 Example plot showing the increase in etch time required to clear a space. Smaller spaces require increased etch time

sometimes occurs in small features as ions and electrons are reflected off the sidewall of the feature. This effect can change the pattern dependence on plasma damage.

This effect can vary widely as a function of the antenna layout, changing with antenna pitch, connections between lines, and variations in topology. As described in Sect. 9.13.3, the antenna geometry can vary from the start of the damaging process step to the end, particularly during a subtractive conductor etch step. For example, on a patterned conductor structure, the etch time for a subtractive metal etch may be a function of the metal space, as shown in Fig. 9.54. The metal etched in wide spaces is removed entirely first, while the metal in small spaces is removed entirely later. Therefore, the antenna area and edge change as different features complete the etch step with the removal of the conductor being etched. The worst-case antenna layout may not occur at a minimum pitch, and multiple test structures with different pitches are needed to determine the worst-case layout.

Additionally, the conducting lines do not have to be connected to the final antenna to impact the process damage, as shown in Fig. 9.55 (Simon et al., 2000). Figure 9.55 shows the drive current shift for a layout with the metal lines connected to the antenna in different configurations. Figure 9.55 shows that damage still occurs as a function of space even though the metal lines are not connected to the final antenna pattern. The isolated unconnected patterns form temporary island antennas during the etch process, as shown in Fig. 9.56 (Simon et al., 2000).

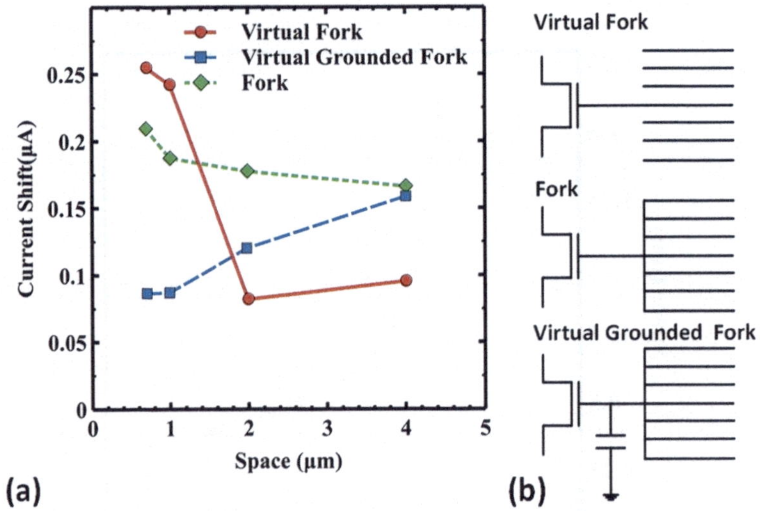

Fig. 9.55 (**a**) Plot of the drive current shift as a function of space between metal fingers compared to a transistor with no antenna for various antenna structures. (**b**) Sketch showing the structures compared. The virtual fork has no connection between the fingers. The virtual grounded fork has a capacitor connected to ground. The fork structure has connected fingers

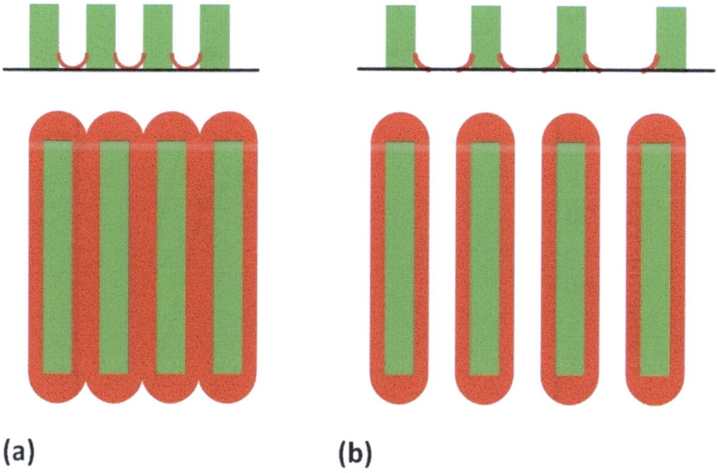

Fig. 9.56 A top down sketch of the dynamic antenna shows an isolated island antenna formed during the etch process. (**a**) Temporary antenna in a tight pitch formed by remaining material to be etched. (**b**) Only a single line antenna is formed in a larger pitch

9.13.4 Etch Rate Process Damage

CMOS process recipes are generally optimized for maximum efficiency and throughput. The etch rate of the material is increased for efficiency. Unfortunately, high throughput creates the worst-case conditions for process damage, as the power, bias, etc., are increased to achieve high efficiency, also increasing the current flow of the damaging charge onto the wafer. Additionally, fast processing leaves less time for the surface charge to leak away, increasing the damage probability. The etch rate can often be correlated with process damage, giving a strong clue pointing to the root cause of related problems. Figure 9.57 plots yield loss for a capacitor with a large via antenna increasing with increasing etch rate (Cheung, 1999).

9.13.5 Spacer Etch Damage

CMOS processes commonly use dielectric spacers to offset source-drain implants from the gate to improve electrical characteristics and reliability. The spacer offsets edge damage from the gate edge during the spacer etch. The spacers used to optimize the CMOS transistor electrical performance require optimum etch processing. The gate, source, and drain may be directly exposed to the plasma during the spacer etch, creating a high probability of plasma damage. The spacer etch may directly attack the semiconductor substrate, which can cause physical damage to the source and drain regions of the transistor (see Sect. 9.10.8). The spacer etch process is

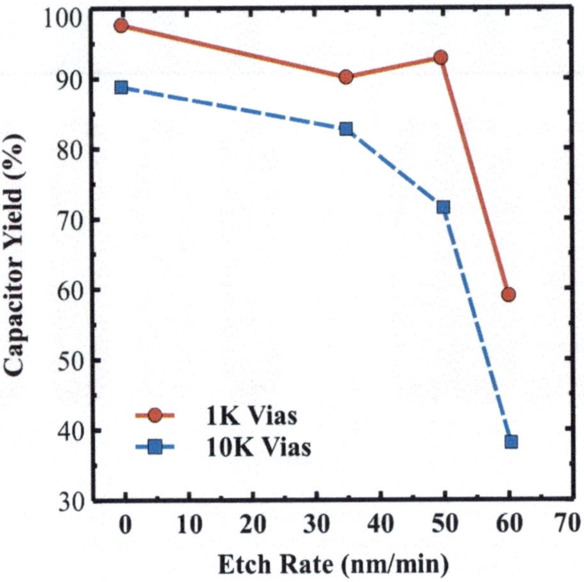

Fig. 9.57 Plot showing decreasing capacitor yield with increasing via etch rates for antenna structures with a large number of vias

critical as it must remove the dielectric spacer anisotropically with minimal silicon loss and damage. The overetch at the end of the spacer etch recipe is a critical step that must be optimized to minimize plasma damage.

The metal gate is commonly covered with a dielectric, which may protect it from process damage. Process damage in this type of structure commonly occurs in the source-drain regions. Damage to structures without a dielectric on top of the gate can also suffer damage in the gate region. Figure 9.58 shows sketches of a planar CMOS structure and a finfet CMOS structure showing the spacers after the gate is processed and formed (in green). The transistor in this section of the process flow is sensitive to damage as the transistor is open to damage.

For CMOS finfet processes, a replacement gate flow (gate-last) is typically used to be compatible with high-k dielectrics, optimize the gate metallurgy, etc. (Lai et al., 2009). First, a dummy gate is formed, and the spacer is deposited and formed on the dummy gate. For finfets, the spacer etch must be longer to remove the material along the fin sidewall. The dummy gate is subsequently removed, and a gate dielectric and metal gate are deposited in the open crevice left by the dummy gate removal. Figure 9.59 shows a cartoon of a finfet gate formed with a replacement gate process flow (Gulshan, 2019). A replacement gate flow has multiple steps where damage can occur, including forming the spacer on the dummy gate, removing the dummy gate, replacing the gate with metal, etc.

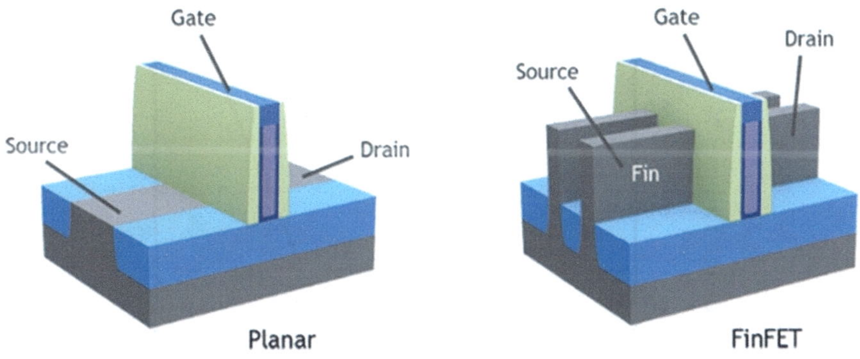

Fig. 9.58 Sketch showing a planar transistor and a finfet

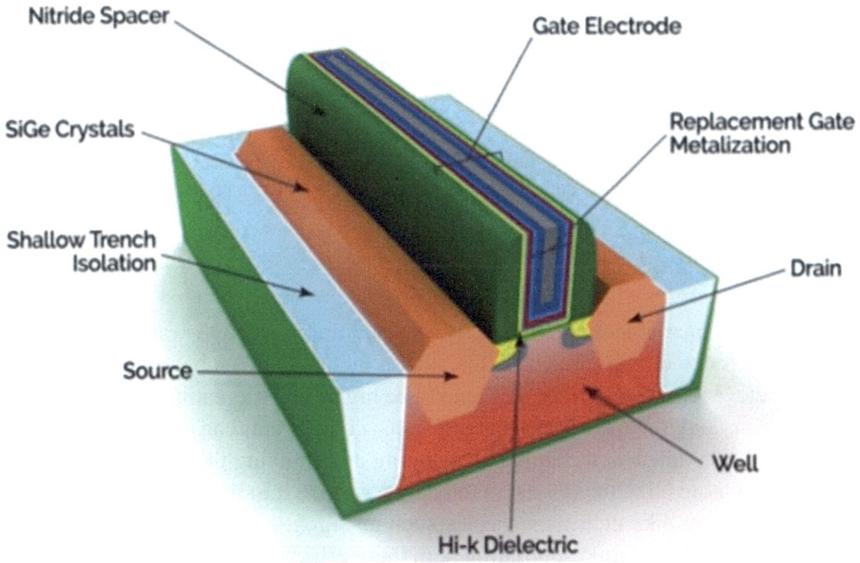

Fig. 9.59 Sketch showing a replacement gate finfet

Figure 9.60 shows the gate leakage failure rate at a specific voltage and current as a function of the spacer over etch (Gabriel & Weling, 1994). A dielectric is deposited on top of the gate metal before gate patterning, and spacer etch. The purpose of this deposited layer is to protect the gate metal from plasma charging during the spacer etch. Typically, a thicker layer on the gate protects better against plasma damage.

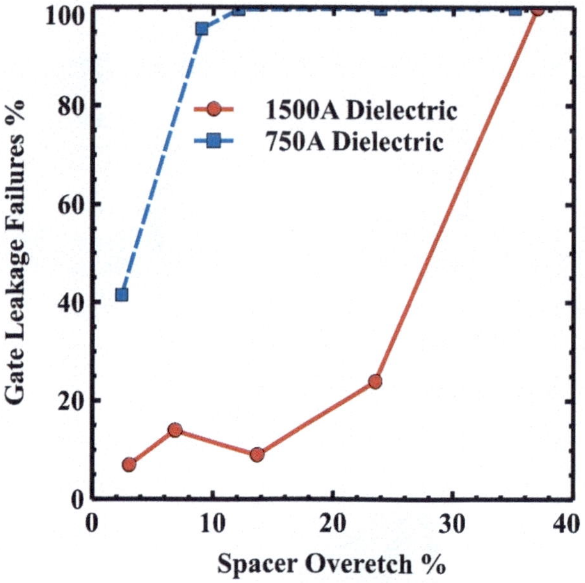

Fig. 9.60 Gate dielectric leakage failure percentage as a function of spacer over etch percentage for different dielectric thickness deposited on top of the gate metal as a protective layer

9.13.6 Surface Etch Damage

The silicon surface is inevitably directly exposed during CMOS wafer fabrication. Plasma etch steps directly touching the silicon include gate etch, spacer etch, contact etch, etc. Modern plasma etches tend to be high energy, creating highly accelerated electrons and ions. The high-energy plasma components can damage the silicon surface through various mechanisms, including amorphizing the crystalline lattice, knocking on atoms like oxygen from the native surface into the wafer, and implanting foreign atoms like etch ions species such as argon, halogens, etc. Figure 9.61 shows a molecular dynamic simulation with these effects (Nakakubo, 2015, pg. 33).

The wafer silicon surface becomes structurally damaged and contaminated by the etch process, impacting the CMOS electrical parameters. CMOS processes are optimized to minimize or eliminate surface damage. The etch recipe can operate with reduced power as the etch intersects the surface to minimize damage. Wet or dry cleans are added to gently remove the damaged layers.

9.13.7 Contact and Via Etches

The dielectric etch for contacts and vias connecting metal layers can also cause plasma damage. A dielectric typically has good isolation properties, protecting any underlying metal or semiconductor layer until all the dielectric is removed from the

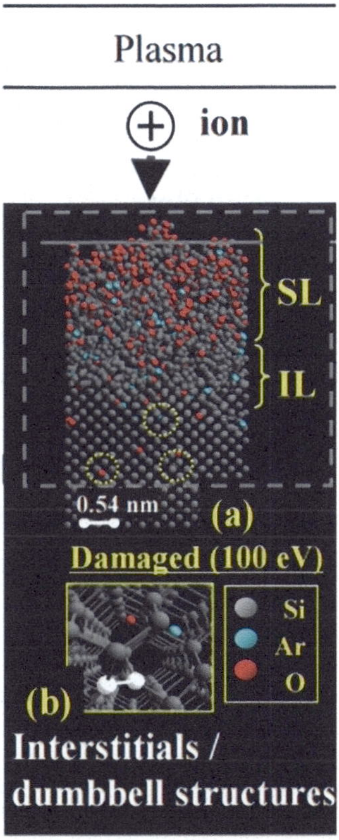

Fig. 9.61 (**a**) Molecular dynamic simulation showing silicon surface damage caused by exposure to an Ar plasma showing an etched surface where the surface layer (SL) and interfacial layers (IL). (**b**) Interstitial contamination with argon, silicon, and oxygen. Yellow circles show ions knocked deep into the substrate

hole being etched. At this point, the ions in the etch can cause process damage, beginning as the ions from the plasma contact the conductor at the bottom of the contact hole. The amount of damage is strongly dependent on the overetch time. Additionally, there can be a capacitive coupling component of the damage similar to the one described in Sect. 9.11.1 that can cause a voltage to be coupled onto the underlying conductor before the plasma contacts the conductor. This effect is typically worse for structures with many contacts or vias.

The process damage increases with increased numbers of vias or contacts connected to a common conductor. Each via or contact collects charge and transfers it to the underlying metal, eventually reaching the transistor after being transmitted through the underlying metal layers. Therefore, design rules sometimes limit the maximum number of contacts and vias connected to a common metal line to minimize damage.

The conductors in a CMOS process can be at different distances from the semiconductor surface, requiring substantial amounts of overetch to remove the dielectric for the deepest contact. For example, the same contact level may be used to contact both the gate conductor and the substrate. However, the gate contacts are usually located farther from the substrate than the substrate contacts. As a result, the

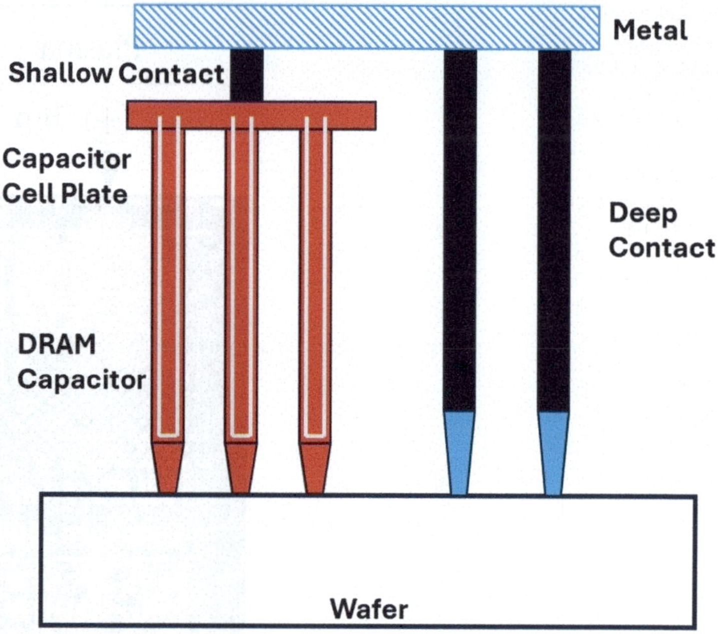

Fig. 9.62 Cross-section of a DRAM process flow showing a contact used to connect both the bitline (BL) and the cell plate

gate contact undergoes plasma damage while the etch continues to remove the additional material necessary to clear the substrate contact, impacting the health of the critical gate dielectric.

CMOS technologies with three-dimensional structures can have extremely high aspect ratio contacts and vias. For 3D technologies such as NAND or DRAM processes require high aspect ratio contacts to reach the CMOS devices below the array structures. Figure 9.62 shows a cross-section cartoon of a CMOS DRAM with the interconnects (Xiao, 2012). The same contact connects the DRAM capacitor cell plate and the bitline that runs beneath the capacitor to reduce manufacturing costs. The cell plate capacitor contact is much farther from the silicon than the bitline contact. Therefore, the cell plate completes etching and is exposed to the plasma longer than the bitline contact. The cell plate is exposed to the plasma during the time required to etch the sizeable additional amount of dielectric to expose the bitline contact, increasing the propensity for plasma damage. NAND technology, especially 3D NAND, requires different depth contacts, as shown in Chap. 8, creating a propensity for damage.

Figure 9.63 shows some examples of process damage caused by plasma damage during the contact etch (Tsui, 2000). Figure 9.63a shows a transistor's gate leakage with different contact numbers connected to the gate. The transistor with the larger number of contacts shows increased variability in gate dielectric leakage and variation. The reduction in the gate current for about 40% of the distribution is likely due

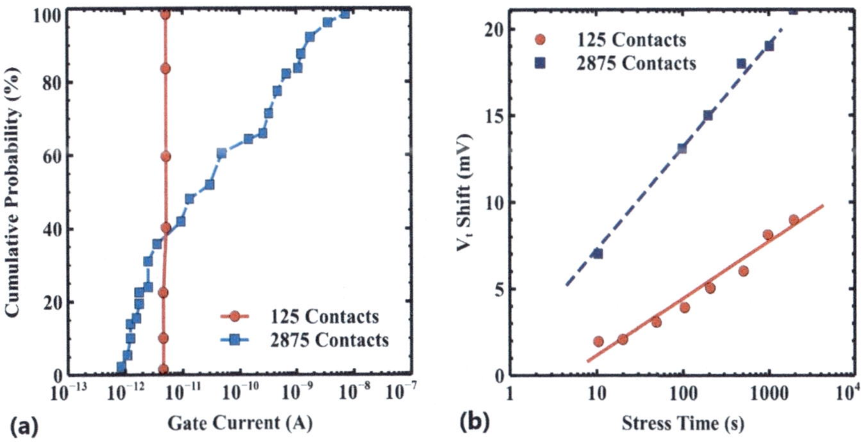

Fig. 9.63 Plots showing increasing process damage with an increasing number of contacts connected to the gate. (**a**) Degraded gate current and variation with increasing number of contacts. (**b**) Latent reliability damage is observed in pmos transistors during transistor hot electron electrical stress with $V_{ds} = -0.5$ V and $V_{gs} = -0.8$

to trapped charge reducing the gate current. The damage can shift time-zero electrical results and create latent reliability problems that can not be detected at time zero and can only be detected with reliability stress. Figure 9.63b shows that a hot electron reliability test can observe latent damage undetectable at time zero with worsening reliability with increasing numbers of contacts connected to the gate.

For example, Fig. 9.64 shows a degradation in dielectric breakdown for a gate antenna structure for a directional in situ resist strip in the same chamber as the contact etch (Ma et al., 2000). The dielectric breakdown is degraded with an increasing number of contacts, indicating increased charge flow onto the gate through more contacts. The degradation increases with increased power during the resist strip.

Process damage in contact to silicon surfaces can occur through other mechanisms in addition to antenna charging. Contact etches remove the dielectrics over the wafer surface and exposes the surface directly to the plasma as the etch completes the dielectric removal at the bottom of the contact hole. Exposing the wafer surface at the bottom of the contact creates the damaged region described in Chap. 8. Surface damage in a contact hole typically degrades the contact resistance, as the critical silicide formation is disrupted.

9.13.8 Surface Charge Creation

The dielectric films deposited on the wafer surface can adversely impact the CMOS devices by trapping charge that electrically modify the surface properties (Li et al., 2007). Dielectric films are used as stressors to increase channel mobility in CMOS

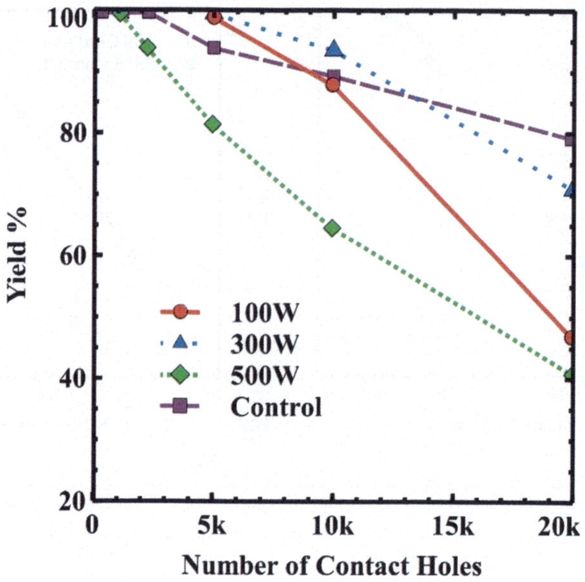

Fig. 9.64 Plot showing dielectric breakdown yield of a mos antenna capacitor structure as a function of the number of contact holes and the contact etch related insitu resist removal power showing decreased yield for increasing power and the number of contacts. The control did not have the insitu resist strip

technology. The stress induced by the dielectric film can change the electrical properties of the film, including trapping and hydrogen sensitivity. Plasma etch removal of surface films can cause surface charge by creating charged defects. Figure 9.65 shows an example of this phenomenon. Different transistor layouts have a different V_t shift response to a hydrogen alloy caused by a plasma etch removal of a surface dielectric used to block salicide formation. An experiment replacing the plasma etch with a wet etch eliminated the problem. This experiment revealed that the stressor dielectric interacted with the plasma etch.

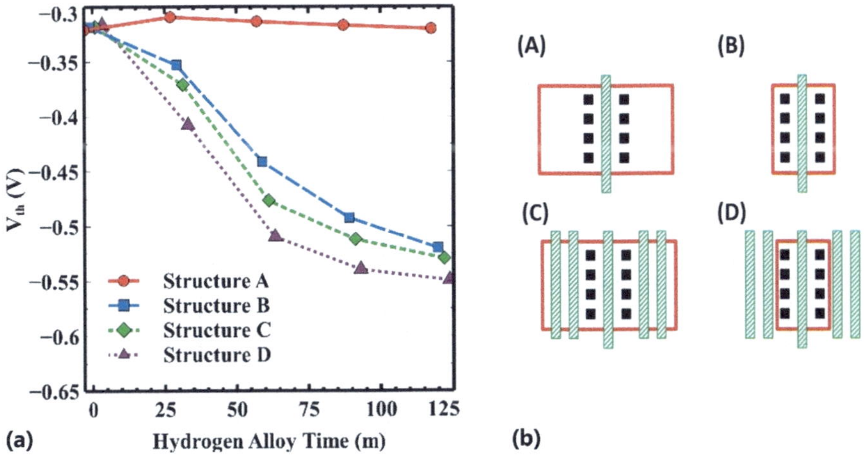

Fig. 9.65 Threshold voltage shifts of different layout structures caused by a plasma etch of a silicon surface dielectric used to block salicide formation. (**a**) V_t as a function of hydrogen alloy times. (**b**) Sketch showing the different transistor layout used in the experiment

9.14 Through-Silicon Vias, Backside Power, and Advanced Packaging Damage

Modern CMOS devices often use advanced packaging techniques to improve CMOS chips' packing density, speed, power dissipation, etc. Unfortunately, advanced packaging techniques have the propensity for process damage because the large features required for the interface to the external world can form antennas to the tiny transistors in the CMOS chip. In addition, the trend toward backside power distribution may increase the likelihood of damage. For example, through-silicon vias (TSVs) enable bonding to the backside of a wafer by forming a conductor through the entire wafer (Lee et al., 2014). Wafers are thinned before or after the formation of the TSV to minimize processing time. A large hole is etched through the wafer, connecting the front side of the wafer to the backside. The hole is typically filled with copper to form a low-resistance electrical connection. A connection from the CMOS circuitry to the TSV is formed. An assembly process completes the packaging by connecting the backside through-silicon vias to the package, forming an interface with the external world through the package. Figure 9.66 shows a cross-section sketch of a wafer with a TSV connecting to the backside (Li et al., 2018). The TSV process flow is typically lengthy and complex, consisting of multiple steps with process damage risk. Common steps that can cause damage include TSV etch, TSV dielectric liner deposition, TSV dielectric etch to clean theliner, Cu barrier metal, TSV sputter clean, etc. Process steps after the TSVs are formed can also cause damage as the chip completes the assembly process.

Figure 9.67 shows an example of process degradation from a TSV process to a 40 nm dielectric thickness MOS capacitor (Li et al., 2018). The No TSV group has

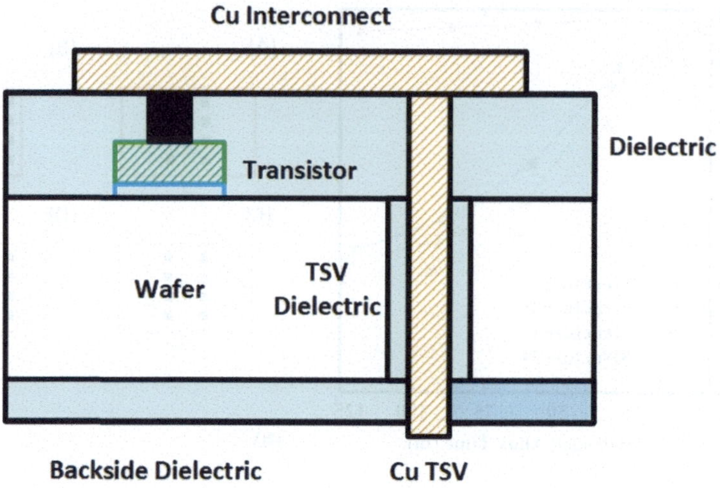

Fig. 9.66 Sketch showing a through silicon via (TSV) connecting the back of the wafer to the surface of the wafer

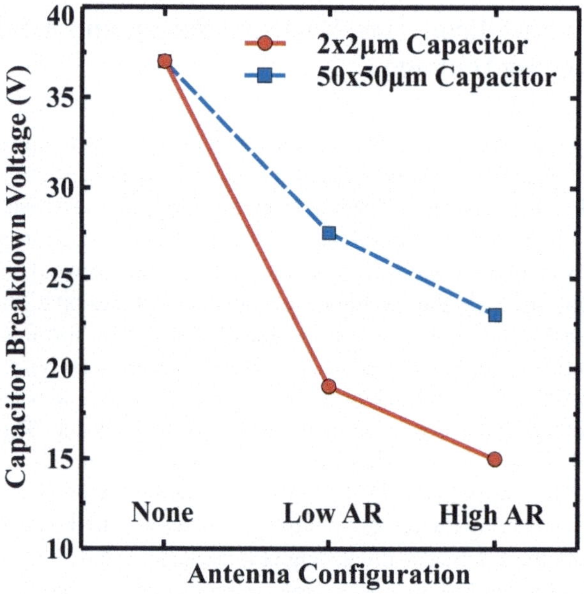

Fig. 9.67 Plot showing p-channel capacitor breakdown voltage with through silicon vias (TSV) antennas consisting of 12 TSVs for the low antenna ratio (Low AR) and 26 TSVs for the high antenna ratio (High AR). A control without TSVs is shown. The capacitor dielectric is 40 nm PECVD oxide

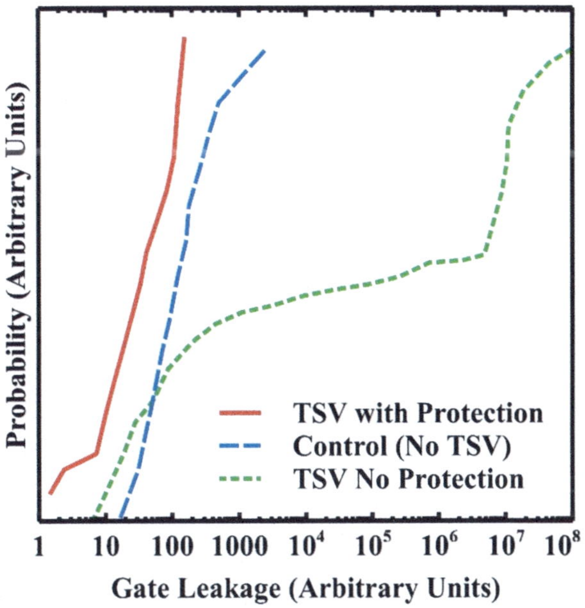

Fig. 9.68 Probability plot of gate leakage demonstrating protection from process damage from a TSV process. A control without a TSV is compared to an ESD protected TSV circuit and an unprotected TSV connection, demonstrating the benefit of the ESD protection structure against process damage

a normal breakdown for a thick capacitor at about 10MV/cm. The TSV groups have antennas formed by a small number of TSVs. The low antenna ratio (AR) with 12 TSVs connected to the capacitor gate shows a degraded breakdown voltage. The high AR capacitor with 26 TSVs connected to the capacitor shows further degradation of the capacitor dielectric.

TSV Design rules limiting the types of devices allowed to connect to a TSV can reduce the probability of process damage. Commonly, similar electrostatic discharge (ESD) protection circuits to bond pads can be placed between sensitive circuits and the TSV to partially protect the chip from damage during TSV processing, with a layout area penalty. Figure 9.68 shows a plot comparing gate dielectric leakage, showing the benefits of TSV ESD protection (Premachandran et al., 2020). Most TSV and backside processes occur after a wafer is completely processed, making ESD structures effective.

9.15 Exposure to High-Energy Light

Exposure to light can cause process damage if the photon energy is sufficient to break atomic bonds, eject electrons, etc. The damage increases with increasing photon energy. Figure 9.69 shows a sketch of some of the damage mechanisms for photon damage (Li et al., 2004). High-energy light exposure is a common intentional component of CMOS processing used in process steps such as plasma processing, laser annealing, etc.

The temporary or permanent films on the wafer interact with the photons, absorbing or blocking the light. Therefore, a detailed analysis of the materials on the wafer is needed to determine the propensity for damage at a particular step in the process flow for a particular wavelength of light.

High-energy radiation, such as ultraviolet light, has sufficient energy to break the atomic bonds in the materials present on the wafer, such as dielectrics. Defects can be created in dielectrics, causing fixed charge, traps, etc. Transient tracks of created electron–hole pairs can be formed in the silicon regions, creating temporary low-resistance conduction paths. These effects can interact with ongoing plasma damage or create defects with light alone.

Semiconductor dielectrics are sensitive to light exposure at higher energies. Electrons can be ejected from their bonded positions and be accelerated in any electric field. The free electrons have both the propensity to cause damage and drain accumulated charge from antennas. Figure 9.70a shows the quantum electron yield where free electrons are generated as a function of photon energy (Cheung, 1999, pg. 179). Free electrons can be generated in SiO_2 with photon energies greater than 2.5 eV, where a bond is occasionally broken when a photon is absorbed.

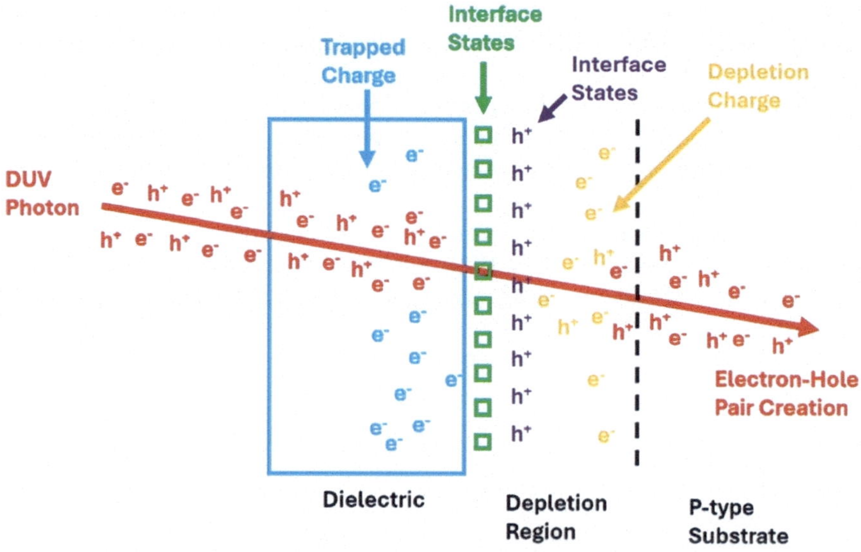

Fig. 9.69 Sketch showing different types of damage to dielectrics and semiconductors caused by exposure to high-energy photons

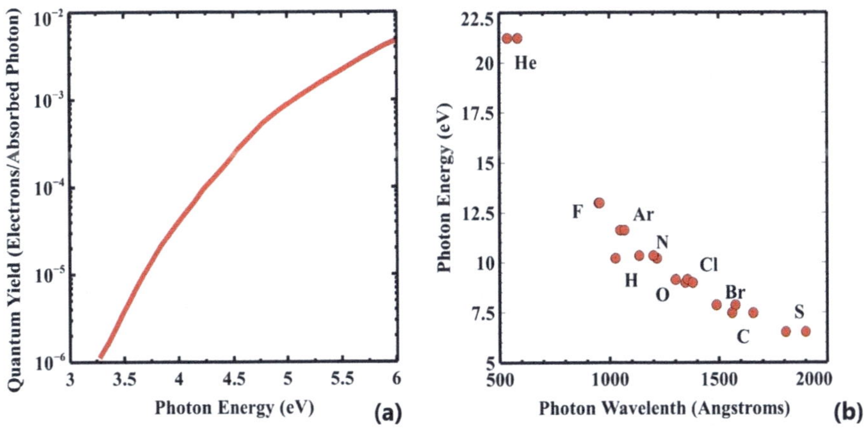

Fig. 9.70 (**a**) Photoconductivity quantum yield of silicon dioxide as a function of photon energy. (**b**) Photon energy as a function of wavelenth for common atoms used in plasma processing

In addition to the increased diode leakage when the wafer is exposed to light in a plasma, dielectric conduction is increased when the wafer is exposed to high-energy photons during processing. Figure 9.70a shows a plot of the photoconduction electron generation quantum yield of silicon dioxide as a function of the wavelength. The conduction of SiO_2 dramatically increases for photon energies greater than about 9.0 eV, which is the bandgap of SiO_2. Photoconduction differs for high-k dielectrics with lower bandgaps of about 6 eV (Huff & Gilmer, 2005). Many common gases used in plasma processing emit high-energy photons capable of dielectric conduction and damage. Figure 9.70b shows a plot of photon energy as a function of wavelength for common ions used in plasma processing (Cheung, 1999, pg. 176–177). Energies greater than the gate dielectric bandgap are frequently encountered in CMOS processing. The photoinduced dielectric leakage may mitigate process damage during plasma processing. The combination of high-temperature plasma processing and light exposure further increases the dielectric leakage by more than one factor alone.

All dielectrics exposed to high-energy light from plasma exposure can be expected to have high leakage relative to ambient conditions. Measurements have shown a resistance level of 50-200 $k\Omega/cm^2$ for oxides in a plasma (Cismaru et al., 1999). For comparison, an oxide dielectric resistance level under dark ambient conditions was greater than 10^{12} Ω/cm^2. The six orders of magnitude decrease in the dielectric resistivity increases charge loss from the antennas, potentially mitigating damage.

The higher energy photons shown in Fig. 9.70b can cause substantial damage to dielectrics. Figure 9.71 shows a plot of C–V threshold voltage shifts caused by 8–10 eV photons in the range of common plasma ions (Fleetwood, 2013). Substantial shifts in the order of tens of volts can occur if the gate is biased during exposure. These conditions can occur if a dielectric is charged by damage and exposed to UV light simultaneously, creating a worst-case damage situation without the light being blocked by films on the wafer. The thickness of the 200 nm dielectric is similar to that of shallow trench isolation dielectric thicknesses.

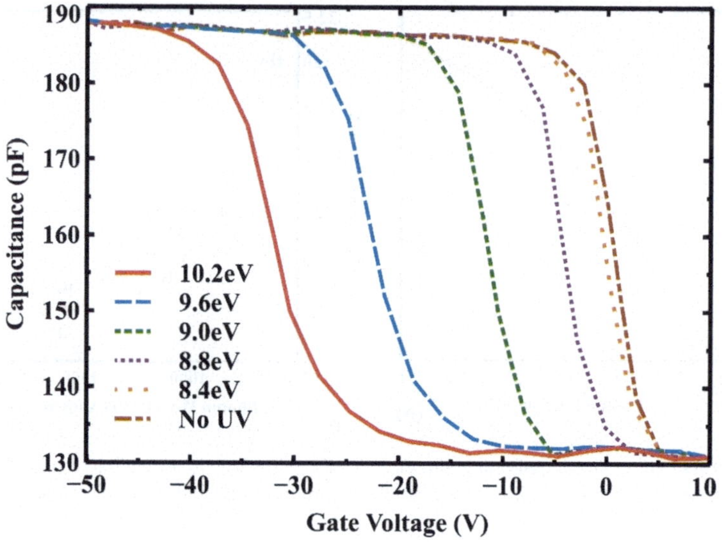

Fig. 9.71 High-frequency capacitance voltage curves for a 200 nm silicon dioxide dielectric exposed to various UV energies while under gate bias. The gate electrode is transparent at UV wavelengths. The undamaged device is labeled No UV

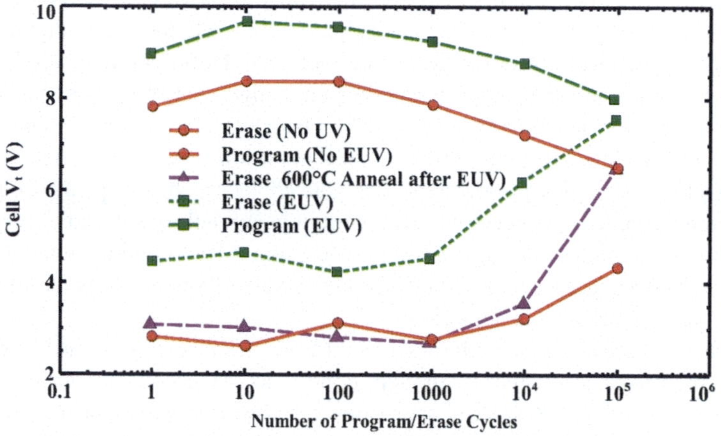

Fig. 9.72 Example of high energy photon damage to a silicon oxide nitride oxide silicon (SONOS) nonvolatile memory. V_t vs. cycling performance for a control cell not exposed to extreme ultraviolet radiation (No EUV) is shown in red. V_t vs. cycling performance for a cell exposed to extreme ultraviolet (EUV) radiation of 13.5 nm (91.85 eV) is shown. Partial recovery with an anneal after EUV exposure is shown

High-energy photon damage can be observed in diverse types of structures. For example, Fig. 9.72 shows an example of a nonvolatile memory cell cycling degradation caused by 13.5 nm wavelength photons (Tsui et al., 2011). The entire Vt window is shifted upward by the EUV exposure. The window position partially recovers with a 600 °C anneal.

9.16 Damage Recovery by Exposure to High-Energy Light

Exposure to high-energy light, such as UV radiation, can also cause healing or recovery from process damage in addition to damage, confounding the electrical behavior. This opposite result from UV damage can confound the understanding of plasma damage. Exposure to UV light can reduce process damage by ejecting trapped electrons or holes from locations where they are trapped. Energy damage in a CMOS process can also occur from X-rays created during processing by secondary emission during processes such as sputtering (Pearton et al., 1992). Figure 9.73 shows the dark current recovery in a charge-coupled device after X-ray damage with exposure to UV light (Janesick, 2001). The UV light was turned off at 100 s, and the change in slope shows that UV light played a role in the recovery.

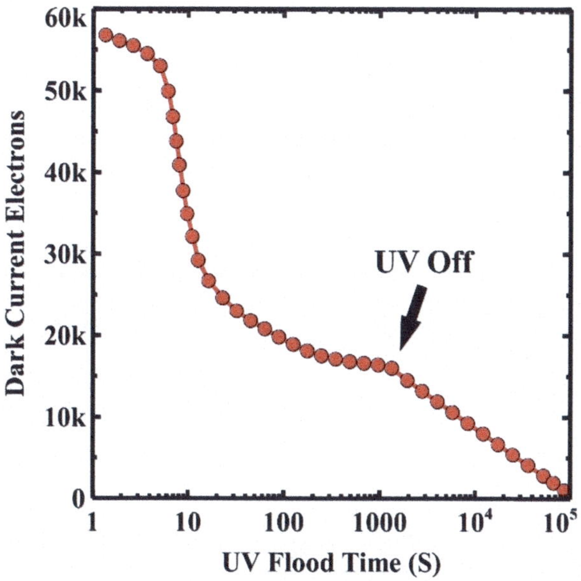

Fig. 9.73 Dark current recovery by exposure to UV in a charge coupled device (CCD) after x-ray radiation exposure. The CCD is exposed to x-rays and then exposed to UV radiation at 5 eV. The UV radiation is later turned off at the indicated point on the plot

References

Ackaert, J., et al. (2000). Prevention of plasma induced damage on thin gate oxide damage of HDP oxide deposition. In *5th international symposiyum on plasma induced damage*. https://doi.org/10.1109/PPID.2000.870600

Alba, S., et al. (1997). Study of plasma charging and photoresist coating effects using UV-erased FLASH cell as charging sensor. In *2nd International symposiyum on plasma process induced damage*. https://doi.org/10.1109/PPID.1997.596759

Cambell, A. N., et al. (1997). Effects of focused ion beam irradiation on MOS transistors. In *IEEE international reliability physics symposiyum*. https://doi.org/10.1109/RELPHY.1997.584241

Carrere, J. P., et al. (2003). Plasma damage in ultra-thin gate oxide induced by dielectric deposition processes: An overview on main mechanisms and characterization techniques. In R. E. Sah et al. (Eds.), *Silicon nitride and silicon dioxide thin insulating films VII*. Electrochemical Society.

Cheung, K. P. (1999). Plasma Charging Damage. Springer https://doi.org/10.1007/978-1-4471-0247-2.

Cheung, K. P. (2001). Charging damage during plasma enhanced SiN/SiO thin film deposition. In *silicon nitride and silicon dioxide thin insulating films electrochemical society proceedings Vol. 2001-7*.

Cho, H. J., et al. (2007). Study of plasma damage at Recess-Channel Gate (RG) structure during plasma nitridation. In *IEEE international reliability physics symposium*. https://doi.org/10.1109/RELPHY.2007.369917

Cismaru, C., et al. (1999). Plasma vacuum ultraviolet emission in a high density etcher. In *4th international symposiyum on plasma and process induced damage*. https://doi.org/10.1109/PPID.1999.798846

Current, M. I. (2014). New applications for ion implantation: Life in a vertical CMOS world, materials modification, and deep proton implants. *Ion Implantation Technology, 2014*.

Current, M. I. (2019). Shallow and high-dose implants for IC and PV devices.. In *Unknown Title*.

Dirnecker, T., et al. (2002). Influence of photoresist pattern on charging damage during high current ion implantation. In *7th international symposiyum on plasma and process induced damage*. https://doi.org/10.1109/PPID.2002.1042620

Dixon, W., et al. (1997). Photoresist-enhanced wafer charging during high current ion implantation. In *IEEE international conference on ion implantation*. https://doi.org/10.1109/IIT.1996.586134

Egerton, R. F., Li, P., & Malac, M. (2004). Radiation damage in the TEM and SEM. *Micron, 35*(6), 399–409. https://doi.org/10.1016/j.micron.2004.02.003

Electronics, W. (2015). Basics: Electrostatic discharge (ESD). In *Wavelength electronics*.

Fang, S., et al. (1992). A new model for thin oxide degradation from wafer charging in plasma etching. *IEEE IEDM*. https://doi.org/10.1109/IEDM.1992.307309

Fleetwood, D. M. (2000). *Focused ion beam damage to MOS integrated circuits*. Sandia Nat. Labs.

Fleetwood, D. M. (2013). Total ionizing dose effects in MOS and low-dose-rate-sensitive linear-bipolar devices. *IEEE Trans. on Nuclear Science, 60*(3), 1706–1730. https://doi.org/10.1109/TNS.2013.2259260

Fletch, S. (2014). *Doping FinFETs: Trade-offs for beamline implantation, plasma doping, and diffusion from doped epi/CVD films*. NCCAVS User Group Meeting.

Gabriel, C. T., & Weling, M. G. (1994). Gate oxide damage reduction using a protective dielectric layer. *IEEE EDL, 15*(8), 269–271. https://doi.org/10.1109/55.296212

Gulshan, M. B. S. (2019). Advanced VLSI technology: FinFet technology. *Journal of Emerging Technologies and Innovative Research, 6*(6), 64–68.

Hagimoto, Y., et al. (2009). Defects of silicon substrates caused by electrostatic discharge in single wafer cleaning processes. *Solid State Phenomena, 145-146*, 185–188. https://doi.org/10.4028/www.scientific.net/SSP.145-146.185

Huff, H. R., & Gilmer, D. C. (2005). *High dielectric constant materials VLSI MOSFET applications*. Springer. https://doi.org/10.1007/b137574

Istsumi, M., et al. (2002). *SiO2 in Si microdevices*. Kodansha and Springer.

Ito, H. (2012). New ion implantation solutions for advanced image sensor devices. In *Applied materials*.

Janesick, J. R. (2001). *Scientific charge-coupled devices*. The International Society for Optical Engineering.

Jank, M. P. M., et al. (2000). Gate oxide damage due to through the gate implantation in MOS-structures with ultrathin and standard oxides. In *International conferance on ion implantation technology*.

Koffel, S., et al. (2009). End of range defects in Ge. *Journal of Applied Physics, 105*, 126110. https://doi.org/10.1063/1.3153985

Krishnan, S., et al. (1996). Inductively coupled plasma (ICP) metal etch damage to 35-60A gate oxide. *IEEE IEDM*. https://doi.org/10.1109/IEDM.1996.554084

Lai, C. M., et al. (2009). A novel "hybrid" high-k/metal gate process for 28nm high performance CMOSFETs. In *IEEE International Electron Devices Meeting*. https://doi.org/10.1109/IEDM.2009.5424256

Lee, K. W. (2020). Damage control of ion implantation for advanced doping process by using in-situ temperature control. *Materials Science in Semiconductor Processing, 117*, 105164. https://doi.org/10.1016/j.mssp.2020.105164

Lee, S. W., et al. (2004). Thin oxide degradation from HDP-CVD oxide deposition in 300mm process. In *IEEE international conference on integrated circuit design and technology*. https://doi.org/10.1109/ICICDT.2004.1309980

Lee, J., et al. (2009). Study on tunnel oxide degradation by metal pad etch-induced plasma damage in 0.04 μm flash memory technology. *Journal of Korean Physical Society, 55*(1), 346–351. https://doi.org/10.3938/jkps.55.346

Lee, J., Kim, D.-K., Min, M., & G-J., & Chung, I. (2012). Characterization of tunnel oxide degradation due to plasma field oxide recess in flash memory devices. *Thin Solid Films, 520*, 5007. https://doi.org/10.1016/j.tsf.2012.03.005

Lee, M., Pak, J. S., & Kim, J. (2014). *Electrical design of through Silicon via*. Springer. https://doi.org/10.1007/978-94-017-9038-3

Li, F. M., Nixon, O., & Nathan, A. (2004). Degradation behavior and damage mechanisms of CCD image sensor with deep-UV laser radiation. *IEEE Transactions on Electron Devices, 51*(12), 2229–2236. https://doi.org/10.1109/ted.2004.839758

Li, R., Kong, W.-R., Tao, K., Yu, L.-J., Huang, K., Ning, J., Geng, C.-Q., & Wang, C.-D. (2007). Threshold voltage shift due to mechanical stress-enhanced plasma process-induced damage in 0.13 Micron pMOSFET. *IEEE Electron Device Letters, 28*(5), 360–362. https://doi.org/10.1109/led.2007.894644

Li, Y., et al. (2018). TSV process-induced MOS reliability degradation. In *IEEE international reliability physics symposiyum*. https://doi.org/10.1109/IRPS.2018.8353610

Ma, S. (2003). Backend dielectric etch induced wafer arcing mechanism and solution. *International Symposiyum on Plasma and Process-Induced Damage*. https://doi.org/10.1109/PPID.2003.1200951

Ma, S. (2004). *Plasma induced charging damage: From a semiconductor equipment vendor point of view* July 8, 2004 Plasma Etch User Group (PEUG).

Ma, S., et al. (1995). Effects of wafer temperature on plasma charging induced damage to MOS gate oxide. *IEEE EDL, 16*(12), 534–536. https://doi.org/10.1109/55.475578

Ma, S., et al. (2000). Plasma damage evaluation of an integrated in-situ directional resist stripping process in magnetically enhanced RIE etcher for dual damascene application Int. *Symposiyum on Plasma Process Induced Damage*. https://doi.org/10.1109/PPID.2000.870631

Mack, M. E., et al. (1985). Wafer charging and beam interactions in ion implants. *Nuclear Instruments and Methods in Physics Research, 405-411*, 405. https://doi.org/10.1016/0168-583X(85)90665-2

McPherson, J. W., et al. (2002). Proposed universal relationship between dielectric breakdown and dielectric constant. *IEEE IEDM*. https://doi.org/10.1109/IEDM.2002.1175919

Nakakubo, Y. (2015). *A study of defect generation phenomena in single crystalline silicon substrate during plasma processing and the characterization techniques*. Kyoto University.

Nojiri, K. (2012). *Dry etching technology for semiconductors*. Springer. https://doi.org/10.1007/978-3-319-10295-5

Pearton, S. J., Corbett, J. W., & Stavola, M. (1992). *Hydrogen in crystalline semiconductors*. Springer. https://doi.org/10.1007/978-3-642-84778-3

Pipes, L. C., McGill, L., & Jahagirdar, A. (2014). NMOS source-drain extension ion implantation into heated substrates. In *2014 20th international conference on Ion Implantation Technology (IIT)*.

Premachandran, C. S., et al. (2020). Efficient bidirectional protection structure for Plasma Induced Damage (PID) and Electrostatic Discharge (ESD) for 3D IC integration. In *International reliability physics symposiyum*. https://doi.org/10.1109/IRPS45951.2020.9129158

Pyo, S. G., & Kim, S. (2013). Plasma damage mechanism of electron beam curing process for spin on dielectrics. *IEEE Electron Device Letters, 34*(5), 575–577. https://doi.org/10.1109/led.2013.2250246

Rao, K. V. (2014). *Ion implant applications to enable advances in semiconductor technologies*. NCCAVS Junction Tech. User Group Meeting.

Rimini, E. (1995). *Ion implantation basics to device fabrication*. Kluwer Academic Publishers.

Romano, L., Privitera, V., & Jagadish, C. (2015). *Defects in semiconductors*. Academic Press.

Rudawski, N. G., Elliman, R. G., & Jones, K. S. (2008). Influence of as on the formation of mask-edge defects during stressed solid phase epitaxy in patterned Si wafers. *Journal of Vacuum Science and Technology B, 26*, 435. https://doi.org/10.1116/1.2775459

Rudawski, N. G., et al. (2009). Stressed multidirectional solid-phase epitaxial growth of Si. *Journal of Applied Physics, 105*(081101). https://doi.org/10.1063/1.3091395

Sacchettini, Y., et al. (2019). Plasma antenna charging in CMOS image sensors. *IEEE International Reliability Physics Symposium*. https://doi.org/10.1109/IRPS.2019.8720518

Shin, H., & Hu, C. (1993). Monitoring plasma-process induced damage in thin oxide. *IEEE Trans. on Semi. Manufacturing., 6*(2), 96–102. https://doi.org/10.1109/66.216927

Shiraki, K. (2008). Study of ion implantation induced gate oxide quality problem and solutions. In *2008 international symposiyum on semiconductor manufacturing*.

Shul, R. J., & Pearton, S. J. (2000). *Handbook of Advanced Plasma Processing Techniques Springer*. https://doi.org/10.1007/978-3-642-56989-0

Simon, P., et al. (2000). Identification of plasma-induced damage conditions in VLSI designs. *IEEE Trans on Semiconductor Manufacturing, 13*(2), 136–144. https://doi.org/10.1109/66.843628

Siu, S., et al. (2002). Utility of CHARM-2 in diagnosing sources of plasma charging damage in high density etchers and in assisting hardware development. In *7th international symposiyum on plasma and process induced damage*. https://doi.org/10.1109/PPID.2002.1042614

Strong, A. W., et al. (2009). *Reliability Wearout mechanisms in advanced CMOS technologies*. Wiley. https://doi.org/10.1002/9780470455265

Sze, S. M., & Lee, M. K. (2010). *Semiconductor devices physics and technology* (3rd ed.). Wiley.

Tsai, Y. P., et al. (2018). 7nm FinFET plasma charge recording device. *IEEE International Reliability Physics Symposium*. https://doi.org/10.1109/IEDM.2018.8614520

Tsui, B. Y. (2000). Plasma charging damage during contact hole etch in high-density plasma etcher. *Microelectronics Reliability, 40*, 2039–2046. https://doi.org/10.1016/S0026-2714(00)00023-8

Tsui, B. Y., Yen, C. C., Li, P. H., & Lai, J. Y. (2011). Effects of EUV irradiation on poly-Si SONOS NVM devices. *IEEE Electron Device Letters, 32*(5), 614–616. https://doi.org/10.1109/led.2011.2121052

Tung, C. H., Sheng, G. T. T., & Lu, C. Y. (2003). *ULSI semiconductor technology atlas*. Wiley-Interscience. https://doi.org/10.1002/0471668796

Ushiki, T., et al. (1998). Reduction of plasma-induced gate oxide damage using low-energy large-mass ion bombardment in gate-metal sputtering deposition. *IEEE International Reliability Physics Symposium*. https://doi.org/10.1109/RELPHY.1998.670661

Wang, Z. (2004). *Detection of and protection against plasma charging damage in modern IC technology*. Twente University.

Wolf, S. (2002). Silicon processing for the VLSI era. In *Deep-submicron process technology* (Vol. 4). Lattice Press.

Wood, B., et al. (2013). Fin doping by hot implant for 14nm FinFET technology and beyond. *ECS Transactions, 58*(9), 249–256. https://doi.org/10.1149/05809.0249ecst

Wu, C. P., & Kolondra, F. (1991). Wafer charging control in high-Current implanters. *Journal of the Electrochemical Society, 138*(10), 3100–3107.

Wu, Y. W., & Sune, J. (2013). Recent advances in dielectric breakdown of modern gate dielectrics IEEE Int. *Integrated Reliability Workshop*. https://doi.org/10.1109/IIRW.2013.6804141

Wu, C. H., et al. (2015). Mapping of wafer-level plasma induced charge contour by novel on-chip in-situ recorders in advanced FinFET technologies. *IEEE IEDM*. https://doi.org/10.1109/IEDM.2015.7409644

Xiao, H. (2012). *Introduction to semiconductor manufacturing technology* (2nd ed.). SPIE Press. https://doi.org/10.1117/3.924283

Yang, K. C., Park, S. W., Lee, H. S., Kim, D. W., & Yeom, G. Y. (2017). Effect of structure and material variation of focus ring for enhanced etch resistance. *Nanoscience and Nanotechnology Letters, 9*(1), 24–29. https://doi.org/10.1166/nnl.2017.2296

Ye, B., Zhu, H., Chen, L., Li, J., Sun, Q. Q., & Zhang, D. W. (2021). Suppression of stress-induced defects in FinFET by implantation and STI co-optimization. *IEEE Transactions on Electron Devices, 68*(5), 2587–2589. https://doi.org/10.1109/ted.2021.3068241

Chapter 10
Inline Process Damage Measurements

10.1 Introduction

The capability to collect in-process electrical and damage-related data is advantageous. It allows rapid detection of process damage information, enabling troubleshooting and improvement, avoiding waiting months for full CMOS flow wafer processing completion. The same parametric structures on product wafers used to monitor process damage at the final parametric test step at the end of the process can be modified to allow inline parametric testing with simple inline bond pads, as described in Chap. 11.

Additionally, other inline methods can provide an indirect method of detecting process damage. These methods do not provide direct quantitative measurements of process damage but can provide valuable qualitative information.

10.2 Surface Potential Measurements

Plasma, ion implantation, and similar processes can inject charge into the surface of a wafer. This charge can be retained on the wafer surface after processing for some time if the wafer surface has suitable material conditions. The wafer surface must be highly insulating with high-resistivity materials to prevent charge from migrating across the surface of the wafer. After processing, the wafer must be maintained in an environment with little moisture, low light, and over a specific temperature range to keep charge in place on the wafer surface until the measurement is performed. Measurements typically have to be performed quickly after exposure to plasma to avoid charge migration or loss. Deionized water rinses (see Chap. 9) and thermal alloys can neutralize the wafer surface charge, allowing the test wafer to be reused.

K. Prall, *CMOS Plasma and Process Damage*,
https://doi.org/10.1007/978-3-031-89029-1_10

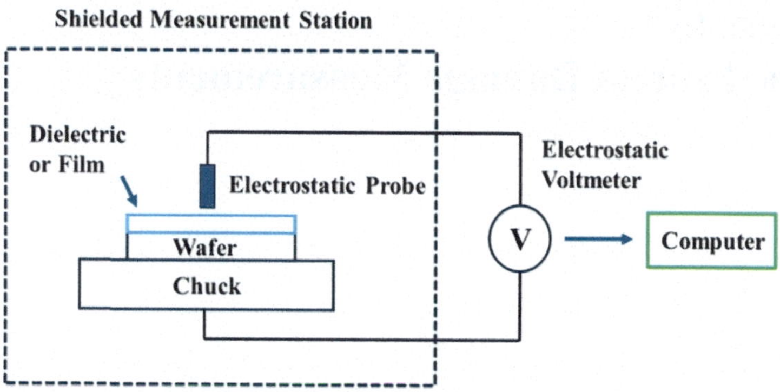

Fig. 10.1 Sketch of a wafer potential measurement system and a sketch showing the instrument configuration for surface charge measurements on a wafer

Charge on the surface of a wafer can be detected electrically by placing an electrical probe close to the wafer surface connected to an electrostatic voltmeter (Shibata et al., 2011). The voltmeter is sensitive enough to detect the potential induced on the surface of the wafer by the charge due to processing. Figure 10.1 shows a sketch of a measurement system. The probe tip is placed near the wafer surface. The expected voltage can be calculated using a simple area capacitor formula.

As an example, the potential of a patterned resist layer with a 10^{13} ions/cm^2 40 keV phosphorous implant is shown in Fig. 10.2 (Shibata et al., 2011). Charge from the ion implant remains embedded in the resist and can be detected by electrostatic potential measurements. The amount of charge is dependent on the type of resist. The resist is optimized for different lithography wavelengths and performance, with little or no consideration for the film's resistivity. A smaller amount of charge remains in the i-line resist, while the krF resist retains a significant amount of charge. Excessive charge build-up in the resist can discharge in a destructive arc to silicon or other structures, as shown in Chap. 5.

The charge embedded in the resist can be a function of the design rules and the underlying layers, potentially resulting in process damage in specific patterns. Figure 10.3 shows the surface potential across a resist and isolation pattern, with some locations reaching 200 volts. No charge is observed on the gate oxide silicon surface, as it would have been discharged into the wafer or implanted through the gate dielectric (Shibata et al., 2011). This effect is the mechanism causing resist pattern sensitivity described in Chap. 9.

10.3 Surface Parametric Measurements

Sophisticated instrumentation can enable measurement of many parameters inline between process steps as the wafer is manufactured using very simple test wafers (KLA-Tencor, 2002). An example of advanced inline electrical capability is the

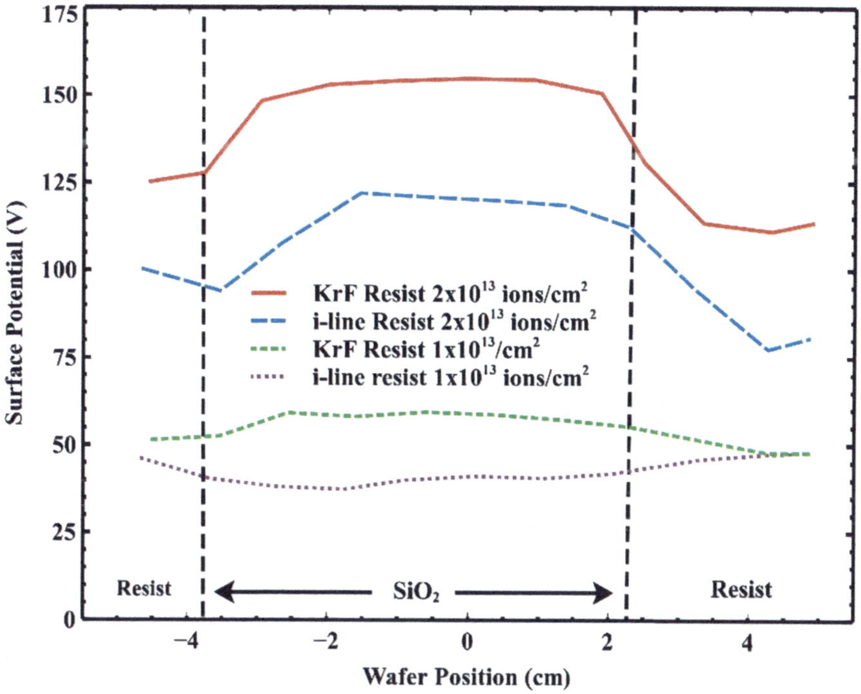

Fig. 10.2 Surface potential measurement of implanted resist charge for different resist types and implant doses measured across a region with no resist

Quantox™ system shown in Fig. 10.4 (Wu et al., 2007). This system contains three methods to enable inline measurements (Schroder, 2001). A corona bias is used to bias the surface of the wafer. The corona bias uses a high-voltage probe to inject low-energy ions onto the surface of the wafer in a room atmosphere. The low-energy ions stick to the wafer surface and tend to be temporarily immobile. The combination of the corona, dielectric, and substrate results in a structure that behaves as a metal oxide semiconductor (MOS) capacitor. The surface potential can be driven to specific bias conditions, allowing characterization of the capacitor. A light source in the system enables the generation of minority carriers in the substrate. A vibrating Kelvin probe creates an AC signal, allowing a current measurement where the current is given by Eq. (10.1), where I is the current calculated from the measurement, V_c is the forced voltage, C is the capacitance generated by the measurement, and t is time.

$$I = V_C \frac{dC}{dt} \tag{10.1}$$

Surface photovoltage (SPV) measurements enable the capability to measure minority carrier lifetime.

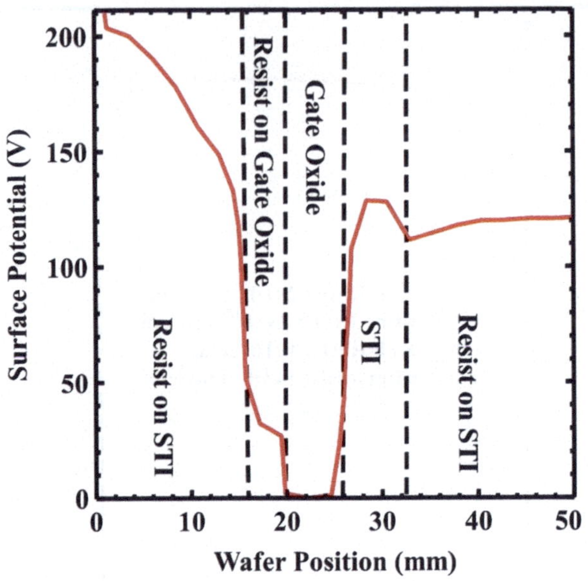

Fig. 10.3 Surface potential measured across different wafer surface conditions with shallow trench isolation (STI), gate oxide, and resist

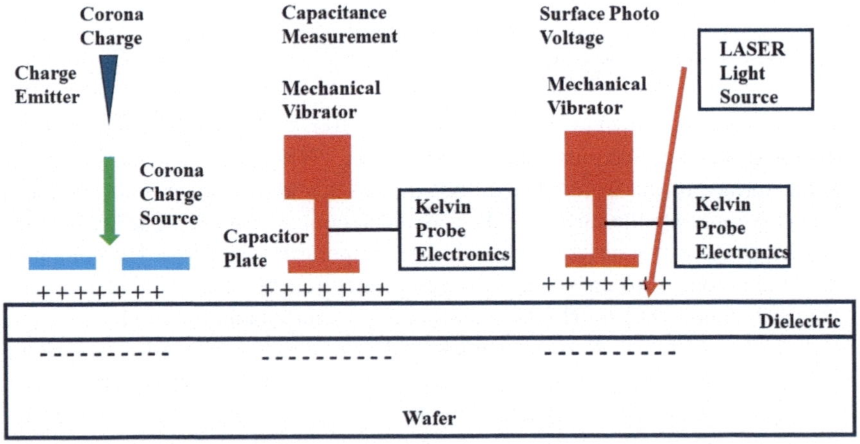

Fig. 10.4 Sketch showing the operation and capabilities of the Quantox™ system

Surface parametric measurement capability can enable detailed rapid electrical characterization of simple test wafers (Gupta & Horner, 2000). Table 10.1 shows a flowchart for a simple process flow for inline measurements. Almost all of the measurements that can be performed on a MOS capacitor can be performed on a test wafer using these techniques, albeit slower. The short loop flow does not create an antenna but can detect process damage. These measurements include trapped

Table 10.1 Short loop process flow for inline test wafers

Wafer start and clean
Dielectric deposition
Resist coat, backside dielectric removal, resist clean (optional)
Plasma or process damage experiment
Measurements with minimum delay
DI rinse and 400 °C alloy to remove charge
Recycle the test wafer for another experiment

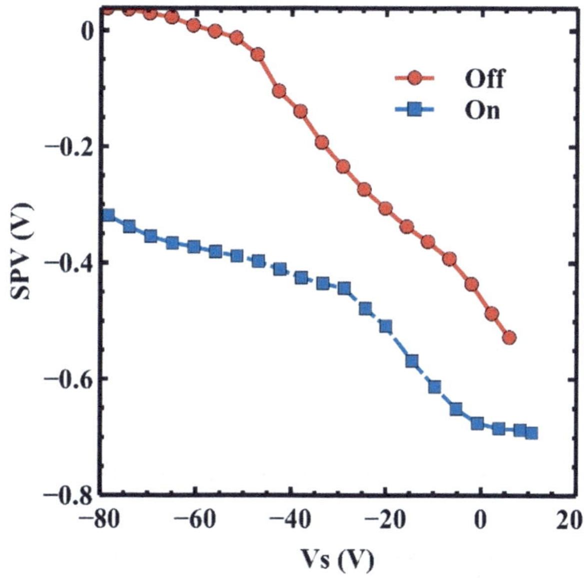

Fig. 10.5 Example of recipe component isolation by simple inline measurements showing a power lift step turned on and off. The power lift is a step in the termination of the plasma process. More negative voltage indicates more plasma damage. Vs is the forced voltage, and SPV is the measured surface voltage

charge, surface voltage, surface barrier height, flat band voltage, dielectric thickness, dielectric leakage, interface trap density, mobile ions, dielectric integrity, minority carrier lifetime, doping density, etc. This capability provides a valuable tool for process damage diagnostics (Schroder, 2001).

Simple inline process tests can be used to optimize processes to minimize process damage by allowing the testing of recipes and isolation of individual components of the recipe. Figure 10.5 shows an example of recipe component isolation (Gupta & Horner, 2000). The surface photovoltage technique is used in this example. At the end of the process, a recipe step is required to turn off the plasma, remove the static charge holding the wafer to the chuck, and remove the wafer from the chuck. The figure shows SPV measurements for two different methods of terminating the process. The larger negative value indicates that the "Powerlift" recipe results in more process damage.

Simple test wafers can be used to determine the fundamental process and equipment differences. For example, Fig. 10.6 compared different etchers where the test wafer was exposed to a 30 s etch step. This technique provides a valuable tool for process optimization. In this example, a more negative SPV voltage indicates more injected charge in the dielectric, which is symptomatic of increased process damage. The control (E) has a typical capacitance–voltage characteristic, while the experimental groups have distorted characteristics (Gupta & Horner, 2000).

Another example of the usefulness of measurements on simple test wafers is shown in Fig. 10.7, which were performed on elements of a CMOS imager process (Sacchettini, 2019). The dielectric on the pixel is critical for CMOS imagers. Three different dielectric films are compared in this experiment by exposing test wafers and product-level wafers to plasma damage. The plasma damage injects charge into the dielectric, changing the electrical surface properties. The test wafer is measured using the corona technique, which detects a change in the surface voltage, as shown in Fig. 10.7. The dark current is measured on CMOS imager pixels, as described in Chap. 8, comparing the performance of the three films. There is a good correlation between the test wafer results and the product-level dark current, highlighting the inline technique's advantages. The oxide–nitride–oxide (ONO) film has positive charge trapped in the as-deposited film, likely due to the charge trapping described in Chap. 8.

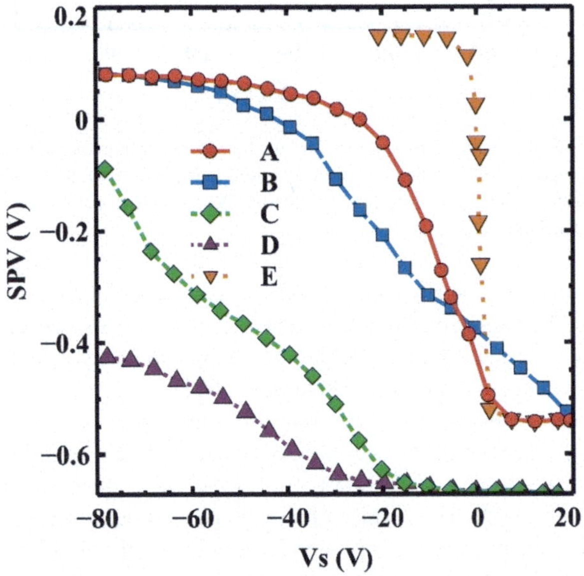

Fig. 10.6 Surface photovoltage comparison of different etchers. Vs is the forced voltage, and SPV is the measured surface voltage

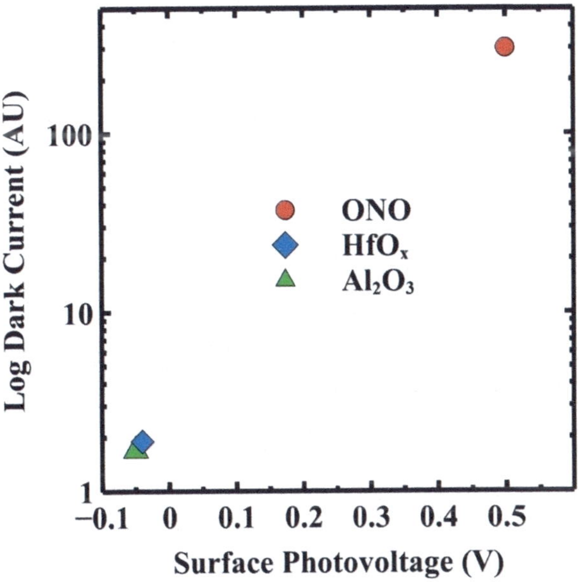

Fig. 10.7 Comparison of corona data on dielectric test wafers with product level imager dark current performance data for different dielectric films on the silicon surface after plasma damage

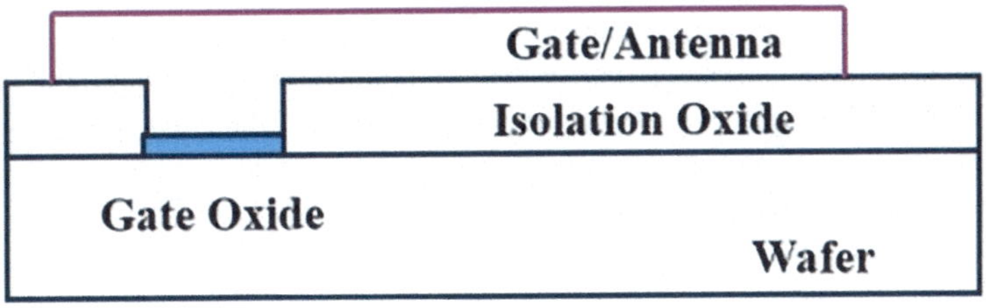

Fig. 10.8 Sketch showing a cross-section of a capacitor short loop

10.4 Short Loop Flows

Simple short loop process flows can be used to construct capacitors, transistors, etc., to allow quick and easy experimentation of some sources of process damage. For example, a short loop capacitor flow with a gate antenna can be used to investigate etch damage during the gate etch process. Figure 10.8 shows a cross-section cartoon of a short loop structure used to test process damage on a flash memory tunnel oxide (Lee et al., 2009). The short loop flow starts with a thick oxide. A hole is cut in the thick oxide, where the tunnel dielectric is formed. A gate material is deposited and patterned, forming a capacitor structure. Different antenna structures, including area and edge antennas, are formed over the thick oxide. Limited process damage experiments can be performed on wafers formed by this short loop process flow. For

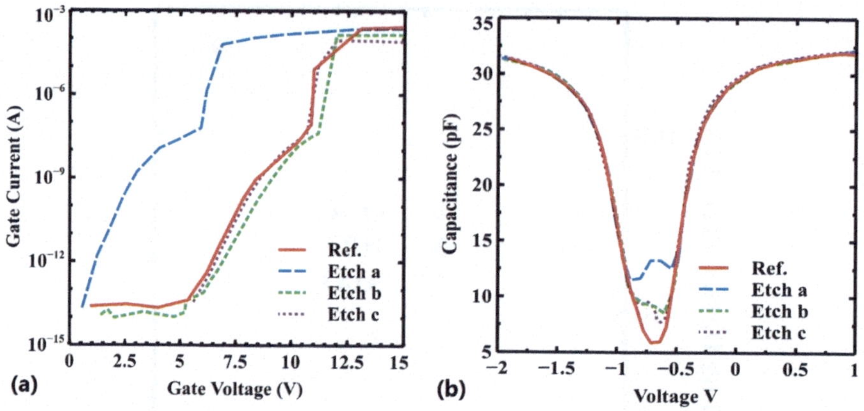

Fig. 10.9 Electrical results from the capacitor short loop flow. (**a**) 10.1% uniformity gate etch with a 30% $O_2/(C_2F_6 + O_2)$ gas ratio. (**b**) 6% uniformity gate etch with a 20% $O_2/(C_2F_6 + O_2)$ gas ratio. (**c**) 4.9% uniformity gate etch with a 10% $O_2/(C_2F_6 + O_2)$ gas ratio

example, gate etch, resist strip, etc., can be performed on the short loop, allowing experimental optimization of process damage without consuming expensive full-flow CMOS wafers.

The results of an experiment performed with this short loop are shown in Fig. 10.9 (Lee et al., 2009). Three different gate etch recipes are compared, revealing a significant difference in damage with variation in the recipe gases. Figure 10.9a shows the breakdown characteristics of the tunnel dielectric as a function of the etch recipe. Figure 10.9b shows a low-frequency C–V plot of the capacitor. The abnormal shape of the C–V plot, at the minimum near the flat-band voltage, is associated with excessive interface generation caused by process damage. The damage correlates with the nonuniformity of the etch across the wafer, indicating possible current flow through the wafer.

References

Gupta, I. J., & Horner, G. S. (2000). A comprehensive assessment of contact oxide etch damage using device monitor and in line noncontact testing. In *5th international symposiym on plasma and process induced damage*. https://doi.org/10.1109/PPID.2000.870599

KLA-Tencor. (2002). *Quantox user's guide*. KLA Tencor.

Lee, J., et al. (2009). Study on tunnel oxide degradation by metal pad etch-induced plasma damage in 0.04 μm flash memory technology. *Journal of the Korean Physical Society, 55*(1), 346–351. https://doi.org/10.3938/jkps.55.346

Sacchettini, Y. (2019). A highly reliable Back side illuminated pixel against plasma induced damage. *IEEE IEDM*. https://doi.org/10.1109/IEDM19573.2019.8993561

Schroder, D. K. (2001). Surface voltage and surface Photovoltage: History, theory, and applications. *Measurement Science and Technology, 12*, R16–R31. https://doi.org/10.1088/0957-0233/12/3/202

Shibata, S., Kamiyanagi, H., Kawase, F., Kaneki, R., Kawashima, T., & Kitamura, A. (2011). Charging phenomena during medium current ion implantation of carbonized photo-resist surface layers. *IEEE Transactions on Semiconductor Manufacturing, 24*(3), 455–463. https://doi.org/10.1109/tsm.2011.2128352

Wu, S. H., et al. (2007). *Inline electrical monitor of plasma-enhanced silicon Oxynitride for 70mn node or beyond*. IEEE ASMC.

Chapter 11
Process Damage Test Structures

11.1 Introduction

Parametric test structures added to a CMOS product or test wafers are necessary to characterize and understand plasma and process damage. This chapter summarizes the structures necessary to cover the mechanisms described in this book. The structures need to be optimized for each CMOS technology. A comprehensive set of test structures can require a great deal of effort. The test structure design should be consistent with the relevant design rules and have structures with larger antennas, etc., above the design rule requirements to allow margin verification of the design rules. All the structures should have a control structure without antennas for reference. Structures should be duplicated with and without diodes to enable definitive data to prove that plasma damage is occurring. The layout of structures should exactly match the product layout, including dummy gates, proximity correction patterns, etc. Table 11.1 lists potential parametric test structures for plasma and process damage organized by the process module discussed in this chapter.

11.2 Bond and Probe Pads

Bond or probe pads form an antenna on every structure probed through the connection to the pad. Therefore, every structure connected to a pad has the possibility of additional plasma damage during the pad and passivation opening process modules. The addition of duplicate test structures with different bond pad sizes gives the capability of separating the damage from these steps from damage caused by the rest of the process. A mixture of different pad sizes and passivation opening sizes can be used to isolate damage during the passivation process module. The structures can be measured with and without passivation.

K. Prall, *CMOS Plasma and Process Damage*,
https://doi.org/10.1007/978-3-031-89029-1_11

Table 11.1 List of plasma and process damage-related test structures

Process module	Description of the test structure
Transistor formation-related process steps	Transistors with large and small bond pads with various configurations on the gate, drain, source, and substrate terminals for all types of CMOS devices present in the technology. Dummy gates, proximity, etc., should be included
	Gate area and edge antennas consistent with the design rules
	Alternative gate configurations, such as donut devices, should be included if allowed by design rules to detect field edge leakage problems
	Source and drain diffusion area antennas from minimum to maximum allowed design rules
	Transistors with and without light-blocking layers to detect high-energy light damage
	Transistor mismatch structures with and without antennas
	Ring oscillators with and without antennas, with and without diodes
Well and triple well	Well or triple-well antennas connected to transistor terminals to detect well-related damage
Implant steps	Resist antennas consistent with the design rules for implanted layers
Metal-related	Metal area and edge antennas connected to a gate for each metal layer consistent with the design rules. Structures for individual metal layers and cumulative metal layers should be included. Metal structures should cover a range of allowed metal spaces
	Include metal comb and serpentine structures used for the backend of line to allow detection of metallization and dielectric damage. Comb and serpentine structures with one side grounded for worst-case stress
	Metal reliability structures with and without antennas
Contact and via-related	Contact and via antennas connected to gates for each layer consistent with the design rules. Structures for individual contact or via layers and cumulative contact or via layers should be included
	Contact and via chains with metal antennas
Metal–insulator–metal capacitor	MIM structures to detect MIM capacitor damage (if allowed) connected to the terminals of MOS devices to detect damage to the MIM capacitor or to the transistor. MIM structures connected to the wafer substrate as a hard ground for the worst-case condition
Bond pads	Many structures with large and small bond pads with and without diodes
Assembly-related structures	Through silicon via structures should be added if used in the technology
Memories	Word-line-related structures

11.3 Inline Probe Pads

Parametric test structures allow for optimizing a CMOS process to minimize the detrimental effects of process damage. The test structures should be designed to allow inline measurements after completing specific process steps. This method allows the process damage to be quickly quantified as processing progresses, enabling the detection of the most destructive process steps. Metal, contact, and via layers can serve as inline probe pads, enabling inline measurements. The design rules generally forbid large area bond pads on intermediate layers to avoid excess topology interfering with chemical mechanical polishing (CMP), deposition, and lithography processes.

CMP dishing can occur for a metal pad formed with a damascene process, creating a low spot in a metal feature in the center of the pad. The subsurface dishing feature can interfere with subsequent processing, creating defects and uneven topology that may kill chips. Modification of the pad layout can avoid this problem. For example, Fig. 11.1a shows an inline pad layout formed with metal stripes (Kahng & Samadi, 2008). If the metal is formed by a damascene process, the dielectric between

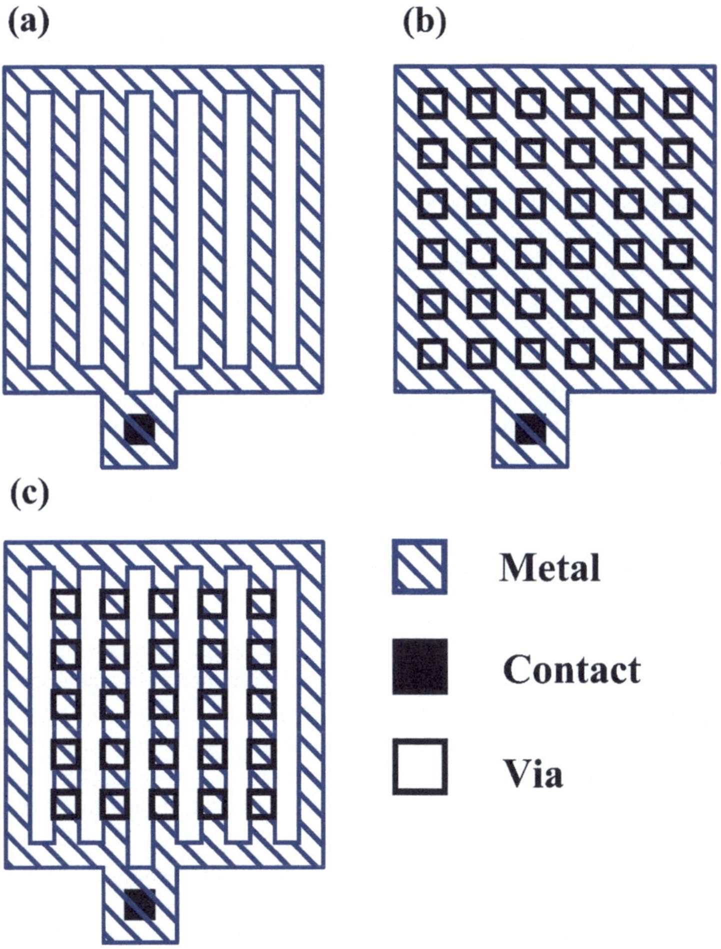

Fig. 11.1 Sketch showing the layout of bond pads to allow inline testing during CMOS processing. (**a**) Bond pad formed with alternating lines and spaces of metal lines. (**b**) Bond pad formed with a sea of vias on top of a solid metal pad. (**c**) Bond pad formed with alternating metal lines and spaces with a sea of vias on top

the lines prevents or minimizes the dishing. This style of bond pad can be formed in the copper, tungsten, and cobalt layers used in intermediate metal layers. The probe tip contacting the pad is much larger than the metallization pitch. The probe tip shorts many metal lines together, connecting the pad to the underlying structures and forming a low-resistance connection.

The striped pads shown in Fig. 11.1a can also be used in subtractive metallization processes, where metal is formed on the wafer surface and then patterned. However, a subtractive solid metallization probe pad can form the opposite problem of damascene metallization, where subsequent process steps like dielectric deposition can form a high spot or doming, interfering with subsequent processing.

An inline pad can be formed in contact and via layers, allowing measurement between the metal routing layers and further partitioning the process flow to detect problem process steps. Figure 11.1b shows an example probe pad formed by placing a sea of contacts or vias on top of a solid metal pad. The probe tip is much larger than the via or contact dimension, shorting many contacts or vias and creating an electrical connection to the underlying devices. Figure 11.1c combines the previous structures.

Different types of pads can be stacked on top of each other, allowing the same device to be probed multiple times during CMOS wafer processing. This technique allows monitoring of the progression of process damage throughout the process. However, the probe pads can also create antennas with the propensity for process damage, which must be accounted for in the test structure design.

The probe tips contacting the wafer inline may leave a mark on the probe pad surface due to mechanical damage. As a result, particles may be generated by the damaged region, degrading yield and requiring caution when performing inline probing.

11.4 Gate Area and Edge Antennas

As described in the previous chapters, process damage occurs with increasing antenna area or perimeter. Simple test structures with appropriate antennas can be used to detect process damage. Figure 11.2 shows sketches of different antenna structures. Figure 11.2a shows a transistor layout with no antenna used as a control test structure to compare with the antenna structures. Figure 11.2b shows a transistor connected to an area antenna in the gate metal layer of the transistor. Figure 11.2c shows a perimeter antenna structure.

Typically, a minimum area transistor allowed by the design rules is used in antenna structures to increase the antenna ratio, as described in Chap. 2. Structures with different antenna areas and perimeters are used to vary the degree of process damage and determine the design rule capability for a process. For example, antenna ratios of 100, 1000, 10,000, etc., are used to explore orders of magnitude changes in the antenna ratio. Structures similar to the sketch in Fig. 11.2 can be repeated in any conducting layer in a CMOS process and connected to a small area transistor.

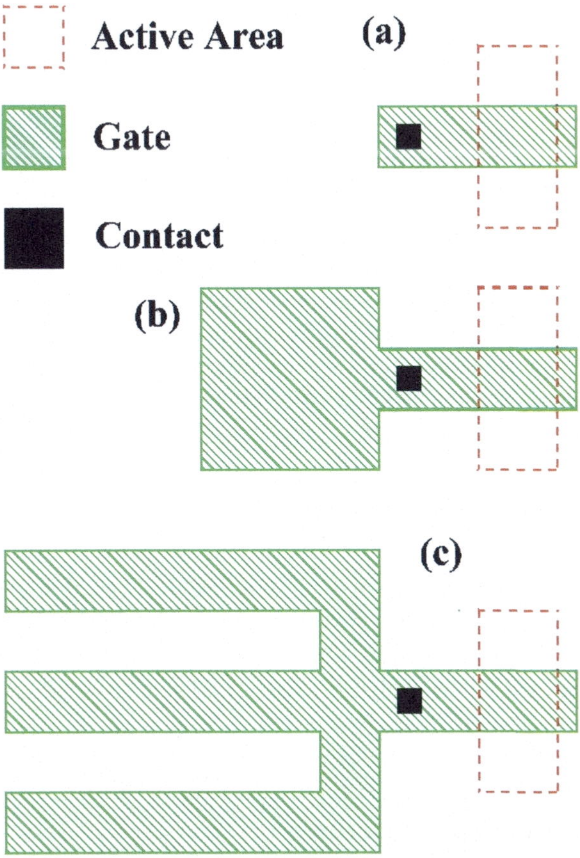

Fig. 11.2 Layout sketch showing gate antenna test structures with (**a**) no antenna, (**b**) area antenna, and (**c**) edge antenna. The antennas are not drawn to scale

11.5 Well Antenna Structures

Plasma damage due to coupling through CMOS wells, as described in Chap. 2, can be characterized by the addition of transistors connected to well antennas. Structures such as well antennas (Chap. 2) can be configured to match the design rules varying from the minimum antenna dimensions and beyond.

11.6 Source-Drain Antennas

Damage from source-drain antennas is described in Chap. 2. Structures can be designed to create source-drain antennas with different metal antennas, as shown in this section. Differential antennas have different metal-to-gate antenna ratios and different source-drain antenna ratios. These structures can generate the results shown in Chap. 2.

11.7 Matching Transistor Structures

Matching transistors are crucial devices for critical circuits. Matching transistors are typically constructed using the best-known methods, and many design rules are optimized for minimum transistor variation (Hastings, 2005). Matching test structures can include plasma damage experiments that are consistent with the design rules to analyze potential problems. Specifically, matching metal patterns and coverage of the transistor with metal play critical roles in matching transistor variation.

11.8 Light Damage Structures

High-energy light can damage devices, as described in Chap. 9. Light damage test structures consisting of structures covered and not covered by large metal features can be used to detect light damage. The light damage detection capability of these structures can be confounded with hydrogen blocking or absorption, as described in Chap. 14.

11.9 Ring Oscillator Structures

Ring oscillators are commonly used to characterize and monitor circuit speeds. Ring oscillators are susceptible to plasma and process damage, as described in Chap. 5. Ring oscillators with antenna structures are not very practical as the antenna-related capacitive loads slow down the oscillator. However, structures with and without gate diodes can detect degradation in performance because of the small diode capacitance. The difference in speed with the addition of the diode can be predicted with design simulations enabling the detection of performance degradation.

11.10 Multiple Metal Layer Detection

The area available on a wafer for parametric test structures is often limited, requiring optimization of the structures for maximum learning and minimum area. Antennas in multiple process steps can be combined into one structure. The

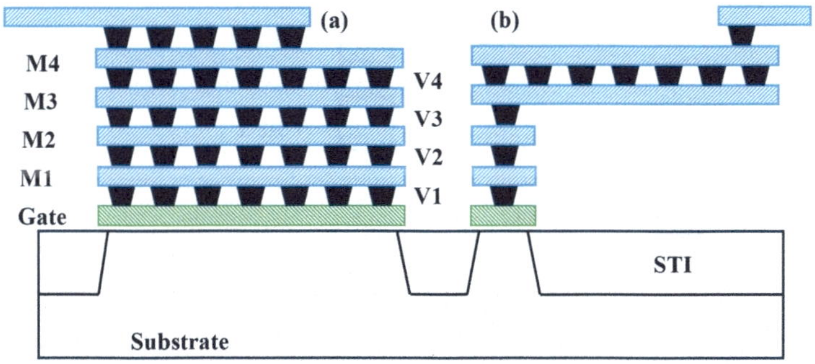

Fig. 11.3 Cross-section sketch showing parametric antenna structures. (**a**) Multilevel antenna for detection of plasma damage on many levels. (**b**) Single level antenna for the detection of process damage on metal3 M3, via4 V4, and metal4 M4

composite structure enables the detection of plasma damage on as many levels as possible with fewer test structures but creates difficulty in isolating the exact damaging step. Figure 11.3 shows an example of combined and isolated structures (Martin, 2020). Figure 11.3a shows a combined test structure cross-section sketch. Each of the metal and via layers is replicated in the antenna layout and is stacked on top of each other. Parametric measurements on these structures at the completion of the process reveal the cumulative effects of process damage on the stacked layers. Figure 11.3b shows an isolated structure where only process damage on metal 3 M3, via 3 V3, and metal 4 M4 is measured. Connections in the layers not measured use the minimum areas to avoid creating small antennas. This type of structure allows quantitative measurements of process damage on these specific levels.

11.11 Contact and Via Structures

Specialized contact and via antenna structures can electrically isolate process damage that occurs during the fabrication of these process modules. Since all contacts and vias are usually designed with the same exact sizes of contacts and vias in a CMOS process, the process damage increases with increasing numbers of connections to a single node. Instead of using area or perimeter to increase process damage, via and contact antennas use different numbers of contacts to accelerate damage. Figure 11.4 shows a layout sketch of contact and via antennas. Figure 11.4a and b show contact antennas connected to the gate metal of a transistor. One structure has a single contact, and the second structure has an extensive array of contacts, allowing comparison of the transistor characteristics. A similar structure for vias is shown in Fig. 11.4c and d, where one structure has a single via and the second structure has an array of vias. Thus, the extent of the damage caused by contact and via processes can be determined by an electrical comparison of the two structures. The

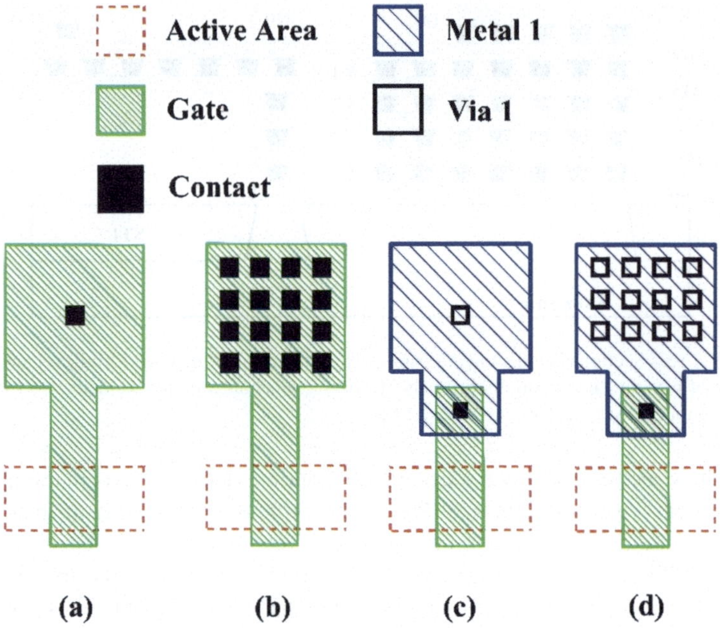

Fig. 11.4 Layout sketch showing contact and via antennas. Not to scale. (**a**) single contact used as a control structure for comparison. (**b**) array of contacts creating an antenna on the contact layer. (**c**) single via used as a control structure for comparison. (**d**) array of vias creating antennas on the via layer. Additionally, the resist can completely cover the antenna

numbers of contacts and vias in this type of structure typically vary by orders of magnitude to span the requirements for a CMOS chip.

11.12 Resist Antenna Structures

Resist structures can play a role in process damage, as described in Chap. 9. Resist or other masking materials can modify the behavior of any antenna with a resist patterning layer over it. Figure 11.5 shows some example layout sketches of resist-related structures (Dirnecker et al., 2002). Figure 11.5a shows a gate area antenna structure used as a control. Figure 11.5b shows a gate antenna with a large resist area overlapping it. This structure determines if the resist conducts charge, increasing the effective antenna area. Figure 11.5c shows a gate antenna with a nearby resist antenna with a gap between the resist antenna and the gate antenna. These types of structures typically are used in implant layers with a preference for the implantation conditions more likely to cause damage, as described in Chap. 9. This type of structure is used to measure the damage caused by charge flow or arcing between the resist and the gate.

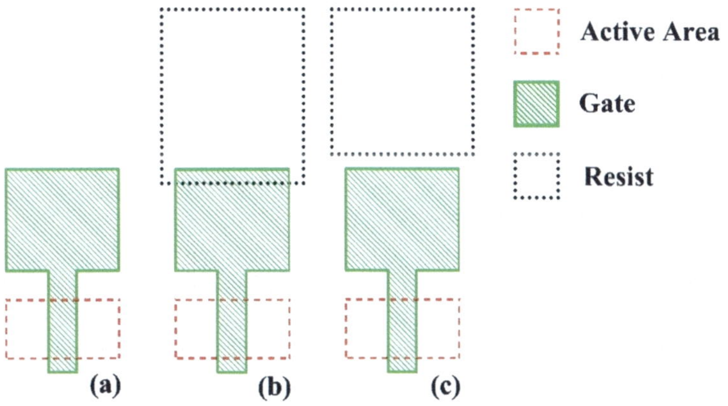

Fig. 11.5 Layout sketch showing resist antenna test structures. (**a**) Control antenna with no resist. (**b**) Resist antenna connected to a gate antenna. (**c**) Resist antenna with a gap to the gate antenna

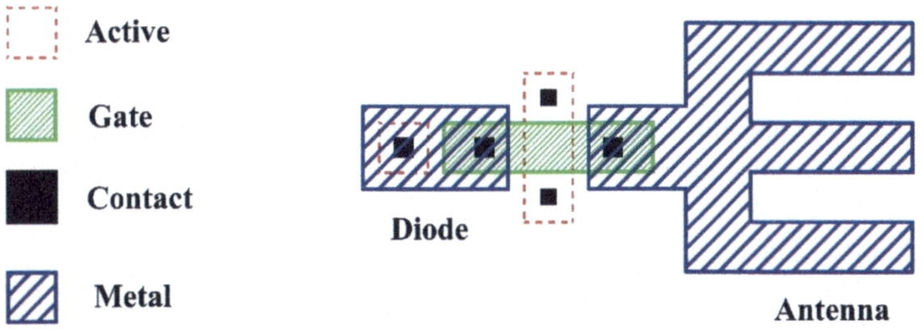

Fig. 11.6 Layout sketch of a metal antenna protected by a diode

11.13 Diode Protection

Diodes can be used to protect structures from process damage, as described in Chap. 3. Diodes protect sensitive process elements across multiple processing steps, starting from the process step, where the diode is connected in the metal to the end of the process. Diodes can be selectively added to process damage test structures. Identical test structures with and without diodes connected at various points in the process can provide critical data to isolate the specific process steps causing process damage. The diodes can be easily added or removed from a structure by adding or removing a contact or modification of a metal layer. The comparison of structures with and without diodes allows careful experimentation to understand the source of process damage. Figure 11.6 shows an example sketch of an antenna protected by a diode.

11.14 Comb Structures

Process damage on an edge or comb structure can be a function of the pitch or space in addition to the perimeter of the comb (see Chap. 9). Differences in etch rates or plasma composition as a function of pitch or space are responsible for this variation. Therefore, test structures with varying spaces or pitches are necessary to determine the worst-case design rules for process damage.

The fingers of perimeter antennas do not necessarily have to be electrically connected to affect process damage (see Chap. 9). The unconnected fingers add to the antenna perimeter or area until they are disconnected, as all the conductor material is removed from the space during the etch. Figure 11.7 shows a layout sketch of connected and unconnected perimeter antennas (Simon et al., 2000).

11.15 Alternative Transistor Layout

11.15.1 Introduction

An alternative transistor layout for bulk CMOS can provide some improvement in process damage resistance by reducing the probability of failure along the shallow trench isolation edge described in Chap. 5. These types of structures are also used in radiation-hardened applications. However, these types of layout are not practical for finfets. The key feature of these alternative transistor layouts is preventing leakage along the isolation edge by eliminating the leakage path or increasing the leakage path distance to reduce the parasitic current. These techniques are valuable in circuits such as critical analog modules such as sense amplifiers. These types of layout may also improve transistor parameters such as breakdown. In addition, the layout penalty for these types of layout is minor, and the layout can be used selectively.

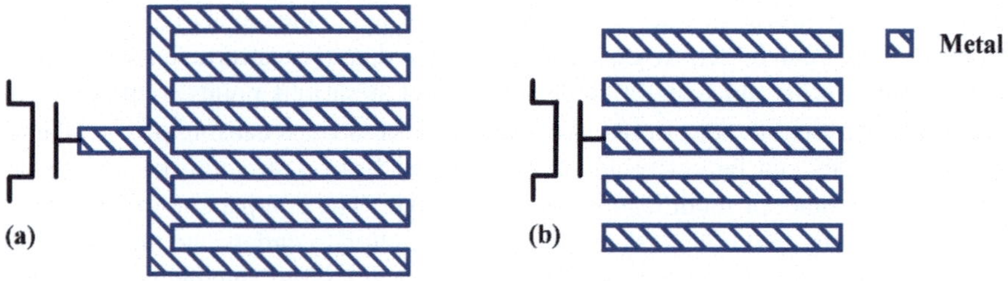

Fig. 11.7 Layout sketch showing two different comb antennas. (**a**) Connected metal comb. (**b**) Unconnected comb

11.15.2 Dog Bone Field Edge Layout

One layout option is to place a small rectangle at the gate edge to increase the gate length overlapping the field edge, as shown in Fig. 11.8, resulting in a dog bone-shaped gate layout (Esqueda et al., 2005). The field edge leakage path is increased, and the leakage path is placed under gate control. The plot shows improved leakage with increasing dog bone size.

11.15.3 Donut Devices to Suppress Field Edge Leakage

Donut devices can be used to suppress field or STI edge leakage by eliminating the parallel field conduction path between the drain and source (Bezenova et al., 2018; Lee & Lee, 2013; McLain et al., 2009; Wang et al., 2009). These types of devices are often used in radiation-hardened CMOS to prevent field edge leakage after radiation damage. Figure 11.9 shows several options for field edge leakage prevention. Figure 11.9a shows the gate placed entirely inside the active area, with the drain in the center of the donut and the source outside of the donut. This donut device uses a gate and active area layout to place all the field edge regions on the source side of the transistor to prevent any possibility of field edge leakage. All field edges are on the source side, which is often connected to the ground. The octagon shape of the donut is used to reduce the peak electric fields by eliminating the gate corners shown in the other devices. Figure 11.9b–d shows alternative layouts where the length of the conduction path along the field edge is increased but not eliminated. These devices do not completely eliminate the field edge leakage path but may increase the path length and resistance to the point where the STI edge leakage is negligible.

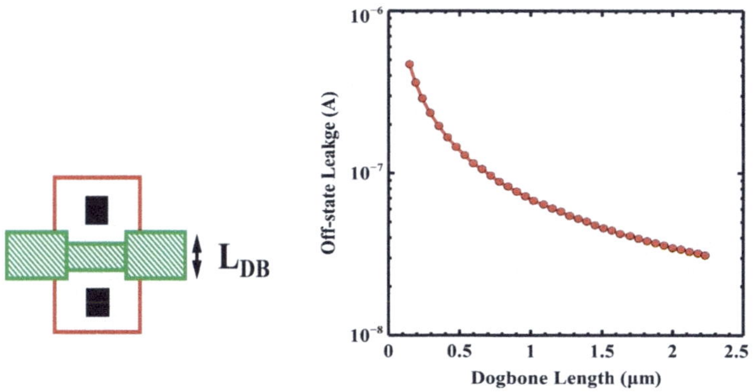

Fig. 11.8 Dog bone layout to improve field edge leakage and plot showing drain to source leakage as a function of dog bone length

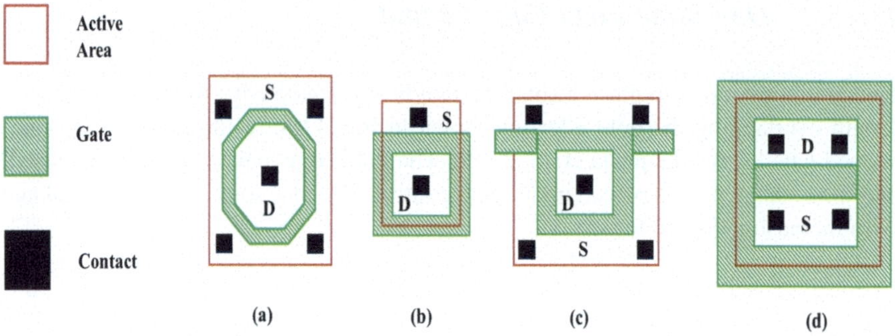

Fig. 11.9 Layout examples for a donut gate transistor

Design rules may or may not allow a contact to the gate over the active area. Traditionally, gate contacts over active areas were forbidden because metallization, such as aluminum, would diffuse into the polysilicon gate and damage the gate dielectric. Modern CMOS processes are generally capable of forming gate contacts over active areas as the metallization prevents any interaction of the contact and gate materials. The structure in Fig. 11.9a requires a gate contact over the active area, and the rest of the structures can be contacted over the isolation. The device in Fig. 11.9a is the best solution for STI edge leakage if the contact to the gate over the active area is allowed in the design rules.

11.16 Through-Silicon Via Test Structures

Through-silicon vias (TSVs) are a potential source of plasma damage due to the complex structure and the large amount of plasma exposure during the processing of the very deep structure, as described in Chap. 9. Device structures connected to TSVs can suffer process and plasma damage. The TSV serves as an antenna, with the number of TSVs connected to a gate serving as the antenna ratio. Different antenna ratios can be created by varying the TSV numbers connected to varying gate sizes. Figure 11.10 shows an example layout for a TSV antenna structure (Li et al., 2018). TSVs are connected to a small area transistor or capacitor. The metallization used in the chip to connect the TSVs is used for the test structure.

11.17 Structures for Metal Insulator Metal Capacitors

Plasma damage of metal–insulator–metal capacitors is described in Chap. 8. As shown in Chap. 8 figures, several options are needed to fully characterize the capacitor, including a lower plate connected to the ground, a metal line, or a gate. These three devices need to be configured with different metal sizes and via antennas for

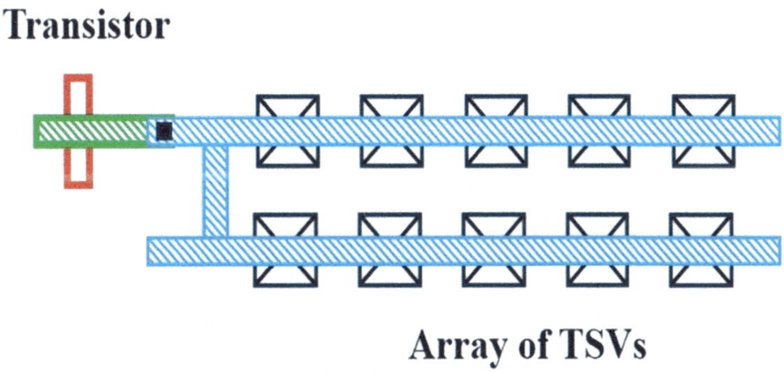

Fig. 11.10 Layout sketch of TSV plasma damage test structure

the different metal layers or a multilevel antenna, as shown in Chap. 8. The antenna ratio between the metallization and the MIM capacitor can be explored with different MIM capacitors and antenna sizes consistent with the design rules.

11.18 Memory Array Wordline Structures

The wordlines of memory arrays such as DRAM or flash are susceptible to plasma damage, as described in Chap. 8. Parametric structures can be configured to place various antennas connected to wordlines, along with skipping or adding layers, as shown in Chap. 3. Structures with and without diodes can be used if the layout is possible. Parametric structures are helpful for the observation of more catastrophic problems such as dielectric leakage, breakdown, or fundamental cell failure. The complete characterization of the reliability problems described in the memory section typically requires the layout of wordline damage structures which are placed inside the CMOS memory product chip so that full product-level functional and reliability testing can be performed.

11.19 Structures for Metallization Damage

Plasma damage can occur to the interconnect metallization in a CMOS chip, degrading the intermetal dielectrics, causing leakage, and other degraded electrical properties. Low-k dielectrics can be more susceptible to damage because of the weak bonds relative to silicon dioxide. The metallization reliability may be degraded, affecting electromigration and stress migration and creating field reliability hazards. Plasma damage in metals is commonly measured using the simple comb and serpentine test structures shown in Fig. 11.11. Comb structures are commonly placed on each metal layer and stacked, allowing multiple metal layers to be tested with

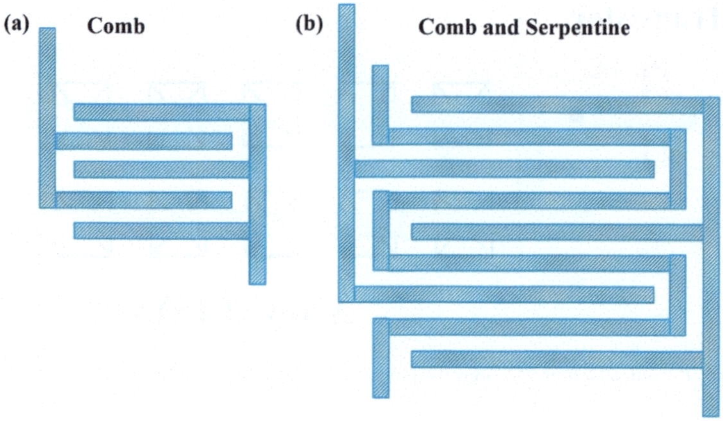

Fig. 11.11 Sketch showing common metallization structures. (**a**) Interdigitated comb structure. (**b**) Interdigitated comb structure with a serpentine continuity structure between the combs

one structure and emulating an actual chip layout (Baklanov et al., 2012, pg. 380). A worst-case layout can be implemented by connecting one terminal of the comb to the substrate through contacts and vias. The connection to the substrate creates a high electric field during plasma damage compared to a structure with both terminals of the comb floating, creating the test structure shown in Chap. 6. The comb and serpentine structures can also be connected to transistors to form an antenna.

Additionally, damage can be detected with contact and via chains (Baklanov et al., 2012; Tsai et al., 2018). These types of structures may fail more frequently due to the interfaces between different metals and process-related interfaces.

References

Baklanov, M. R., et al. (2012). *Advanced interconnects for ULSI technology*. Wiley. https://doi.org/10.1002/9781119963677

Bezenova, V., et al. (2018). Modeling of the equivalent channel dimensions of integrated MOS transistors in radiation-hardened enclosed layout. *e & i Elektrotechnik und Informationstechnik, 135*(1), 61–68.

Dirnecker, T., et al. (2002). Influence of photoresist pattern on charging damage during high current ion implantation. In *7th international symposium on plasma- and process-induced damage*. https://doi.org/10.1109/PPID.2002.1042620

Esqueda, I. S., et al. (2005). Two-dimensional methodology for modeling radiation-induced off-state leakage in CMOS technologies. *IEEE Transactions on Nuclear Science, 52*(6), 2259–2264. https://doi.org/10.1109/TNS.2005.860671

Hastings, A. (2005). *The art of analog layout* (2nd ed.). Pearson.

Kahng, A. B., & Samadi, K. (2008). CMP fill synthesis: A survey of recent studies. *IEEE Transactions on Computer-Aided Design of Integrated Circuits and Systems, 27*(1), 3–19. https://doi.org/10.1109/TCAD.2007.907061

Lee, M. S., & Lee, H. C. (2013). Dummy gate-assisted n-MOSFET layout for a radiation-tolerant integrated circuit. *IEEE Transactions on Nuclear Science, 60*(4), 3084–3091. https://doi.org/10.1109/tns.2013.2268390

Li, Y., et al. (2018). TSV process-induced MOS reliability degradation. *IEEE International Reliability Physics Symposium*. https://doi.org/10.1109/IRPS.2018.8353610

Martin, A. (2020). Plasma induced charging damage: From appropriate MOS test structures to antenna design rules, a comprehensive process qualification procedure. In *2020 IEEE International Integrated Reliability Workshop (IIRW)*. https://doi.org/10.1109/IIRW49815.2020.9312854

McLain, M., et al. (2009). Reliability of high performance standard two-edge and radiation hardened by design enclosed geometry transistors. *IEEE International Reliability Physics Symposium*. https://doi.org/10.1109/IRPS.2009.5173247

Simon, P., et al. (2000). Identification of plasma-induced damage conditions in VLSI designs. *IEEE Transactions on Semiconductor Manufacturing, 13*(2), 136–144. https://doi.org/10.1109/66.843628

Tsai, Y.-P., Liou, P.-C., Lin, C. J., & King, Y.-C. (2018). Plasma charging effect on the reliability of copper BEOL structures in advanced FinFET technologies. *IEEE Journal of the Electron Devices Society, 6*, 875–883. https://doi.org/10.1109/JEDS.2018.2859365

Wang, R.-V., et al. (2009). Shallow trench isolation edge effect on random telegraph signal noise and implications for flash memory. *IEEE Transactions on Electron Devices, 56*(9), 2107–2113. https://doi.org/10.1109/TED.2009.2026116

Chapter 12
Design Rules Related to Process Damage

12.1 Introduction

Design rules and common sense naturally minimize some worst-case process damage layout configurations. For example, it makes very little sense to connect an extensive array of contacts to the gate of a transistor. It wastes chip area with no added benefit. Designers usually avoid nonsense situations like this. However, design rules must be carefully modified with consideration for process damage to reduce the probability of process damage. There is a trade-off between mitigating process damage and complex, burdensome design requirements. Design practices cannot eliminate process damage. For example, plasma damage can vary as a function of the metal pitch. It is likely too costly from a design perspective to attempt to manage this issue. The fundamental responsibility for mitigating process damage lies with the process technology, with design practices assisting in mitigation.

12.2 Process Module-Specific Design Rules

Antenna-specific design rules are commonly added to CMOS design rules to prevent the formation of the worst-case circuits and layout. Table 12.1 shows example design rules for the prevention of charging. The design rules typically mitigate charging by limiting the size of the antennas by setting an upper limit on the antenna ratio based on the antenna area or perimeter. Design rules also attempt to minimize the worst-case conditions. The management of process damage risk by design practices can be challenging to implement in design rule checkers, increasing the complexity of the design methodology. The suggested rules in Table 12.1 are provided to create a list for consideration. The actual rules are very specific to the underlying

K. Prall, *CMOS Plasma and Process Damage*,
https://doi.org/10.1007/978-3-031-89029-1_12

Table 12.1 Design rules for the prevention of process damage

Module	Rule number	Description	Considerations	References
Well	W1	Maximum well area for all wells (nwell, pwell, triple-well)	Limit well to adjacent active area antenna ratio	Beckmeier and Ring (2021), Martin and Kamp (2020), Chu et al. (2018)
Well	W2	Well to adjacent connected active area ratio for all wells (nwell, pwell, triple-well)	Limit well to adjacent active area antenna ratio	Beckmeier and Ring (2021), Martin and Kamp (2020), Chu et al. (2018)
Gate	G1	Maximum gate area on isolation	Limit maximum gate antenna ratio	Cheung (1999)
Gate	G2	Maximum gate perimeter on isolation	Limit maximum gate antenna ratio	Cheung (1999)
Gate	G3	Maximum gate area on isolation to gate area ratio	Limit maximum gate antenna ratio	Cheung (1999)
Gate	G4	Maximum gate perimeter on isolation to gate area ratio	Limit maximum gate antenna ratio	Cheung (1999)
Gate	G5	No direct gate connection to voltage supply or ground without passing through a transistor	Prevent gate connection to large power supply or ground metal antenna requiring the formation of a diode connection to the gate	Cheung (1999)
Source Drain	S1	Maximum source-drain area for an individual transistor	Limit S/D to gate antenna ratio Sets maximum W of transistor	Lin and Sery (2003)
Metal	M1	Maximum metal area to gate, source, or drain area ratio for metal X layer	Limit maximum metal to gate, source, and drain antenna ratio by individual metal layer for all metal layers	Martin (2013)
Metal	M2	Maximum cumulative metal area to gate, source, or drain area ratio for all metal layers combined	Limit maximum cumulative metal to gate antenna ratio for all metal layers	Martin (2013)
Metal	M3	Maximum metal to gate, source, or drain area ratio for metal antennas greater than a distance from the gate	Reduce metal antenna ratios for distant antennas across the chip to reduce damage risk	Lai et al. (2006)

(continued)

Table 12.1 (continued)

Module	Rule number	Description	Considerations	References
Metal	M4	Maximum differential metal antenna ratios between gate, source, and drain ratios for each metal layer	Limit differential antenna risk	Lai et al. (2006)
Metal	M5	Cumulative maximum differential metal antenna ratios between gate, source, and drain ratios for all metal layers	Limit differential antenna risk	Lai et al. (2006)
Contact	C1	Maximum number of contacts connected to a gate	Limit contact antenna by limiting the maximum number of contacts to a gate	Tsui (2000)
Contact	C2	Maximum number of contacts connected to a source or drain as a function of transistor W	Limit contact antenna by limiting the maximum number of contacts to a source or drain	Tsui (2000)
Via	V1	Maximum number of vias indirectly connected to a gate by via layer X	Limit via antenna by limiting the maximum number of vias to a gate for a via layer	Tsui (2000)
Via	V2	Maximum cumulative number of vias indirectly connected to a gate for all via layers.	Limit via antenna by limiting the cumulative maximum number of vias for all via layers	Tsui (2000)
MIM	MIM1	No MIM connected directly to the gate of a transistor	Eliminate circuits with high damage potential	Weng et al. (2009)
MIM	MIM2	Maximum MIM area to gate area ratio	Limit MIM antenna area ratio	Weng et al. (2009)
Pad	P1	Maximum pad area	Limit the scope of pad antennas	Lin (2016)
TSV	TSV1	Maximum number of TSVs connected to a sensitive structure like gates, MIM caps, etc.	Limit damage from TSVs	Li et al. (2018)
TSV	TSV2	Forbid TSVs from connecting directly to a sensitive structure like gates, MIM caps, etc. without a diode or protection circuit	Limit damage from TSVs	Li et al. (2018)

CMOS technology. Diode protection-related rules are not included, as they can usually be added using the regular design rules. A sound design rule methodology may prevent a few instances of extreme layout that can risk the entire chip.

12.2.1 Well Design Rules

Well design rules are needed for bulk CMOS to limit the well coupling-related damage described in Chap. 2. Wells are usually configured in banks containing many transistors to conserve the chip area, as well-related rules tend to be large. Therefore, any limitation to the well size limits the maximum number of transistors in a single well. Rules for all types of wells in a given CMOS technology must be considered. Special attention needs to be paid to triple-wells as they tend to have less leakage and are more susceptible to plasma damage. Large wells are more conducive to practically using diode protection, as the design penalty can be small, with one diode for a large well. Therefore, diode protection, such as the ones described in Chap. 3, can often be implemented without penalty. Rules W1 and W2 in Table 12.1 are suggested to protect against well-related damage. Design rules should prevent an unnecessarily large well from being used, with the minimum well necessary to enclose the transistors being implemented.

12.2.2 Gate-Related Design Rules

Gate-related plasma protection design rules are commonly used to prevent a large gate antenna from being formed. Gate protection design rules typically do not cause any design burden because there is no benefit to running large gate runs on top of the isolation in modern CMOS. Maximum gate antenna rules typically prevent the inadvertent layout of antennas. The gate layer is not often used as a routing layer as any design edits require mask revisions early in the process, degrading the cycle time for design revisions. The gate protection rules can limit the maximum gate routing on the field isolation (rules G1 and G2) or control the maximum antenna ratios (rules G3 and G4). Direct gate connection to a power supply or ground should be discouraged as power supply buses tend to form large antennas (rule G5). Gates connected to power supplies or ground should be connected through the drain and source of a transistor, avoiding a direct connection.

12.2.3 Source-Drain-Related Design Rules

Similar to well-related plasma damage protection, the maximum source-drain active areas should be limited to avoid the formation of an unnecessarily sizeable active area region. There are typically no good design reasons for an abnormally large active area. Historically, the active area was sometimes used as a routing layer, but modern performance requirements prohibit this use. The active area should be restricted to the minimum necessary for the source-drain junction formation and the related contacts. Rule S1 shows an example of this type of rule. Also, the minimum source-drain area can be enforced by rules such as a maximum contact to gate distance or a maximum active area surround of the contact. Advanced CMOS processes have aggressive design rules to achieve high gate density, making it unlikely that large source-drain areas are used.

12.2.4 Metal-Related Design Rules

Metallization fundamentally creates antennas due to the dense routing requirements of modern CMOS technology. In addition, advanced logic devices commonly have more than ten metal layers to achieve the wiring density necessary to connect high-density transistor arrays. As a result, large metal antennas can be naturally created while achieving the necessary wiring density.

Process damage design rules limit the magnitude of metal antennas, typically achieved by limiting the maximum antenna area, perimeter, or the antenna ratio of the metal area or perimeter to the gate, source, drain area, or perimeter. This limitation limits the maximum length of a metal line on an individual layer, requiring a given signal routed in metal to jump between metal layers through via connections to spread the antenna among multiple metal layers. Figure 12.1 shows a layout sketch of this modification (Sicard & Bendhia, 2007, pg. 175). Figure 12.1a shows a metal line running across the chip in a single layer without plasma damage design rules, and Fig. 12.1b shows the impact of the implementation of antenna design rules. The metal antenna limitations by design rules have to be sufficiently large to be practical, allowing the chip design to achieve the necessary wiring density without choking the routing by requiring many signals to be routed in many metal layers.

The maximum metal area and perimeter are limited by design rules to prevent plasma damage. These rules can significantly impact the layout because metal routing can run large distances across the chip. The metal routing for long runs must jump from one metal layer to another to break up the antenna into smaller

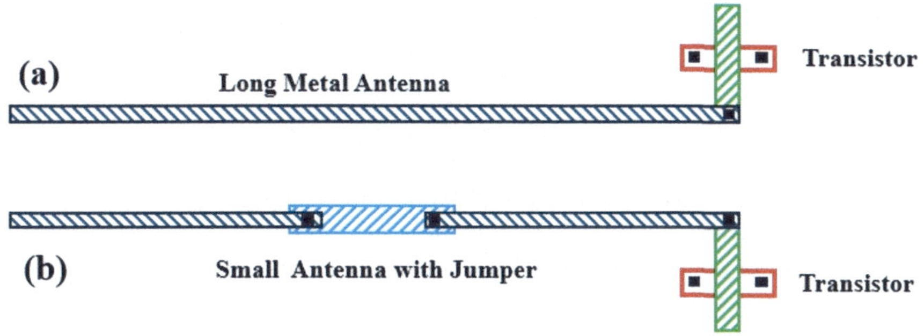

Fig. 12.1 Top down layout sketch showing an example of (**a**) worst case layout creating a large antenna and (**b**) reduction in the antenna ratio by adding a metal jumper, routing the signal to another metal layer to break up the large antenna

independent sections to avoid violating the design rules. An isolated metal line not connected to the gate at the process step where charging damage occurs mitigates the charging problem due to the reduction in the antenna size (rules M1 and M2). The metal is connected to the circuit at subsequent process steps, with each interconnect mask layer antenna size being below the design rule limits.

Connections to power and ground tend to form massive antennas that are practically difficult to limit. Large antennas connected to gates are easier to manage by preventing a direct connection to power or ground (rule G5). On the other hand, large antennas to sources and drains are more challenging to manage, as a significant fraction of the sources and drains of the transistors must be connected to the power supply and ground. Typically, antenna rules can be more strict for gate connections than for sources or drains.

It would be helpful to reduce the metal antenna sizes for antennas far from the transistor to reduce remote antenna plasma damage problems described in Chap. 2. Usually, circuit blocks are placed in the chip layout, with the metal being routed later by computer-aided design tools such as place and route. As a result, design rules to limit remote antennas (rule M3) are challenging to implement as they would require shifting the placement of transistor blocks to avoid remote antennas.

The differential antennas described in Chap. 2 are problematic to mitigate with design rules (rules M4 and M5). Attempting to force similar antenna sizes on both the source and drain requires increasing the amount of metallization on one diffusion antenna while reducing the amount of metallization on the other diffusion. Forcing more wiring than is needed for design reasons may result in an area and performance penalty.

12.2.5 Contact and Via-Related Design Rules

Contacts or vias can also create charging problems, usually during the etch or metallization steps, which form the contact or via. Therefore, the design rules limit the maximum number of contacts or vias to reduce the antenna area (rules C1, C2, V1, and V2). Practically, only one or two contacts are needed for a good electrical connection to the gates, so the design rules can forbid any large contact arrays connected to the gate. The number of practical contacts to the drain is limited by the width of the transistor for planar CMOS technology, so design rules can be written to require as many contacts in a line that can fit in the active area parallel to the gates as possible. Additional contacts can be forbidden by the design rules.

For finfet transistors, the very small width of the fins allows only a maximum of a single drain or source contact per fin, limiting the need for multiple contacts. Normally, a single rectangular drain or source contact shorts several fins together, forming a single structure. Finfet gate contacts are usually limited because of the limited area.

Limiting via antennas with design rules is more complicated than contacts. The number of vias placed between two perpendicular crossing metal lines is a function of the width of the metal lines. Usually, as many vias as possible are placed between the two lines in an orthogonal regular array. This practice achieves the lowest resistance interconnect and the best reliability for a given metallization technology. Limiting the size of via antennas requires a reduced number of vias between large metal busses, impacting resistance, performance, and reliability. Via antenna rules (rules V1 and V2) are generally specified to restrict the maximum upper numbers of vias to avoid interfering with low-resistance bus requirements by eliminating the worst-case conditions.

12.2.6 Metal-Insulator-Metal Capacitor-Related Design Rules

Metal-insulator-metal (MIM) capacitors also need process damage-related design rules to prevent the problems described in Chap. 8. The design rules can prevent a direct connection between a MIM capacitor terminal and a transistor gate by forbidding the connection and avoiding the worst-case situation (rule MIM1). Optionally, design rules can limit the maximum antenna ratio of the MIM capacitor to the gates (rule MIM2). However, limitations in MIM antenna ratios to source-drain regions are problematic because of the heavy use of MIM capacitors connecting power supplies and ground to improve noise and supply stability.

12.2.7 Bond Pad-Related Design Rules

Bond pads can be particularly problematic because the large area of the bond pad is a natural antenna. As a result, charging can occur during the bond pad formation and the passivation process modules. The electrical probing and assembly requirements set minimum bond pad areas, setting an upper limit on the pad size and antenna ratio. Pad sizes larger than the dimensions necessary for testing and assembly are usually unnecessary and can be restricted by the design rules (rule P1). CMOS product chip pads are usually heavily protected by electrostatic discharge protection circuits that also protect against process damage. Parametric test structures are susceptible because the bond pad commonly has no electrostatic discharge circuits to protect it like the pads in the product chip. Special pad rules and configurations for parametric structures are usually necessary. For example, a duplicate test structure layout with tiny pads can be used to experimentally determine damage from pad antennas.

12.2.8 Through-Silicon Via-Related Design Rules

The use of through-silicon vias (TSVs) described in Chap. 9 can cause process damage that can be mitigated through design rules. The design rules can limit the maximum number of TSVs connected to a single sensitive node. The design rules may forbid TSV connections to sensitive nodes such as gates or MIM capacitors without a diode or an electrostatic discharge protection circuit.

12.2.9 Technology-Specific Design Rules

Plasma damage-related design rules for technology-specific structures such as DRAM and NVM rowlines should be considered to mitigate potential damage problems. The primary area of concern is often damage on the wordlines or gate connections for technologies like DRAMs, nonvolatile memories, and CMOS imagers. Typically, there is little or no layout space available to add diodes, as described in Chap. 8. Damage mitigation usually involves careful optimization and minimization of metallization-related antennas and placing a transistor as close as possible to the worldline or gate connection with metal antennas on the opposite side of the transistor from the wordline connections.

References

Beckmeier, D., & Ring, M. (2021). Relation of PID well to gate antenna charging effects. IEEE International Integrated Reliability Workshop, https://doi.org/10.1109/IIRW53245.2021.9635620.

Cheung, K. P. (1999). *Plasma charging damage*. Springer. https://doi.org/10.1007/978-1-4471-0247-2

Chu, Y.-L., et al. (2018). A new mechanism of signal path charging damage across separated power domain deep N-well Interface. *IEEE International Reliability Physics Conference*. https://doi.org/10.1109/IRPS.2018.8353631

Lai, W., et al. (2006). Ultra-thin gate dielectric plasma charging damage in SOI technology. *IEEE International Reliability Physics Symposium*. https://doi.org/10.1109/RELPHY.2006.251246

Li, Y., et al. (2018). TSV process-induced MOS reliability degradation. *IEEE International Reliability Physics Symposium*. https://doi.org/10.1109/IRPS.2018.8353610

Lin, W. (2016). Bond-pad charging protection design for charging-free reference transistor test structures. *IEEE Transactions on Electron Devices, 63*(1), 272–279. https://doi.org/10.1109/TED.2015.2501418

Lin, W., & Sery, G. (2003). Multiple-terminal gate charging effect—Competing/compensating charging behavior. *IEEE Electron Device Letters, 24*(7), 481–483. https://doi.org/10.1109/led.2003.814997

Martin, A. (2013). Circuit relevant well charging from metal antenna and its degradation on digital MOS transistor reliability. *IEEE International Integrated Reliability Workshop*. https://doi.org/10.1109/IIRW.2013.6804156

Martin, A., & Kamp, A. (2020). A new implementation approach for reliability design rules against plasma induced charging damage from well configurations of complex ICs. *IEEE International Reliability Physics Symposium*. https://doi.org/10.1109/IRPS45951.2020.9128866

Sicard, E., & Bendhia, S. D. (2007). *Basics of CMOS cell design*. McGraw-Hill.

Tsui, B. Y. (2000). Plasma charging damage during contact hole etch in high-density plasma etcher. *Microelectronics Reliability, 40*, 2039–2046. https://doi.org/10.1016/S0026-2714(00)00023-8

Weng, W.-T., Lee, Y.-J., Lin, H.-C., & Huang, T.-Y. (2009). Effects of plasma damage on metal–insulator–metal capacitors and transistors for advanced mixed-signal/radio-frequency metal–oxide–semiconductor field-effect transistor technology. *Japanese Journal of Applied Physics, 48*(8), 086001. https://doi.org/10.1143/jjap.48.086001

Chapter 13
Parametric Damage Testing Strategy and Procedures

13.1 Introduction

Process and plasma damage effects are complicated and deserve careful attention to the definition of the parametric testing strategy. Data collection is needed for many parameters, followed by reliability testing to determine the reliability degradation. Different types of test structures need to be measured to isolate process damage from other parametric shifts.

13.2 Plasma Damage Detection Strategy

The widely varying damage caused by process and plasma damage requires a comprehensive and well-thought-out testing strategy to detect time zero and reliability degradation. Figure 13.1 shows an example flowchart for a process damage parametric test flow. The suite of process damage-related parametric tests is typically a subset of the standard tests performed during wafer acceptance testing at the end of the process flow. The test results are compared with the process specifications established for the CMOS technology. Each parametric test is usually controlled with process control charts, enabling the statistical detection of process damage. A diagnostic analysis of failing parameters is performed to determine the cause of the deviation and if process damage is a possible cause of the problem (Martin, 2020).

K. Prall, *CMOS Plasma and Process Damage*,
https://doi.org/10.1007/978-3-031-89029-1_13

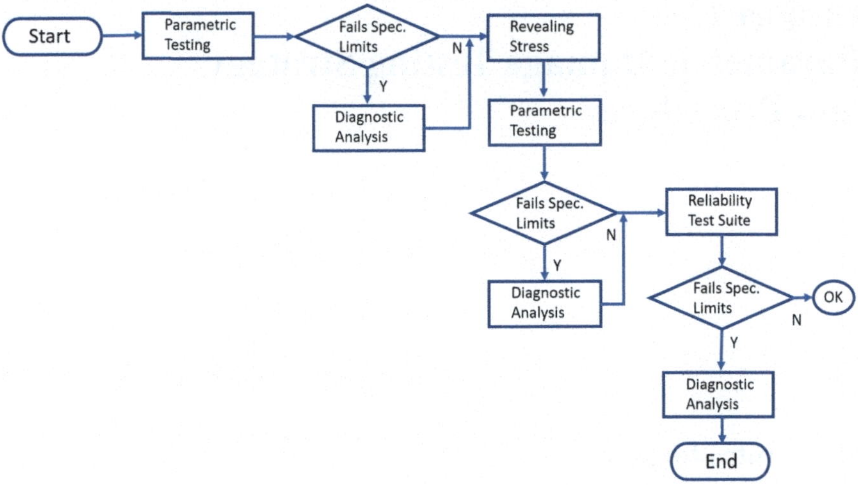

Fig. 13.1 Example flow chart showing process and plasma damage testing strategy

13.3 Parametric Testing

The basic function and integrity of the measured devices and structures need to be checked before extensive, time-consuming measurements are performed. Integrity checks are designed to detect fundamentally defective structures and devices that do not function properly to shorten test times and avoid contaminating the test database with erroneous results. In addition, defective devices need to be screened and analyzed to determine if the failure is associated with process damage or some other cause. This section describes simple checks to ensure that the measured devices are functioning by using extreme limits that would not be expected and failing devices that exceed the extreme limits. The values of the limits set are typically orders of magnitude higher than the expected values during regular device operation. These tests also eliminate catastrophically misprocessed wafers. Table 13.1 shows a list of commonly used integrity checks.

Shorts tests need to be performed to ensure that the different terminals are electrically isolated. Typical test conditions force a small voltage, which is applied between two terminals, and the resulting current is measured. The test results are compared with the expectation of continuity or isolation, and defective devices are screened. For example, the gate should not leak to the drain and source at a high level, and the drain should not leak to the source at a high level.

Continuity tests can be performed to ensure that proper connections exist between the probe tips and the devices to be measured. For example, probe tips connected to the source and drain should conduct current when the junction is forward-biased, and a small current should flow when the junctions are reverse-biased. The well continuity should also be checked.

Table 13.1 List of integrity and function checks

Shorts check (terminal-to-terminal checks)
Junction continuity (source, drain, diodes, and well)
Junction isolation (source, drain, diodes, and well)
Dielectric leakage (gate dielectric, MIM capacitor, intermetal dielectrics, etc.)

Table 13.2 Common parametric measurements related to process damage

Threshold voltage (V_t)
Subthreshold slope (SV_{tm} or S)
Conductance (g_m)
Punch-through (BV_{ds})
Drain current in the linear region of operation (Id_{lin})
Drain current in the saturation region of operation (Id_{sat})
Breakdown voltage (BV)
Off-state drain-to-source current (I_{off})
Drain-induced barrier lowering (DIBL)
Gate leakage (I_{gate})
Gate breakdown (BV_{ox})
Gate-induced drain leakage (GIDL)
Transistor matching
Contact resistance (R_c)
Ring oscillator frequency and supply voltage sensitivity

Dielectrics should be screened for gross leakage to eliminate shorted devices.

Parametric testing is widely used in CMOS technology to develop CMOS technology and monitor the production line health. Parametric testing for process and plasma damage can reuse the same measurement methods used for regular parametric testing with little or no modifications. The regular methods used in parametric testing need to be applied to the specialized process damage structures described previously. Table 13.2 shows a list of the more common tests performed on process damage-related structures.

The parameters shown in Table 13.2 are transistor-centric and related to the process damage of the CMOS transistors. As described in Chap. 6, the backend of line (BEOL) metallization used to construct the interconnect may also be susceptible to plasma damage. Therefore, a different suite of parametric tests is needed to analyze BEOL damage-related parameters. Table 13.3 shows a list of metallization parameters related to process damage. Most of the parameters shown are commonly measured during parametric testing. The electromigration and stress migration reliability tests tend to be very time-consuming, requiring specific test flows and high-temperature testing or stress. Fast tests such as the breakdown energy of metal (BEM) can be used for routine monitoring (Hong & Crook, 1985).

Structures with diodes protecting the gates need to be tested differently from devices without diodes, as the diodes may overwhelm other unrelated leakages. For example, gate leakage measurements can be overwhelmed by diode leakage, etc. In

Table 13.3 Common metallization-related parameters associated with process damage

Low-k interconnect parameters (see Chap. 8)
Intermetal dielectric capacitance
Intermetal dielectric leakage
Metal and via resistance
Electromigration
Stress migration

addition, the diode may limit the gate voltages through breakdown or forward bias current.

Inline parametric testing can be performed during processing, as described in Chap. 11, while the wafer is partially completed using specialized bond pads with a pattern similar to a waffle. Inline probing can be very useful in isolating the exact process step responsible for process damage.

The history of parametric testing performed on a given structure and location on the wafer must be considered when reprobing structures. Some tests can damage or destroy the structure, causing damage unrelated to processing. These tests include dielectric tests such as breakdown, leakage, etc. Higher voltage tests, such as transistor or junction breakdown tests, can degrade the devices. Tests that can cause the transistor to operate in a negative resistance region, causing snap-back, can damage the transistor.

The parametric health and capability of a CMOS process are heavily characterized and optimized during the technology development. There is a need for ongoing monitoring during manufacturing to ensure that damage to the production process remains under control. This monitoring is necessary to detect and remedy excursions such as those described in Chap. 4. Wafers are typically sampled periodically from the outgoing production volume and tested using a process damage-specific parametric test flow. Statistical methods are used to optimize the sampling with cost and capability constraints.

13.4 Diagnostic Revealing Stress

After the initial parametric test is performed, reliability tests are performed to determine if latent process damage has occurred. These tests can be very time-consuming and are impractical for large-scale routine parametric testing. A revealing or diagnostic stress can be performed on transistors as a proxy for dielectric-related transistor damage. The advantage of this type of stress is that it can be done very quickly, consumes a small amount of test capacity, and allows for increased process damage testing. This stress is a valuable technique that cannot replace process damage-related reliability testing. A revealing stress cannot be used for reliability lifetime prediction. It only shows the presence of process damage.

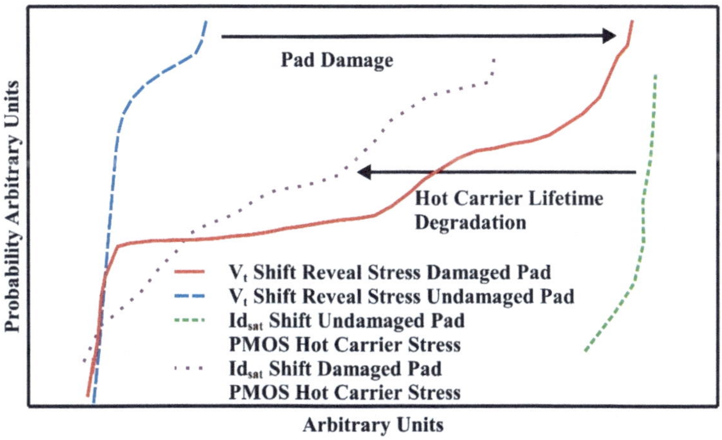

Fig. 13.2 Probability plot showing the threshold voltage shift after a 1 s constant current revealing stress for a pmos transistor with and without a bond pad compared with the results of hot carrier stress. The large change in threshold with the revealing stress correlates with a large degradation in hot carrier reliability. This indicates sucessful detection of the reliability problem with the reavealing stress

The revealing stress is performed by forcing a constant current into the gate of the transistor for a short time. The constant current stress ensures that current flows through the dielectric, filling neutral electron and hole traps and revealing the presence of hidden damage. On the other hand, a constant voltage stress does not ensure that current flows through the dielectric. The forced current level must inject carriers into the dielectric with a sufficient bias to place the capacitor into direct or Fowler-Nordheim tunneling leakage. Additionally, the forced current level must be well below the amount that causes dielectric breakdown. Therefore, the exact current level needs to be determined empirically. Various levels are used in the literature (Cheung, 1996; Martin, 2009). The revealing stress may shift all transistors, including undamaged devices without antennas. The key metric is the difference between structures with and without antennas.

As an example of the usefulness of revealing stress, Fig. 13.2 shows an example of a revealing stress performed on two different structures (Beckmeier & Martin, 2016). The standard parametric or reliability tests are performed before and after the revealing stress. One structure has a large bond pad, and the other structure has a minimal bond pad. The shift in threshold voltage from before and after the revealing stress is plotted on a probability plot. The minimal bond pad structure has a nearly normal distribution with a slight V_t shift. The structure with a large bond pad has a much larger V_t shift with a wide variation. Figure 13.2 shows the result of hot carrier stress for the two pad experiments performed on identically processed wafers without a revealing stress. The hot carrier stress results show that the revealing stress correctly and qualitatively predicted that the large pad structure would have degraded reliability relative to the control small pad structure. The results indicate that the

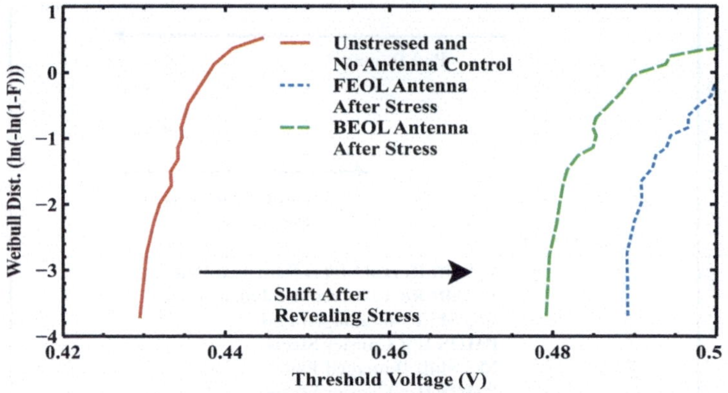

Fig. 13.3 Example of a 2 s revealing stress for transistors with 5.2 nm gate oxide. The probability plot shows the initial threshold prestress voltage distribution of transistors with various antenna experiments on the left side of the plot, which are indistuinguishable for the various antenna sizes. The right side of the plot shows the shifted distribution with the different antenna experiments separated with a 2 s stress

revealing stress successfully detected the damage that degraded the hot-electron lifetime.

Figure 13.3 shows an example of revealing or diagnostic stress (Martin, 2009). An antenna experiment shows no difference at time zero with and without plasma damage for various antenna sizes. However, after a 2 s stress, the distribution of all transistors is shifted, and the curves are separated, with the control distribution showing the least amount of shift, while transistors with antennas shift by varying amounts. For this type of measurement, the time between the stress and the measurement needs to be tightly controlled as a recovery process occurs, shifting the distributions back toward the original threshold, as described in Chap. 6.

The revealing stress can usually correctly predict that a wafer may have reliability problems (Martin, 2022). However, a revealing stress can have false positives, showing a sizeable revealing shift when minimal reliability problems exist. Figure 13.4 shows a diagram comparing the experimental results of hot carrier and NBTI reliability (*y*-axis) with the reveling stress results (*x*-axis). Points along the diagonal represent experiments with a good qualitative correlation between the reliability tests and the revealing stress results. Points next to the *x*-axis represent experiments with a strong revealing stress response and minimal reliability stress response, indicating a false-positive result for the revealing stress.

Revealing stress is a valuable technique for detecting the presence of traps unoccupied with charge carriers and other damage, such as interface states. The impact of traps is a strong function of the dielectric thickness, as shown in Chap. 5.

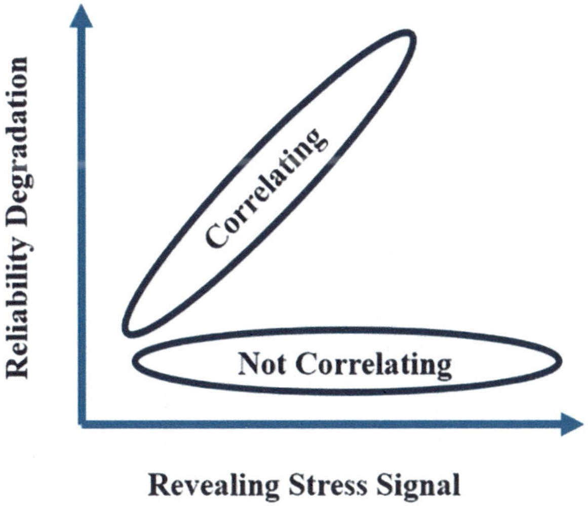

Fig. 13.4 Plot showing a qualitative correlation between a revealing stress with hot carrier and NBTI tests

13.5 Reliability Testing

Process and plasma damage complicates CMOS reliability testing by confounding the inherent reliability with the associated reliability degradation due to process damage. A suite of different transistor-level parametric reliability tests are normally performed on CMOS technologies to guarantee a reliable process. These tests include time-dependent dielectric breakdown, hot carrier lifetime, off-state lifetime, bias instability testing, electromigration, stress migration, etc. Some of these reliability tests have industry-standard procedures defined by the Joint Electron Device Engineering Council (JEDEC).

Reliability test parametric structures need to be duplicated with and without antennas to allow the qualification and verification of the technology reliability with and without process damage. The worst-case damage likely occurs with antennas present, requiring duplication of testing to ensure reliability in worse-case situations.

13.6 Diagnostic Analysis

An investigation of the failure characteristics can provide a positive conclusion that process or plasma damage occurred, requiring extensive data analysis. Plasma damage has a set of unique symptoms that allows the diagnosis of a plasma damage problem, separating it from a plethora of other potential problems. The unique

Table 13.4 Unique symptoms of plasma damage

Typically, it is much worse on structures with antennas. The problem can be mitigated with diode connections to the substrate
Damage occurs on specific wafer locations, design rules, and circuits. Similar structures are not affected. It may affect only one of the identical structures on the chip
Damage varies with process conditions, especially parameters related to power, bias, throughput, voltage, etc.
The problem correlates with specific fabrication tools
Sensitive to final alloy conditions with more time, temperature, and hydrogen typically improving symptoms
Recovery with time at room and elevated temperatures
It may improve with post-testing thermal bake (200–500 °C)
It may improve with high-temperature processing (500+°C)
Exposure to UV light may improve symptoms
The parametric data shift between inline and final testing
Damage can be detected with parametric test structures. It can be worse on parametric test structures depending on design rules. Parametric data can be bad, with product results being good
Damage is detected with a diagnostic stress. See Chap. 13
Across wafer variation that may or may not correlate with the wafer physical characteristics
Reliability degradation in structures with antennas
Failure analysis reveals no physical defect or discharge-type defects (see Chap. 5)

symptoms are described in Chap. 4 and are shown in Table 13.4. Typically, a plasma damage problem occurs with several, but not all, of the symptoms shown in the table. A high-confidence plasma damage conclusion can be reached by understanding how all of the symptoms match the failure mode in a coherent manner.

Once a problem is determined not to be plasma damage, reaching a root cause conclusion may be more difficult, requiring exhaustive investigation, failure analysis, and experimentation to select among the potential causes of failure described in Chap. 4.

References

Beckmeier, D., & Martin, A. (2016). Plasma process induced damage detection by fast wafer level reliability monitoring for automotive applications. *Microelectronics Reliability, 64*, 189–193. https://doi.org/10.1016/j.microrel.2016.07.129

Cheung, K. P. (1996). On the use of fowler-Nordheim stress to reveal plasma-charging damage. *International Symposium of Plasma Process-Induced Damage*. https://doi.org/10.1109/PPID.1996.715192

Hong, C. C., & Crook, D. L. (1985). Breakdown Energy of Metal (BEM) a new technique for monitoring metallization reliability at wafer level. *IEEE International Reliability Physics Symposium*. https://doi.org/10.1109/IRPS.1985.362084

Martin, A. (2009). Review on the reliability characterization of plasma-induced damage. *Journal of Vacuum Science & Technology B: Microelectronics and Nanometer Structures, 27*(1), 426. https://doi.org/10.1116/1.3054356

Martin, A. (2020). Plasma induced charging damage: From appropriate MOS test structures to antenna design rules, a comprehensive process qualification procedure. *2020 IEEE International Integrated Reliability Workshop (IIRW)*. https://doi.org/10.1109/IIRW49815.2020.9312854

Martin, A. (2022). Plasma Processing Induced Charging Damage (PID) assessment with appropriate fWLR stress methods ensuring expected MOS reliability and lifetimes for automotive products (invited). *IEEE International Reliability Physics Symposium*. https://doi.org/10.1109/IRPS48227.2022.9764556

Chapter 14
Role of Hydrogen

14.1 Introduction

Hydrogen plays a critical role in CMOS technology, passivating and neutralizing defects. CMOS technology would be slower, less capable, and less reliable without the hydrogen mechanisms that passivate defects. In addition, hydrogen plays a significant role in process and plasma damage, confounding the data analysis and problem-solving. Typically, a large fraction of process damage problems are observed interacting with hydrogen processing, shifting electrical parameters, and changing the impact of process damage. Exposure to hydrogen can partially recover the impact of plasma damage, hiding the damage and creating the potential for latent damage to be detected as a reliability problem.

Hydrogen is ubiquitous in CMOS processing; preventing the exposure of the wafer during manufacturing to hydrogen is nearly impossible. Most plasma processes crack gases, often containing hydrogen components, creating atomic hydrogen that rapidly diffuses into the wafer. Hydrogen containing acids, resists, and other films expose the wafer to hydrogen. Even water used in wafer processing exposes the wafer to a hydrogen source (see Chap. 9) (Pankove & Johnson, 1991).

Most CMOS processes intentionally expose the wafer to a 400–500 °C hydrogen alloy in a furnace tube at or near the final process steps to reduce interface states by hydrogen passivation. The alloy furnace is partially filled with hydrogen molecules that penetrate the wafer, interacting with it through multiple mechanisms. The molecular hydrogen breaks apart into more reactive atomic hydrogen through interaction with defects in the wafer. In some cases, the wafer is directly exposed to atomic hydrogen through a remote plasma process to provide a high-concentration source of atomic hydrogen.

There are many complex hydrogen mechanisms in silicon and its dielectrics that occur independently or interact. These hydrogen effects are very complicated and not completely understood. Table 14.1 shows a list of important CMOS-related

K. Prall, *CMOS Plasma and Process Damage*,
https://doi.org/10.1007/978-3-031-89029-1_14

Table 14.1 Summary of lower temperature hydrogen mechanisms in CMOS wafers

Mechanism	Reaction	Activation energy eV Temperature range	References
Diffusion/drift			
H_2 diffusion in SiO_2 H^0 (neutral) diffusion in SiO_2 H^+ diffusion in SiO_2	N/A	0.3–0.5 0.1–0.2 0.7–0.9 (25 °C)	Baumvol (1999, pg. 56) Fleetwood et al., (2024)
H_2 diffusion in Si Through interstitial sites	N/A	0.78	Fleetwood et al. (2009, pg. 43)
Hydrogen diffusion accelerated by dopants in Si	N/A		Pearton et al. (1992, pg. 226)
H_2O diffusion in SiO_2	N/A	0.86	Baumvol (1999, pg. 56)
H^+ proton drift in SiO_2	N/A	0.7–0.9 (25 °C)	Fleetwood et al. (2024) Fleetwood et al. (2009, pg. 390)
H^+ drift in n-type Si H^- drift in n-type Si	N/A	~0.5 ~0.7	Fleetwood et al. (2024)
Enhanced hydrogen migration along interfaces	N/A	~0.3	Fleetwood et al. (2024)
Dihydrogen formation in SiO_2	$2H \rightarrow H_2$	0.75	Baumvol (1999, pg. 56)
Dihydrogen disassociation in SiO_2	$H_2 \rightarrow 2H$	4.22	Baumvol (1999, pg. 56)
Interface states			
Interface state passivation—P_b center passivation H_2 reaction	$H_2 + P_b^+ \rightarrow$ Si-H + H^+	1.51–1.57 1.5–1.7 0.5 with holes	Fleetwood et al. (2009, pg. 386) Fleetwood et al. (2024)
Interface state passivation—P_b center passivation H reaction	H + $Pb^+ \rightarrow$ Si-H		Baumvol (1999, pg. 56)

Interface state depassivation—P_b center restoration H reaction	Si-H → Pb^++H	2.5–2.8 (>400 °C) 2.6 without holes 2.1–2.3 with holes	Baumvol (1999, pg. 56) Fleetwood et al. (2009, pg. 387) Fleetwood et al. (2024)
Interface state depassivation—P_b center restoration H_2 reaction	Si-H + H → Pb^++H_2	1.0 1.0–1.2	Baumvol (1999, pg. 56) Fleetwood et al. (2009, pg. 387) Tsetseris et al. (2005, pg. 2267)
Dopants			
Acceptor passivation in p-type silicon	H + B → HB (boron)	(60–140 °C) >150 °C >40 °C	Fleetwood et al. (2009, pg.32) Pearton et al. (1992, pg. 63) Pankove and Johnson (1991, pg. 96)
Acceptor depassivation in p-type silicon	H-B → H + B (boron)	1.28 1.1–1.3 without holes 0.55–0.65 with holes	Fleetwood et al. (2009, pg. 38) Pearton et al. (1992, pg. 74) Tsetseris et al. (2005, pg. 2267) Pankove & Johnson (1991, pg. 96) Fleetwood et al. (2024)
P donor passivation in n-type silicon	H + P → H-P (phosphorus)	(100–150 °C) (100–150 °C)	Fleetwood et al. (2009, pg. 38) Tsetseris et al. (2005, pg. 2266) Pearton et al. (1992, pg. 75)
As Donor passivation in n-type silicon	As+H → H-As (Arsenic)		Fleetwood et al. (2009, pg. 38) Tsetseris et al. (2005, pg. 2266) Pearton et al. (1992, pg. 75)
P donor depassivation in n-type silicon	P-H → P^++H^- (phosphorus)	1.2,1.12 1.2–1.35 without electrons 0.35–0.45 with electrons	Fleetwood et al. (2009, pg. 38) Tsetseris et al. (2005, pg. 2266) Fleetwood et al. (2024) Tsetseris et al. (2005, pg. 2266)
As donor depassivation in n-type silicon	H-As→As^++H^- (Arsenic)	1.5 without electrons 0.5–0.6 with electrons	Fleetwood et al. (2009, pg. 38) Tsetseris et al. (2005, pg. 2266) Fleetwood et al. (2024)

(continued)

Table 14.1 (continued)

Mechanism	Reaction	Activation energy eV Temperature range	References
Interaction with silicon and oxygen			
Hydrogen-silicon vacancy formation	$H + V \rightarrow H\text{-}V$	0.4–0.5 (<300 °C)	Fleetwood et al. (2009, pg. 38) Fleetwood et al. (2024)
Dihydrogen-silicon vacancy formation	$H_2 + V \rightarrow H_2V$		Fleetwood et al. (2009, pg. 43)
Hydrogen-silicon vacancy dissociation	$H\text{-}V \rightarrow H + V$	0.18 (>300 °C)	Fleetwood et al. (2009, pg. 39)
Dihydrogen-silicon vacancy dissociation	$H + H\text{-}V_o \rightarrow H_2 + V$	0.26	Fleetwood et al. (2009, pg. 43)
Acceleration of thermal donor formation by hydrogen		450 °C	Pankove & Johnson (1991, pg. 52)
Hydrogen passivation of thermal donors formed by oxygen in silicon		~450 °C, 550–800 °C	Fleetwood et al. (2009, pg. 389) Pearton et al. (1992, pg. 38,48)
Hydrogen release from thermal donors		1.15–1.25	Fleetwood et al. (2009, pg. 389)
H binding to interstitial O	$H + O_i \rightarrow H\text{-}O_i$	(<250 K)	Fleetwood et al. (2009, pg. 36)
Hydrogen-interstitial O dissociation	$H\text{-}O_i \rightarrow H + O_i$	0.65–0.75 (>250 K)	(Fleetwood et al., 2009)pg. 36
Interaction with contaminants			
Carbon-hydrogen passivation	$H + C \rightarrow HC$		Fleetwood et al. (2009, pg. 40) Pearton et al. (1992, pg. 44)
Carbon-hydrogen dissociation	$H\text{-}C \rightarrow H + C$	0.73	Fleetwood et al. (2009, pg. 40) Pearton et al. (1992, pg. 44)
Transition-metal hydrogen passivation	$T_m + H \rightarrow T_m\text{-}H$		Fleetwood et al. (2009, pg. 41)
Transition-metal hydrogen depassivation	$T_m\text{-}H \rightarrow T_m + H$	1.1–2.5 (300 °C+)	Fleetwood et al. (2009, pg. 41) Pearton et al. (1992, pg. 31,48)
Hydrogen defects			
H cluster and platelet formation	$[2Di\text{-}H]_n + H_2 \rightarrow [2Di\text{-}H]_{n+1}$	(<100 °C)	Fleetwood et al. (2009, pg. 44)
H cluster and platelet disassociation	$[2Di\text{-}H]_{n+1} \rightarrow [2Di\text{-}H]_n + H_2$	(200–400 °C)	Fleetwood et al. (2009, pg. 44)

Isolated free hydrogen bandgap state introduction	H^o H^+ H^-	0.1–0.2 0.48 0.7	Fleetwood et al. (2009, pg. 388
Hydrogen donor level in silicon bandgap (E_c-0.4)	N/A		Pearton et al. (1992, pg. 48)
Bias temperature instability	N/A	0.35	Fleetwood et al. (2009, pg. 392) Strong et al. (2009)
Silicon embrittlement, ductility change with hydrogen	N/A		Pearton et al. (1992, pg. 319) Pankove & Johnson, (1991, pg. 51)
Hydrogen aids the motion of dislocations in silicon	N/A		Pankove and Johnson, (1991, pg. 51)
Hydrogen passivation of dislocations in silicon	N/A		Pankove and Johnson, (1991, pg. 41
Hydrogen-induced defects	N/A		Pankove and Johnson, (1991, pg. 129)
Miscellaneous			
Enhanced low dose rate sensitivity (ELDRS) to radiation	N/A		Fleetwood et al. (2009, pg. 392)
Hydrogen degrades noise	N/A		Fleetwood et al. (2024)
Deuterium substitution for hydrogen		1.84 1.1	Fleetwood et al. (2009, pg. 393) Pearton et al. (1992, pg. 48)
Grain boundary passivation		2.5	Pearton et al. (1992, pg. 45,48)
Sandia proton memory			Fleetwood et al. (1998)
Silicon etching with hydrogen			Lee et al. (2006)
Hydrogen release and motion from SiN_4		0.35–0.6	Fleetwood et al. (2024)
Hf to Si hydrogen transfer	Hf-H → Hf^- + H^+ → H + Si → Si-H	0.2–0.4	Fleetwood et al. (2024)

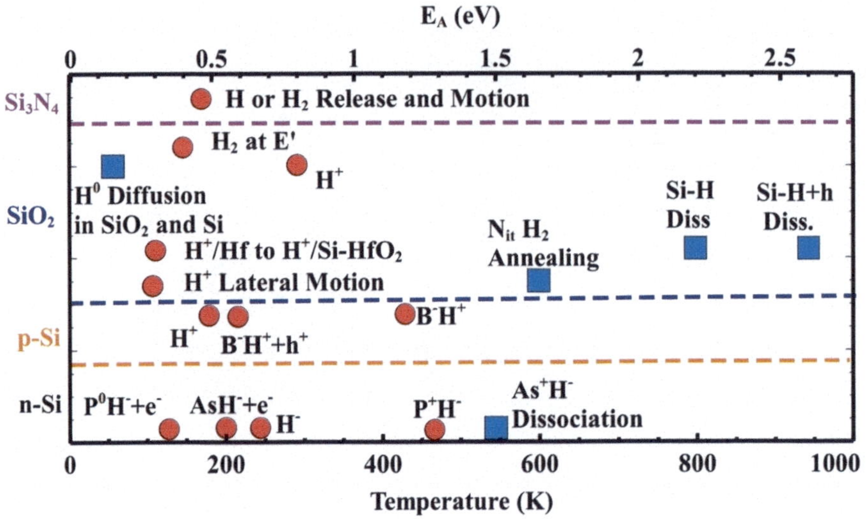

Fig. 14.1 Plot showing activation energies and temperatures for hydrogen-related action in CMOS technology for silicon nitride, silicon dioxide, p-type silicon, and n-type silicon

hydrogen mechanisms along with the related chemical equation. In addition, the activation energies and temperature ranges of interest are shown for some of the mechanisms. Unfortunately, there are conflicting activation energies in the literature. Some older activation energies have been excluded as they conflict with modern values (Baumvol, 1999). This chapter provides a concise description of hydrogen mechanisms in CMOS technology.

Temperature plays an enormous role in hydrogen action in CMOS technology, driving the mechanisms shown in the table. The temperatures of interest in CMOS manufacturing range from near room temperature to around 1200 °C. The temperatures of interest during packaging and reliability testing range from −25 °C to around 200 °C. The temperatures shown are from large area structures involving large numbers of hydrogen reactions. Single atom effects covered in this book can be expected to be observed at lower temperatures at a part per million or billion level. Figure 14.1 shows a chart of the hydrogen-related effects in the silicon substrate, gate dielectric, and silicon nitride as a function of temperature and activation energy discussed in this chapter (Fleetwood et al., 2024).

14.2 Hydrogen Diffusion and Drift

The smallest atom, hydrogen, is highly mobile in silicon and most dielectrics, moving interstitially or hopping between sites, quickly moving in the temperature range of CMOS operation, causing electrical changes in CMOS devices. In addition, hydrogen can drift due to the electric field present in operating transistors and other

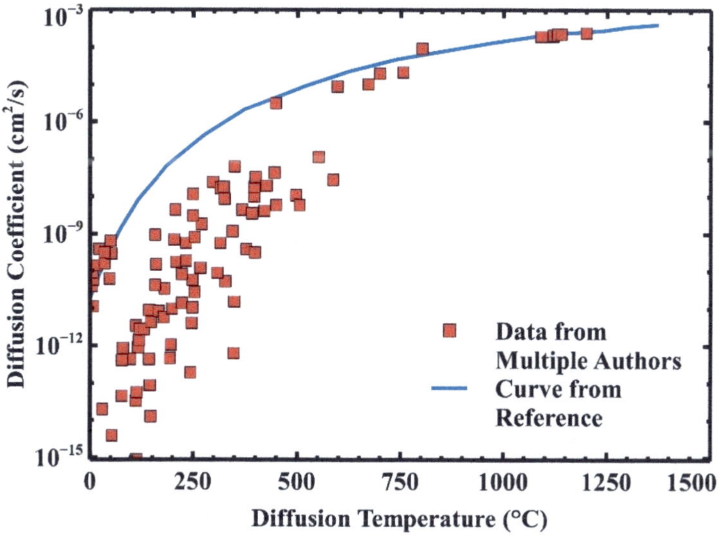

Fig. 14.2 Hydrogen diffusion coefficient in silicon plotted against temperature compiled from various sources

devices. Hydrogen can drift and diffuse in several forms, including molecular hydrogen (H_2), neutral atomic hydrogen (H^0), positively ionized atomic hydrogen (H^+), and negatively ionized atomic hydrogen (H^-). Hydrogen easily changes between these forms as it drifts and diffuses. Atomic hydrogen can combine to form molecular hydrogen and dissolve later, reforming atomic hydrogen, transitioning back and forth with time and position in the semiconductor and dielectrics.

Hydrogen migrates in two distinct methods. At high temperatures, hydrogen migrates predominately as an atomic species by interstitial diffusion. At low temperatures, hydrogen diffuses by a hopping mechanism between stable or metastable defects (Fleetwood et al., 2009). Figure 14.2 shows a diffusion coefficient plot versus temperature for hydrogen in silicon (Hallam et al., 2018).

Free unbound hydrogen is very mobile under electric fields and can be expected to drift in CMOS devices under typical operating conditions, changing the electrical characteristics of the devices. Hydrogen drift or diffusion can be expected during the operation of CMOS devices. The hydrogen mechanisms described in this chapter can impact the functionality of the CMOS chip, performance, and reliability. Figure 14.3 shows the results of a drift experiment (Herring et al., 2001). A Schottky diode is reverse-biased, and a laser light pulse is applied to the back of the diode. The light pulse supplies holes to the hydrogen-phosphorous complex where the hydrogen is located, reducing the energy for the proton required to escape the phosphorous-hydrogen complex where it is trapped. The free hydrogen drifts under the electric field. The plot shows the change in diode capacitance after the laser pulse for a sample with and without hydrogenation. Without hydrogenation, the plot shows no hydrogen drift, only a transient caused by the laser light. For a diode that

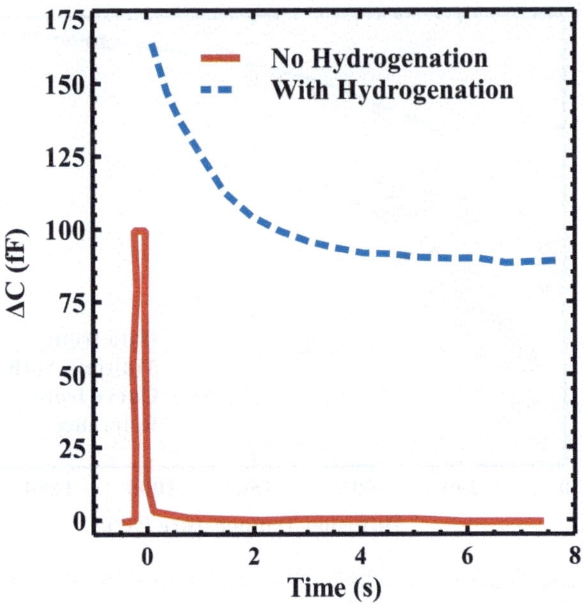

Fig. 14.3 Change in reverse-biased Schottky diode capacitance with a hole injection laser pulse with and without a hydrogen plasma treatment. The hydrogenation was performed with a remote hydrogen plasma for 20 min at 120 °C

has been hydrogenated by a remote hydrogen plasma at 120 °C for 20 min, the plot shows a transient decay over seconds as the hydrogen drifts through the diode's depletion region. Similar effects can be expected during CMOS operation.

14.3 Passivation and Depassivation of Interface States

Interface state passivation of hydrogen is a critical process enabling successful CMOS technology. The interface quality where the crystalline silicon semiconductor meets the amorphous or nanocrystalline gate dielectric controls the transistor performance and reliability. Unfortunately, there are a large number of defects at this interface. Therefore, a CMOS process flow usually contains one or more hydrogen passivation steps near the last steps of the flow just before the wafer processing is completed. For example, a hydrogenation step may consist of a thermal treatment or alloy in a partial hydrogen ambient consisting of molecular hydrogen. Also, the hydrogen treatment may involve exposing the wafer to an atomic hydrogen plasma.

Additionally, a hydrogen treatment may consist of a thermal treatment, driving hydrogen from a hydrogen-containing film such as silicon nitride into the wafer. Hydrogen can also be driven from the wafer. The hydrogenation is usually performed at temperatures of around 400–500 °C, which is an optimal temperature for

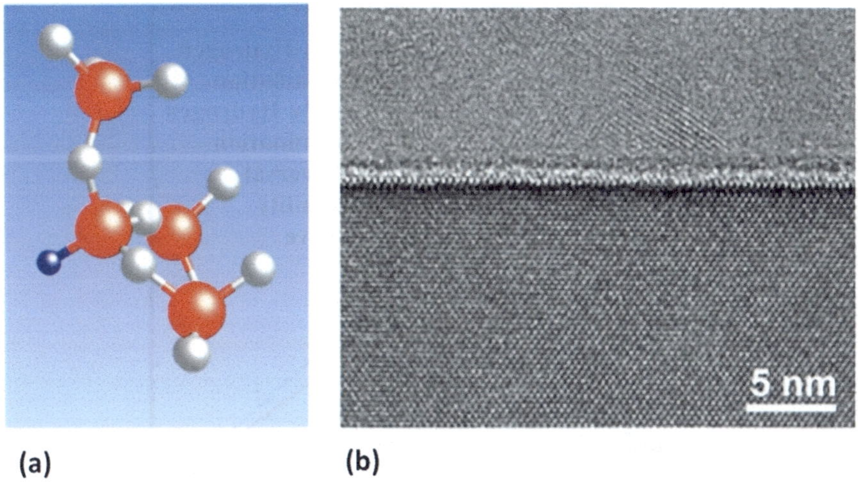

Fig. 14.4 (**a**) Sketch of ball and stick pattern of SiO_2-Si with a hydrogen atom passivating a missing bond with the following colors: Si—red, O—white, and H—blue. (**b**) Cross-section TEM of silicon-silicon dioxide interface

passivation of interface states (See Fig. 14.7). For temperatures greater than 500 °C, the metallization on the wafer may be damaged, so the maximum temperature is limited. The optimum alloy temperature is shown in Fig. 14.11.

The most common interface state defect is a dangling bond where one or more of the unpaired silicon valence bonds create an electrically active defect known as a P_b center. Hydrogen can bond with the P_b center, forming an electrically inactive center. Figure 14.4a shows a ball and stick diagram of the Si-SiO_2 interface where a hydrogen atom has passivated a dangling bond (Gos, 2011). Figure 14.4b shows a transmission electron microscope (TEM) image of the Si-SiO_2 interface where the crystalline silicon meets the amorphous silicon dioxide (Tao et al., 2024). The rough and nonideal interface can be observed, showing the complex structure involved in interface states.

Hydrogen passivates interface states by chemically reacting with the dangling silicon bond at the P_b center, with the following reactions forming an electrically neutral silicon hydrogen bond. The electrically neutralized defect does not affect the electrical function of the CMOS devices.

$$H_2 + P_b^{\,+} \rightarrow Si - H + H^+$$

or

$$H + P_b^{\,+} \rightarrow Si - H$$

Hydrogen can also depassivate passivated interface states through the reverse processes.

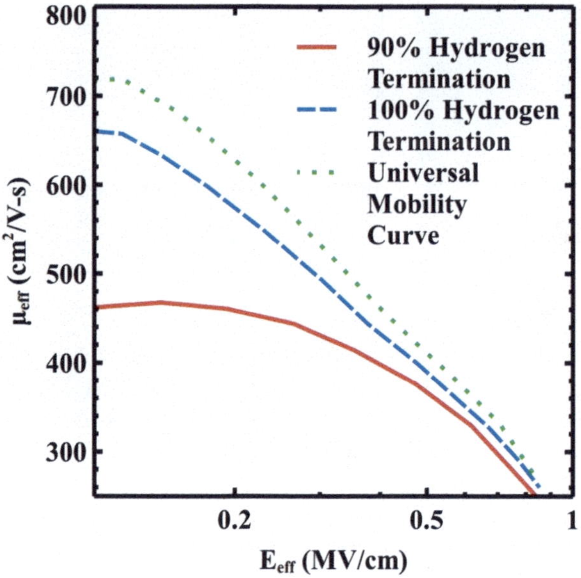

Fig. 14.5 Computer simulation of the silicon surface mobility with hydrogen termination for different levels of hydrogen passivation

$$Si-H+H \rightarrow P_b^{\ +}+H_2$$

or

$$Si-H \rightarrow P_b^{\ +}+H$$

Hydrogen passivation of interface states improves CMOS devices. Interface states degrade the movement of the electrical channel carriers, reflecting or temporarily trapping the electrical carriers. This effect degrades the mobility of the transistor, limiting its speed and drive capabilities. Hydrogen passivation neutralizes the electrical activity of the interface states. Figure 14.5 shows a computer simulation of the channel mobility with 90% and 100% of the interface states passivated (Ishihara et al., 2020). The transistor mobility is severely degraded with unpassivated interface state levels of 10^{11} cm^{-2}. This level is larger than that achieved in CMOS technology (~10^{10} cm^{-2}).

Hydrogen-passivated interface states can become depassivated through exposure to atomic hydrogen. The depassivation increases with increased temperatures. A catalytic reaction can occur where a single proton (hydrogen ion) can depassivate multiple passivated interface states, as shown in Fig. 14.6. A proton can diffuse to the passivated interface state, combining with the passivating hydrogen atom forming H_2, which diffuses away, depassivating the interface state. The hydrogen molecule diffuses through the dielectric and can be cracked into two atoms by dielectric

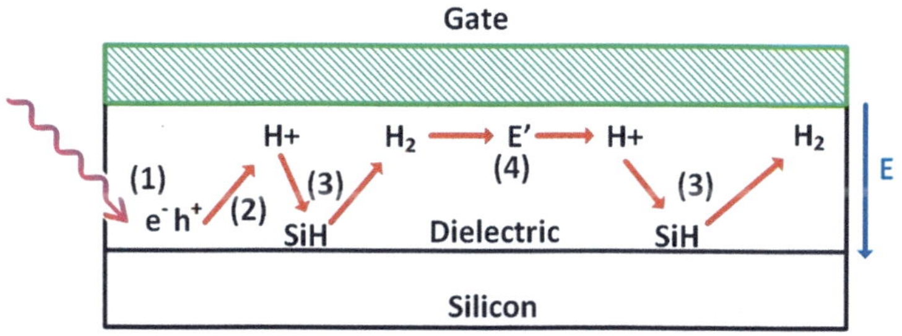

Fig. 14.6 Sketch of silicon dioxide-silicon interface showing a catalytic reaction passivating and depassivating interface states. (1) Energy from a photon or electric field creates an electron-hole pair. (2) The hole drifts toward the gate and reacts with a neutral hydrogen atom, creating a positively charged hydrogen ion. (3) The hydrogen reacts with a Si-H bond at the silicon interface, depassivating an interface state and creating a hydrogen molecule. (4) The hydrogen molecule is captured by a defect such as an E′ center cracking the molecule, creating hydrogen ions. Steps 3 and 4 create a catalytic reaction where a single hydrogen ion can depassivate multiple interface states

defects such as a fixed oxide charge E' center. The cracked hydrogen atom can diffuse to another interface state and repeat the process (Rashkeev et al., 2001).

Hydrogen treatment of CMOS wafers is heavily optimized to obtain the best possible overall results. The final hydrogen treatments in the process flow control the overall hydrogen passivation of defects. The temperature of the treatment is important, as shown in Fig. 14.7 (Ishihara et al., 2020). The two types of P_b centers responsible for interface states are measured using an electron spin resonance method (ESR). The P_b center concentration is measured as a function of the anneal and alloy temperature. Thermally treating the wafer in an inactive neutral nitrogen ambient reduces the interface state density by reducing the Pb center density. The reduction in interface states is further improved by adding hydrogen to the alloy. The lowest interface state density occurs near 425 °C in this experiment. For lower temperatures, the passivation is limited by the hydrogen-silicon reaction rate. For temperatures above 450 °C, the interface state density increases as the hydrogen is thermally ejected from the P_b center (Ishihara et al., 2020).

14.4 Mobile Hydrogen

In addition to the poor interface state passivation performance, hydrogen alloys at higher temperatures can leave unreacted ionized hydrogen (protons) in the dielectrics. The protons are mobile in the dielectrics, behaving as a mobile ion, affecting the MOS device stability. Figure 14.8a shows a current-voltage measurement on a silicon on insulator structure with a 400 nm silicon dioxide layer, changing the threshold voltage. The voltage is swept in opposite directions while measuring the current through the dielectric. The hysteresis is caused by protons drifting back and

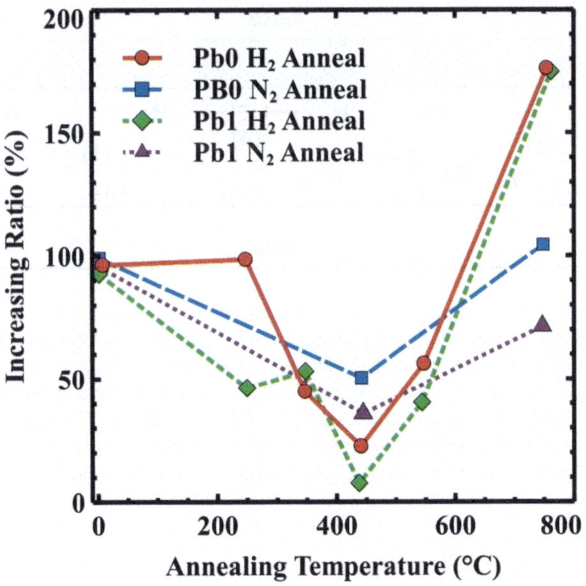

Fig. 14.7 Interface defect related Pb0 and Pb1 density plotted as a function of temperature for a hydrogen alloy or nitrogen anneal process for a 2.5 nm oxide grown at 850 °C. The interface state density is plotted as a ratio from the initial low temperature unannealed interface state concentration

forth between the top and bottom interfaces. Figure 14.8b shows the width of the hysteresis voltage as a function of the alloying temperature with and without hydrogen. The maximum hysteresis occurs around 600 °C. Hydrogen exposure at these temperatures is generally avoided.

Hydrogen can be repeatably mobile under certain conditions. A hydrogen-based proton memory has been demonstrated, showing protons behaving as mobile ions in a SiO_2 dielectric dielectric (Fleetwood et al., 1998). The proton is trapped in the dielectric between the silicon substrate and a polysilicon or aluminum gate. The positively charged protons are attracted or repelled by the gate voltage, transiting between the top and bottom of the dielectric and modifying the threshold voltage. As shown in Fig. 14.9, the threshold voltage modification is reflected as a hysteresis in the C-V curve (Vanheusden, 2011). The hysteresis could be observed through millions of cycles of the C-V plot, indicating that the hydrogen was not trapped and remained mobile. This hydrogen behavior is abnormal and not generally observed in CMOS technology, indicating that special processing conditions are needed to generate this behavior, such as a dielectric with high concentrations of oxygen vacancies (Warren et al., 1997).

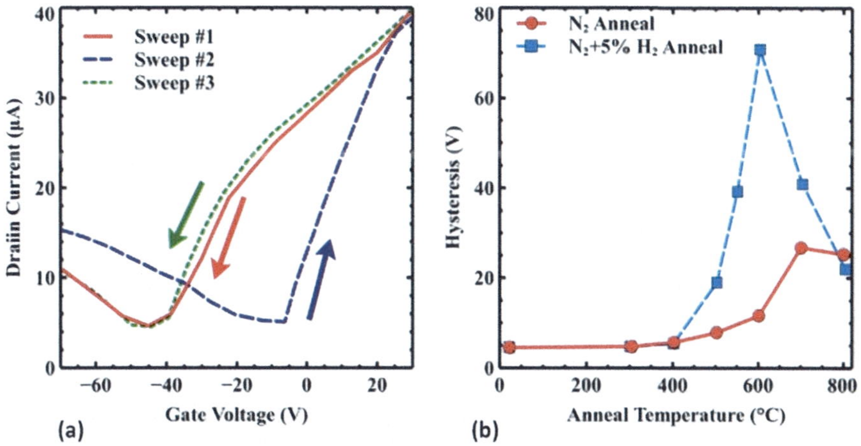

Fig. 14.8 Proton motion observed in a silicon on insulator (SOI) mos capacitor structure. (**a**) current-voltage curve of the SOI capacitor showing hysteresis. Sweep 1 is a positive to negative voltage sweep on the gate. Sweep 2 is a subsequent negative to positive sweep. Sweep 3 follows curve 2 with a positive to negative sweep. (**b**) plot showing the width of the hysteresis versus annealing temperature with and without hydrogen

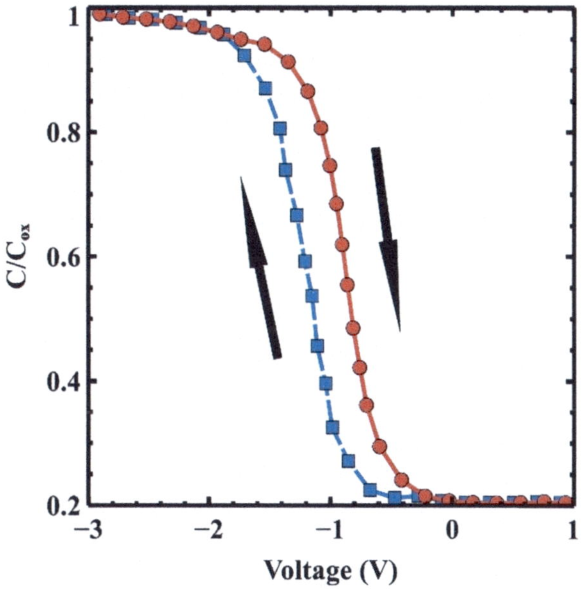

Fig. 14.9 Capacitance voltage plot of a mos capacitor showing hysteresis behavior due to mobile hydrogen

14.5 Dopant Passivation and Depassivation

14.5.1 Acceptor Passivation and Depassivation by Hydrogen

Hydrogen can passivate the dangling silicon bonds created by introducing dopants into the silicon lattice (Pankove & Johnson, 1991). Acceptors such as boron are trivalent atoms that substitute for silicon in the lattice, creating a hole by capturing an electron in the dangling bond. As a result, one of the silicon bonds is left vacant. Hydrogen can be attracted to the acceptor site and be captured by the silicon lattice at the location of the missing bond, as shown in Fig. 14.10. The A-H pair is neutral and does not capture an electron from the valence bond, eliminating the electrical activity of the acceptor dopants.

Hydrogen can heavily passivate acceptors at low temperatures around 100 °C (Pankove & Johnson, 1991). At temperatures higher than 100 °C, the hydrogen has enough thermal energy to escape the dangling bond site. Above 100 °C, very few acceptors are neutralized. However, at temperatures lower than 100 °C, the hydrogen does not have enough thermal energy to penetrate to the acceptor site. These competing mechanisms create a maximum acceptor neutralization temperature of around 100 °C, as shown in Fig. 14.11. The higher resistivity near 100 °C is direct evidence of boron neutralization by hydrogen.

A large fraction of the acceptors in p-type silicon can be neutralized by hydrogen. The fraction neutralized can exceed 90% with heavy plasma hydrogenation, changing the resistivity orders of magnitude. The passivation can be reversed by heating the wafer at temperatures significantly higher than the peak at 100 °C, as shown in Fig. 14.12 (Pearton et al., 1992). The ratio plotted tracks the recovered dopant density (N_R) as a fraction of the initial acceptor density (N_0). There is a difference in the recovery temperature with different acceptor dopants.

The acceptor passivation by hydrogen tracks the hydrogen diffusion into the silicon ((Pankove & Johnson, 1991) pg. 93). Figure 14.13 shows a plot of the resistivity

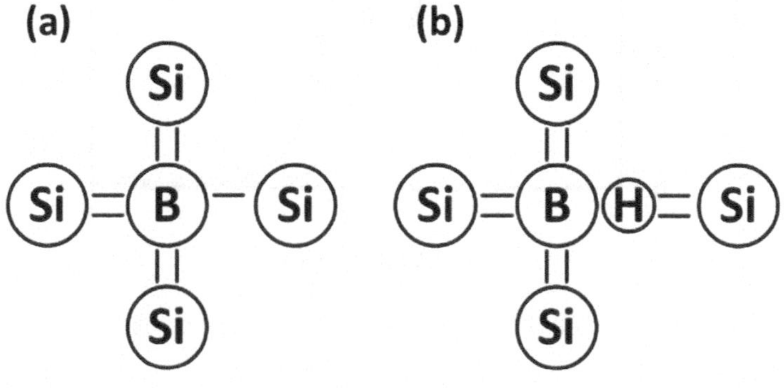

Fig. 14.10 Sketch showing hydrogen passivation of acceptor dopants. (**a**) the acceptor dopant (boron) is bonded into silicon lattice with missing bond. (**b**) passivated acceptor with hydrogen

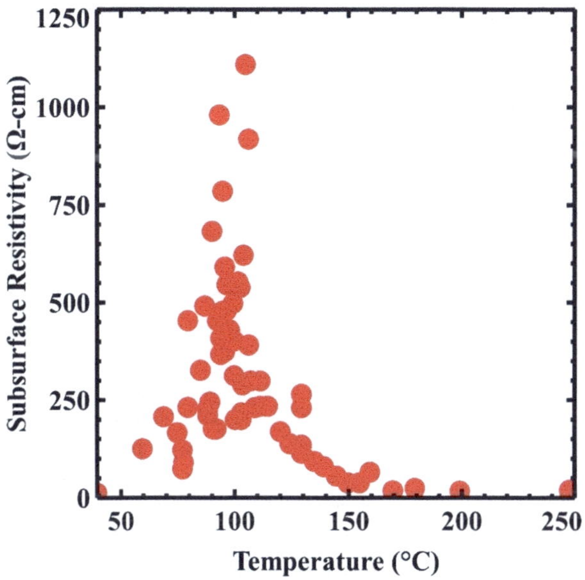

Fig. 14.11 Plot showing silicon resistivity as a function of hydrogenation temperature for a 4 h treatment in a hydrogen plasma

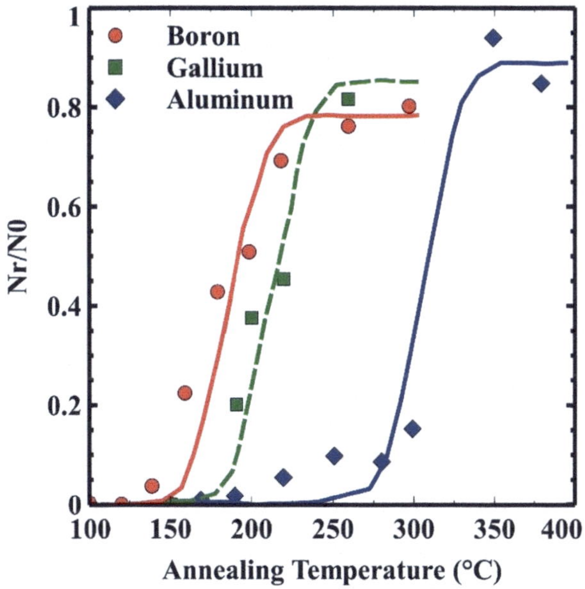

Fig. 14.12 Plot showing the recovery of acceptor electrical activity by thermal annealing in a non-hydrogen ambient for B, Ga, and Al in silicon

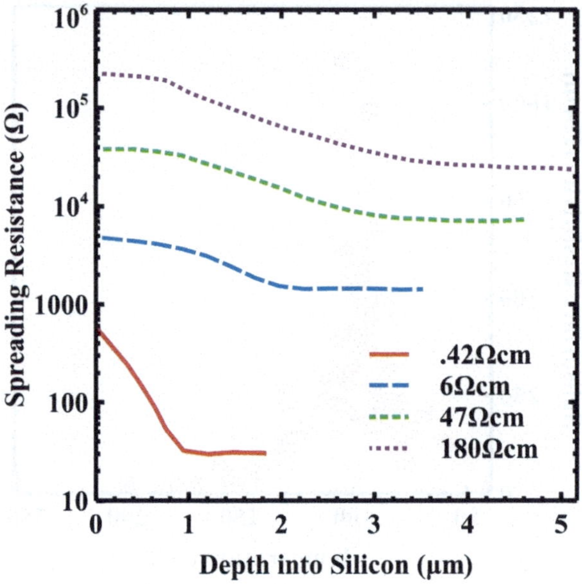

Fig. 14.13 Spreading resistance measurements as a function of depth for various uniformly boron doped substrates exposed to a hydrogen plasma. The top 3 curves were plasma hydrogenated for 1 h at 122 °C. The most heavily doped bottom sample was plasma hydrogenated for 4 h at 122 °C

measured by spreading resistance for several uniformly boron-doped silicon samples. The resistivity is high near the silicon surface, where hydrogen has diffused from the surface. The hydrogen also tracks the acceptor dopant distribution created by ion implantation as the hydrogen is trapped at the acceptor location.

14.5.2 Donor Passivation and Depassivation by Hydrogen

Donor dopants can also be passivated by hydrogen. The hydrogen passivation of donors follows similar physics to the acceptor passivation previously described. The fraction of donors passivated is typically less than the fraction of acceptors passivated. However, in certain cases, such as highly doped n-type silicon, more than 90% of donors can be neutralized by hydrogen (Pearton et al., 1992). Figure 14.14 shows the electrically active implanted donor dopant profile measured by spreading resistance. The electrically active dopant concentration is reduced by an order of magnitude with plasma hydrogenation.

The depassivation of donor-hydrogen complexes can occur similarly to the depassivation of acceptors. Depassivation of donor-hydrogen can occur at temperature ranges of 24–110 °C that are in the range of operation and testing of chips (Seager & Anderson, 1990).

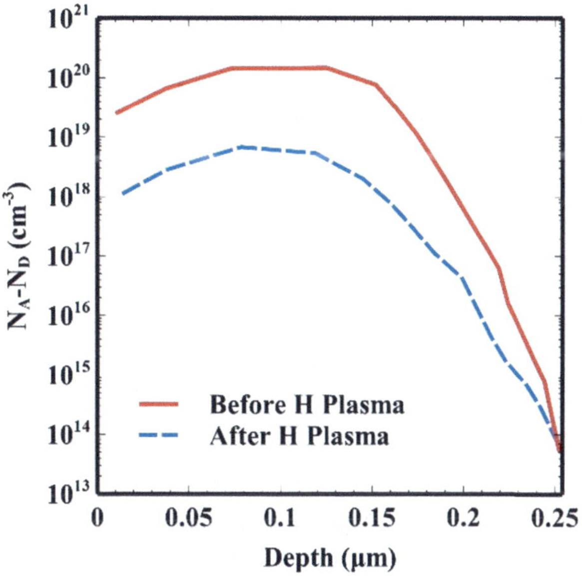

Fig. 14.14 Plot of electrically active implanted Sb donor concentration in silicon measured by spreading resistance with and without a hydrogen plasma treatment

14.6 Hydrogen Interaction with Silicon and Oxygen Vacancies

Hydrogen can react with both silicon and oxygen vacancies in the wafer, creating electrically active defects in the silicon bandgap. These defects can create shallow energy levels in the bandgap below the conduction band, causing changes in device behavior.

14.7 Hydrogen Interaction with Thermal Donors

Thermal donors are oxygen clusters formed in the silicon lattice. Thermal donors are electrically active defects primarily created in silicon by exposing a wafer to thermal processing steps in the temperature range of 300–550 °C. About 25 different thermal donor species have been detected in silicon (Singh & Nagabalasubraminian, 2019). Thermal donors are created when oxygen from the wafer substrate manufacturing process and other chip manufacturing processes are in the silicon lattice. Thermal donor formation requires the presence of oxygen in the wafer. Thermal donors can affect the chip operation when formed in the electrically active wafers near the surface of the wafer. Modern wafer manufacturing uses various techniques to mitigate

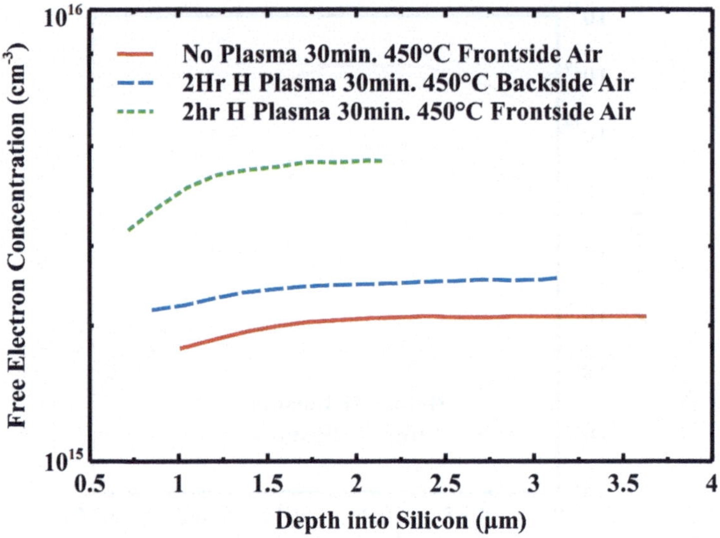

Fig. 14.15 Free electron concentration as a function of depth in the silicon derived from capacitance-voltage measurements on schottky barrier diodes. The thermal donors are formed during the 30 min. 450 °C treatment in air. Three hydrogenation experiments are performed before the thermal donor anneal. The bottom curve has no hydrogen exposure. The middle curve had a backside plasma hydrogen treatment. The top curve had a frontside hydrogen exposure

the existence of oxygen at the wafer surface. These techniques include hydrogen and argon annealing of the wafer before chip manufacturing begins.

Hydrogen can enhance the migration of oxygen clusters. In addition, hydrogen can bind to oxygen clusters, enhancing thermal donor formation (Fleetwood et al., 2009). Figure 14.15 shows a plot demonstrating accelerated thermal donor formation by plasma hydrogen exposure (Simoen et al., 2006). Schottky diodes are formed on the treated wafers and measured using a capacitance-voltage technique. The free electron concentration is calculated and plotted as a function of depth into the silicon. The electron concentration due to the created thermal donors is strongly enhanced by plasma hydrogen exposure before thermal annealing. N-type wafers are used to avoid hydrogen acceptor passivation.

14.8 Hydrogen Interaction with Contaminants

14.8.1 Metals, Chalcogenides, Etc.

Impurities in a CMOS wafer, such as transition metals, chalcogenides, etc., create deep-level traps in the center of the bandgap that can degrade device performance, as described in Chap. 15 (Claeys & Simoen, 2018). Hydrogen can passivate these

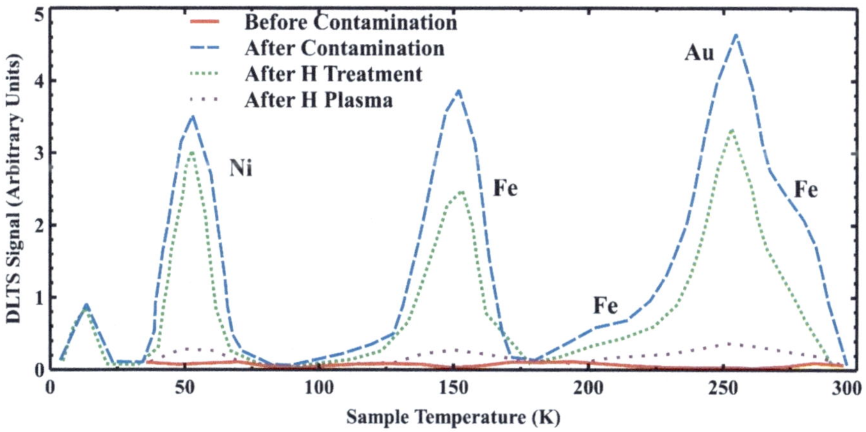

Fig. 14.16 Deep level transient spectroscopy (DLTS) plot showing transition metal contamination in a silicon wafer. Samples are shown for an uncontaminated silicon sample, after contamination in a contaminated furnace (2 h 900 °C in Ar), after passivation with hydrogen (2 h 700 °C H_2), and after passivation with a hydrogen plasma (2 h 700 °C). The metal contaminants signal peaks are identified

deep-level traps, eliminating their electrical activity. Figure 14.16 shows a plot of the results of hydrogen treatment of samples contaminated with metal impurities (Pearton et al., 1984). An electrical technique called deep-level transient spectroscopy (DLTS) is used to measure the electrically active metal impurities in the wafer. A silicon wafer was annealed in a contaminated furnace, contaminating the wafer with Ni, Fe, and Au. The wafer was then passivated with hydrogen by annealing in a clean furnace or exposing the wafer to a hydrogen plasma in a clean tool. The treatment partially electrically deactivates the metal defects, as shown in the figure.

Unfortunately, the hydrogen passivation of contaminants is not thermally stable and can be depassivated at low temperatures in the range of 200–300 °C, making hydrogen passivation an impractical solution to metal contamination (Fleetwood et al., 2009; Pearton et al., 1984).

14.8.2 Carbon

Carbon is a common silicon contaminant introduced into the wafer through SiC quartzware used during the crystalline silicon growth in the prevalent Czochralski wafer fabrication method. Also, carbon can be introduced into the wafer through common semiconductor processes such as plasma etching, etc. Hydrogen can interact with carbon in the wafer, forming hydrogen-carbon complexes and creating electrically active donor and acceptor defects in the silicon bandgap (Fleetwood et al., 2009; Pearton et al., 1992).

14.9 Bias Temperature Instability

Bias temperature instability is one of the most severe reliability risks in CMOS technology, particularly negative bias temperature instability (NBTI) for PMOS devices (Strong et al., 2009). A detailed discussion of NBTI is outside the scope of this book. However, a short description of the hydrogen behavior affecting NBTI follows. Figure 14.17 shows a sketch of hydrogen mechanisms affecting NBTI (Tsetseris et al., 2005). During negative voltage stress, thermal energy and the electric field break a donor-hydrogen complex in the silicon (1a,1b). As a result, the hydrogen diffuses to the interface, where it can depassivate the hydrogen-passivated interface state (2). In addition, the hydrogen can diffuse into the dielectric, where it can become trapped, forming a positive fixed oxide charge and shifting the threshold voltage (3). Finally, the hydrogen in the dielectric can form a hydrogen molecule that drifts back to the interface and repassivates the interface state (4,5), completing the cyclic process. The dynamic change in fixed oxide charge and interface states in the NBTI degradation produces shifts in the transistor parameters capable of causing circuit failure.

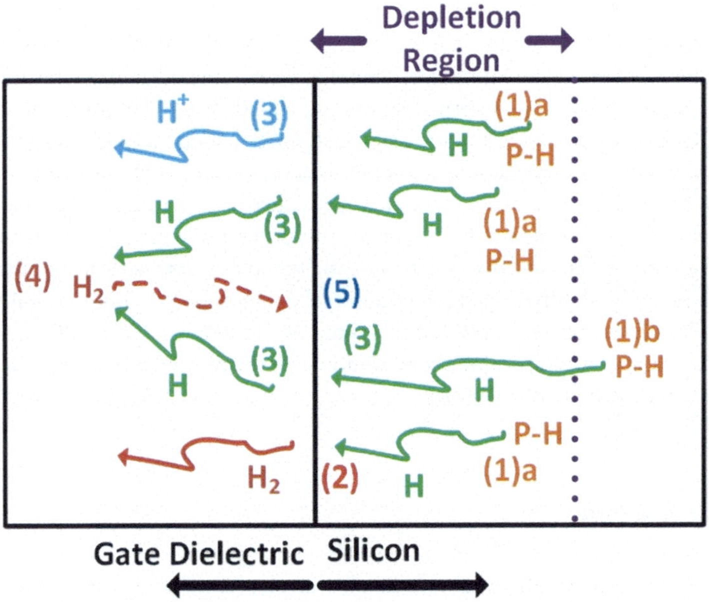

Fig. 14.17 Sketch showing the role of hydrogen in NBTI. (negative bias on n-type Si): (1a)–(1b) release of H from donor-H complexes in the silicon, (2) depassivation of interface states, (3) entry of H into silicon dioxide, (4) formation of H_2 molecule, (5) repassivation of interface state

14.10 Hydrogen Degrades Noise in CMOS

The hydrogen behavior in CMOS is very dynamic, constantly moving, interacting with defects, forming and breaking bonds at the temperatures where CMOS chips operate. This behavior, as shown in this chapter, can be observed as transistor noise. For example, Figs. 14.6 and 14.17 show the dynamic interaction of hydrogen with the interface states that create random telegraph noise (RTS) and 1/f noise as the interface states are passivated and depassivated by the mobile hydrogen. RTS noise is quantized transitioning between discrete observable states as single atomic level defects, as described in Chap. 7. The 1/f noise occurs when many defects are involved in the noise (Fleetwood et al., 2024). Figure 14.18 shows increasing noise as a function of the change in threshold voltage caused by interface states after radiation damage and annealing. Careful experimentation attributed the noise to hydrogen action (Aigier et al., 1992; Fleetwood et al., 2024).

Figure 14.19 shows an example of hydrogen-related noise in CMOS. A noise metric is plotted as a function of temperature before and after radiation. The three peaks in the after-radiation curve correspond to hydrogen action in the dielectric of lateral hydrogen motion in the dielectric, H_2 diffusion or disassociation, and H_2 drift. The corresponding activation energies are approximately 0.3, 0.45. and 0.75 eV. The advantage of studying radiation-damaged transistors is that large

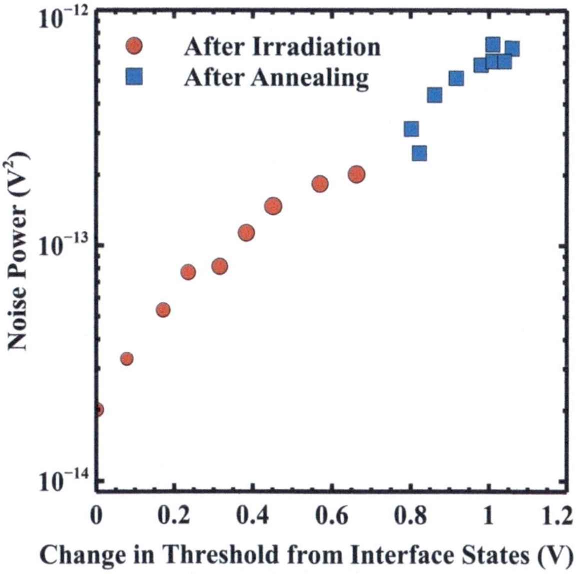

Fig. 14.18 Noise power plotted as a function of the shift in threshold voltage caused by interface states for power transistors irradiated and annealed after irradiation. The anneal is performed at room temperature with the transistor biased. The threshold voltage shift is attributed to hydrogen activity

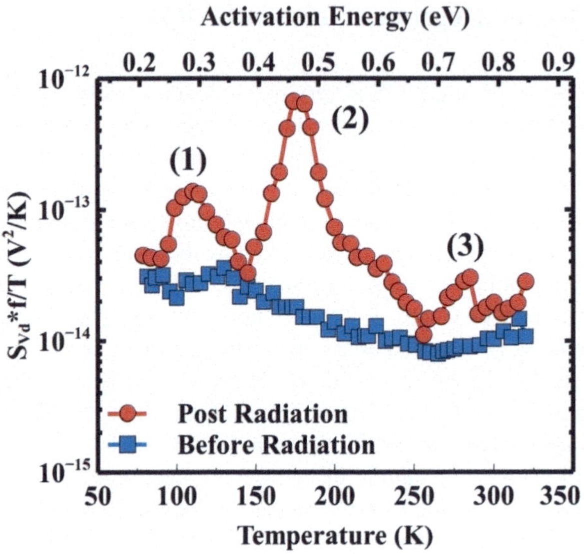

Fig. 14.19 Plot showing hydrogen related noise behavior at room and low temperatures before and after radiation damage. Peak (1) is associated with lateral hydrogen movement in the gate dielectric. Peak (2) is associated with H_2 diffusion or dissociation in the gate dielectric. Peak (3) is associated with H_2 drift in the dielectric

numbers of dielectric defects are created, significantly reducing the measurement difficulty. Plasma-damaged structures should behave similarly.

14.11 Hydrogen-Based Donors and Acceptors

Isolated hydrogen in the silicon lattice can act as a donor or acceptor depending on the charge state of the proton. Hydrogen can behave both as a donor or acceptor, varying with the charge on the atom. Figure 14.20 shows a plot of the energy level of hydrogen in the silicon bandgap varying with the hydrogen location in the lattice (Johnson et al., 1994). The plot shows the energy levels for hydrogen as a function of hydrogen position in the lattice. As the hydrogen is trapped interstitially, the stability of the hydrogen-based donor and acceptors can be expected to be weak with the energy level unstable with temperature and electric field.

14.12 Physical Defects Caused by Hydrogen

Hydrogen can cause structural and electrical defects in a CMOS wafer by itself without interacting with other atoms (Fleetwood et al., 2009; Pearton et al., 1992). Hydrogen can form clusters and platelets where high hydrogen concentrations

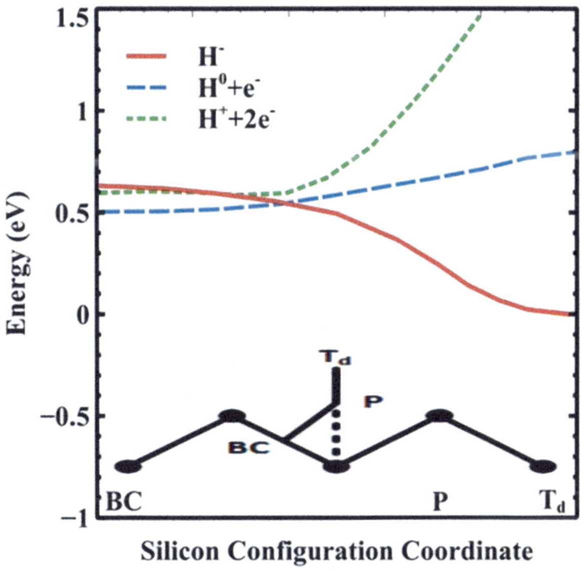

Fig. 14.20 Calculated hydrogen energy levels in the bandgap as a function of the hydrogen position in the silicon lattice for hydrogen in phosphorus doped silicon. BC is bond center and Td is the tetrahedral location

create an electrically active defect at low temperatures. These concentrated defects can form or dissolve depending on the temperature.

Hydrogen can play other roles in the silicon structure, including silicon embrittlement and modifying ductility (Pankove & Johnson, 1991; Pearton et al., 1992). In addition, hydrogen can impact structural defects, such as aiding in the growth of dislocations (Pankove & Johnson, 1991).

Hydrogen is an essential component of plasma etch processes. Also, hydrogen can thermally etch silicon at high temperatures without using plasmas. For example, silicon exposure to hydrogen at 670 °C can etch silicon (Lee et al., 2006).

14.13 Hydrogen Barriers

14.13.1 Introduction

Some materials commonly used in semiconductor processing can effectively block hydrogen from diffusing through them. This effect can be advantageous or disadvantageous depending on the engineering strategy for controlling the hydrogen reaching the semiconductor surface where the active devices exist. This section describes some examples of the common hydrogen barrier films used in CMOS fabrication. Unfortunately, an exhaustive survey of hydrogen barriers was not found during the preparation of this book. Literature from tritium barrier experiments and hydrogen metal protection experiments can provide insight into hydrogen barriers (Nemanic, 2019).

The hydrogen barrier capability of a film is strongly dependent on its material properties, including crystallinity, density, grain structure, porosity, solubility, diffusivity, etc. (Nemanic, 2019). An effective hydrogen barrier should be capable of preventing hydrogen penetration with a film thickness compatible with CMOS processing. Generally, the thinner the film, the easier it is to add to a CMOS process flow.

Many films used in CMOS processing impact the hydrogen behavior in a CMOS chip by design or inadvertently. For example, common metals such as copper or aluminum used in metallization can block hydrogen, preventing the passivation of the interface states (Fukamoto et al., 2004). Other metals, such as titanium, can absorb hydrogen, preventing it from reaching the semiconductor surface (Popegen et al., 2013). Hydrogen diffuses through dielectric films depending on the properties of the materials. Hydrogen can be trapped in the grain boundaries of films such as polysilicon (Pearton et al., 1992).

14.13.2 Alumina

Aluminum oxide (Al_2O_3) is a high dielectric constant material that is one of the more effective hydrogen barriers. It is commonly used in applications where the prevention of hydrogen encroachment is needed. One such application is for the manufacturing of ferroelectrics. Ferroelectrics such as lead zirconate titanate (PZT) or strontium bismuth tantalate (SBT) and their associated electrodes are extremely sensitive to hydrogen. Exposing these materials to hydrogen can lead to catastrophic electrical failure (Chowdhury et al., 2007). Hence, ferroelectric memories have heavily optimized hydrogen barriers, with alumina commonly used as a barrier. Alumina films as thin as 10 nm have been used as hydrogen barriers. Figure 14.21 shows the results of hydrogen barrier experiments on alumina (Helmich et al., 2020). Various thicknesses of alumina films are placed on a silicon wafer with a hydrogen emitting silicon nitride layer on top of the alumina. The wafers are exposed to a rapid thermal anneal (RTA), which drives hydrogen out of the nitride film into the wafer and atmosphere. The hydrogen concentration in the wafer was measured by monitoring the change in the p-type wafer resistivity caused by the boron neutralization by hydrogen described in Sect. 14.5.1. The plot shows increasing hydrogen concentration with increasing RTA temperature and decreasing hydrogen concentration with alumina thickness, demonstrating alumina's effectiveness as a hydrogen barrier.

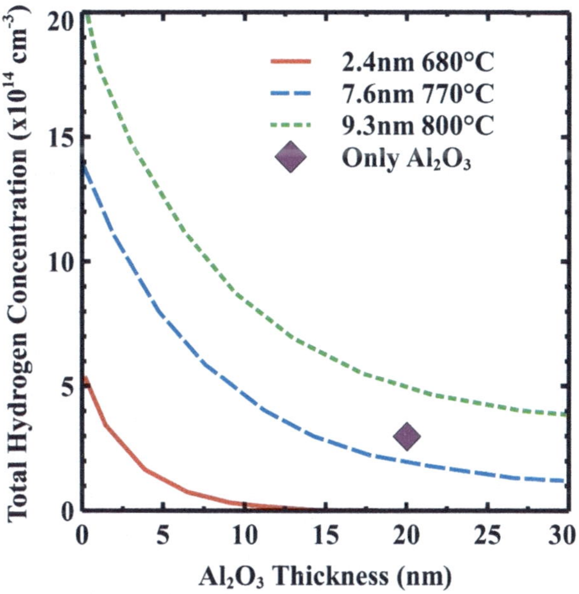

Fig. 14.21 Hydrogen concentration in silicon after a rapid thermal process heat treatment at 680, 770, 800 °C. The wafer is protected by an alumina layer on the silicon. A hydrogen rich silicon nitride layer above the alumina is used as a hydrogen source. The diffusion length L is fit to the data and shown in the legend. The Only Al_2O_3 point shows the background hydrogen concentration in the film

14.13.3 Silicon Nitride

Silicon nitride is a common hydrogen barrier. It can perform as a hydrogen barrier or as a hydrogen-emitting film, depending on the properties of the film and post-deposition processing conditions. There are many different methods of silicon nitride deposition, including thermal chemical vapor deposition (CVD), plasma-enhanced chemical vapor deposition (PEVCD), atomic layer deposition (ALD), physical vapor deposition (PVD), etc. (Kaloyeros et al., 2020).

The silicon nitride properties vary widely depending on the deposition method and process conditions. For example, some silicon nitride films can be optimized through processing conditions to be a hydrogen-emitting film or a hydrogen barrier. PECVD silicon nitride is one of the most common deposition methods due to its low cost and film properties. This section describes the performance of PECVD nitride as an example.

Figure 14.22 shows a plot of the hydrogen barrier properties for selected nitride films with the properties shown in Table 14.2 (Benoit et al., 2007). After the nitride films are deposited, the films are exposed to a deuterium alloy (400 °C 100% D_2) to measure the barrier properties. The use of deuterium alloy as a tracer element allows

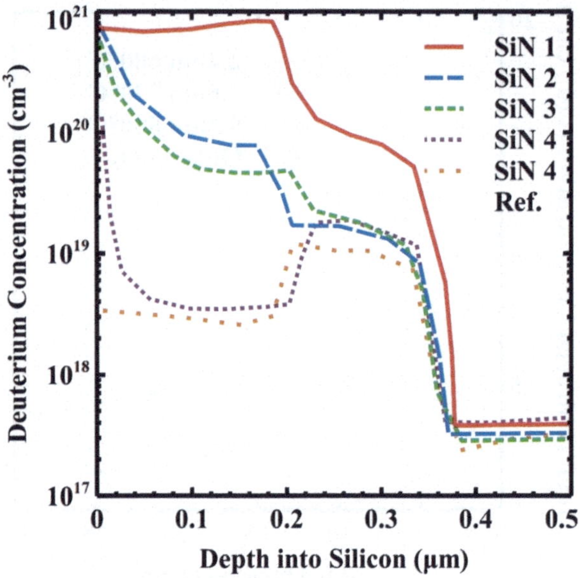

Fig. 14.22 Sketch showing a comparison of hydrogen barrier properties for different silicon nitride films (recipes) by performing a deuterium alloy step after the nitride is deposited. The deuterium concentration is measured by secondary ion mass spectroscopy (SIMS)

Table 14.2 Properties of the nitride films measured in Fig. 14.22

Properties	SiN 1	SiN 2	SiN 3	SiN 4
Refractive index	2.23	1.86	2.01	1.92
Stress (MPa)	166	381	12	−70
Si-H bond density (10^{21} cm^{-3})	21.5	10.5	15	7
N-H bond density (10^{21} cm^{-3})	2.8	8.5	5.5	8
Density (g/cm^3)	2.28	2.36	2.45	2.60
Mean concentration deuterium (10^{20} cm^{-3})	7.4	1.7	1.0	0.09

the identification of the penetrating atoms and separation from the hydrogen already present in the nitride films. There is a significant difference in the barrier performance between the films. The film SiN 1 has the worst performance due to its low density. The film SiN 4 has the best barrier performance with no significant deuterium penetration. The observed deuterium in the plot is the naturally occurring background deuterium in the films. The line labeled SiN 4 (Ref) shows that the deuterium concentration of film SiN 4 without a deuterium alloy is close to the alloyed sample, with the exception of increased deuterium concentration near the nitride film's surface. The best barrier properties can be obtained with a more amorphous film, as shown in Chap. 16 for a mobile ion barrier.

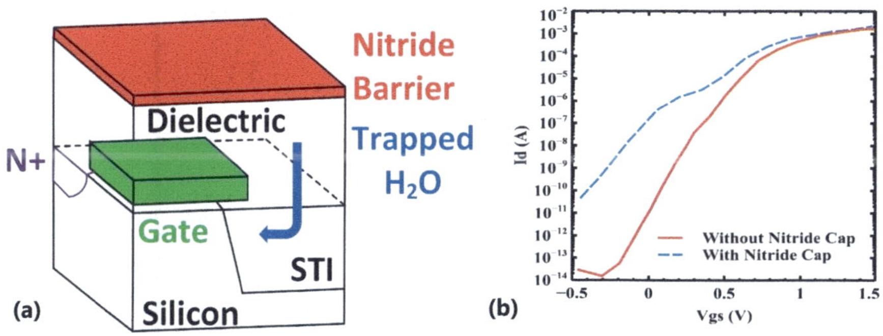

Fig. 14.23 Example of hydrogen related degradation at a shallow trench isolation edge due to moisture trapped in a dielectric layer. (**a**) sketch showing moisture related degradation at the shallow trench isolation edge. (**b**) subthreshold plot of log drain current versus gate voltage with and without a nitride capping layer

14.13.4 Hydrogen and Water Trapped Beneath Barriers

The electrical behavior of hydrogen depends on the materials present on the wafer, which affect the drift and diffusion properties. Hydrogen barriers and gettering materials can change the severity of mobile hydrogen problems (Park et al., 2000). For example, a silicon nitride hydrogen barrier can trap moisture in a wafer, causing isolation degradation along the shallow trench isolation edge, as shown in Fig. 14.23. The subthreshold conduction is degraded by creating a positive charge, as described in Chap. 5. The problem can be resolved in several ways, such as vacuum annealing immediately before the nitride deposition. The proposed mechanism in this example is oxygen-enhanced diffusion of the boron dopants, lowering the field threshold voltage at the edge of the channel. Additionally, a fixed oxide charge may be created by the water-based oxidation of the wafer from the trapped moisture.

14.14 Hydrogen Emission from Silicon Nitride Films

Silicon Nitride is commonly deposited using silane (SiH_4) as the silicon source for the film and ammonia (NH_3) as the nitrogen source. These hydrogen-based precursors result in films with a large hydrogen concentration, up to 30% (Braun, 2021; He, 1995). The atomic or molecular hydrogen can be emitted from the silicon nitride films at film deposition temperatures and higher temperatures. The emitted hydrogen can diffuse through films on the wafer's surface, reaching the silicon surface and causing changes in the electrical characteristics of the CMOS devices. These electrical changes can be beneficial or detrimental to the transistors. Examples of beneficial changes include passivating interface states, improving DRAM refresh,

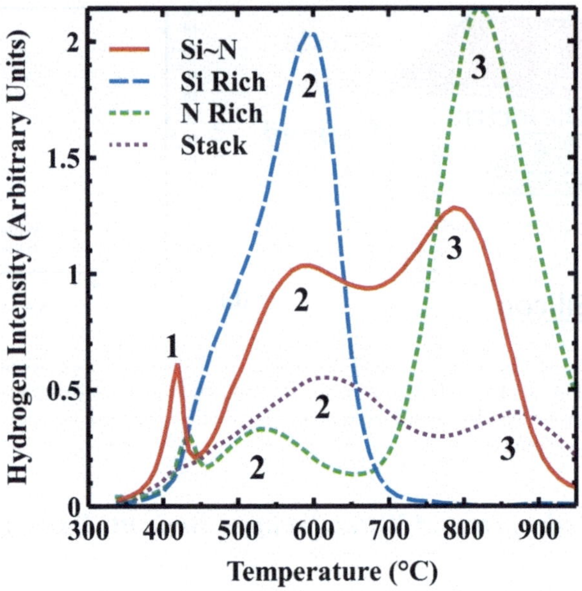

Fig. 14.24 Plot showing the amount of hydrogen emitted from a silicon nitride layer as a function of temperature for four different films. The label Si ~ N is an intermediate silicon-nitride ratio film. Si-rich is a silicon rich silicon nitride film. N-rich is a nitrogen rich nitride film. Stack is a three layer film consisting of the previous three films. Peak 1 is the surface desorption of H_2. Peak 2 is H_2 effusion from Si-H bonds. Peak 3 is hydrogen effusion from Si-N bonds. Note that there is no high temperature effusion from the silicon rich film, indicating a lack of Si-N bonds

etc. (Lee et al., 2012). Examples of detrimental changes include degradation of hot electron lifetime and negative bias temperature instability (Gasner, 1990).

Hydrogen effusion from different silicon nitride films is shown in Fig. 14.24 (Jarafi et al., 2019). The nitride film is exposed to a temperature ramp in a vacuum, causing outgassing, and the emitted gases are analyzed with a mass spectrometer. This technique is called thermal desorption spectroscopy (TDS). The amount of hydrogen at each temperature step is measured, producing the plot. There is a significant difference in the hydrogen emission as a function of film properties.

The film with the silicon and nitrogen concentrations nearly equal (Si~N) has a prominent peak below 400 °C where unreacted unbonded H is emitted from the films. The films have two emission peaks near 500–550 °C and 750–800 °C. The lower temperature peak is caused by thermal energy breaking Si-H bonds, resulting in H emission from the film. The higher temperature peak is caused by thermal energy breaking N-H bonds, resulting in H emission from the film. The higher temperature hydrogen emission for the N-H bond is due to the higher bond strength of the N-H bond. The disassociation energy for the H-Si bond is 293.3 Kj/mol, and the disassociation energy for the H-N bond is 358.8 Kj/mol, indicating the H-N bond breaking occurs at higher temperatures (Haynes, 2016–2017, pg. 9–76).

Figure 14.25 shows a sketch explaining the hydrogen emission dependence on the silicon nitride deposition recipe conditions (Wang, 2001). Higher deposition

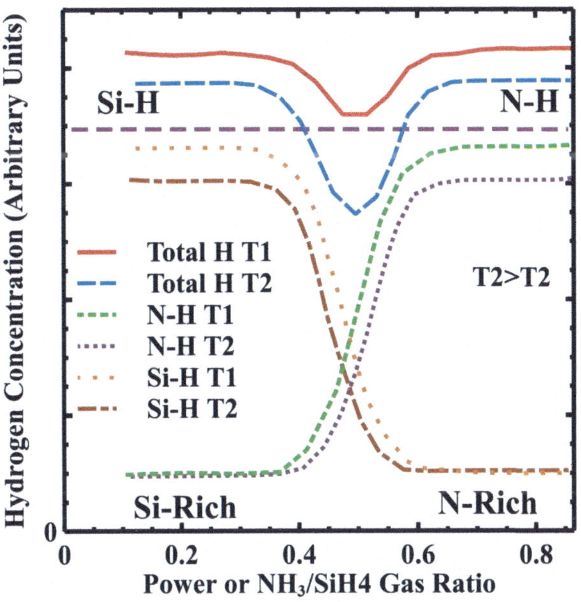

Fig. 14.25 Sketch showing emitted hydrogen as a function of silicon nitride process deposition conditions. There are two recipe conditions that can achieve a hydrogen target level as shown by the pink line

temperatures result in less hydrogen emission due to less hydrogen incorporation into the film. The minimum hydrogen incorporation occurs when the deposited film is stoichiometric with a nitrogen-to-silicon ratio of 1.33 for the Si_3N_4 film. More hydrogen is trapped in the film for silicon or nitrogen-rich films as the hydrogen bonds with the excess silicon or nitrogen atoms.

The film can be optimized for low or high hydrogen emission with recipe modifications. Higher hydrogen emission at low temperatures can be obtained with a silicon-rich film by increased silane flow or lower power. Higher emission at high temperatures can be obtained with a nitrogen-rich film. The minimum emission can be obtained with a stoichiometric film deposited at high temperatures.

Figure 14.26 shows some important properties of silicon nitride as a function of the refractive index and silicon-to-nitrogen ratio (Bredemeier et al., 2019). The ratio of silicon-hydrogen and nitride-hydrogen bonds in the film determines important hydrogen-related properties. The film density increases with increasing nitride concentration, limiting hydrogen diffusion and limiting hydrogen emission.

The hydrogen concentration as a function of the refractive index is shown in Fig. 14.27, peaking near a refractive index of 2.4 (Bredemeier et al., 2019). The hydrogen emission from a silicon nitride film exposed to a 900 °C rapid thermal anneal as a function of the refractive index is also shown in the figure. Increasing the refractive index and increasing the silicon-to-nitrogen ratio leads to more hydrogen loss from the film. Low silicon concentration films contain low concentrations of

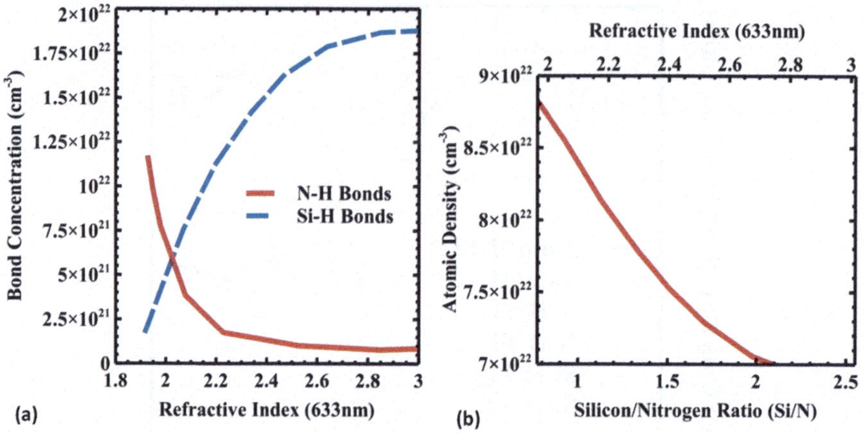

Fig. 14.26 Properties of silicon nitride films as a function of refractive index. (**a**) Si-H and N-H bond concentration as a function of refractive index. (**b**) film density as a function of refractive index and silicon to nitrogen ratio

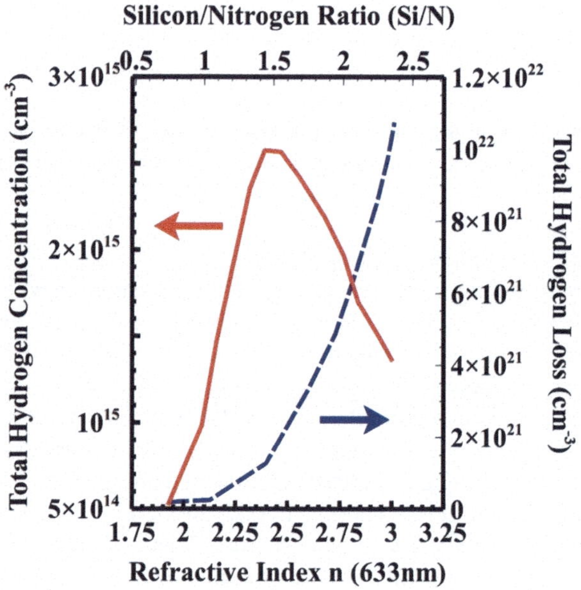

Fig. 14.27 The initial hydrogen concentration in the film and the emitted hydrogen concentration from a silicon nitride film as a function of refractive index and silicon to nitrogen ratio in the film. The hydrogen loss is measured as a combination of the film hydrogen and the silicon hydrogen losses accounting for the hydrogen lost from the film and the hydrogen diffusion through it from the substrate

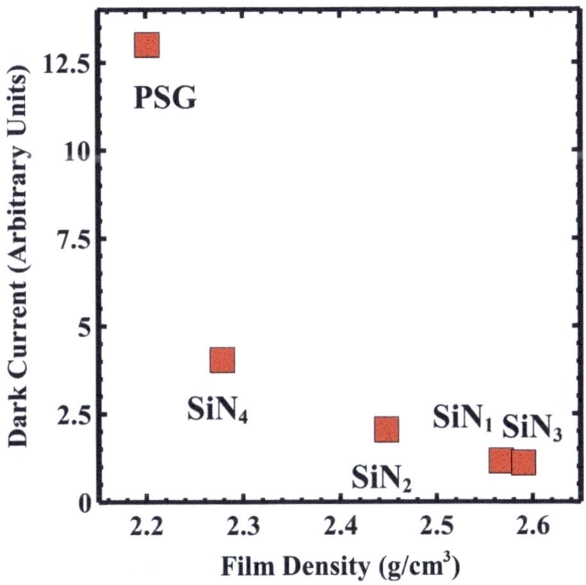

Fig. 14.28 Plot showing CMOS imager dark current improvement as a function of passivation density

hydrogen and emit lower amounts of hydrogen. Therefore, this type of film performs better as a hydrogen barrier.

Nitride films can significantly impact sensitive CMOS devices, improving or degrading performance depending on process condition details. Figure 14.28 shows a plot of CMOS imager dark current as a function of the passivation for silicon nitride and phosphorous silicate glass films (Regolini et al., 2007). There is a strong improvement in the imager dark current with increasing film density. The improvement with increasing density is attributed to the higher density passivation, which prevents the out-diffusion of hydrogen from the wafer as well as hydrogen emission from the film.

The different silicon nitride barrier deposition conditions described previously also affect the crystallinity of the film. The effectiveness of the barrier is improved with a more amorphous film and degraded with a more crystalline film, as shown for sodium in Chap. 16, where silicon nitride is demonstrated as a mobile ion barrier.

14.15 Hydrogen Interaction with Metals

Hydrogen can react with the metals commonly used in CMOS wafers, changing its behavior. For example, hydrogen is commonly gettered with titanium, cobalt, or nickel silicides, becoming trapped in the metal temporarily or permanently

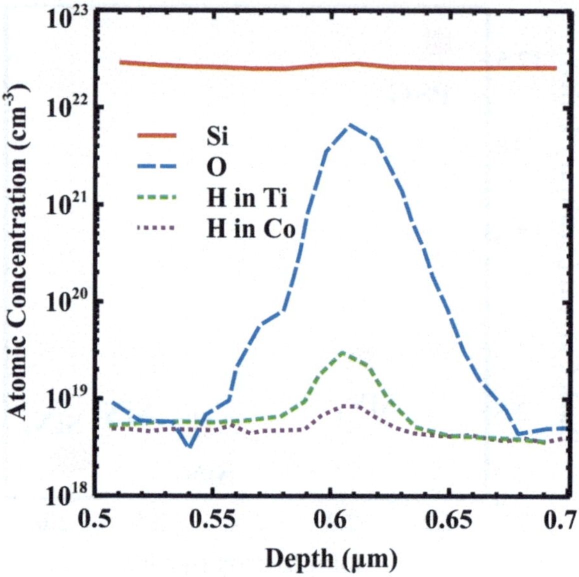

Fig. 14.29 Secondary Ion Mass Spectroscopy (SIMS) of a titanium or cobalt silicide layer exposed to hydrogen on top of a silicon dioxide layer showing hydrogen trapped in the silicide layers

depending on the subsequent temperatures on the wafer. Figure 14.29 shows the hydrogen profile in a cobalt or titanium silicide layer on top of a silicon dioxide layer (Wang, 2001). The titanium silicide layer traps more hydrogen than the cobalt silicide layer.

Other metals, such as aluminum, can function as a catalyst, cracking a water molecule and creating H+ and OH- products that change the hydrogenation of the wafer (Liou et al., 2001). The use of aluminum in processing equipment can increase hydrogen exposure.

Other common materials, such as the plastic used in wafer boxes and front-opening universal pod (FOUP) used to transport wafers between process steps, can absorb hydrogen and other gases, remitting them on the wafer. FOUPs are commonly purged with nitrogen to limit wafer exposure.

14.16 Deuterium

The rare hydrogen isotope deuterium is commonly used as a tracer for hydrogen-related experiments because its natural rarity makes detection easy. Deuterium can be used as a process gas, and the location of the isotope in the wafer can be tracked with common analytical techniques such as secondary ion mass spectroscopy (SIMS), providing insight into hydrogen behavior during semiconductor processing. The use of deuterium allows the isolation of the hydrogen from specific sources from the ubiquitous general background hydrogen. Deuterium, with a higher mass, diffuses slightly slower but otherwise matches the behavior of hydrogen.

Deuterium can also improve CMOS reliability under certain conditions, improving CMOS reliability for problems such as hot-carrier injection, stress-induced leakage current, and bias temperature instability (Jung et al., 2022). The additional mass due to the extra neutron is believed to make the Si-D bond more robust, enhancing reliability.

14.17 Enhanced Low-Dose Rate Radiation Damage

Hydrogen can play an important role in radiation damage for CMOS devices. The radiation damage can be increased for lower radiation exposure rates when compared to high dose rate exposure. This phenomenon is called enhanced low dose rate sensitivity (ELDRS). The hydrogen-related physics behind ELDRS is believed to be very closely related to the physics of bias temperature instability (Tsetseris et al., 2005).

Figure 14.30 shows a plot demonstrating ELDRS (Zimin et al., 2018). The shift in threshold voltage for a NMOS transistor as a function of total radiation dose is plotted for a high and low dose rate. The low dose rate threshold voltage shift is

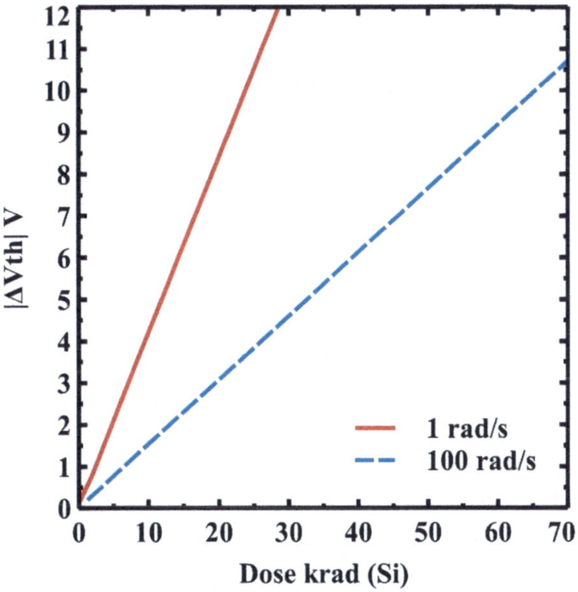

Fig. 14.30 Plot showing enhanced low dose rate sensitivity for radiation damage. The threshold voltage of nmos devices is plotted as a function of total dose. Two dose rates of 1 rad/s and 100 rad/s are compared with the lower dose rate showing more damage and a larger shift in threshold voltage

substantially worse. This phenomenon is attributed to the formation of hydrogen molecules from diffusing hydrogen atoms (Tsetseris et al., 2005). This phenomenon can lead to an underestimation of the radiation sensitivity of CMOS technologies measured at high dose rates.

References

Aigier, P., et al. (1992). Comparison of l/f noise in irradiated power MOSFETs measured in the linear and saturation regions. *IEEE Transactions on Nuclear Science, 39*(6), 2012. https://doi.org/10.1109/23.211398

Baumvol, I. J. R. (1999). Atomic transport during growth of ultrathin dielectrics on silicon. *Surface Science Reports, 36*, 1–166. https://doi.org/10.1016/S0167-5729(99)00006-0

Benoit, D., Regolini, J., & Morin, P. (2007). Hydrogen desorption and diffusion in PECVD silicon nitride. Application to passivation of CMOS active pixel sensors. *Microelectronic Engineering, 84*, 2169–2172. https://doi.org/10.1016/j.mee.2007.04.071

Braun, J. L. (2021). Hydrogen effects on the thermal conductivity of delocalized vibrational modes in amorphous silicon nitride (a-SiN x:H). *Physical Review Materials*, 5, 035604. https://doi.org/10.1103/PhysRevMaterials.5.035604

Bredemeier, D., et al. (2019). Impact of silicon nitride film properties on hydrogen in-diffusion into crystalline silicon. In 36th *European photovoltaic solar energy conference and exhibition.*

Chowdhury, A., Kim, J., & Wallace, R. M. (2007). Alumina as a hydrogen barrier layer for FeRAM devices. In *Non-volatile memory technology symposium.* https://doi.org/10.1109/NVMT.2007.4389944

Claeys, C., & Simoen, E. (2018). *Metal impurities in silicon- and germanium-based technologies origin, characterization, control, and device impact.* Springer. https://doi.org/10.1007/978-3-319-93925-4

Fleetwood, D. M., et al. (1998). Nonvolatile memory based on mobile protons. In *International non-volatile memory technology*. https://doi.org/10.1109/NVMT.1998.723227

Fleetwood, D. M., Pantelides, S. T., & Schrimpf, R. D. (2009). *Defects in microelectronic materials and devices*. CRC Press Taylor and Francis Group. https://doi.org/10.1201/9781420043778

Fleetwood, D. M., et al. (2024). Effects of Interface traps and hydrogen on the low-frequency noise of irradiated MOS devices. *IEEE Transactions on Nuclear Science, 71*(4), 555–568. https://doi.org/10.1109/TNS.2023.3323548

Fukamoto, J., et al. (2004). Impact of asymmetric metal coverage on high performance MOSFET mismatch. *Solid-State Electronics, 48*, 1767. https://doi.org/10.1016/j.sse.2004.05.060

Gasner, J. (1990). An investigation of device instabilities arising from the encapsulation material and composition. In *VLSI multilevel interconnection conference*. https://doi.org/10.1109/VMIC.1990.127895

Gos, W. (2011). *Hole trapping and the negative bias temperature instability*. TU Wien.

Hallam, B. J., et al. (2018). Development of advanced hydrogenation processes for silicon solar cells via an improved understanding of the behaviour of hydrogen in silicon. *Progress in Photovoltaics, 1217–1238.* https://doi.org/10.1002/pip.3240

Haynes, W. M., et al. (2016–2017). *Handbook of chemistry and physics*. CRC.

He, S. S. (1995). Hydrogen diffusion and redistribution in PECVD Si-Rich silicon nitride during rapid thermal annealing. *International Conference on Solid-State and IC Technology*. https://doi.org/10.1109/ICSICT.1995.500083

Helmich, L., et al. (2020). Atomic-layer-deposited Al2O3 as effective barrier against the diffusion of hydrogen from SiNx:H layers into crystalline silicon during rapid thermal annealing. *Physica Status Solidi Rapid Research Layers, 14*, 2000367. https://doi.org/10.1002/pssr.202000367

Herring, C., et al. (2001). Energy levels of isolated interstitial hydrogen in silicon. *Physical Review B, 64*, 125209. https://doi.org/10.1103/PhysRevB.64.125209

Ishihara, T., et al. (2020). Interface defect engineering for high performance MOSFETs with novel carrier mobility model: Theory and experimental verification. *AIP Advances, 10*(055020), 1–15. https://doi.org/10.1063/5.0005813

Jarafi, S., et al. (2019). Composition limited hydrogen effusion rate of a-SiNx:H passivation stack. *AIP Conference Proceedings*. https://doi.org/10.1063/1.5123853

Johnson, N. M., et al. (1994). Inverted order of acceptor and donor levels of monoatomic hydrogen in silicon. *Physical Review Letters, 73*(1), 130–133.

Jung, D.-H., et al. (2022). High pressure deuterium annealing for improved immunity against stress-induced threshold voltage degradation. *IEEE Transactions on Devices and Materials Reliability, 22*(3), 457. https://doi.org/10.1109/TDMR.2022.3194504

Kaloyeros, A. E., et al. (2020). Review—Silicon nitride and silicon nitride-rich thin film technologies: State-of-the-art processing technologies, properties, and applications. *ECS Journal of Solid State Science and Technology, 9*, 063006. https://doi.org/10.1149/2162-8777/aba447

Lee, J.-H., et al. (2006). Low-pressure, low-temperature hydrogen annealing for nanoscale silicon fin rounding. In *IEEE nanotechnology materials and devices conference*. https://doi.org/10.1109/NMDC.2006.4388941

Lee, C.-Y., Lai, C.-S., Yang, C.-M., & Wang, D. H. L. (2012). DRAM data retention and cell transistor threshold voltage reliability improved by passivation annealing prior to the deposition of plasma nitride layer. *IEEE Transactions on Device and Materials Reliability, 12*(2), 406–412. https://doi.org/10.1109/TDMR.2012.2188895

Liou, J.-W., et al. (2001). Plasma charging and mobile ions on the data retention of 0.25um flash memory. In *6th international symposiyum on plasma process induced damage*. https://doi.org/10.1109/PPID.2001.929982

Nemanic, V. (2019). Hydrogen permeation barriers: Basic requirements, materials selection, deposition methods, and quality evaluation. *Nuclear Materials and Energy, 19*, 451–457. https://doi.org/10.1016/j.nme.2019.04.001

Pankove, J. I., & Johnson, N. M. (1991). *Hydrogen in semiconductors*. Academic Press.

Park, S.-K., et al. (2000). CMOSFET characteristics induced by moisture diffusion from interlayer dielectric in 0.23um DRAM technology with shallow trench isolation. In *International reliability physics symposium*. https://doi.org/10.1109/RELPHY.2000.843908

Pearton, S. J., et al. (1984). Deep level impurities in germanium and silicon: Low temperature passivation or removal techniques. *IEEE Transactions on Nuclear Science, NS-31 No. 1*, 326–330. https://doi.org/10.1109/TNS.1984.4333269

Pearton, S. J., Corbett, J. W., & Stavola, M. (1992). Hydrogen in crystalline semiconductors. Springer https://doi.org/10.1007/978-3-642-84778-3.

Popegen, G., Nelhiebel, M., & Grasser, T. (2013). Detrimental impact of hydrogen passivation on NBTI and HC degradation. In *IEEE international reliability physics symposium*. https://doi.org/10.1109/IRPS.2013.6532125

Rashkeev, S. N., et al. (2001). Proton-induced defect generation at the Si–SiO2 interface. *IEEE Transactions on Nuclear Science, 48*(6), 2086–2092. https://doi.org/10.1109/23.983177

Regolini, J. L., Benoit, D., & Morin, P. (2007). Passivation issues in active pixel CMOS image sensors. *Microelectronic Reliability, 47*, 739–742. https://doi.org/10.1016/j.microrel.2007.01.067

Seager, C. H., & Anderson, R. A. (1990). Minority carrier induced debonding of hydrogen from shallow donors in silicon. *MRS Online Proceedings Library, 209*.

Simoen, E., et al. (2006). Thermal donor formation in direct-plasma hydrogenated n-type Czochralski silicon. *Materials Science and Engineering B, 134*, 189–192. https://doi.org/10.1016/j.mseb.2006.06.041

Singh, S., & Nagabalasubraminian, P. B. (2019). A review on thermal donors in CZ-silicon and their impact on electronic industry. *International Journal of Engineering Research and Applications, 14*(2), 43–47. https://doi.org/10.37591/rrjophy.v12i3.3808

Strong, A. W., et al. (2009). *Reliability wearout mechanisms in advanced CMOS technologies*. Wiley. https://doi.org/10.1002/9780470455265

Tao, Y., et al. (2024). Tunnel oxide passivating contact enabled by polysilicon on ultra-thin SiO2 for advanced silicon radiation detectors. *Scientific Reports, 14*, 17307. https://doi.org/10.1038/s41598-024-68368-w

Tsetseris, L., et al. (2005). Common origin for enhanced low-dose-rate sensitivity and bias temperature instability under negative bias. *IEEE Transactions on Nuclear Science, 52*(6), 2265–2271. https://doi.org/10.1109/TNS.2005.860670

Vanheusden, K. (2011). *Direct observation of Mobile proton in SiO2 thin films: Potential application in a novel memory device*. Materials Research Society Symposia Proceedings.

Wang, C. Y. (2001). *Hydrogen concentration analysis in PECVD and RTCVD silicon nitride thin films and it's impact on device performance*. Materials Research Society.

Warren, W. L., et al. (1997). Protonic nonvolatile field effect transistor in Si/SiO2/Si structures. *IEEE Transactions on Nuclear Science, 44*(6), 1789–1798. https://doi.org/10.1109/23.658944

Zimin, P. A., et al. (2018). ELDRS in p-MOS and p-MNOS based RAD FETs with thick gate insulators: Experiment and simulation. In *European conference on radiation and its effects on components and systems*. https://doi.org/10.1109/RADECS45761.2018.9328654

Chapter 15
Metallic Defects

15.1 Introduction

Metal impurities in CMOS wafers can cause many electrical and physical defects detrimental to the yield and operation of a CMOS chip. A single metal atom can destroy sensitive devices such as a DRAM cell or a CMOS imager pixel, resulting in yield loss or degraded quality. Advanced logic chips can fail to yield due to a single atomic metal defect in one finfet, killing the entire chip. Higher amounts of metal contamination can degrade all types of CMOS chips. The mitigation and elimination of metal defects during CMOS processing have been heavily studied and are essential to successful CMOS manufacturing. CMOS manufacturers have minimized metal contamination through rigid control of the contamination levels of incoming consumable materials, optimization of the processing tools to prevent the wafer from coming into contact with harmful contaminants, etc.

Figure 15.1 shows a periodic table highlighting the risk of device degradation or damage for each element (Shamazaki et al., 2008). The elements with the highest propensity for damage are on the right side of the transition metals (in Red) in the periodic table and include gold, which is probably the worst element for killing the minority carrier lifetime. Copper is highly mobile and is a significant risk because of its wide use in chips for metallization. Cobalt and nickel are widely used in silicides to reduce the resistance of semiconductor regions. Finally, iron is a common contaminant due to its ubiquitous use in metals used in modern industry.

The orange-colored elements denote elements with medium risk. Titanium and tungsten are widely used in CMOS metallization. Yellow highlights the elements associated with a lower risk for device degradation. Zirconia and hafnia are widely used in gate dielectrics where limited thermal processes prevent damage to the underlying silicon. The alkali metals lithium, sodium, and potassium cause the mobile ion problems described in Chap. 16.

K. Prall, *CMOS Plasma and Process Damage*,
https://doi.org/10.1007/978-3-031-89029-1_15

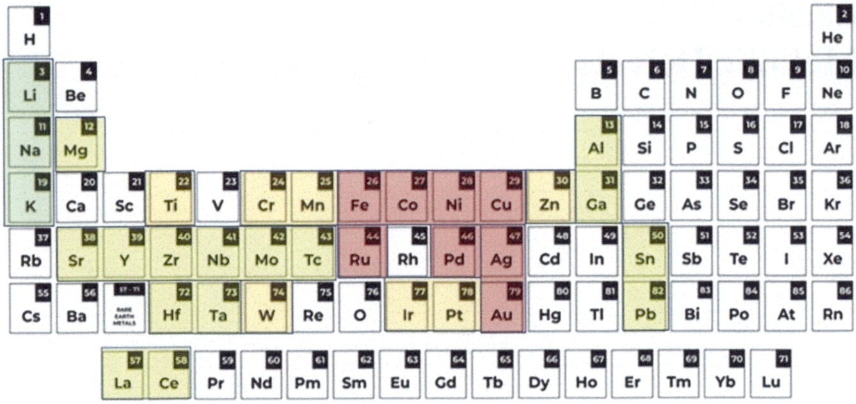

Fig. 15.1 Periodic table showing the propensity of an element for causing damage or electrical degradation to CMOS chips. Red color indicates a large propensity for damage or degradation. Orange color indicates a medium propensity for damage or degradation. Yellow color indicates a smaller propensity for damage or degradation. Green color indicates damaging elements that typically cause mobile ion problems

An excellent book on this subject is highly recommended (Claeys & Simoen, 2018).

The probability of an element contaminating a wafer is related to the likelihood of the wafer being exposed to the element somewhere along the manufacturing flow. For example, the use of stainless steel is minimized in semiconductor equipment to avoid wafer contamination by the steel components (Fe, Cr, Ni, Mo, Ti, Al, Cu, etc.). Gold is a severe contaminant but rarely causes problems because its use in wafer manufacturing is nonexistent in CMOS processes (excluding wire bonding). Commonly used metals in CMOS metallization, such as Ti, Co, Ni, W, Cu, Hf, etc., have increased risk because they are used in many CMOS processes. The consumables used in manufacturing are heavily optimized to minimize metal contamination from the incoming wafer, gases, chemicals, etc.

Metal contamination can be detected through electrical testing of the affected silicon devices, as described in the following sections. However, chemical analytical techniques have substantial advantages in troubleshooting metal ion contamination in a wafer fab. Chemical techniques allow exact, unambiguous determination of the exact contaminating atoms. Additionally, chemical techniques are very useful for inline troubleshooting, allowing equipment, chemicals, etc., to be isolated and tested as the sources of the contamination. Chemical techniques use a simple test wafer instead of the complex devices needed for electrical testing.

The semiconductor industry generally uses a few analytical techniques for metal contamination detection. The most common method is vapor phase decomposition (VPD), in which vapor hydrogen fluoride is used to remove the surface native oxide, forming water droplets, followed by the collection of the water droplets for chemical techniques. The droplets can be analyzed by total reflection X-ray fluorescence (TXRF) and other techniques, resulting in detection levels of around 10^8 to

10^{10} atoms/cm^3 (Claeys & Simoen, 2018). Other common analytical techniques include time-of-flight secondary ion mass spectrometry (TOF-SIMS), which has a sensitivity of approximately 6×10^8 atoms/cm^3. Energy-dispersive X-ray (EDX) analysis is useful for analyzing particles visible in a cross section with a sensitivity of 0.1–1.0 atomic percent (see Chap. 16 for a discussion of analytical chemistry).

15.2 Common CMOS Contamination Sources

Virtually any process step and equipment can introduce metallic impurities into the wafer. Therefore, semiconductor consumable materials, such as wafers, resists, chemicals, gases, water, etc., are heavily filtered and purified to eliminate metallic and other impurities. Wafer manufacturers grow the silicon crystal and fabricate the starting material wafers using methods to minimize metal contamination. Fabrication tools are designed not to introduce metallic materials into the wafer by using specific alloys, ceramics, etc., in the chamber exposed to the wafer and external to the chamber. Any component of the fabrication tools that comes into direct contact with a wafer is a potential source of contamination.

The metallization steps necessary to create a CMOS chip may add metal impurities in unintended locations, including the wafer substrate, gate dielectrics, etc. The copper metallization used in CMOS processes has a high propensity for metallic contamination. The copper back end of line processes use metal and dielectric barriers to minimize copper migration. Gate metallization, such as tungsten, can contaminate the gate dielectric and substrate. Salicide formation using titanium, cobalt, nickel, etc., in direct contact with the silicon substrate can contaminate the wafer. Assembly-related processes such as through-silicon vias (TSVs) can also contaminate the wafer. High-k gate transition-metal oxides are a potential source of metal contamination. Table 15.1 shows a partial list of common sources of metallic contamination in wafer fabrication. However, virtually, all process steps have the propensity to contaminate the wafer with metals (Claeys & Simoen, 2018).

15.3 Common Effects of Metal Contamination

Metal contamination in a CMOS wafer can cause many distinct types of physical and electrical defects. A partial list of first-order effects is shown in Table 15.2 in rough order of the contamination level necessary to observe the effects, decreasing from top to bottom. Heavy contamination can cause physical structural defects, and low contamination levels can destroy DRAM or imager chips. Multiple observable types of physical defects can be caused by metal contamination. Metal contamination can induce wafer haze and roughness on the surface where metal precipitates on the wafer surface, modifying the reflectivity. Haze is typically associated with for high levels of metal contamination. Metals that induce haze include iron, cobalt,

Table 15.1 Common sources of metal contamination in CMOS wafers

Crystal growth and wafer manufacturing
Wafer cleaning, deionized water
Resist and masking material deposition, strip, and removal
Wafer handling and backside contact with chucks, etc.
Ion implantation
Gate dielectric (high-k)
Rapid thermal processing and furnace processing
Silicidation or salicide processing
Gate metallization
Contact metallization
Metallization
3D integration, through-silicon vias, etc.

nickel, copper, rhenium, and palladium (Graff, 1995). Sometimes, special etchants are used to highlight haze, making it more visible.

Metal contamination on the wafer surface can prevent normal oxide growth, creating dielectric defects. In addition, metal defects can induce crystalline defects like stacking faults in silicon. Metal atoms can precipitate into clusters of metals in the wafer bulk or on the surface. Metal atoms can create silicon defects such as dislocations or become trapped at defect locations, possibly enlarging them. The propensity for cluster formation is a strong function of the thermal history and ramp-down conditions during processing (Graff, 1995).

Metal defects can severely degrade semiconductor junction diodes and the source and drain junctions of CMOS transistors, resulting in excessive leakage, reduced breakdown, shorted junctions, etc. In addition, metal defects can have a subtle effect on the device parameters of CMOS devices, shifting threshold voltages, creating Vt hysteresis, mobility reduction, degraded subthreshold slopes, short channel effects, and increased noise (Claeys & Simoen, 2018).

Single-atom metal defects can cause failure in sensitive devices when one of the first-order effects becomes large enough to cause the device to fail. For example, a DRAM cell fails when the junction or dielectric leakage becomes sufficient to drain the stored charge from the capacitor node, causing the cell to lose the digital state before the required refresh interval. Another example is that excessive leakage in a CMOS imager photodiode may cause the pixel to lose the stored image information. Details on these types of failures may be described in the technology-specific chapter (Chap. 8).

15.4 Electrical Activity of Metal Ion Contaminants

Metal impurities in a silicon wafer create generation-recombination (G-R) centers where electrons and holes can be trapped, created, recombine, and transition between the bands. The defect centers are commonly called Shockley-Read-Hall

Table 15.2 Common physical and electrical defects caused by metal contamination in a CMOS wafer

Physical defects
Wafer haze
Surface roughness
Stacking faults
Metal precipitates
Gate dielectric thinning
Junction electrical defects
Minority carrier lifetime degradation
Reduction in minority carrier diffusion length
Bulk carrier generation
Junction leakage
Junction weak or soft breakdown
Low activation energy for junction leakage
Spiked junction and salicide spikes
MOS dielectric and capacitor electrical defects
Gate leakage
Trap centers
Mobile charge
Fixed charge
Gate dielectric leakage
Gate dielectric breakdown
Low-k dielectric degradation
Surface recombination velocity degradation
Back surface generation
Intermetal dielectric degradation
CMOS transistor electrical defects
V_t shift
V_t hysteresis
Degraded subthreshold slope
Mobility reduction
Short-channel effects
Increased noise
Chip specific defects
DRAM refresh failures
CMOS imager single pixel failures

(SRH) centers. Figure 15.2 is a band diagram sketch of the electrical processes that occur at an SRH defect, showing the before and after conditions for each transition (Sze & Lee, 2010). An electron in the conduction band can become trapped in the center (Fig. 15.2a), remain in the center for a while, and be re-emitted back into the conduction band (Fig. 15.2b). A hole in the valence band can become trapped in the center (Fig. 15.2c), remain in the center for a while, and be re-emitted back into the valence band (Fig. 15.2d). An electron and hole can also recombine in the metal defect, annihilating both carriers (Fig. 15.2e).

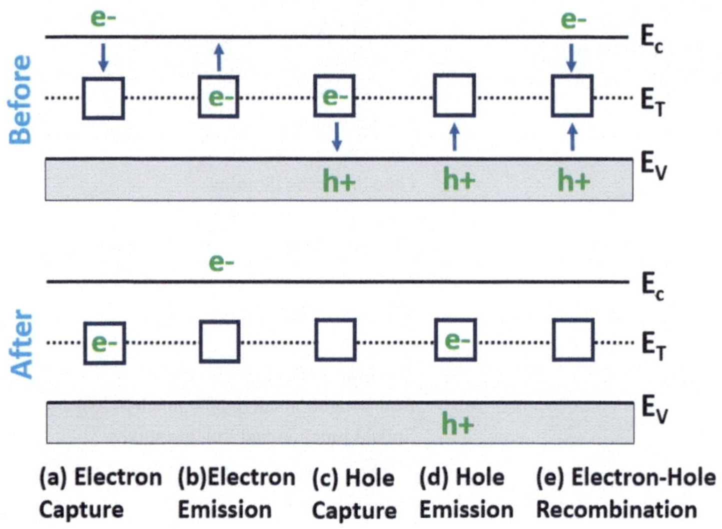

Fig. 15.2 Generation-recombination mechanisms that occur at a metal SRH defect: (**a**) electron capture, (**b**) electron emission, (**c**) hole capture, (**d**) hole emission, and (**e**) electron-hole recombination

The SRH mechanisms created by metal and other defects degrade the electrical device characteristics of contaminated wafers, as summarized in Table 15.2. Increased leakage in junction diodes or mos capacitors is one of the primary symptoms of SRH metal-related leakage. SRH-related leakage typically dominates near room temperature, while other mechanisms, such as carrier thermal generation, dominate at elevated temperatures. Other mechanisms, such as radiative and Auger, are neglected here as they are less important in silicon. Radiative effects can be neglected in indirect bandgap silicon. Auger mechanisms only become significant at high doping or current density.

Figure 15.3 shows some of the G-R effects in a one-sided n+/p− reverse-biased diode with the depletion region (dashed line) extending into the silicon. Nearly identical effects occur in a mos capacitor in depletion. SRH centers are shown at the wafer surface or inside the n+ region (Fig. 15.3a), inside the depletion region (Fig. 15.3b), at the STI edge (Fig. 15.3c), inside the wafer bulk (Fig. 15.3d), and at the backside wafer surface (Fig. 15.3e). Carriers generated by the thermal processes create an electron-hole pair at the G-R centers. The generated carriers respond to the electric field with the negative electrons drifting toward the positive biased n + region and the positively charged holes drifting toward the grounded backside contact. The current created by the carrier flow in the electric field creates a leakage current in the diode with electrons collected by the positive supply and the holes collected by the substrate ground.

Figure 15.4 shows a plot of diode leakage as a function of the areal surface metal concentration for a device intentionally contaminated with a wet chemical treatment (Claeys & Simoen, 2018, pg. 307). The metal contaminants diffuse into the wafer

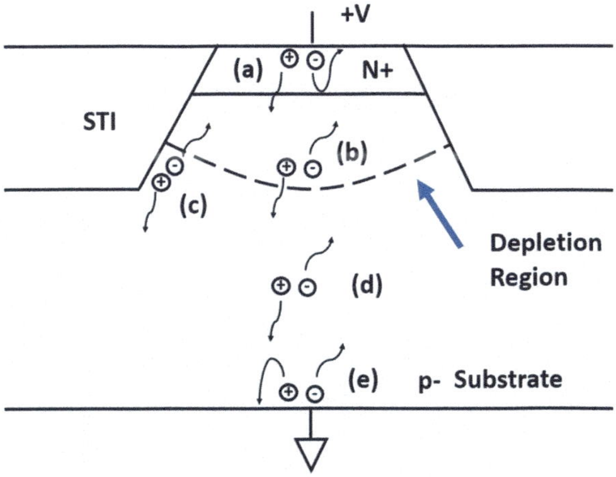

Fig. 15.3 Sketch of SRH leakage mechanisms for a reverse biased n+ junction diode. The arrows show the direction electrons and holes drift with positive bias on the junction. (**a**) SRH leakage at the surface or in the n+ region. (**b**) SRH leakage in the depletion region. (**c**) SRH leakage at the shallow trench isolation (STI) edge. (**d**) G-R leakage inside the silicon bulk. (**e**) G-R leakage at the wafer backside interface

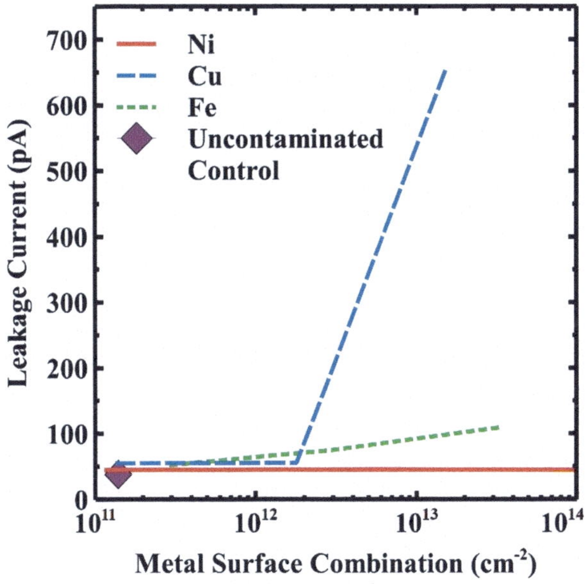

Fig. 15.4 Plot showing reverse bias diode leakage as a function of areal surface metal concentration measured by X-ray fluorescence spectroscopy (TXRF) for three different contaminant metals (copper, iron, and nickel). An uncontaminated measurement labeled is shown on the left side of the plot (diamond)

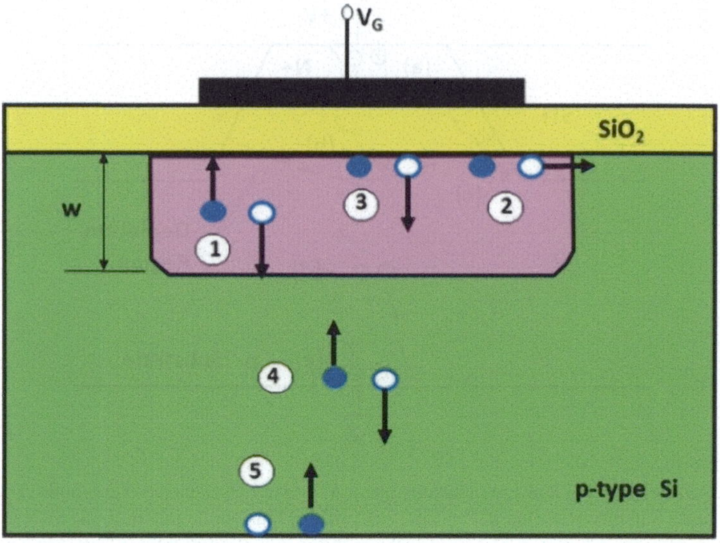

Fig. 15.5 Common effects due to metal contamination in a MOS capacitor or transistor channel. (1) Bulk generation and recombination inside the depletion region (pink region). (2) Surface generation and recombination. (3) Surface generation and recombination inside the gate region characterized by surface recombination velocity. (4) Quasi-neutral bulk generation characterized by minority carrier diffusion length. (5) Backside surface generation

with further processing after the contaminating step. The contaminants form SRH defects in the active region of the diodes. The surface metal concentration cannot directly be converted to a volume concentration because the metal diffusion profile is unknown.

Figure 15.5 shows a similar diagram for a positively biased mos capacitor biased in depletion containing identical defects as the diode shown above in Fig. 15.3 (Claeys & Simoen, 2018, pg. 108). Again, carriers generated in the bulk of the semiconductor (4) and the backside drift (5) and diffuse toward the surface, where they are captured by the MOS capacitor depletion region, increasing the minority carrier population and affecting the device characteristics. Carriers generated in the depletion region (1) separate with the electrons drifting to the surface, and the positive gate field pushes the holes away from the surface. Defects at the surface generate electron-hole pairs inside the gate region (3) or outside the gate region (2), and the electric field also separates the electron-hole pairs. The metal defect-induced generation recombination mechanisms can affect many device characteristics, as shown in Table 15.2.

The structure of a MOS transistor combines the mechanisms of the junction diode and MOS capacitor. During normal operating conditions, the electrical biases inside a CMOS transistor can affect how metals drift, diffuse, and accumulate inside the transistor junctions. Figure 15.6 shows an example sketch explaining how positive copper ions behave inside the transistor. (Claeys & Simoen, 2018; Yeon et al.,

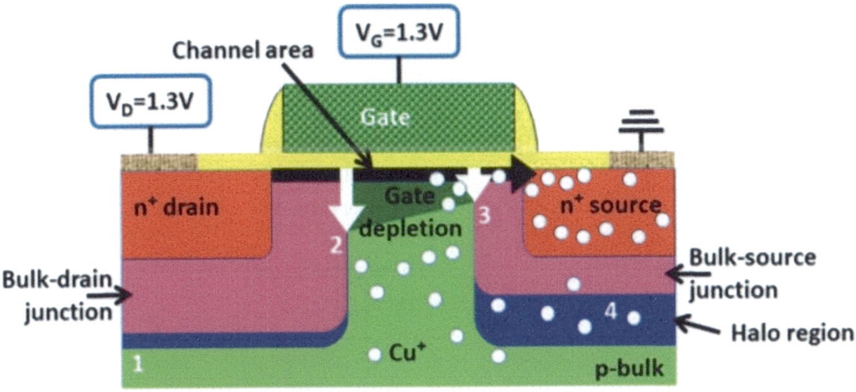

Fig. 15.6 Cross-section sketch of a nmos transistor showing the location of metal contamination in a mos transistor, accumulating in the source of the device

2012, pg. 308). The NMOS transistor is biased with the drain and gate biased positively. The source is grounded. This biasing condition tends to push and pull the positively charged metal ions into the source, concentrating the metal defect density into the source region.

A transistor operated in an unipolar mode with the source and drain relative polarities always biased in the same direction tends to collect metals in the source region (3,4) and drive metals from the drain region (1,2). An example of this biasing type is a pulldown NMOS transistor in a CMOS inverter. On the other hand, a transistor operated in a bipolar mode with the relative source and drain reversed during operation tends to push the metal ions back and forth between the source and drain during operation, resulting in metal distribution throughout the device. An example of this biasing type is a bidirectional CMOS pass gate passing a varying voltage.

Metal ions in a PMOS transistor behave differently from the NMOS transistor behavior described above. The polarities in Fig. 15.6 are reversed for a PMOS device with the gate and drain biased negatively with the source grounded. This biasing scheme attracts the positively charged metal ions toward the drain in a PMOS device instead of the source in a NMOS device.

15.5 Relevant Important Properties of Metals

15.5.1 Introduction

The atomic properties of metals and other contaminants determine the likelihood of chip damage and degradation in addition to the probability of exposure. This section describes the material properties relevant to semiconductor contamination. Table 15.3 lists the properties important in metal contamination, some of which are

Table 15.3 Important metal properties related to electrical contamination in semiconductors

Property
Location of energy levels in the bandgap
Diffusivity
Capture cross section
Lifetime and diffusion length
Solid solubility
Melting temperature
Metallic radius
Segregation coefficient
Formation energy
Number of electrons in d and s orbits

discussed in the following sections (Claeys & Simoen, 2018). The likelihood of the wafer and chip being exposed to metal contaminants plays a significant role, as some potentially damaging metals are not used in chip manufacturing.

The propensity for harm to a CMOS chip cannot be directly predicted from a simple examination of the periodic table and is determined empirically. However, studying the individual fundamental properties can provide some limited insights. This section reviews some of the material properties shown in Table 15.3. The energy levels in the bandgap, diffusivity, capture cross section, lifetime, diffusion length, and solid solubility are reviewed in detail. Other parameters are discussed in the references (Claeys & Simoen, 2018; Graff, 1995).

15.5.2 Trap Energy Location in Bandgap

The bandgap energy barrier is a crucial parameter for a semiconductor. The bandgap barrier height must be large enough to prevent electrons and holes from significant direct tunneling from band to band. Silicon is an indirect semiconductor where it is usually difficult for carriers to tunnel between bands because the bands are not aligned in space and energy. As a result, traps are normally required to allow transitions between bands. Figure 15.7 shows a sketch of the bandgap of a semiconductor (Rubin, 2024). Figure 15.7a shows the energy transition required without a trap. The energy gap for silicon is 1.1 eV.

The energy level of the metal contamination trap defect in the bandgap is an important metal defect property. The worst-case trap energy level occurs when the trap is located near the center of the bandgap, as shown in Fig. 15.7c. The tunneling or emission probability is higher when the defect is located in the center of the bandgap, causing more generation, recombination, and a higher trapping probability. The center of bandgap traps significantly impacts leakage and other forms of device degradation. Traps located away from the center of the bandgap, as shown in Fig. 15.7b, have a lower tunneling or emission probability and have less impact on device degradation.

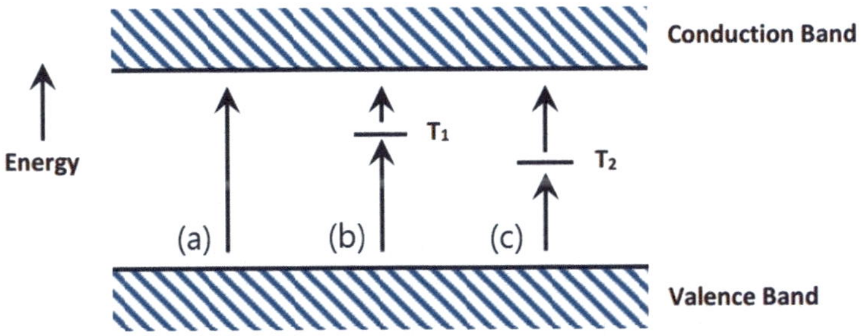

Fig. 15.7 Sketch showing trap levels in the silicon bandgap. (**a**) No trap present. (**b**) Trap T_1 located away from the center of the bandgap. (**c**) Trap T_2 located at the center of the band gap

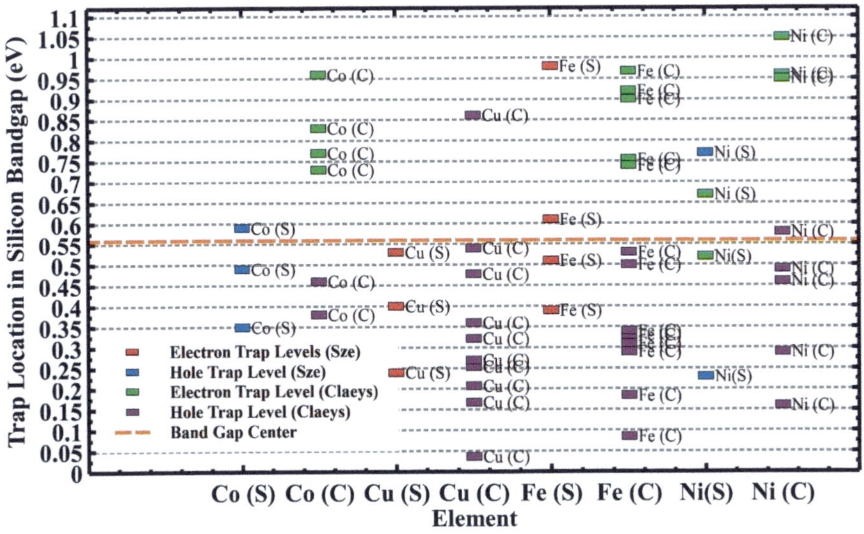

Fig. 15.8 Plot showing metal defect position in the band gap for some of the worst contaminants in silicon. The dashed line marks the center of the band gap. Two sources of data are compared with (S) denoting and old source (Chen et al., 1987) and (C) denoting a modern source (Claeys & Simoen, 2018)

The technical literature commonly includes graphs and tables of the defect energy of metal contaminants that are useful in determining the severity of a contaminant by its location in the bandgap. For example, Fig. 15.8 shows a chart showing the position of energy levels for common silicon metal contaminants in the bandgap (Sze & Lee, 2010). The bandgap center at 0.56 eV is marked with a dashed line in the plot. Elements such as copper, iron, etc., have levels close to the center of the bandgap, indicating that they create a strong SRH center.

Figure 15.8 compares older and recent data on trap levels for some of the worst contaminants in silicon. Older publications typically show a few levels in the bandgap for each element, labeled (S) in Fig. 15.8 (Sze & Lee, 2010). Newer publications using more advanced techniques tend to show many levels in the bandgap, as shown in the figure labeled (C) (Claeys & Simoen, 2018). The results in the literature can be inconsistent with many different values published by different authors (Rein, 2005). The trap levels are not very useful for identifying unknown contaminants but may provide some insight into the electrical severity of a metal contaminant.

An individual element can create multiple electrical energy levels in the bandgap, each with different electrical characteristics. Additionally, some defect states are capable of trapping more than one charge carrier. Each energy level has an individual defect density, which determines the number of defects per unit volume in the semiconductor. For example, the energy levels for hafnium contamination in the bandgap are shown in Table 15.4 (Claeys & Simoen, 2018; Sachdeva et al., 2006). Hafnium has 20 unique defect levels in the silicon bandgap. The first column in the table denotes the energy level name. The defect energy level shows where the defect exists in the silicon bandgap relative to the valence (E_v) and conduction bands (E_c) for fast and slow diffusing levels. The capture cross section and the measured defect density are shown. Despite hafnium having a large number of levels in the bandgap and a few levels near the center of the bandgap, it is not an important SRH defect center due to its low diffusivity at CMOS processing temperatures after gate formation. The number of levels for each element can vary widely.

Table 15.4 Energy levels for hafnium as a contaminant in silicon

	Defect level (eV)			
Level	Slow	Fast	Capture cross section (cm^2)	Defect density (cm^{-3})
E55		EC-0.09	$\sigma_n = 4.4 \times 10^{-15}$	3.3×10^{14}
E70	EC-0.117		$\sigma_n = 2.94 \times 10^{-11}$	4.6×10^{13}
E81		Ec-0.15	$\sigma_n = 4.28 \times 10^{-14}$	5.92×10^{14}
E84	Ec-0.20		$\sigma_n = 1.61 \times 10^{-11}$	
E92		Ec-0.17	$\sigma_n = 4.14 \times 10^{-13}$	5.8×10^{14}
E109	Ec-0.22		$\sigma_n = 1.08 \times 10^{-13}$	3.4×10^{13}
E127		Ec-0.22	$\sigma_n = 6.8 \times 10^{-15}$	5.5×10^{14}
E146	Ec-0.27		$\sigma_n = 1.53 \times 10^{-14}$	1.08×10^{14}
E146		Ec-0.29	$\sigma_n = 1.21 \times 10^{-14}$	4.7×10^{14}
E178		Ec 0.39	$\sigma_n = 5.05 \times 10^{-13}$	4.8×10^{14}
E202	Ec-0.4		$\sigma_n = 3.26 \times 10^{-14}$	6.59×10^{13}
E232	Ec-0.44		$\sigma_n = 2.25 \times 10^{-14}$	5.26×10^{13}
E232		Ec-0.49	$\sigma_n = 2.8 \times 10^{-14}$	4.17×10^{14}
H64	Ev + 0.13		$\sigma_p = 1.79 \times 10^{-13}$	8×10^{13}
H81	Ev + 0.15		$\sigma_p = 1.49 \times 10^{-14}$	2.5×10^{13}
H109	Ev + 0.25		$\sigma_p = 1.39 \times 10^{-16}$	1.3×10^{13}
H132		Ev + 0.26	$\sigma_p = 4.15 \times 10^{-14}$	2.5×10^{12}
H143	Ev + 0.24		$\sigma_p = 1.89 \times 10^{-15}$ Exp.(0.04 eV/k_BT)	3.4×10^{13}
H191		Ev + 0.43	$\sigma_p = 3.2 \times 10^{-13}$	1.1×10^{13}
H249		Ev + 0.50	$\sigma_p = 3.08 \times 10^{-14}$	1.5×10^{13}

15.5.3 Diffusivity

A contaminant's ability to move through silicon and other materials affects its ability to reach device structures that can be degraded and damaged. Slow diffusing elements are more likely to stay at the location where they were introduced into the wafer and not migrate to the critical device regions. Conversely, elements that diffuse rapidly are highly likely to reach sensitive device regions and degrade them.

Elements diffuse through different mechanisms (Bracht, 2000). Fast diffusing elements diffuse through interstitial diffusion, migrating rapidly through the spaces between atoms in the crystal lattice. Slow diffusing elements diffuse substitutionally, replacing the silicon atom in the lattice and then repeating the replacement process for the next atom as it diffuses. Additionally, some contaminants diffuse by both mechanisms with diffusion rates between pure interstitial and substitutional diffusion rates. Figure 15.9 shows a plot of the diffusion rates versus inverse temperature for several common metal contaminants. The three different types of diffusion are shown in the figure. Note the difference in the temperature slope between the three different types of diffusion.

The slope of the diffusion rate is an essential factor in determining the severity of a contaminant. The diffusion coefficient temperature slope is the lowest for the interstitial diffusers labeled (a) in Fig. 15.9. The shallow slope means that these contaminants can diffuse at temperatures inside the operating temperature range of CMOS chips. Low-temperature diffusion can cause chip operation and reliability degradation during packaging, testing, burn-in, reliability tests, etc., used in chip manufacturing flows.

Most historical diffusion data was taken at high temperatures. Extrapolation to low temperatures is not necessarily accurate, requiring data from low-temperature diffusion measurements.

15.5.4 Carrier Capture Cross Section

The carrier capture cross section is a property that measures the propensity for a contaminant to capture an electron or hole as it nears the defect. A large cross section indicates that the SRH defect is more likely to capture the carrier, which impacts the device characteristics. The capture cross sections are different for electrons and holes, sometimes by many orders of magnitude, creating a defect that traps one dominant carrier. Table 15.5 shows a table of example data for some common silicon contaminants (Aissa et al., 2015). The location of the defect in the bandgap is shown. The cross section of the electrons (σ_n) and holes (σ_p) are shown, along with the ratio of the n-type to the p-type cross section (σ_n/σ_p). High ratios indicate that the defect is less active as an SRH center, as the smallest cross section limits the generation rate. The energy level in the bandgap is shown to affect the defect behavior, as described in Table 15.5. The difference between elements is significant, and the role

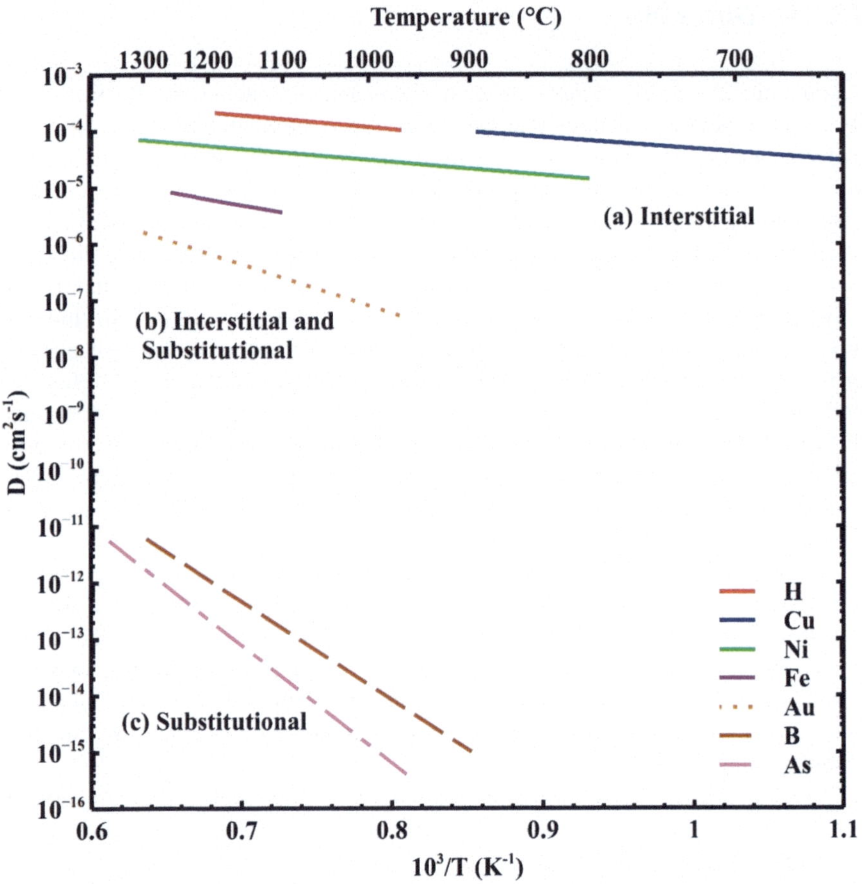

Fig. 15.9 Diffusivity (*D*) of some elements in silicon as a function of inverse temperature. (**a**) Elements with solid lines diffuse mainly through interstitial diffusion. (**b**) Gold with large dotted lines diffuses through both interstitial and substitutional diffusion. (**c**) Elements with dashed lines diffuse mainly through substitutional diffusion

of the cross section in CMOS degradation is not always obvious. For example, copper, a severe contaminant, has a small cross section for both electrons and holes. Other elements, such as iron, chrome, vanadium, titanium, gold, and zinc, have larger cross sections, indicating that they should be electrically active in silicon.

15.5.5 *Solid Solubility*

Solid solubility is an important parameter for metal contamination. The solid solubility of a contaminating element determines the maximum amount of contaminant that can be introduced into a silicon wafer (Graff, 1995). An element with a low

Table 15.5 Capture cross sections for various silicon contaminants in n-type silicon

Defect	Energy (eV)	σ_n (cm^2)	σ_p (cm^2)	σ_n/σ_p
Fe_i	$E_v + 0.38$	5.0×10^{-14}	7.0×10^{-17}	714.3
FeB	E_c-0.26	6.0×10^{-15}	7.0×10^{-17}	85.7
		5.0×10^{-15}	3.0×10^{-15}	1.7
Cr_i	E_c-0.24	2.4×10^{-14}	8.0×10^{-15}	3.0
	E_c-0.24	2.0×10^{-13}	8.0×10^{-14}	2.5
		2.0×10^{-14}	4.0×10^{-15}	5.0
CrB	$E_v + 0.27$	3.8×10^{-14}	7.0×10^{-15}	5.4
	$E_v + 0.28$	0.5×10^{-14}	1.0×10^{-14}	0.5
	$E_v + 0.27$	2.0×10^{-14}	1.0×10^{-14}	2.0
V_i	$E_v + 0.36$	1.0×10^{-16}	2.0×10^{-18}	50.0
		5.0×10^{-14}	3.0×10^{-18}	16666.7
Ti_i	E_c-0.27	1.0×10^{-14}	3.0×10^{-17}	333.3
		3.1×10^{-14}	1.4×10^{-15}	22.0
Mo_i	$E_v + 0.28$	1.6×10^{-14}	6.0×10^{-16}	26.0
Mn_i		3.5×10^{-15}	2.0×10^{-18}	1750.0
Ni_s	E_c-0.40	5.6×10^{-17}	8.0×10^{-17}	0.7
	$E_v + 0.19$	4.0×10^{-12}	1.0×10^{-15}	4000.0
Cu	E_c-0.50	1.6×10^{-18}	1.0×10^{-19}	16.0
Au_s	E_c-0.50	1.4×10^{-16}	7.6×10^{-15}	0.02
Co_s	$E_v + 0.41$	2.0×10^{-15}	5.0×10^{-18}	0.0004
		1.6×10^{-14}	1.0×10^{-15}	16.0
Zn_s	$E_v + 0.33$	1.5×10^{-15}	4.4×10^{-15}	0.34

solubility has a lower maximum number of atoms possible in the wafer compared to an element with high solid solubility when contaminated to a maximum amount. Therefore, a low solid solubility element may cause less degradation of the CMOS devices depending on the influence of the other contamination parameters. Figure 15.10 shows a plot of solid solubility versus inverse temperature for common metal contaminants. For example, zinc has more than two orders of magnitude higher solubility than titanium, making it a worse contaminant.

15.5.6 Minority Carrier Lifetime and Diffusion Length

The minority carrier lifetime and diffusion length are important related metrics for a semiconductor. The minority carrier lifetime and diffusion length play essential roles in CMOS technology, affecting device leakage, DRAM refresh, CMOS imager defects, etc.

Minority carriers are carriers of the opposite type relative to the local silicon doping (Meroli). A minority carrier is created by the SRH generation process, moves through the device through diffusion and drift, and it is eliminated through SRH recombination. The lifetime is the average time the minority carrier exists between its creation through generation and annihilation by recombination, existing in the

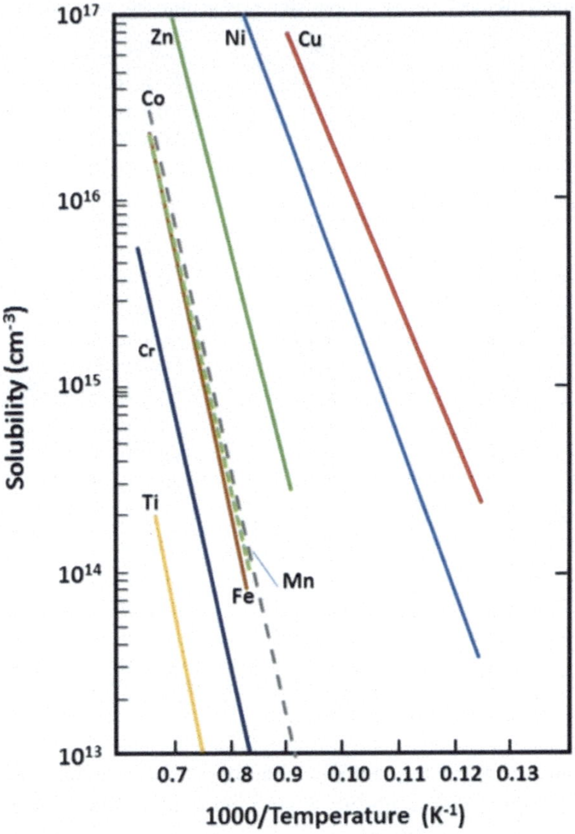

Fig. 15.10 Solid solubility of common metal contaminants in silicon as a function of inverse temperature

conduction band for electrons and the valence band for holes. The lifetime τ and diffusion length L_{diff} are related by Eq. (15.1), where D is the diffusivity for electrons or holes. The equation applies separately for electrons and holes.

$$L_{diff} = \sqrt{D\tau} \tag{15.1}$$

The diffusion length specifies the distance the minority carriers travel before recombining. These two parameters are closely related as they are different methods of measuring the same phenomenon. A carrier with a long lifetime and diffusion length has more time to travel through diffusion for a longer distance. Figure 15.11 shows the relationship between lifetime and diffusion length as a function of dopant concentration for p-type and n-type semiconductors.

Generation is the process of creating mobile electrons in the conduction band and holes in the valence band. The generated charge carriers allow the semiconductor to conduct charge, changing its resistivity. Generation can occur thermally or

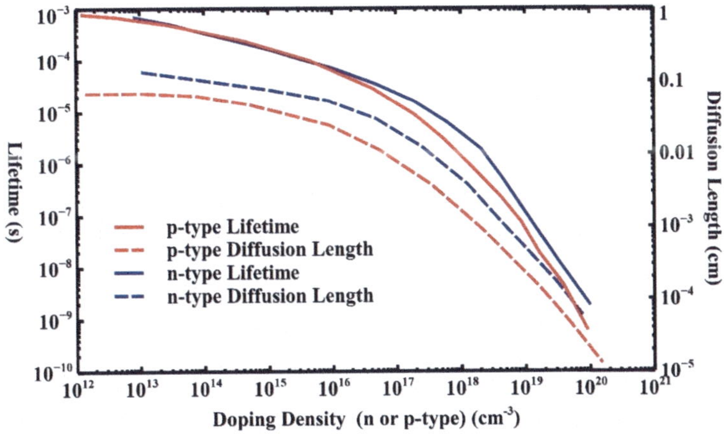

Fig. 15.11 Plot showing the relationship between lifetime and diffusion length as a function of dopant density for both n-type and p-type wafers

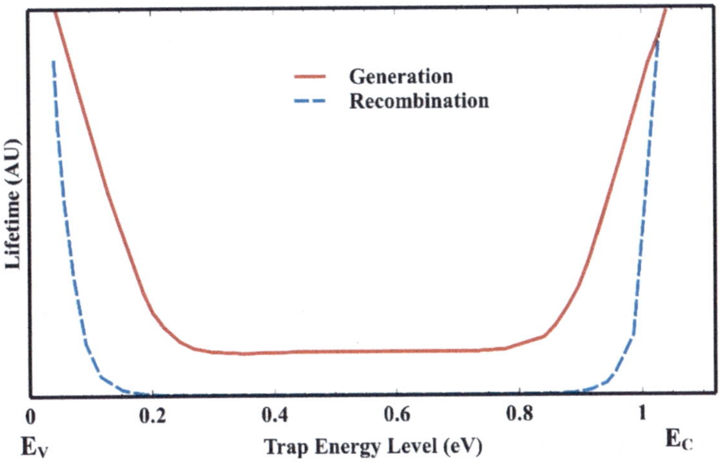

Fig. 15.12 Plot showing the generation and recombination lifetime as a function of the metal defects position in the bandgap as a function of the defect energy level location in the silicon bandgap

through electrical methods such as drift and diffusion. Recombination is the process of annihilating an electron and a hole, causing them to cease to exist as charge carriers removing them from the conduction and valence bands. Lifetime can refer to either the generation or recombination process. There is a significant difference between the generation and recombination lifetimes as a function of the defect location in the bandgap.

Figure 15.12 shows a sketch of the generation and recombination lifetimes as a function of the trap energy level in the bandgap (Rein, 2005; Schroder, 1997). The

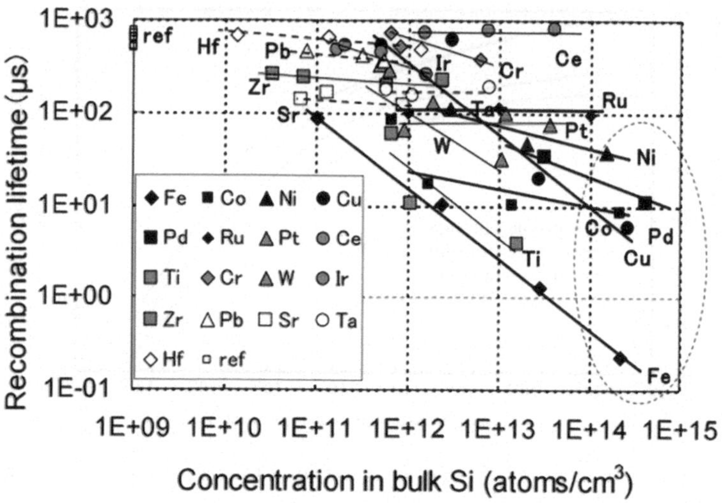

Fig. 15.13 Recombination lifetime as a function of metal concentration in silicon for common metal contaminants. The points labeled ref in the top left corner show the results for an uncontaminated silicon sample. The worst lifetime degradation is on the bottom right of the plot

left side of the plot with a trap energy level of zero denotes the edge of the valence band (E_v). The right side of the plot label E_c denotes the edge of the conduction band. The shape of the curve differs for generation and recombination. Note that the generation lifetime is a stronger function of the trap bandgap location, while the recombination lifetime is a weaker function of the trap location.

Metal contamination degrades minority carrier lifetime and diffusion length by creating SRH centers where the charge carriers become trapped and recombine, annihilating the carriers. More metal-related SRH centers can degrade the lifetime and diffusion length. Figure 15.13 shows a plot of the recombination lifetime versus the contaminant concentration for some metals and other elements as a function of the concentration of the contaminants (Shamazaki et al., 2008). The lifetime of some elements, such as copper and iron, is a strong function of the metal concentration. The concentration of other elements, such as cerium or hafnium, has little or no impact on the lifetime.

15.6 The Worst Contaminants

15.6.1 Copper

Copper is one of the worst silicon contaminants because of its common usage in semiconductor processes where it is used as an interconnect in metal lines and packaging processes such as through-silicon vias. The properties of copper, such as its

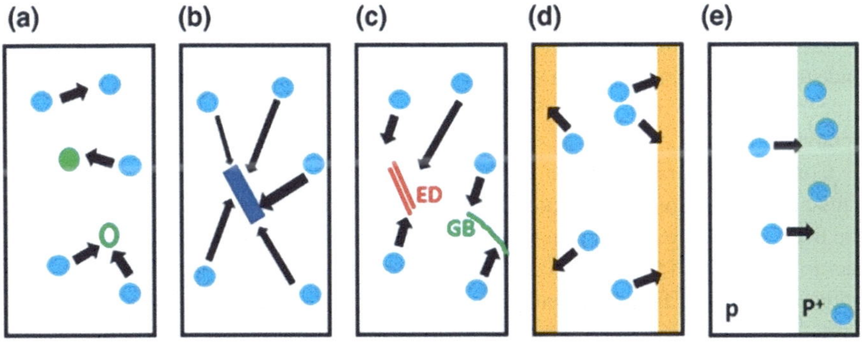

Fig. 15.14 Sketch showing different types of copper defects in silicon. (**a**) Point defects, including individual atoms and point defects. (**b**) Formation of copper silicide precipitates in the silicon bulk. (**c**) Migration to existing defects, including dislocations or grain boundaries. (**d**) Out diffusion to the surface. (**e**) Segregation in p+ acceptor regions

location in the bandgap, high interstitial diffusivity, etc., create a severe risk for CMOS chips contaminated with copper. Copper is one of the fastest diffusing metals, which makes copper particularly dangerous to CMOS chips (Claeys & Simoen, 2018).

Copper precipitates at high concentrations typically greater than 10^{16} atoms/cm^3 into several different forms and locations, including copper silicide precipitates, precipitation into dislocations, oxygen clusters, and other existing defects. Copper is attracted to boron or acceptor-doped regions and tends to precipitate there. Figure 15.14 shows a sketch showing some of the copper mechanisms in silicon.

The behavior of copper in silicon is strongly dependent on the temperature ramp down of thermal steps during wafer fabrication. Slow cooling creates copper silicide precipitates and dislocation loops. At moderate ramp-down rates (~500 °C/s), copper can remain interstitial until it combines with boron (and other acceptors), forming copper-boron pairs. Rapid quenching at 2000 °C/s can form plate-shaped copper silicide defects (Claeys & Simoen, 2018, pg. 152). Copper forms point defects at lower concentrations, resulting in nonvisible and single-atom defects.

CMOS processes have many low-temperature steps throughout the backend of the process flow where the metallization layers are formed, exposing the wafer to long times at around 400 °C. At this temperature range, copper has a propensity to migrate from throughout the bulk of the wafer to the wafer surface, collecting in active transistor regions such as the channel and isolation oxide interface. As a result, the surface copper can degrade device performance by reducing the transistor drive through degradation in mobility through an increase in surface recombination velocity. In addition, transistor and junction leakage can be degraded by the presence of surface copper. Figure 15.15 shows a plot of the lifetime and surface recombination velocity for high and low-temperature anneals. Higher surface recombination values indicate degraded performance. Minority carrier recombination at a semiconductor surface tends to be much higher than recombination in the semiconductor bulk (Claeys & Simoen, 2018, pg. 168). The surface recombination locally depletes the minority carriers. The surface recombination velocity is a metric measuring the minority carrier loss at the surface. Note that the diffusion length and surface recombination velocity behave in opposite directions with the annealing process, as shown in Fig. 15.15. This result indicates a change in surface and bulk behavior.

Copper is one of the worst elements for diffusing at near-room temperatures. This property of copper can cause degradation of CMOS chips during operation in the field and during packaging, testing, and burn-in at the final stages of chip manufacturing. Figure 15.16 shows a plot of minority carrier lifetime measured by an optical technique for chemically copper-contaminated wafers. First, the wafers were coated with a copper-containing chemical, and the copper was diffused with an 850 °C oxidation. Then, a thermal anneal was performed on the wafers at lower temperatures at various times. The lifetime degraded when the wafers were annealed at around 100 °C for 45 min, indicating that the copper's electrical behavior changed with the low-temperature anneal. The control wafers with no copper contamination showed no lifetime degradation. The low-temperature anneal disassociates a Cu-Cu

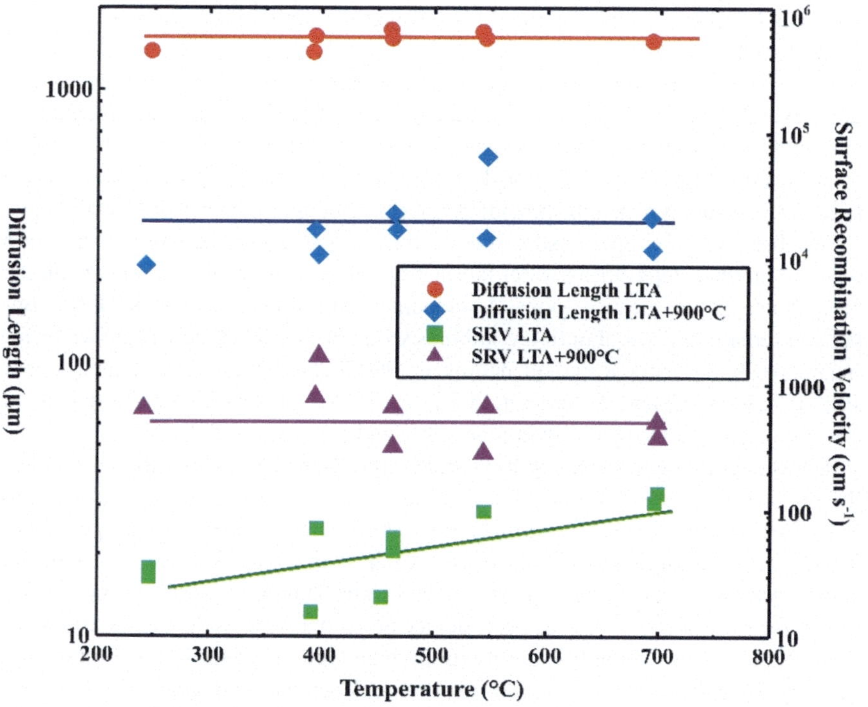

Fig. 15.15 Comparison of diffusion length and surface recombination velocity for copper contaminated silicon annealed at high and low temperatures. (**a**) Diffusion length is degraded less by only low temperature annealing. (**b**) Surface recombination velocity is degraded less with low temperature annealing

pair and creates an electrically active substitutional defect, forming an SRH center and degrading the lifetime (Ramappa et al., 2001).

Copper also migrates easily through many types of materials, including dielectrics. Copper metallization in CMOS chips must be surrounded by metal or dielectric copper barriers to prevent copper migration, which can cause chip failure. The barriers for copper metallization must be optimized to prevent copper migration in addition to other properties such as adhesion. Failure can be caused by copper drifting between two metal lines, migrating through the dielectric during electrical bias where two adjacent copper lines are at different potentials. This degradation can occur under bias and temperature conditions during normal chip operation (Ciofi et al., 2006). The copper migrates into the dielectric until a sufficient number of copper atoms is accumulated in the dielectric. The copper can then form a filament, causing dielectric breakdown (Wu et al., 2018).

Copper in a dielectric is usually a positive ion and may behave like a mobile ion, as described in Chap. 16 (Ciofi et al., 2006; Miyazaki et al., 2008). Therefore, the migration of copper can be detected using the same techniques used to detect mobile ions described in Chap. 16. Figure 15.17 shows the detection of copper migration using the triangular voltage sweep technique (TVS) (Ciofi et al., 2006). The current during a voltage ramp is measured across the capacitor created by the intermetal capacitance. The copper migration is detected by the current in the circuit displacing the current from the copper migration.

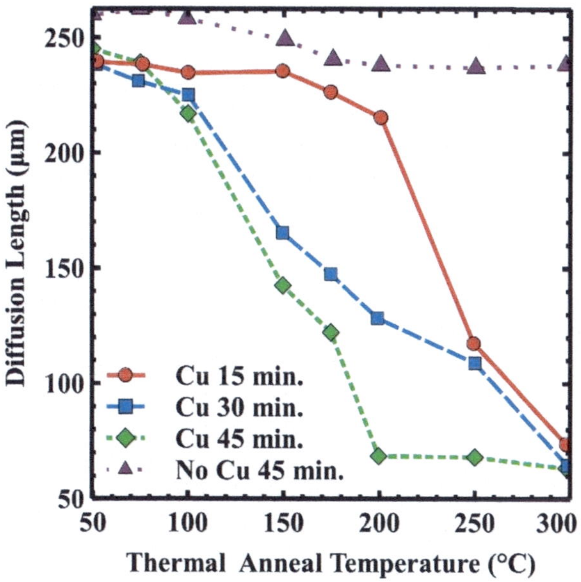

Fig. 15.16 Diffusion length measured by surface photo voltage diffusion length as a function of annealing temperature for copper-contaminated p-type wafer samples annealed at various times. An uncontaminated sample is shown as a reference

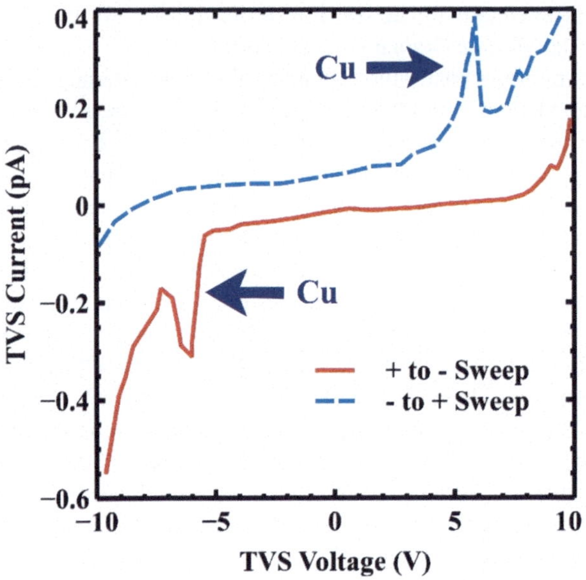

Fig. 15.17 Detection of copper drift and migration using the triangular voltage sweep (TVS) technique. The voltage across a dielectric between two copper lines is swept between opposite polarity voltages, and the ion movement is detected by measuring the current flowing into the structure. The small spike labeled Cu shows the copper movement detected by measuring the displacement current in the circuit as copper ions move from one metal line to the adjacent line

Copper metallization requires a barrier metal or dielectric surrounding the copper interconnect on all sides. Otherwise, the copper can readily migrate through the intermetal dielectrics. As an example, Fig. 15.18 shows a plot of the drift rate of copper ions in various common intermetal dielectrics with no intentional dielectric or metal barriers (Loke et al., 1999). The copper drift rate is plotted as a function of the inverse temperature. There is a significant difference between materials, with some showing high drift near room temperature. The material properties that make low-k materials useful, such as low density and porosity, may make the materials more susceptible to copper migration.

Figure 15.19 shows a triangular voltage sweep plot comparing the copper barrier performance of several metal barriers (Wojick et al., 2011). The poorer performing barrier metals, such as physical vapor deposition (PVD) ruthenium, show a significant spike in capacitance near zero volts where the electrical field changes direction, indicating a large amount of copper migration. PVD tantalum nitride (TaN) shows a tiny spike, indicating low copper migration. A control experiment using only aluminum metalization shows a similar small spike, indicating the signal was from mobile ions and not copper.

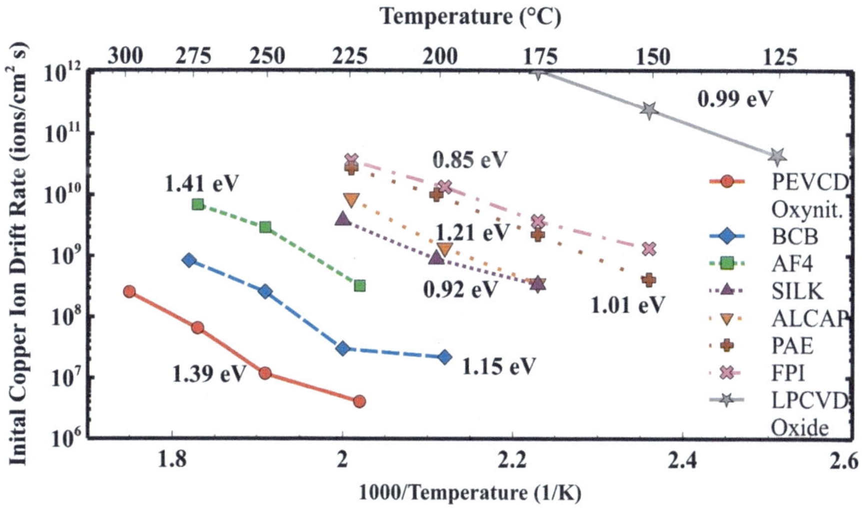

Fig. 15.18 Initial copper ion drift rate of various intermetal dielectrics, including low-k dielectrics as a function of inverse temperature. The activation energies are shown for each dielectric. Abbreviations: *PECVD* plasma enhanced chemical vapor deposition, *BCB* benzocyclobutene, *AF4* parylene-F, *SILK* aromatic hydrocarbon, *ALCAP* poly(arylene ether), *PAE* poly(arylene ether), *FPI* fluorinated polyimide, *LPCVD* low-pressure chemical vapor deposition

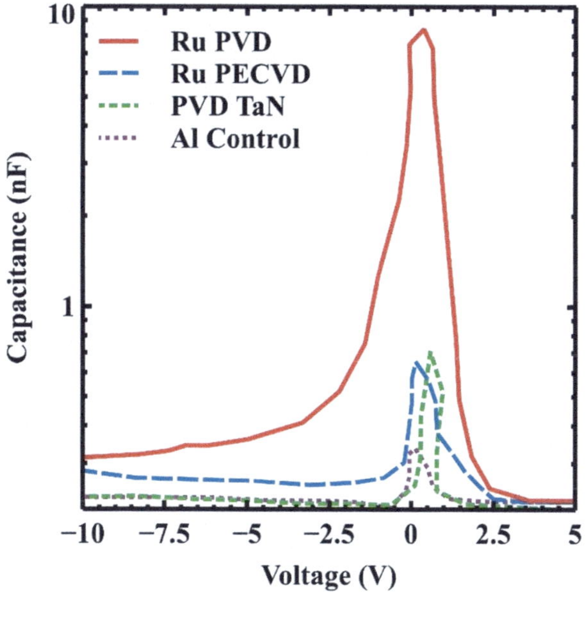

Fig. 15.19 TVS plot comparing different copper barrier metals

15.6.2 Iron

Iron is an ubiquitous contaminant for CMOS technology. It is commonly used in many metals used in processing technology, such as stainless steel. Iron is one of the metals that produce SRH defects with defect states near the center of the bandgap, as shown in Fig. 15.8. Iron contamination has one of the worst impacts on the lifetime and electrical behavior of CMOS devices, as shown in Fig. 15.13. Iron has a high vapor pressure and can evaporate from surfaces during thermal processing, contaminating nearby wafers and equipment.

Iron can form more than many different levels in the bandgap, as shown in Table 15.6. The defect level can be formed in several different configurations, including interstitial iron (Fe_i), substitutional iron (Fe_s), iron-silicon vacancies (V), iron-oxygen interstitials (O_i), and iron-hydrogen (H). The trap level in the silicon bandgap is shown, along with the structural details of the defect. e and h denote if the level is an electron or hole trap (Claeys & Simoen, 2018).

Iron can interact with the dopant boron in silicon in an unstable manner, alternating between two chemical and electrical states (Rein, 2005). Figure 15.20 shows a diagram of the behavior of the two states. The two alternating states are a single iron atom in an interstitial site and an iron-boron pair. The positively charged iron atom is attracted to the negatively charged accepter ion that has captured an electron. As a result, the iron and boron combine and are bound together by colombic attraction, forming a bonded pair. The iron-boron pair is metastable and can be dissociated by a small amount of energy added to the pair. The energy can be added thermally, by exposure to light, or by injection of electrical current. The added energy causes the iron-boron pair to disassociate, creating an electrically active acceptor and an electrically active interstitial iron atom. The two states are unstable, and oscillation

Table 15.6 Table showing trap levels of iron defects in silicon

E_T (eV)	Identity
0.085 ± 0.03	$(Fe_i)_2V_2$,h
0.152	Fe-P-H,e
0.184	$(Fe_i)V_2$,h
0.199	Fe-V-P,e
0.220 ± 0.01	$(Fe_i)_2V_2$,e
0.290	FeV_2,h
0.305 ± 0.01	Fe_iO_i
0.31	Fe_iH,h
0.33	$(Fe_i)V$,h
0.340	VFeV,h
0.365	FeVO,e
0.380 ± 0.02	Fe_s,e
0.410	$(Fe_i)_2$
0.500 ± 0.02	$(Fe_i)V$,h
0.530 ± 0.02	$(Fe_i)_2V$,h

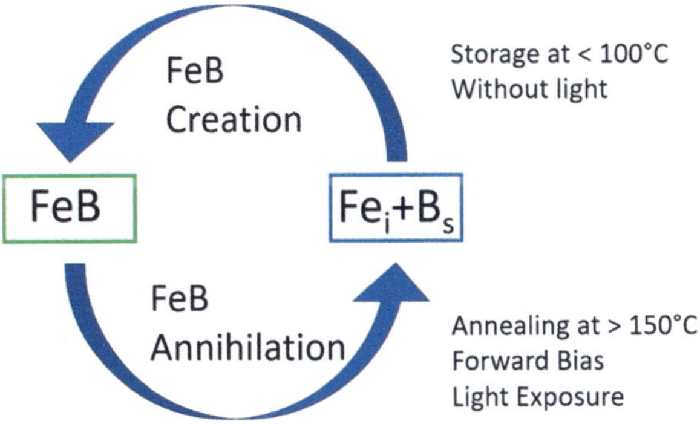

Fig. 15.20 Diagram showing the two metastable electrochemical states and the stimulus necessary to change between the states

between the states can occur over the typical operating temperature range of a CMOS chip, including room temperature.

The metastable iron-boron pair forms in low-energy environments near room temperature. Annealing wafers at near room temperature (in the dark) can induce the formation of iron-boron pairs.

The minority carrier lifetime for silicon containing iron and boron states is dependent on the current level of injection, with the longest lifetime depending on the injection level and the presence of the two states.

The minority carrier lifetime is a function of the fraction of metastable iron-boron and interstitial states (Rein, 2005, pg. 347). A plot of the lifetime as a function of inverse temperature is shown in Fig. 15.21. Below 100 °C, the iron-born pair formation dominates, as shown in Fig. 15.21b. From 100–150 °C, iron-boron pair dissociation dominates, and beyond 150 °C, interstitial iron dominates. The fraction of iron-boron pairs and interstitial iron as a function of temperature is shown in Fig. 15.21a. Over the temperature range of CMOS device operation, transitions can be expected between the two states, impacting the electrical behavior of devices. The iron-boron metastable behavior may be a source of metastable electrical behavior in CMOS, such as random telegraph signal (RTS) noise.

15.6.3 Distinguishing Between Copper and Iron Contamination

As copper and iron are common silicon contaminants, it is helpful to be able to determine if a particular problem is caused by copper or iron. The metastable behavior of iron related to the iron-boron pair can be used to identify iron contamination, as this behavior does not occur in copper-contaminated samples.

Figure 15.22 shows a technique to distinguish copper and iron contamination (Ramappa et al., 2001). Two wafers with copper or iron contamination are measured

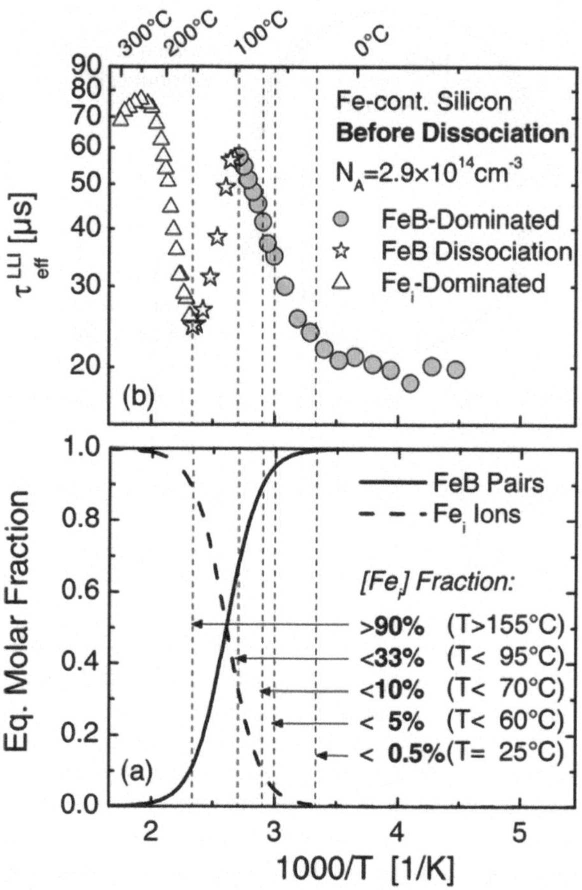

Fig. 15.21 Plots showing the behavior of the low level injection lifetime for iron-boron pairs and interstitial iron as a function of inverse temperature. The top plot (**b**) shows the dominant mechanism as a function of temperature for iron-boron combination (filled circles), Iron-boron pair dissociation (stars), and iron interstitials (triangles). The bottom plot (**a**) shows the fraction of iron interstitials and iron-boron pairs as a function of inverse temperature

to determine the initial minority carrier lifetime. Next, they are exposed to light, which causes the disassociation of iron-boron pairs and the dissociation of Cu-Cu pairs in the iron and copper-contaminated samples, respectively. Both disassociation processes degrade the lifetime with light exposure, with a larger degradation shown for the copper wafer. Then, the lifetimes of the wafers are measured as a function of time at room temperature. The lifetime of the iron-contaminated wafer recovers through the iron-boron combination process, as described in Fig. 15.20. The copper-contaminated wafer does not recover as there is no recombination process. The light exposure severely degrades the diffusion length semi-permanently at near room temperatures.

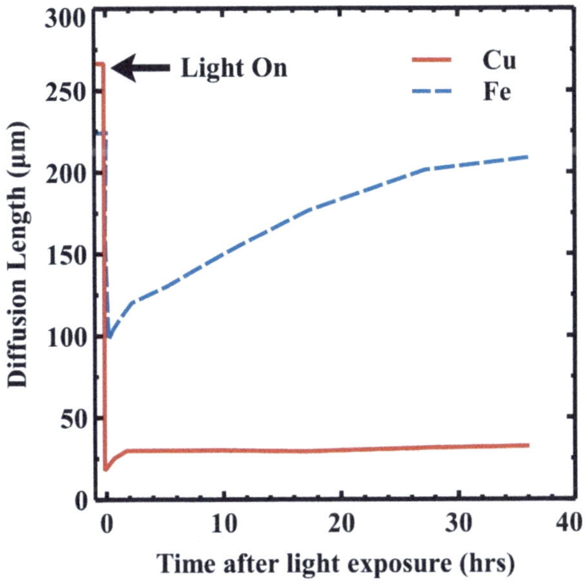

Fig. 15.22 Comparison of the time-dependent diffusion length evolution after exposing copper and iron contaminated samples to halogen light. The lifetime is measured near the surface. Photovoltage. technique

15.7 Case Studies for Metal Contamination

15.7.1 DRAM Refresh

DRAM memory chips are particularly sensitive to metal contamination as the refresh time is sensitive to leakage in the single transistor and single capacitor cell. Metal-related leakage in the cell junction, access transistor, and isolation between cells can degrade the refresh performance and other cell metrics such as variable retention time, row hammer, etc. (see Chap. 8).

The following is a back-of-the-envelope calculation highlighting the impact of low levels of leakage on the DRAM cell performance.

The capacitor in a DRAM cell, as shown in Fig. 15.23, typically has a capacitance of about 8fF in a modern DRAM cell. The voltage V_c used to store the charge in the cell is about 0.5 V (difference between the bitline and cell plate). The charge Q_s stored in the cell is given by Eq. (15.2) and is estimated to be 4 fC. The number of electrons N_e stored in the cell is given by Eq. (15.3), where q is the charge per electron. The calculation shows a small number of electrons stored in the cell (about 2500).

$$Q_S = C_S V_c = 8\times10^{-15} F\times0.5V = 4\times10^{-15} C \quad (15.2)$$

Fig. 15.23 Schematic of a DRAM cell

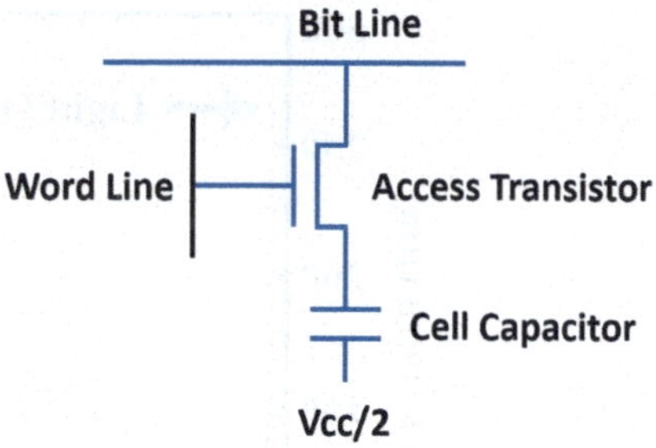

$$N_e = \frac{Q_S}{q} = \frac{4\times10^{-15}}{1.6\times10^{-16}} = 2500\,\text{electrons} \tag{15.3}$$

A typical DRAM cell can retain data for a few seconds, and the DRAM datasheet minimum refresh specification is around 64 ms. Assuming that the DRAM cell can lose 50% of its charge before the cell fails, the DRAM cell can lose 1250 electrons before failing (Schmid et al., 1994). Therefore, the maximum leakage current (I_L) to meet a 1 s refresh is estimated by Eq. (15.4), resulting in a leakage level of about a femptoamp. Most leakage mechanisms are worse at high temperatures, so the refresh specification is more difficult to meet at the upper datasheet temperature specification.

$$I_L = \frac{Q_S}{2\tau_R} = \frac{2\times10^{-15}C}{2\times1\text{S}} = 1\times10^{-15}A \tag{15.4}$$

There are multiple leakage mechanisms in a DRAM cell that can cause a refresh failure, as shown in Fig. 15.24. These mechanisms include transistor-related leakage properties described in Chap. 8, along with the capacitor leakage. Some of these mechanisms can be degraded by metal-related defects, including junction leakage, diffusion currents from the bulk and backside, and gate-induced leakage. The sum of all the leakages from the different mechanisms has to be less than the maximum leakage current, requiring each mechanism to be well below roughly an order of magnitude less than the maximum leakage current for an estimated 1×10^{-16} Amps per mechanism.

A single metal contamination point defect can cause leakage in the range of 0.1 fA at 70 °C. For a defect in the depletion region, the SRH leakage can be estimated from Eq. (15.5), where q is the electron charge, n_i is the intrinsic carrier concentration, C_R is the capture rate of the SRH trap, E_T is the trap level in the bandgap, and E_i is half the bandgap (Vanhellemont & Simeon, 2007). The capture rate ranges around 10^{-10} to 10^{-8} cm^3 s^{-1}.

$$I_{\text{SRH}} = \frac{q\,n_i C_R}{2\cosh\left(\frac{E_T - E_i}{kT}\right)} \tag{15.5}$$

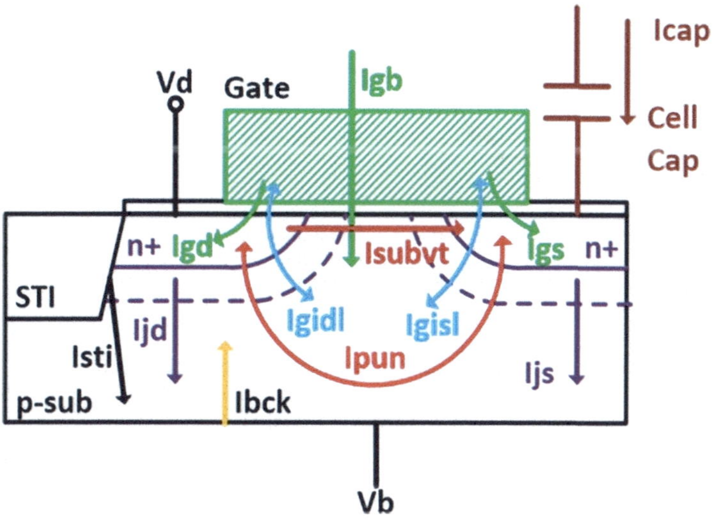

Fig. 15.24 Cross-section sketch of a DRAM cell showing leakage mechanisms affecting refresh

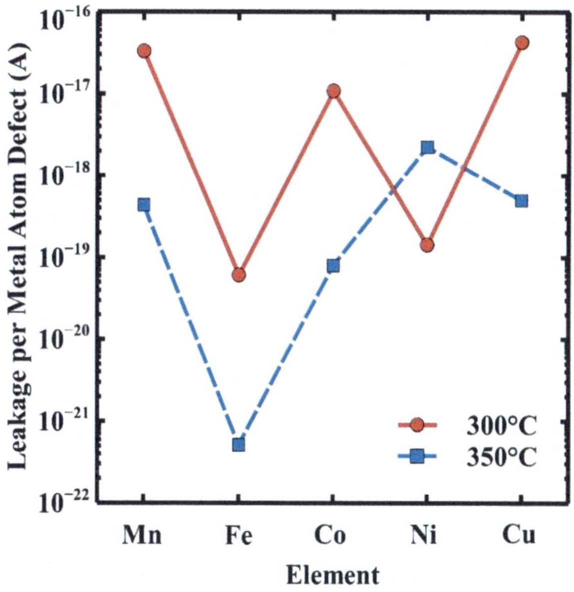

Fig. 15.25 Leakage per metal defect deep level for several metal atoms in silicon at different temperatures

The leakage per SRH state is shown in Fig. 15.25 for several metal contaminants in silicon at different temperatures (Vanhellemont & Simeon, 2007). Leakage levels near 5×10^{-17}A per SRH state can occur, indicating that a DRAM cell can fail or be degraded by an SRH defect containing a single metal atom.

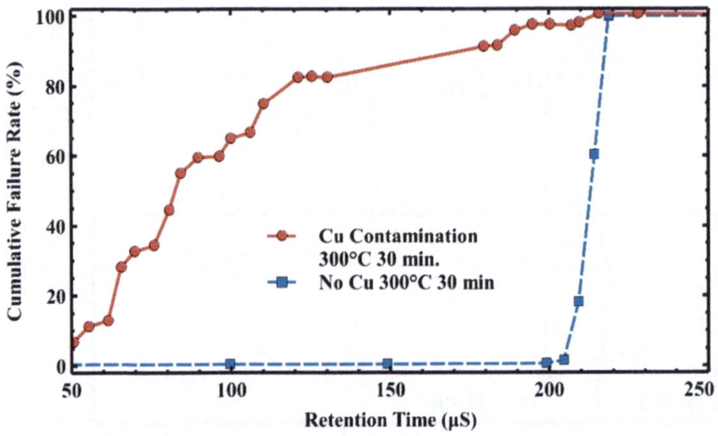

Fig. 15.26 DRAM Refresh degradation due to copper contamination in assembly processes. The right blue curve show the refresh characteristics post anneal without copper contamination. The left most curve shows refresh degradation with backside copper sputter deposition, intentionally contaminating the wafer. Both experimental groups recieved a 300 °C 30 min anneal. The low retention time was due to a unoptimized experimental test chip

Low-temperature processing conditions (200–400 °C) are routinely used in chip assembly processes, such as wafer thinning, silicon vias, etc., which can induce problems due to metal migration. Figure 15.9 shows that copper, nickel, and iron are fast diffusers at low temperatures. Copper is notorious for migrating rapidly at low temperatures, even migrating at room temperature (Istratov & Weber, 2001). Copper contamination is a common cause of refresh degradation during assembly-related processes. As an example, Fig. 15.26 shows the degradation of DRAM refresh with copper contamination during the final assembly process (Lee et al., 2012; Yeon et al., 2012). The DRAM refresh time was degraded when the wafer was intentionally contaminated. The refresh degradation was an experiment that intentionally contaminated the wafer with copper by the deposition of copper on the wafer backside, followed by a 300 °C anneal for 30 min.

15.7.2 Tungsten Contamination in CMOS Imagers

CMOS imagers and charge-coupled devices are another technology that is very sensitive to metal contamination with single-atom defects (Toren & Bisschop, 1995). The sensitivity of a pixel to metal contamination is much higher than that of any other technology, reaching a level as low as 10^5 SRH defects/cm^3. This sensitivity is well below the detection capabilities of most other technologies and measurement techniques.

One imager parameter sensitive to metal contamination is the dark current. The dark current measures the signal coming from the pixel when the pixel is in the dark.

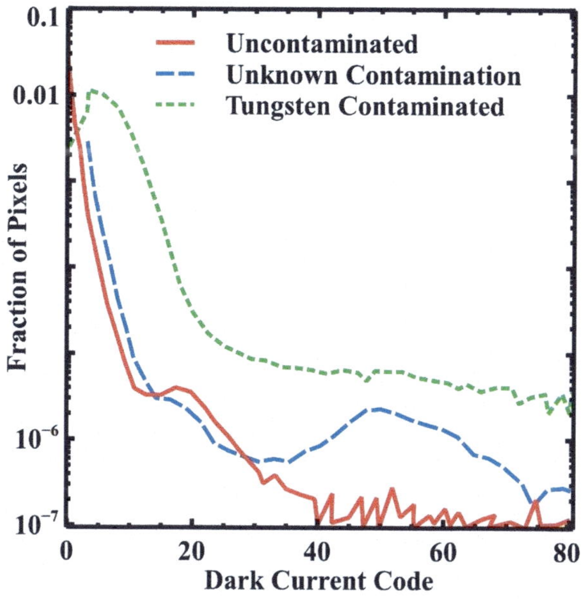

Fig. 15.27 Dark current probability distribution of a CMOS imager for intentionally contaminated and accidentally contaminated imager pixels with an uncontaminated control. One sample was intentionally contaminated by ion implantation of tungsten (2.7E12 ions/cm^2). The *x*-axis is a digital code that measures the dark current from an analog to digital converter

The dark current signal has several components, including noise, thermal generation, and SRH center generation. The pixel signal is transferred to an analog-to-digital converter that converts the dark current to a binary number or code. The distribution of dark current across many pixels can be used to detect metal contamination.

Figure 15.27 shows the dark current distribution for a CMOS imager where an imager was accidentally contaminated with tungsten, along with samples that were intentionally contaminated with tungsten ion implantation (Domengie et al., 2010). An uncontaminated reference imager is also shown. The majority of the pixels have a low digital code number near zero, indicating that the dark current is low. Conversely, defective pixels have a high code number, indicating a high dark current level.

The accidentally contaminated sample has a peak of defective bits near the binary code of 50. The analysis in this study concluded that the accidental sample was contaminated with tungsten at a level of fewer than 1.6×10^{10} atoms/cm^3. The study concluded that a single metal atom SRH center contributed a level of 1.64 dark current codes per defect. The number of SRH centers ranged from 16 to 45 per defective pixel with a dark current code of 50. The large number of centers per defective pixel was attributed to tungsten defect clustering, possibly caused by a cathode gun in a process tool.

The intentionally contaminated ion implanted samples had a peak dark current code of around 6, indicating less leakage per defective pixel. This result was attributed to a lack of tungsten atom clusters. This study highlights the sensitivity of CMOS imagers.

References

Aissa, B., et al. (2015). Emerging frontiers of N-type silicon material for photo voltaic applications: The impurity-defect interactions. *Frontiers in Nanoscience and Nanotechnology, 1*, 2. https://doi.org/10.15761/FNN.1000102

Bracht, H. (2000). Diffusion mechanisms and intrinsic point defect properties in silicon. *MRS Bulletin, 25*, 22. https://doi.org/10.1557/mrs2000.94

Chen, J., et al. (1987). Subbreakdown drain leakage current in MOSFET. *IEEE Electron Device Letters, 8*(11), 515–517. https://doi.org/10.1109/RELPHY.1997.584241

Ciofi, I., et al. (2006). *Water and copper contamination in SiOC:H*. Novel Characterization Methodology based on Triangular Voltage Sweep Measurements. Int. Interconnect Technology Conference. https://doi.org/10.1109/IITC.2006.1648682

Claeys, C., & Simoen, E. (2018). *Metal impurities in silicon- and germanium-based technologies origin, characterization, control, and device impact*. Springer. https://doi.org/10.1007/978-3-319-93925-4

Domengie, F., et al. (2010). Impact on device performance and monitoring of a low dose of tungsten contamination by dark current spectroscopy. In *International Reliability and Physics Symposium*. https://doi.org/10.1109/IRPS.2010.5488821

Graff, K. (1995). *Metal impurities in silicon-device fabrication*. Springer. https://doi.org/10.1007/978-3-642-57121-3

Istratov, A. A., & Weber, E. R. (2001). Physics of copper in silicon. *Journal of the Electrochemical Society, 49*(1). https://doi.org/10.1149/1.1421348

Lee, K., et al. (2012). Impact of cu contamination on memory retention characteristics in thinned DRAM Chip for 3-D integration. *IEEE Electron Device Letters, 33*(9), 1297–1299. https://doi.org/10.1109/LED.2012.2202631

Loke, A. L. S., et al. (1999). Kinetics of copper drift in low-K polymer interlevel dielectrics. *IEEE Transactions on Electron Devices, 46*(11), 2178–2187. https://doi.org/10.1109/16.796294

Meroli, S. The minority carrier lifetime in silicon wafer. *CERN*. https://meroli.web.cern.ch/lecture_lifetime.html

Miyazaki, H., et al. (2008). The observation of stress-induced leakage current of damascene interconnects after bias temperature aging. In *IEEE International Reliability Physics Symposium*. https://doi.org/10.1109/RELPHY.2008.4558877

Ramappa, D. A., Hoff, A. M., & Henley, W. B. (2001). Quantitative analysis of copper contamination in silicon by surface photovoltage minority carrier lifetime analysis. *AIP Conference Proceedings*. https://doi.org/10.1063/1.1354400

Rein, S. (2005). *Lifetime spectroscopy a method of defect characterization in silicon for photovoltaic applications*. Springer. https://doi.org/10.1007/3-540-27922-9

Rubin, L. (2024). *Preventing metallic contamination in image sensors*. Axcelis.

Sachdeva, R., et al. (2006). Electrical properties and diffusion behavior of hafnium in single crystal silicon. *Applied Physics, A, 84*, 351–357. https://doi.org/10.1007/s00339-006-3637-6

Schmid, J. R., et al. (1994). Estimating the effect of contamination-induced leakage current in view of DRAM architectural trends. In *IEEE/SEMI Advanced Semiconductor Manufacturing Conference*. https://doi.org/10.1109/ASMC.1994.588261

Schroder, D. K. (1997). Carrier lifetimes in silicon. *IEEE Transactions on Electron Devices, 44*(1), 160–170. https://doi.org/10.1109/16.554806

Shamazaki, A., et al. (2008). Metallic contamination control in leading-edge ULSI manufacturing. *Ultra Clean Processing of Semiconductor Surfaces, 145-146*, 115. https://doi.org/10.4028/www.scientific.net/SSP.145-146.115

Sze, S. M., & Lee, M. K. (2010). *Semiconductor devices physics and technology* (3rd ed.). Wiley.

Toren, W. J., & Bisschop, J. (1995). Metal contamination characterization in CCD image sensors. In *IEEE International Electron Device Meeting*. https://doi.org/10.1109/IEDM.1995.497205

Vanhellemont, J., & Simeon, E. (2007). Brother silicon, sister germanium. *Journal of Electrochemical Society, 154*, H572–H583. https://doi.org/10.1149/1.2732221

Wojick, H., et al. (2011). Comparison of PVD, PECVD & PEALD Ru(-C) films as Cu diffusion barriers by means of bias temperature stress measurements. *Microelectronic Engineering, 88*, 641–645. https://doi.org/10.1016/j.mee.2010.06.034

Wu, C., et al. (2018). Insights into metal drift induced failure in MOL and BEOL. In *IEEE International Reliaibility Physics Symposium*. https://doi.org/10.1109/IRPS.2018.8353551

Yeon, H. W., et al. (2012). Cu contamination of the nMOSFET in a 3-D integrated circuit under thermal and electrical stress. *Electrochemical Solid State Letters, 15*, H157–H160. https://doi.org/10.1149/2.018205esl

Chapter 16
Mobile Ion Contamination

16.1 Introduction

Mobile ions can be created in a CMOS wafer by contamination with elements that have the propensity to easily lose one or more electrons, forming a positively charged free ion that can move through materials on the wafer. The mobile ions tend to drift and diffuse toward critical circuit elements, damaging and destroying the electrical circuit function.

Mobile ion problems are almost nonexistent in modern CMOS technologies. The rare, infrequent mobile ion problems that do occur are usually catastrophic, taking down a manufacturing plant, contaminating equipment, and contaminating the inline wafers, making them unsellable. The electrical symptoms of mobile ion contamination are strange and nonobvious. Knowledge of mobile ion problems belongs in an engineer's knowledge base, even if there is a low probability of occurrence.

Mobile ions were a severe problem in the early days of metal-oxide-semiconductor (MOS) transistor development, delaying the introduction of the technology by many years (Snow et al., 1965). Early MOS chips could only use PMOS transistors because PMOS devices were somewhat immune to mobile ion problems. Early PMOS chips were slow due to the low mobility, causing a performance penalty to mitigate mobile ion problems. PMOS technology required negative supply voltage. Solutions to the mobile ion problem allowed the manufacture of higher performance NMOS devices, eventually leading to CMOS technology, which combined both types of transistors.

Mobile ions cause time-dependent chip instability where a functioning chip fails and possibly recovers during operation and testing. The failure was commonly caused by threshold voltages of the transistor shifting dramatically with electrical stress at higher temperatures, causing a loss of circuit function. After mobile ions were determined to cause the instability, an intense, focused effort to eliminate the

K. Prall, *CMOS Plasma and Process Damage*,
https://doi.org/10.1007/978-3-031-89029-1_16

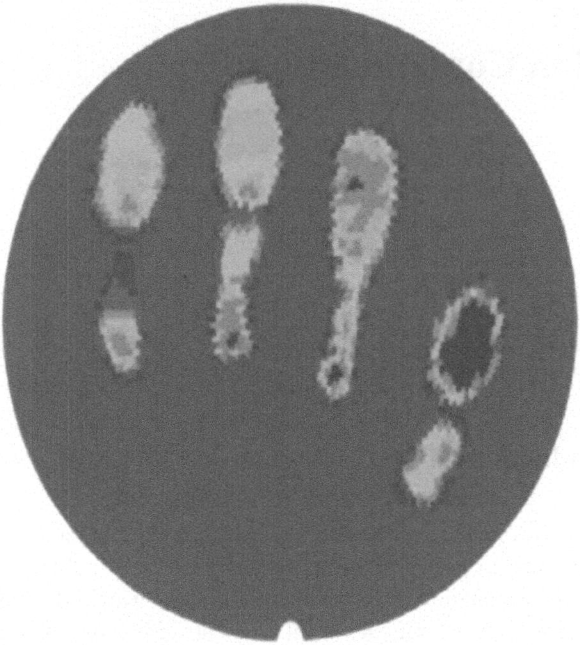

Fig. 16.1 Handprint image formed on a wafer showing the presence of mobile ions measured by a corona discharge system

problem resulted in viable manufacturing solutions enabling the manufacture of NMOS devices, followed by CMOS devices.

The source of the mobile ions was heavily investigated and was determined to be the contamination of the wafers by alkali metals from the first column of the periodic table. Mobile ions can inadvertently be introduced in a CMOS production line through contaminated consumables such as gases, chemicals, slurries, etc. Contaminated manufacturing equipment is not easily repaired as mobile ions can diffuse into glass, quartz, silicon carbide, and other materials, requiring disassembly and replacement of tool components. The recovery of a manufacturing line from mobile ion contamination can be expensive and time-consuming, causing economic pain to the affected company.

The human body was determined to be a source, contaminating the wafers and equipment with primarily sodium and potassium. Figure 16.1 shows a corona measurement of a silicon dioxide coated wafer with a handprint on it (Wilson et al., 2006). The corona measurement can detect the electrical activity of the fingerprint where positively charged mobile ions move under electrical bias at elevated temperatures. Modern fabrication lines protect the wafers from contamination by humans by using extensive protective gear such as bunny suits, gloves, masks, etc.

The author has observed a few mobile ion contamination problems, including contamination caused by workers tracking snow-melting salt into a fab (Ohring, 2015). Nowadays, rigid control of contamination sources protects fabs and keeps them mobile ion free. However, chip manufacturers do not publicly disclose most mobile ion problems as they are highly embarrassing.

16.2 Physics of Mobile Ion Contamination

Many different atoms and some molecules behave as mobile ions, drifting and diffusing in a CMOS chip. The term mobile ion describes elements with a set of properties that are particularly damaging to CMOS chips. Table 16.1 lists these properties. Mobile ions can migrate through the wafer at the typical operating temperatures of CMOS chips and the temperatures used for burn-in and reliability testing. Mobile ions drift during electrical operation and diffuse with thermal energy, reaching critical device locations in the wafer where they cause loss of function and reliability failures.

There are several ways for mobile ions to harm CMOS chips. First, mobile ions can cause threshold voltage shifts, resulting in circuit failures. Mobile ions can cause increased junction leakage and kill minority carrier lifetime. Severe contamination levels can cause dielectric breakdown.

Mobile ions interfere with the operation of CMOS devices by being semi-permanently charged. The mobile ion charges offset the voltage applied to the CMOS transistors' gate, shifting the operating voltages and causing circuit failure. In addition, mobile ions are mobile at low temperatures in the range of CMOS operating temperatures and can move around the wafer, reaching sensitive devices.

Table 16.1 Key properties of mobile ions in semiconductors

Property	Reference
Can diffuse through the chip at CMOS operating temperatures	(Barbottin & Vapaille, 1986)
Can drift under bias during chip operation	(Schroder, 2006)
Shows reversible electrical properties with unbiased bakes, etc.	(Liou et al., 2003)
Can reach critical locations in the chip, causing changes in the electrical properties of devices	(Kuge et al., 2016)
Are permanently or semi-permanently ionized	
Have a severe electrical impact on the function of the CMOS chip, destroying function and reliability.	(Shideler, 1996)
Recovers temporarily with exposure to UV light	(Istsumi et al., 2002) pg. 199
The problem recovers with opposite polarity electrical stress.	

16.2.1 Common Mobile Ions

Most of the mobile ions are positive ions, having lost an electron. The alkali metals in the first column of the periodic have one valence electron, which can easily be lost due to the low electronegativity of the alkali metals. The ions lose the single valence electron and become semi-permanently positively charged. The ions are mobile and move through the wafer through drift and diffusion mechanisms. The ions are strongly affected by electrical fields due to the biases applied to the transistors.

The alkali metals sodium and potassium are the most common mobile ion contaminants due to their universal environmental presence. However, lithium is an unusual mobile ion contaminant as it is seldom present in CMOS manufacturing processes and is occasionally used in lithium grease. Hydrogen can also be a mobile ion, but it is quickly neutralized by interacting with the materials in the wafer. Finally, due to their rarity in CMOS manufacturing, rubidium and cesium are seldom seen as mobile ions. The radioactive element francium is rare, has a short half-life (~22 min), and is not observed in CMOS wafers.

Elements besides alkali metals can be mobile ions. For example, the alkali earth elements in the second column, magnesium, and calcium, are mobile ions in CMOS chips (Williams et al., 1995). Numerous elements can be ionized and behave as a mobile ion in a CMOS wafer. Several atoms, such as those shown in Table 16.2, have been reported in the literature acting as mobile ions.

Transition metals can also be mobile ions. The use of copper is ubiquitous in the semiconductor industry as it is the primary element used for metallization because of its low resistance and other properties (see Chap. 15). Like alkali metals, copper has a single electron in its outer shell, making it highly mobile. Copper behaves as a mobile ion in CMOS chips. CMOS metallization processes are designed to prevent copper migration by using barrier metals and dielectrics to prevent migration. Silver has been reported as a mobile ion in the literature (Snow et al., 1965), and gold is known to be highly mobile in silicon (Coffa et al., 1987). Neither gold nor silver is present in a CMOS process, although gold has historically rarely been used as a minority carrier killer in a few non-CMOS technologies. Historically, gold has sometimes been used in wire bonding.

Table 16.2 Common mobile ion contaminants in CMOS

Element	μ_0 (cm^2/V-s)	E_d (eV)	Reference
H (proton)	40	0.73	(Hofstein, 1967)
Li	4.5×10^{-4} [a]	0.47 ± 0.08	(Schroder, 2006)
Na	3.5×10^{-4} [a]	0.44 ± 0.09	(Schroder, 2006)
K	2.5×10^{-3} [a]	1.04 ± 0.1	(Schroder, 2006)
Mg		0.82–1.07	(Williams et al., 1995)
Ca		0.85–1.12	(Williams et al., 1995)
Cu	4.8×10^{-7} [a]	0.93 ± 0.2	(Schroder, 2006)

[a]Accurate within a factor of 10 (Schroder, 2006).

Hydrogen can also be a mobile ion, but it is quickly neutralized by interacting with the materials in the wafer and can occur with positive, negative, and neutral charges in atomic and molecular configurations. Hydrogen drifts and diffuses very rapidly due to its low mass (a single proton). The behavior of hydrogen in a CMOS wafer is very complex, as described in Chap. 14.

Water can play a role in mobile ions as it is commonly observed to be absorbed in films used in CMOS chips. It can diffuse from films and decompose into hydrogen, becoming mobile (Ciofi et al., 2006b). Water-related signals in triangular voltage sweep measurements are typically observed as mobile protons caused by breaking the water molecules in CMOS wafers, as shown in Fig. 16.23. The prevalent theory is that water can exist as bonded or unbonded forms in films. For example, water can be bonded to hydrophilic silanol. The two different water forms behave differently. Water that diffuses to the dielectric can be fractured, forming atomic hydrogen that reacts with the interface, eliminating the water molecule.

The mobile ions in a CMOS wafer diffuse at relatively low temperatures and drift under electric fields similarly to electrons and holes, albeit at different voltages and temperatures. The element radius determines the mobility of the mobile ion in a CMOS wafer with decreasing mobility down the alkali metal column of the periodic table. Table 16.2 shows the activation energy and mobility for common mobile ion contaminants in SiO_2.

16.2.2 Drift and Diffusion of Mobile Ions

Figure 16.2 shows a plot of potassium and sodium mobility in the presence of an electric field versus inverse temperature in silicon dioxide (Barbottin & Vapaille, 1986, pg. 427). The diffusion mobility is approximately described by a simple Arrhenius equation as shown in Eq. (16.1), where μ_0 is a material constant, E_d is an energy barrier to diffusion, K is the Boltzmann constant, and T is the temperature (Schroder, 2006). The energy barrier is the energy required to emit the mobile ion from traps in the dielectric. The plot shows that sodium is easily mobile inside the operating temperatures of CMOS chips. On the other hand, potassium is much less mobile and requires a slightly longer time to affect CMOS chip operation

$$\mu(T) = \mu_0 \, e^{\left(-\frac{E_d}{KT}\right)} \tag{16.1}$$

An important parameter for mobile ion contamination is the drift time required for a mobile ion contaminant to drift and diffuse from the initial contaminant location to a sensitive location such as the gate dielectric. Figure 16.3 shows a plot of the drift time for a mobile ion to move 100 nm under an electric field in silicon dioxide at 1 MV/cm across various temperatures. The mobile ions Li and Na can transit the dielectric in a few seconds in the operating temperature range of a CMOS chip. The CMOS chip operating temperature ranges vary from 0 to 70 °C for commercial

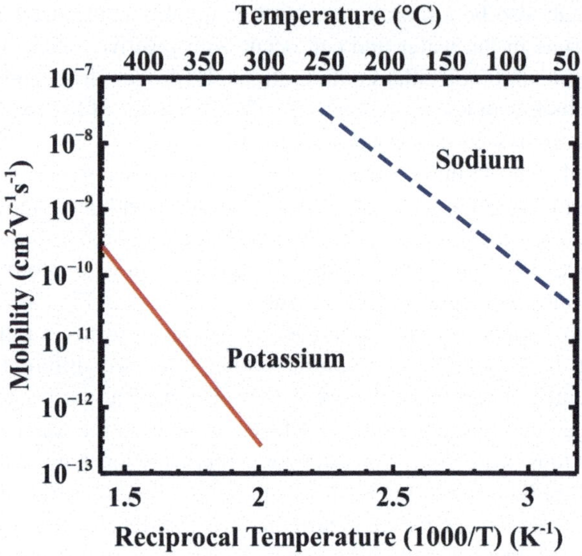

Fig. 16.2 Plot of the mobility of potassium (K+) and sodium (Na+) ions in silicon dioxide as a function of inverse temperature. Different sources of data are used for sodium, as shown in the reference

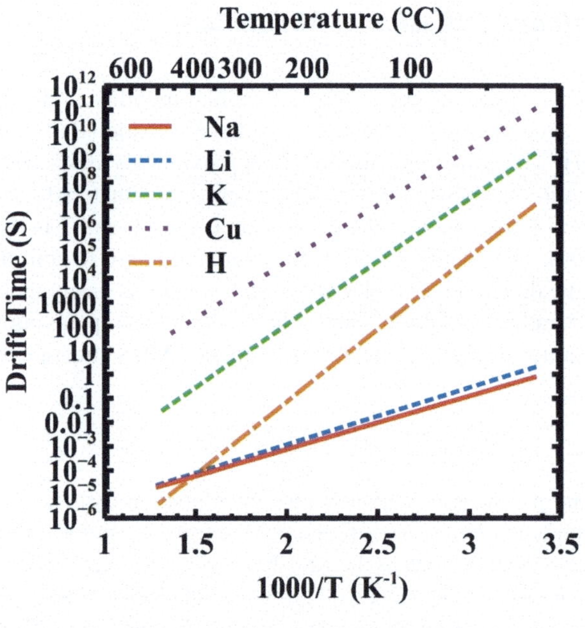

Fig. 16.3 Plot of drift time in silicon dioxide versus inverse temperature for four common mobile ions biased at 1MV/cm for a 100 nm dielectric

temperature ranges to −45 to 125 °C for military, automotive, or extended ranges. This data indicates that Li and Na can cause a chip to fail in a few seconds. The plot also shows the drift time for Cu and K, which are significantly slower but can still cause circuit failure over more extended periods in hundreds or thousands of seconds.

The drift or transit time (t_t) is given by Eq.(16.3), where t_{ox} is the dielectric thickness, V_G is the voltage across the dielectric, μ is the ion mobility, E_d is the ion diffusion activation energy (eV), k is the Boltzmann constant, and T is the temperature (Schroder, 2006, pg. 339). The equation uses a simple drift velocity calculation given by Eq. (16.2) combined with the Arrhenius equation in Eq. (16.1). Note that at chip operating temperatures (100 °C), it takes less than a second for sodium and lithium to transit 100 nm, which is around the thickness of shallow trench isolation (STI). It takes on the order of seconds for these mobile ions to travel through the STI at assembly processing temperatures (100 °C):

$$v_d = \frac{\mu V_G}{t_{ox}} \tag{16.2}$$

$$t_t = \frac{t_{ox}}{v_d} = \frac{t_{ox}^2}{\mu V_G} = \frac{t_{ox}^2}{\mu V_G} \exp\left(-\frac{E_d}{kT}\right) \tag{16.3}$$

16.2.3 Dielectric Properties Affecting Mobile Ion Migration

The material properties of the dielectric film containing the mobile ions strongly affect the mobile ion drift and diffusion properties. The film density is important. Low-density low-k dielectric films are often porous, providing less resistance to mobile ion migration. Dielectrics may contain different characteristics and levels of mobile ion traps that can temporarily delay or permanently stop ion migration. Figure 16.4 shows a plot comparing the copper drift characteristics of different dielectrics as a function of inverse temperature (He & Lu, 2012, pg. 67). The activation energy of drift varies between dielectrics. Copper generally diffuses faster through low-k dielectrics, often orders of magnitude faster than diffusion through silicon dioxide. Crystallinity is important in dielectric barriers, as shown in Sect. 16.14.3.

The drift and diffusion of mobile ions are affected by the local concentration of the ions. Some materials can cause an increased local concentration as mobile ions become concentrated at an interface, increasing the concentration-dependent diffusion component of the migration. Mobile ions may diffuse abnormally fast laterally between and along interfaces of various films.

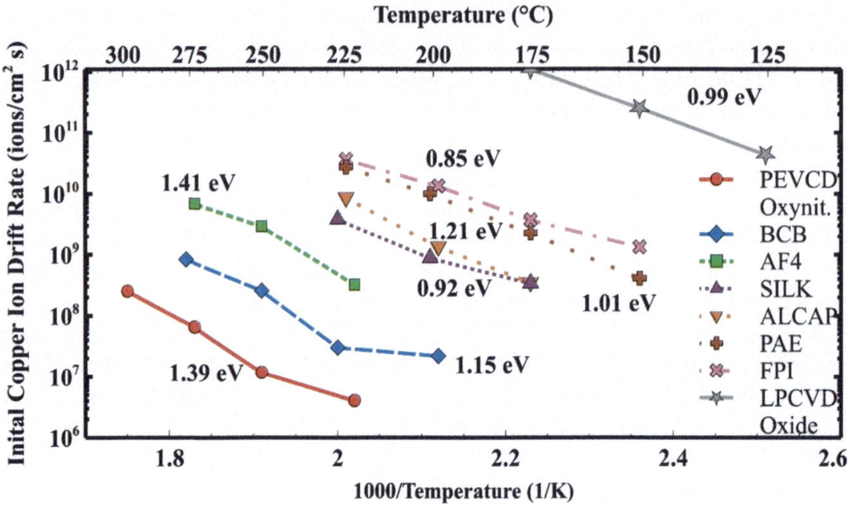

Fig. 16.4 Initial copper ion drift rate of various intermetal dielectrics, including low-k dielectrics as a function of inverse temperature. The activation energies are shown for each dielectric. Abbreviations: PECVD plasma enhanced chemical vapor deposition, BCB benzocyclobutene, AF4 parylene-F, SILK aromatic hydrocarbon, ALCAP poly(arylene ether), PAE poly(arylene ether), FPI fluorinated polyimide, LPCVD low-pressure chemical vapor deposition

16.3 Electrical Symptoms of Mobile Ion Contamination

Mobile ion problems cause several different electrical problems in CMOS devices, as shown in Table 16.3. The top of the table shows the exact device failure mechanism caused by mobile ions. The second half of the table shows electrical symptoms observed parametrically and at the chip level during electrical tests that contain biased and unbiased temperature stresses.

Different mobile ions can damage the chip in different ways. Alkali metals and alkali earth elements typically cause threshold voltage shifts of active and isolation transistors. Copper tends to cause dielectric leakage and dielectric breakdown. For a given mobile ion problem, one or many of the signatures shown are observed. Detailed experimentation often reveals multiple signatures that tend to prove a given problem is caused by mobile ions.

Table 16.3 Common electrical symptoms caused by mobile ions

Product and parametric signatures	Reference
Unstable and shifting threshold voltages	(Hofstein, 1966, 1967; Kothandaraman & Cohen, 2016)
Loss of isolation between transistors	(Gutai, 1997)
Gate dielectric breakdown	(Chen, 2012; Huang et al., 2014)
Junction leakage and breakdown voltage shifts	(Schroen et al., 1965)
Change in junction or gate capacitance as a function of voltage.	(Hofstein, 1966, 1967)
Data retention problems in nonvolatile memories	(Kuge et al., 2016; Liou et al., 2001; Sim et al., 2005)
Signatures during product and parametric reliability testing	Reference
Product failure during burn-in and reliability testing without a visible defect	(Shideler, 1996)
Degradation in standby power with biased or unbiased bakes	(Sabin et al., 1997)
Partially recoverable with unbiased bake	(Snow et al., 1965)
Tracks bias polarity	(Kuhn & Silversmith, 1971)
Faster degradation and failure at higher temperatures tracks Ea of mobile ions.	(Barbottin & Vapaille, 1986)
Signature repeats with pass and fail alternating through cycles of biased and unbiased bakes.	

16.4 Threshold Voltage Shifts

16.4.1 Introduction

The transistor threshold voltage (V_t) shift described in this section is a primary mechanism for chip failure caused by mobile ions, affecting both active and isolation devices. The threshold voltage shifts with time, temperature, and bias. The threshold voltage shift can cause circuit failure, performance degradation, power dissipation degradation, and other failure modes.

Figure 16.5 shows a sketch of the mobile ion impact on threshold voltages for contaminated NMOS and PMOS transistors after positive and negative gate voltage stress performed at elevated temperatures. Initially, as the wafers exit the fab, the distribution of the mobile ions in the gate dielectric is unknown and usually randomized by the thermal energy of the final heat treatments in the process flow. If a positive bias is applied to the gates, the electric field forces the mobile ions to the gate dielectric-silicon interface, where the NMOS V_t is decreased, and the PMOS is increased. If a negative bias is applied to the gates, the mobile ions are forced by the electric field to the gate dielectric-gate interface, where the NMOS V_t is increased, and the PMOS is decreased. After the ions are driven on one interface, an unbiased thermal bake rearranges the ion distribution, diffusing the ions away from the interface. The resultant threshold voltage resides somewhere between the positive and negative threshold voltage points.

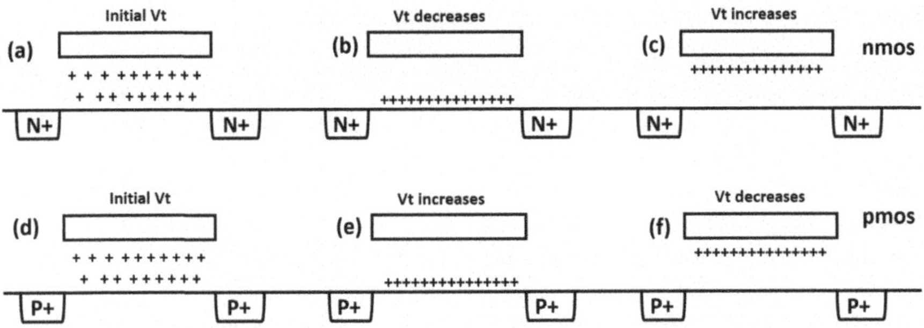

Fig. 16.5 Sketch showing the threshold voltage behavior of contaminated NMOS and PMOS transistors with voltage stress. (**a**) Initial V_t of an NMOS transistor in an unknown state. (**b**) NMOS transistor after a positive gate voltage stress with a reduced V_t . (**c**) NMOS transistor after a negative gate voltage stress with an increased V_t . (**d**) Initial V_t of a pmos transitor in an unknown state. (**e**) PMOS transistor after a positive gate voltage stress with an increased V_t . (**f**) PMOS transistor after a negative gate voltage stress with a decreased V_t

Initially, as the wafers exit the wafer fab, the mobile ions are randomly distributed and widely diffused throughout the wafer by the last thermal steps, such as the final alloy. This condition is shown in Fig. 16.5a and d. If the wafers are stressed with a positive bias on the gate in (Fig. 16.5b and e), the positive mobile ions are repelled by the positive gate potential and are forced to the gate dielectric-semiconductor surface. The positive mobile ions at the semiconductor surface attract electrons that lower the V_t for an NMOS transistor and raise the absolute value of the V_t for a PMOS transistor. If the wafers are stressed with a negative gate bias, the positive mobile ions are attracted to the gate by the negative potential and away from the gate dielectric-semiconductor interface, shifting the threshold voltages in the opposite direction. Mobile ions away from the semiconductor-dielectric interface have a reduced impact on the threshold voltage compared to mobile ions close to the semiconductor-dielectric interface. As a result of the negative gate voltage attracting the ions, the threshold voltage of an NMOS transistor decreases, and the threshold voltage of a PMOS transistor decreases, as shown in Fig. 16.5c and f.

Finally, if an unbiased bake is performed on the transistors after the positive or negative stress is completed, the mobile ions diffuse away from the interfaces, returning closer to the initial state where they are randomly distributed, moving the V_t closer to the initial V_t.

PMOS transistors are normally biased with the gate negative respective to the other transistor terminals. On the other hand, NMOS transistors are normally biased with the gate positive respective to the other transistor terminals. This difference in the polarity of operation for the two devices impacts the device's susceptibility to mobile ions. The PMOS transistor is less susceptible to the positive mobile ions as the negative gate bias attracts the ions to the gate where they have the least impact on the transistor channel. Conversely, the NMOS transistor with the positive gate bias forces the mobile ions close to the channel where they have the most impact on the transistor channel. This phenomenon explains why early MOS chips worked better using PMOS circuits.

16.4.2 Isolation Failure

In addition to the active CMOS transistors, mobile ion problems can cause the isolation that separates transistors electrically to fail, causing shorts and leakage between transistors. The isolation forms a parasitic MOS transistor where the isolation dielectric serves as the gate dielectric. The silicon semiconductor forms the channel like an active transistor, and any gate metal running across the isolation forms the gate of the parasitic device.

The isolation commonly fails first during mobile ion contamination for several reasons, as described in Chap. 5. Additionally, many isolation areas are unprotected by the gate metal, which can block mobile ions. In contrast, the active transistor always has the gate metal protecting it. The isolation has a very thick dielectric relative to the gate dielectric. As a result, the isolation is more sensitive to mobile ions than a CMOS thin transistor dielectric and has a larger threshold voltage shift for an equal amount of charge caused by mobile ions (Chap. 5). For example, the threshold voltage shift for a 2000 A isolation dielectric is 100 times larger than that for a 20 A gate dielectric using the same dielectric constant. However, a much larger threshold voltage shift is needed to cause an isolation device failure relative to the active transistors. Figure 16.6 shows a plot of isolation failure with mobile ion contamination tested with a biased bake (Gutai, 1997). First, the initial V_t before stress is shown.

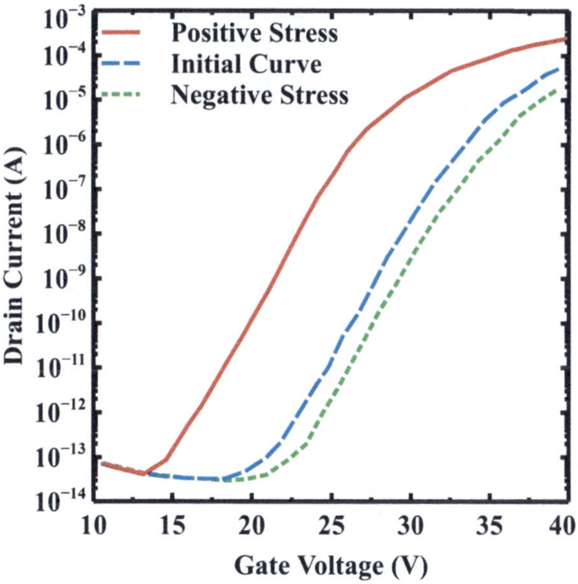

Fig. 16.6 Plot of the drain current versus gate voltage for an nmos parasitic isolation transistor stressed at +/− 20 V at 200 °C 20 s. For a 3000 A isolation dielectric thicknesses. The curves are measured at room temperature. (**a**) After positive stress. (**b**) Initial curve before stress. (**c**) After negative stress

Next, the NMOS isolation device is stressed with +20 V, and the V_t shifts lower. Next, the device is stressed with −20 V, and the V_t shifts higher. Repeating the positive and negative stress results in curves similar to the initial measurements as the mobile ions move back and forth in the isolation dielectric between the gate interface and the semiconductor surface.

As described in Chap. 5, the large isolation threshold voltages are necessary to prevent the isolation devices from leaking at normal CMOS operating voltages. However, mobile ion contamination can cause the threshold voltages to drop to the point where leakage between the active transistors causes the circuits and chip to fail.

16.4.3 CMOS Transistor Degradation

In addition to the isolation field devices, the active transistors can degrade with mobile ion contamination, showing a similar V_t shift, as shown in Fig. 16.6. The CMOS transistor's threshold voltage shifts cause the transistors to operate in a region outside the design space where the circuit was designed to operate, leading to circuit and chip failure. Figure 16.7 shows a sketch of the mobile ion shifts for an NMOS capacitor and transistor, showing the results of mobile ion contamination

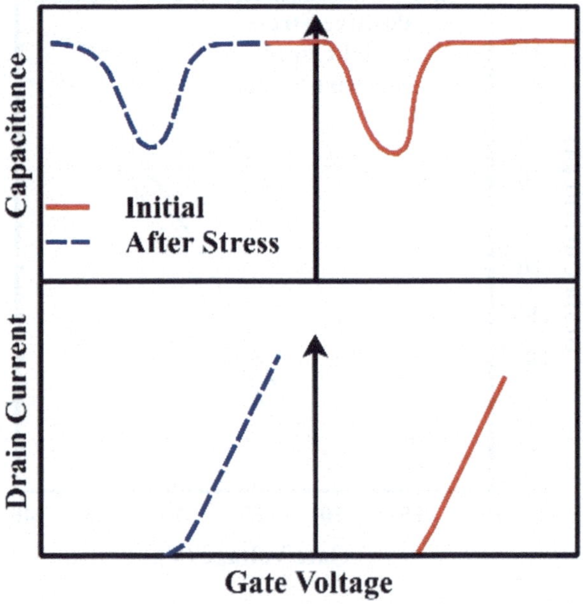

Fig. 16.7 Sketch showing a C-V shift for an nmos capacitor and an I-V V_t shift for an nmos transistor after a positive gate voltage stress at elevated temperatures. The threshold voltage for both the capacitor and the transistor is shifted to the left (more negative)

(Hofstein, 1966). For a positive gate voltage stress at an elevated temperature, the threshold voltage for both devices shifts to the left. Conversely, the threshold voltages shift to the right for a negative gate voltage stress.

Gate dielectrics contaminated by mobile ions tend to have a smaller threshold voltage shift than isolation devices. The gate metallization and gate-related dielectrics (spacers, etc.) can partially block mobile ions from reaching the sensitive gate dielectric. In addition, the orders of magnitude thinner gate dielectric makes lateral diffusion in the gate dielectric more difficult. Migration of mobile ions from the dielectric-gate interface to the dielectric-silicon interface has a lesser effect on the threshold voltage due to the shorter distance traveled. There is some evidence that mobile ions are less mobile in high-k gate dielectrics (Parent et al., 2010). Figure 16.8 shows an example of mobile ion contamination in a 32 nm SOI technology (Kothandaraman & Cohen, 2016).

Through-silicon vias are used to connect the chip to the outside world in advanced packaging applications, as described in Chap. 9. The complex TSV process can create mobile ion contamination problems. Figure 16.8 shows a mobile ion problem caused by a TSV process. A nearby TSV contaminated a PMOS transistor, causing a 100 mV mobile ion V_t shift. The 100 mV shift is sufficient to cause circuit failures in sensitive circuits. Transistors without TSVs are unaffected by mobile contamination.

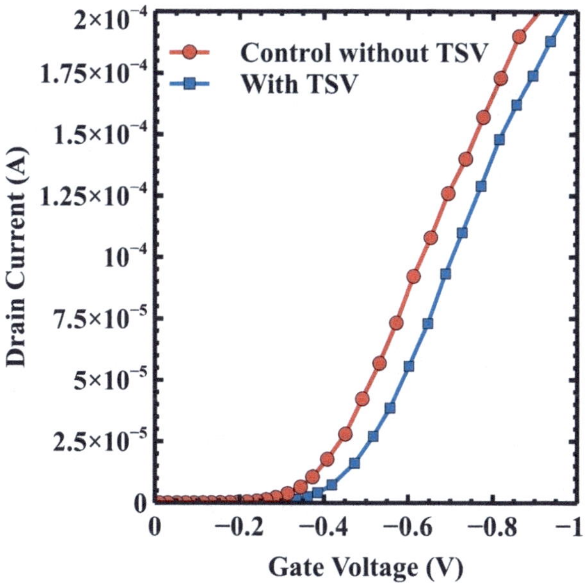

Fig. 16.8 Subthreshold voltage plot showing a 100 mV V_t shift of a 32 nm SOI pmos device contaminated by mobile ions from a nearby through silicon via (TSV)

16.5 MOS Capacitor Degradation

Mobile ion contamination can modify a MOS capacitor's capacitance-voltage (C-V) characteristics, as shown in Fig. 16.7. C-V characterization and analysis is a common method for detecting mobile ion contamination. C-V characterization for mobile ion detection uses a series of biased bakes and electrical measurements.

The shift in MOS capacitor C–V characteristics can also affect the product, causing changes in circuit function. MOS capacitors are often used for circuit design elements, and a change in characteristics may cause chip failure. Mobile ions have been observed to cause additional C-V degradation beyond the lateral shift along the voltage axis shown in Fig. 16.7. Figure 16.9 shows various C-V plots observed on mobile ion-contaminated capacitors for a thick 200 nm dielectric (Snow et al., 1965). The C-V curves are distorted due to various effects, including nonuniform mobile ion contamination across the capacitor, effectively causing multiple different capacitors in parallel with multiple threshold voltages. Also, dielectric leakage degradation may occur for heavily contaminated samples.

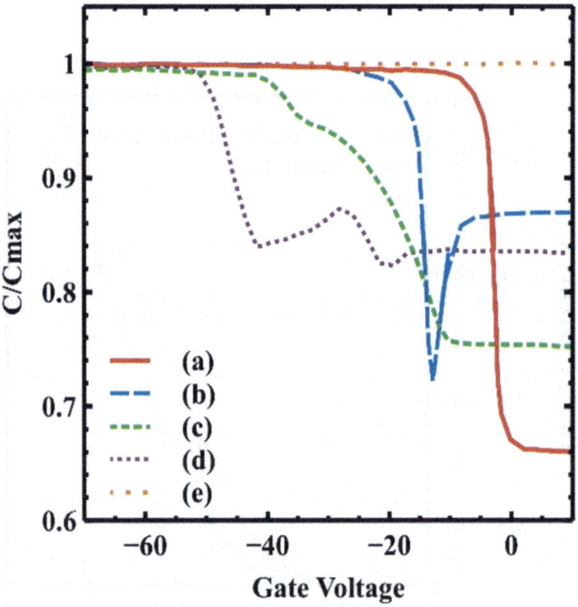

Fig. 16.9 Capacitance-voltage (CV) plots showing anomalous behavior caused by mobile ions. (**a**) uncontaminated C-V curve. (**b**) Shift in V_t and change in inversion region behavior. (**c**)(**d**) C-V curves due to dielectric leakage or multiple threshold voltages. (**e**) Shift in V_t beyond the measurement voltage range

16.6 Diode Degradation

Mobile ions collecting in the dielectrics near a diode junction can affect the diode current-voltage properties by changing the electric field near the junction, similar to a gated diode (Schroen et al., 1965). The positive electric ions change the electric field lines in this region, changing the diode forward and reverse bias currents. The diode characteristics that can be modified by mobile ion contamination include junction breakdown, leakage, and capacitance.

16.7 Gate Dielectric Degradation with Mobile Ions

Gate dielectric properties can be degraded by contamination with many elements, including mobile ions, correlating with increasing degradation with increasing contamination ion density. The affected dielectric properties include leakage, breakdown voltage, and reliability measured by time-dependent dielectric breakdown. Figure 16.10 shows a plot of time-dependent dielectric breakdown (TDDB) as a function of mobile ion contamination concentration (Huang et al., 2014).

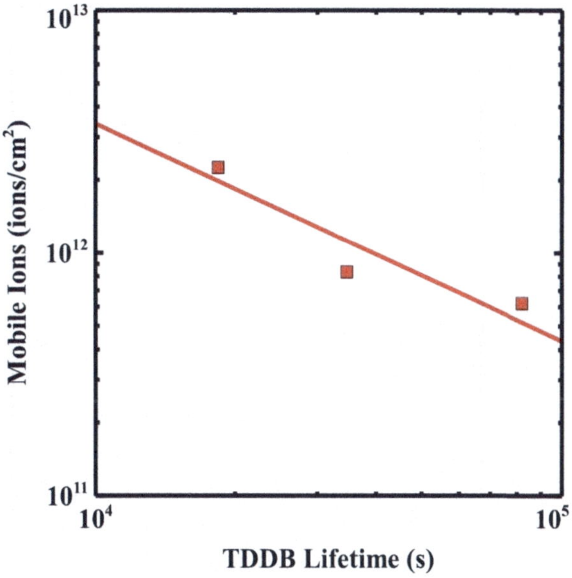

Fig. 16.10 Plot showing degradation in time-dependent breakdown with increasing copper contamination

16.8 Intermetal Dielectrics Degraded by Mobile Ions

Copper is one of the mobile ions listed in Table 16.2, and its mobile ion behavior is described in Chap. 15. Due to low resistance, it is the primary metallization used in most metal layers in modern CMOS technology. Copper is highly mobile in dielectrics (see Chap. 15), migrating by drift and diffusion out of metal lines into the surrounding dielectrics, where it can degrade the dielectrics. Copper dielectric contamination may cause dielectric leakage, reduced breakdown, and degraded reliability.

Much optimization is required to develop a successful copper metallization process, including mitigation of copper migration. A copper line is typically surrounded by a refractory metal barrier metal in a damascene trench, which helps mitigate copper mobile ion problems (Chen, 2012). The metal barrier prevents copper from diffusing and drifting under bias into the surrounding dielectrics. Figure 16.11 shows a current-voltage plot of the intermetal high-k dielectric breakdown for Cu metal lines with and without a barrier metal. The group without the barrier liner shows a decreased breakdown by about 10 V, indicating a degraded dielectric caused by copper. The reference shows both a degradation in breakdown and reliability.

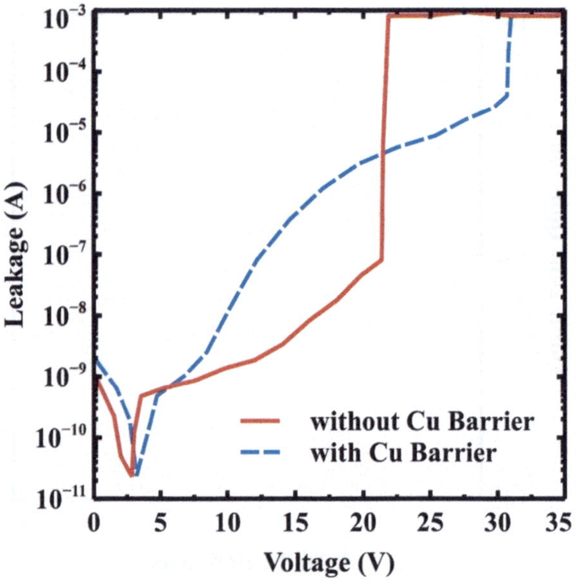

Fig. 16.11 Breakdown of I-V curve of low-k dielectric between copper lines with and without a metal liner

16.9 Product-Level Reliability Failures

The electrical effects described previously in this section can be observed in individual parametric test structures. Additionally, mobile ion-related electrical reliability problems can be observed during product life tests and as reliability failures in the field. CMOS products undergo extensive quality and reliability testing at accelerating voltages and temperatures for extended periods to ensure that products do not fail during their life in the field. The mobile ion-related failure time is governed by the amount of contamination, distance from the contamination location to the sensitive failing circuit, circuit sensitivity to mobile ion contamination, voltage, temperature, and time. Table 16.4 shows an example of foundry qualification data showing a single failure at 500 h in high-temperature operating life reliability testing due to mobile ion contamination. A single chip failed for sodium mobile ion contamination out of hundreds tested. All other qualification tests had no failures.

Product-level failure analysis of mobile ion-related problems uses tests that examine some of the unique electrical properties of mobile ions, including full or partial failure recovery in an unbiased bake and an ability to reverse the failure with a reverse polarity biased bake (See Table 16.3). Additionally, analytical measurements, such as secondary ion emission spectroscopy (SIMS) etc., can be used to locate mobile ions in a CMOS chip.

One possible product mobile ion failure mode is a shift in the power consumption current of the chip (Sabin et al., 1997). This shift may be observed in standby current measurements, where there is maximum sensitivity to isolation or threshold voltages. Few circuits are active in standby mode, resulting in a low current level, enabling the detection of small current shifts. Also, a chip in standby is largely inactive with steady-state biases on circuits, creating a constant field to attract or repel mobile ions. Figure 16.12 shows a plot of the standby current drifting through time measured at 175 °C with and without passivation. The reference concluded that mobile hydrogen protons were the cause of the drift.

Table 16.4 Example foundry qualification data showing a mobile ion related failure in an operating life test at 500 h of testing (Shideler, 1996)

Stress	48 h	168 h	500 h	1000 h
Operating life (6.5 V, 125 °C)	0/900	0/300	1/300 *	0/299

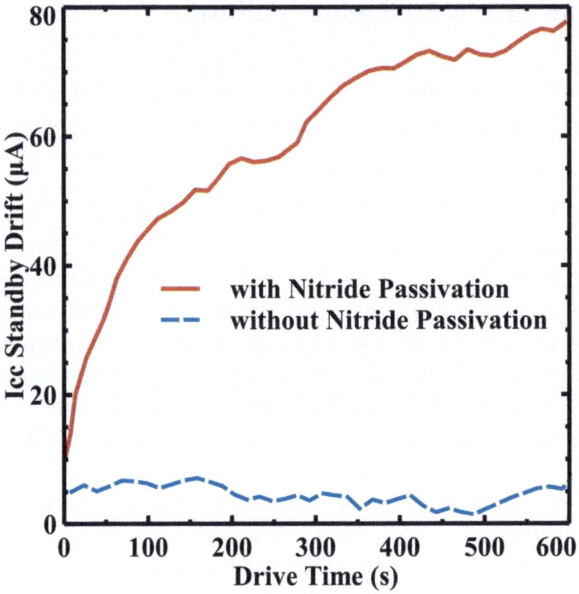

Fig. 16.12 Plot of product level standby current drift as a function of time measured at 175 °C with and without nitride passivation

16.10 Electrical Detection of Mobile Ions

16.10.1 Electric Measurement Introduction

Mobile ion contamination is often first detected electrically on failing production wafers that have escaped all the standard prevention measures. Once isolated, mobile ion problems can be detected chemically on simple test wafers or short loop reduced flow wafers where MOS capacitors are formed.

Electrical detection is an essential diagnostic tool for mobile ions, providing a high level of proof of contamination. Table 16.5 lists some common ways mobile ions can be detected using electrical techniques on parametric test structures. Some methods, such as triangular voltage sweep (TVS) and thermally stimulated ionic current (TSIC), directly measure the current due to ionic motion through dielectrics. Other methods, like threshold voltage device stresses, measure the results of ion motion, such as threshold voltage shifts. This section describes the more common techniques in detail.

Table 16.5 Common electrical methods for the detection of mobile ions

Electrical method	Reference
Triangular voltage sweep (TVS)	(Kuhn & Silversmith, 1971)
Capacitance-voltage analysis (CV)	(Nauta & Hillen, 1978)
Active device and field device stress	(Hofstein, 1966) (Gutai, 1997)
Thermally stimulated ionic current (TSIC)	(Nauta & Hillen, 1978)

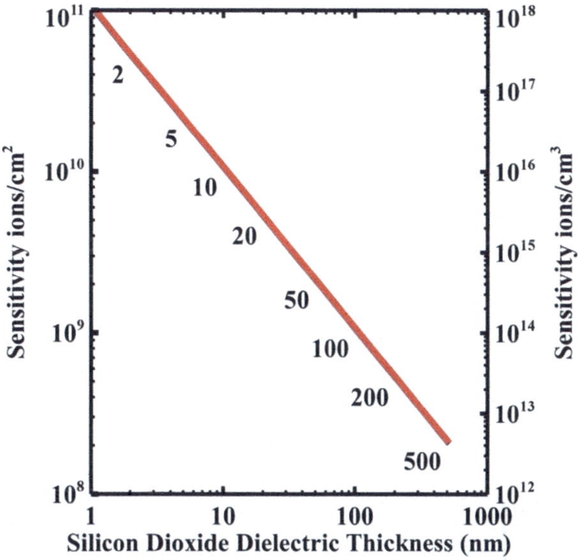

Fig. 16.13 Plot showing the mobile ion measurement sensitivity for transistor measurements and capacitance-voltage measurements plotted against silicon dioxide thickness assuming a 5 mV threshold voltage measurement capability. The left y-axis shows areal sensitivity, and the right y-axis shows volume density sensitivity. This plot is derived from Eq. (16.5)

16.10.2 Measurement Sensitivity and Capability

The minimum detection level is an important capability of each measurement technique. The lower limit of TVS measurement sensitivity has been reported to be around 10^9 ions/cm^2 (Schroder, 2006). TVS measurement capability is limited by the low current capability of the instrumentation, hot chuck, and probe station in the femtoampere to picoampre range. TVS sensitivity can be increased by increasing the size of the test structure, creating a larger chip area for detection, and raising the measurable mobile ion current. The sensitivity is a function of the ramp rate, and the optimum rate must be determined empirically.

C-V analysis, active transistors, and field devices have a sensitivity limited by the instrumentation's capability to resolve a slight shift in the threshold voltage or flat-band. C-V analysis has a reported sensitivity of around 10^{10} ions/cm^2 (Schroder, 2006). However, the threshold voltage shift (ΔV_t) due to mobile ions is a strong function of the dielectric thickness and capacitance, as shown by Eq. (16.4), where Q is the mobile ion concentration, C is the dielectric capacitance.

$$\Delta V_t = \frac{Q}{C} \tag{16.4}$$

Equation (16.4) can be rearranged to calculate the measurement sensitivity per unit area, yielding Eq. (16.5), where N_m is the mobile ion concentration per unit area, ε is the dielectric constant of the dielectric times the permittivity of free space, ΔV_t is the threshold voltage resolution of the instrument, q is the electron charge, and t_{ox} is the dielectric thickness.

$$N_m = \frac{\varepsilon \Delta V_t}{t_{ox} q} \tag{16.5}$$

After estimating the minimum resolution of the measurement system (ΔV_t), including the instruments, hot chuck, and probe station, the mobile ion resolution can be estimated from this equation. Figure 16.13 shows a plot of the measurement sensitivity assuming 5 mV V_t resolution capability for mobile ion detection as a function of the dielectric thickness for silicon dioxide ($\varepsilon = 3.9$). At dielectric thickness near 2 nm, close to typical CMOS dielectric thicknesses, it has a sensitivity of about 10^{11} ions/cm^2. For isolation, a typical dielectric thickness of around 200 nm

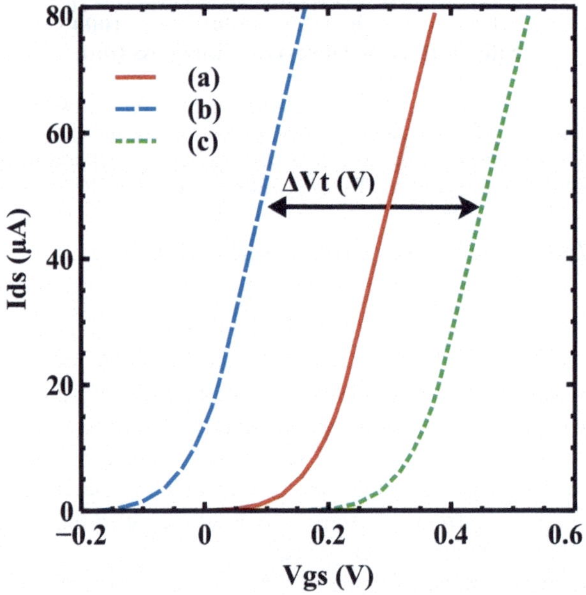

Fig. 16.14 Example nmos threshold voltage measurements used to detect mobile ions on a nmos transistor. (**a**) Initial measurement before stress. (**b**) V_t after a positive voltage high-temperature stress. (**c**) V_t after a negative voltage high-temperature stress

has a sensitivity of around 10^9 ions/cm^2. The increased sensitivity for isolation is one of the reasons the isolation tends to fail first in mobile ion contamination situations.

16.10.3 Detection of Mobile Ion Contamination Using a CMOS Transistor

A CMOS transistor or an isolation transistor can be used to detect mobile ions. The initial detection of mobile ion contamination in a failing chip circuitry often occurs in a CMOS transistor or an isolation device. These measurements take advantage of the amplification properties of a CMOS transistor that can magnify a slight change in gate charge into an easily measurable current.

Figure 16.14 shows an example of the behavior of an NMOS device with mobile ion contamination. The device's threshold voltage is measured by plotting drain current versus gate voltage, typically measured at a low drain voltage (0.05–0.1 V). The threshold voltage is extracted by extrapolating the curve to the x-axis. The threshold voltage can be measured at either room or the high temperature where the device is stressed. First, the initial threshold voltage is measured (Fig. 16.14a), where the threshold voltage is affected by the initial location of the mobile ions. Then, after high-temperature voltage stress with positive gate bias, the positive mobile ions shift to the dielectric-silicon interface, lowering V_t (Fig. 16.14b). After high-temperature voltage stress with negative gate bias, the positive mobile ions shift to the gate dielectric-silicon interface, increasing V_t (Fig. 16.14c). The mobile ion concentration can be calculated by the simple formula $Q = C\Delta V_t/q$, where C is the gate capacitance of the capacitor in accumulation, ΔV_t is the difference in threshold voltage after the positive and negative stress, and q is the electronic charge (1.602×10^{-19} coulombs).

Figure 16.5 shows a sketch of the mobile ion impact on threshold voltages for contaminated NMOS and PMOS transistors after positive and negative gate voltage stress performed at elevated temperatures. Initially, as the wafers exit the fab, the distribution of the mobile ions in the gate dielectric is unknown and usually randomized by the thermal energy of the final heat treatments in the process flow. If a positive bias is applied to the gates, the electric field forces the mobile ions to the gate dielectric-silicon interface, where the NMOS V_t is decreased, and the PMOS is increased. If a negative bias is applied to the gates, the mobile ions are forced by the electric field to the gate dielectric-gate interface, where the NMOS V_t is increased, and the PMOS is decreased. After the ions are driven to one interface, an unbiased thermal bake rearranges the ion distribution, diffusing the ions away from the interface. The resultant threshold voltage resides somewhere between the positive and negative threshold voltage points.

16.10.4 Triangular Voltage Sweep

Because of its unique properties, the triangular voltage sweep (TVS) technique is the most common electrical technique for mobile ion detection. TVS measurements can be performed on several different types of structures, including MOS capacitors, simple parallel plate capacitors, comb, and serpentine capacitors. C-V measurements require an MOS capacitor, while TVS can be performed on any capacitor type. Mobile ion motion can be detected in these structures as the ions transition from one terminal of the capacitor to the other. In addition, TVS has some capability of identifying the exact problem element causing mobile ions through the temperature and voltage dependence of the mobile ion behavior.

The measurement system used for triangular voltage sweep is shown in Fig. 16.15 and consists of a voltage waveform generator, femtoampere meter, and hot chuck in a wafer measurement station (Barbottin & Vapaille, 1986). The voltage waveform generator ramps the voltage applied to the capacitor at a constant programable ramp rate to apply an electric field to the capacitor that causes the mobile ions to drift through the dielectrics. A highly sensitive femtoampere (or picoampere) meter measures the displacement current flow between the two plates of the capacitor that compensates for the mobile ion motion. The voltage source is typically connected to the chuck, and the femtoampere meter is connected to the probe tips to minimize noise. The wafer sits on a low-noise hot chuck that can control and ramp the wafer temperature. The entire system needs to be electrically isolated and shielded to provide a low-noise environment to enable current measurements in the femtoampere to picoampere range. The capacitor size, amount of mobile ion contamination, and mobility of the ions in the dielectric determine the current measurement capability requirements.

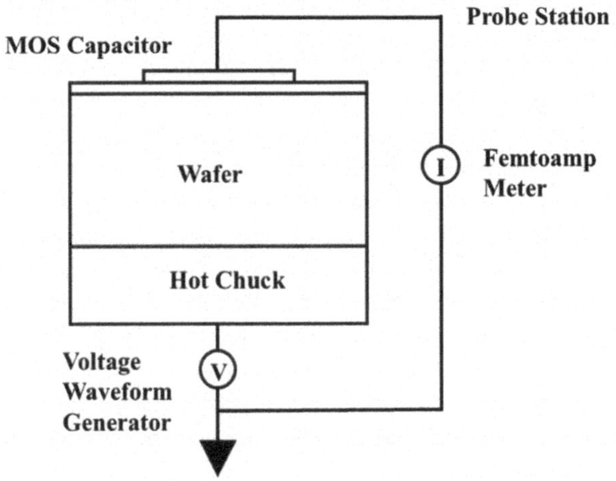

Fig. 16.15 Sketch showing TVS measurement system

Initially, the wafer temperature is ramped to a selected temperature with the capacitor unbiased. The temperature is selected empirically by testing a range of temperatures with each measurement on a virgin capacitor. Typically, ranges from 100 °C to 350 °C are used. An alternative measurement technique prestresses the dielectric at a higher temperature, applying a constant voltage and driving the mobile ions to one interface. This stress maximizes the TVS signal, capturing the maximum measurement current from one polarity to the other.

The voltage is swept at a defined ramp rate in one polarity while the current is continuously measured as the mobile ions drift to the negative electrode. Typically, a maximum electric field of a few MV/cm stress is used to drive the ions while avoiding higher electric fields where direct or Fowler-Nordhiem tunneling currents can interfere with the measurements. Then, the measurement is repeated, ramping the voltage in the opposite direction, and moving the ions to the opposite plate. Figure 16.16 shows an example TVS measurement measured at 350 °C (Kuhn & Silversmith, 1971). The current spikes shown in the plot are the mobile ion signal, while the flat regions of the measurement indicate no mobile ion movement. An uncontaminated TVS measurement is flat and has no peak. Two different mobile ions are detected in the measurement. The more prominent peak is sodium, and the smaller peak is potassium. The area under the mobile ion peaks should be equal for both polarity sweeps if all of the mobile ions are swept back and forth between the plates. If the areas of both polarity sweeps do not match, it indicates that ions are being trapped at one of the interfaces.

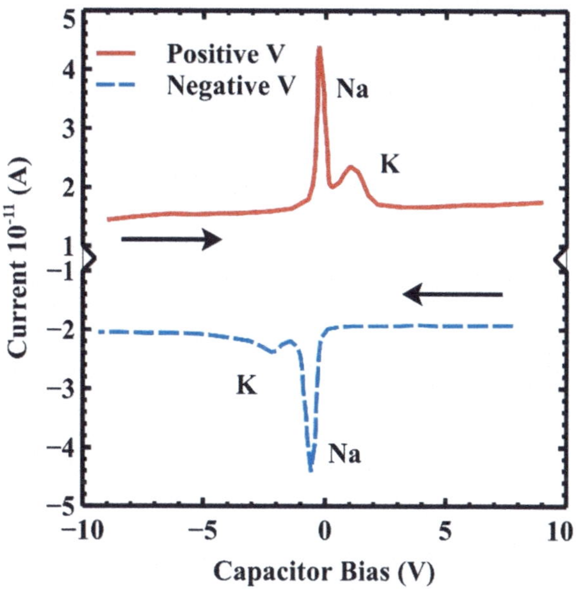

Fig. 16.16 Example triangular voltage sweep detecting the presence of mobile ions at 350 °C. (**a**) The mobile ion current is measured as the voltage is swept in both directions. The arrows show the directions of the voltage sweeps

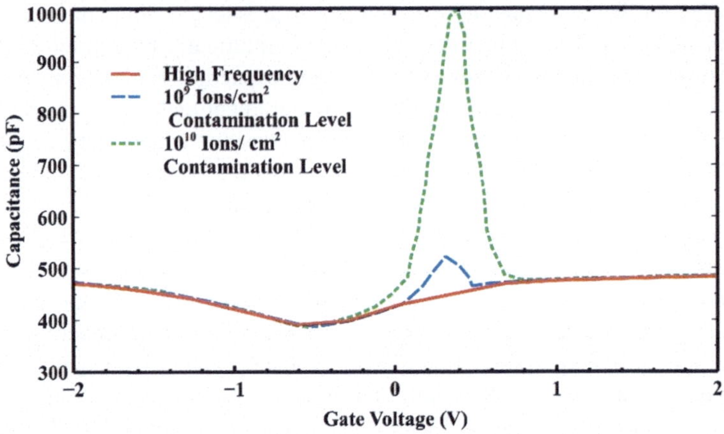

Fig. 16.17 Example plots showing the detection of mobile ions. (a) Increased mobile ion concentrations detected at low frequency compared to a high frequency measurement at 250 °C

The mobility of the mobile ion in the dielectric determines the peak location on the voltage axis. In this example, the potassium peak occurs at a higher positive or more negative voltage due to the lower mobility of the larger mobile ion, requiring a higher field to drive it through the dielectrics.

Figure 16.17 shows some additional examples of TVS signals on a MOS capacitor ((Schroder, 2006) pg. 341). High-frequency and low-frequency plots are compared, highlighting the importance of a low ramp rate in the TVS measurement to obtain the maximum signal. The high-frequency plot does not detect the presence of mobile ions as the ions as the ions do not move with the rapidly alternating voltage. The MOS C-V inversion charge characteristics are still observable as the downward dip in the trace.

Figure 16.18 compares TVS signals for different temperatures (Stauffer et al., 1995). No mobile ion signal is observed at lower temperatures (10 and 40 °C) because the mobile ions are either immobile at these temperatures or not observable with the measurement conditions and instrument capability. The ions may be mobile over time periods longer than the measurement time. The quasistatic C-V characteristics with the dip in capacitance due to the depletion region can be observed as the dip in the plot at lower temperatures. At higher temperatures, the quasistatic C-V disappears as the thermal generation of carriers inverts the channel for all voltages. At higher temperatures, the mobile ion peak increases with increasing temperatures. At some higher temperatures, all ions are mobile, and the peak will no longer increase. The mobile ion charge density can be calculated by integrating the charge observed in the measurements.

The TVS signal peak's position and width provides information on the mobile ions. Figure 16.19 compares two TVS measurements for a thermal oxide and a plasma-enhanced tetraethyl orthosilicate glass (PE-TEOS) that were intentionally contaminated with sodium (Lifshitz & Smolinsky, 1989). The higher-quality

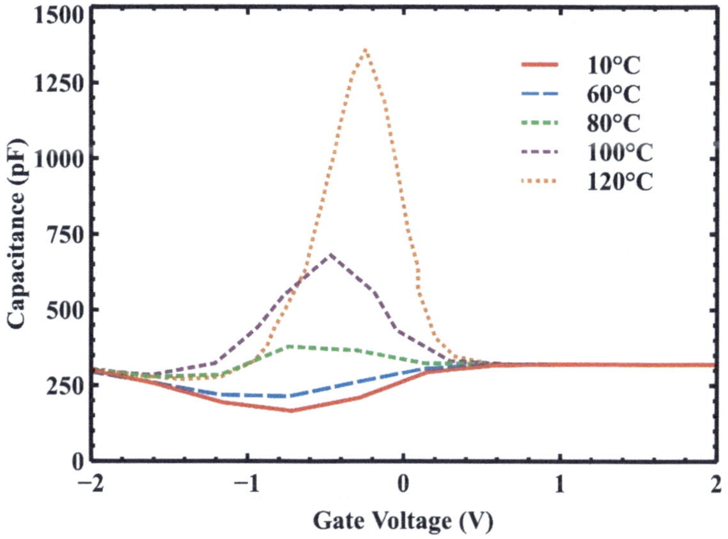

Fig. 16.18 Comparison of TVS measurements at different temperatures

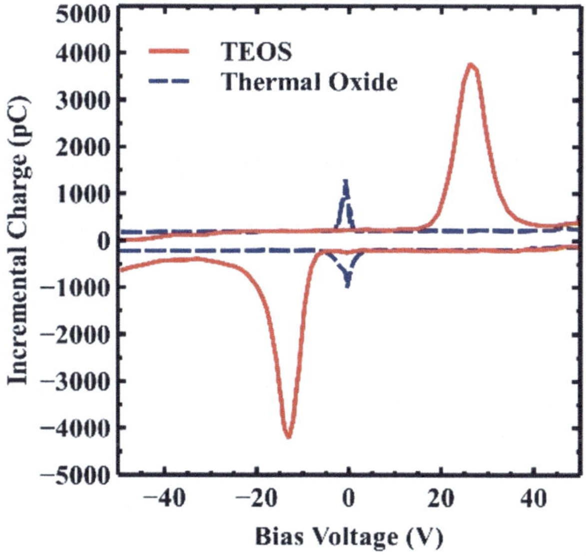

Fig. 16.19 TVS measurements comparing two different sodium contaminated dielectrics measured at 300 °C. (**a**) Measurement on a thermal oxide. (**b**) Measurement on a plamsa-enhanced tetraethyl orthosilicate (PE-TEOS) glass

thermal oxide has a narrow peak, indicating that the mobile ion movement occurred at similar electric fields for all ions. The narrow peak indicates that dielectric consists of uniform ion trap depths, suggesting a homogenous material. The broader

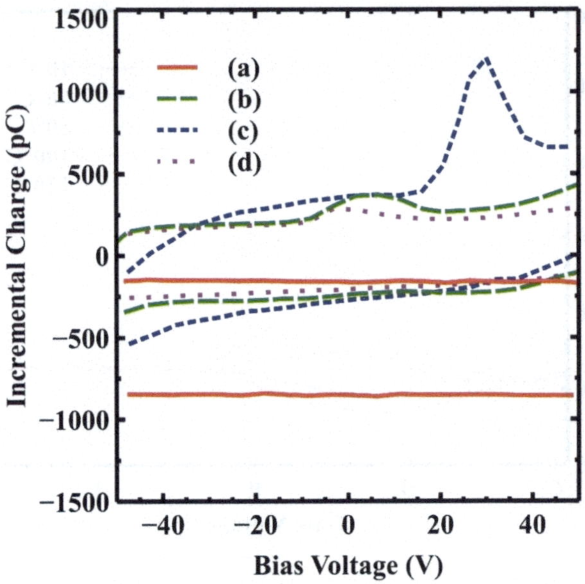

Fig. 16.20 Plots showing a comparison of different dielectrics uncontaminated by alkali metals. (**a**) PE-TEOS deposited at 400 °C. (**b**) PE-TEOS deposited at 300 °C. (**c**) SOX polysilicate glass deposited at 400 °C. (**d**) Accuglass™ 211 polysilicate glass deposited at 400 °C

peak for the PE-TEOS dielectric indicates a large spread of trap energies, indicating a more inhomogeneous film.

The peak's distance from the zero-volt origin indicates the difficulty in moving mobile ions in the film. The thermal oxide peak near zero volts indicates that the ions move easily in the thermal oxide. The large distance from the origin of the PE-TEOS dielectric indicates that the ions are more difficult to move relative to the thermal dielectric ions. The difference in the two traces from the x-axis origin in Fig. 16.19 indicates that one interface has slightly deeper traps that slow the ion migration.

The TVS signal can also show the behavior of mobile ions at the interface. If the TVS peak disappears after a sweep in one direction, it indicates that the mobile ions are trapped at that interface and cannot escape as the polarity is reversed. If hydrogen is the mobile ion under study, the mobile hydrogen may become immobile, as described in Chap. 14. Figure 16.20 compares four different dielectrics free of alkali metal mobile ion contamination (Lifshitz & Smolinsky, 1989). Figure 16.20a shows a PE-TEOS film deposited at 400 °C with no discernable mobile ion contamination. Figure 16.20b shows a PE-TEOS film deposited at 300 °C with a mobile ion peak in one sweep direction. Figure 16.20c and d show TVS peaks in one direction for two low-temperature deposition polysilicate glass films.

The single peak TVS signal in three plots observed in only one direction in Fig. 16.20 is attributed to the motion of hydrogen protons created from water trapped in the films. The TVS signal outside the peak has a slope and is not flat in these films, indicating dielectric leakage or a small percentage of the protons moving at electric fields unrelated to the peak motion. In this example, the protons are trapped at the aluminum gate electrode. The shape of the TVS curve for proton contamination is strongly dependent on the voltage ramp conditions due to the high mobility of the protons.

The less-dense low-temperature films have a high propensity to trap water, creating a film with mobile hydrogen ions. The film's water concentration increases as wafers are exposed to the atmosphere, increasing the TVS signal with time. The water concentration can be decreased by baking or alloying the wafer, reducing the TVS proton signal. In this example, the TVS signal disappears in the SOX dielectric when the film is annealed at 900 °C.

The negative hydroxyl (OH-) molecules remaining after the proton is emitted are not mobile and are usually not observed in the TVS sweep as negative ion motion but can rebond with protons, eliminating the mobile protons.

16.10.5 Detection of Mobile Ions in Intermetal Dielectrics

Triangular voltage sweep (TVS) is a valuable technique for measuring mobile ions in MOS capacitors on the wafer's surface. Additionally, TVS is useful for measuring mobile ions in other structures that do not involve the semiconductor surface. It is commonly used for measuring back-end-of-line contamination in intermetal dielectrics. This feature is a key advantage of the TVS technique over threshold voltage and capacitance-voltage techniques.

Common metallization test structures used for metallization characterization can be used for TVS measurements if they are sufficiently large to generate a detectable mobile ion current. The necessary size of the test structure for mobile ion detectability depends on the severity of the mobile ion contamination, resulting in a more extensive test structure requirement for lower amounts of contamination. Figure 16.21 shows a metal comb and a metal comb with interleaved serpentine (He & Lu, 2012). These structures are commonly used in parametric and yield characterization for metallization process modules. These structures can be used to measure contamination between metal lines and between metal layers. The metal structures are simply connected as a two-terminal device, as shown in Fig. 16.15, and the same measurement methods are used.

Simple flat parallel plate capacitors formed from two different metal layers can also be used for metallization TVS test structures. However, most metals can block or impede mobile ion migration, making the comb and serpentine a more sensitive structure as the mobile ions can move between the metal lines in the surrounding dielectrics. Typically, a comb or serpentine yields a much larger TVS peak compared to a plate capacitor (Chonko et al., 1994).

An example of a TVS measurement using a comb structure is shown in Fig. 16.22 (Huang et al., 2014). Figure 16.22a shows a plot of TVS current versus voltage with different voltage ramp rates, showing that a faster ramp rate gives a more significant peak. Figure 16.22b shows a similar plot showing the TVS signal with different temperatures. No TVS peak is observed below 200 °C in this example. The mobile

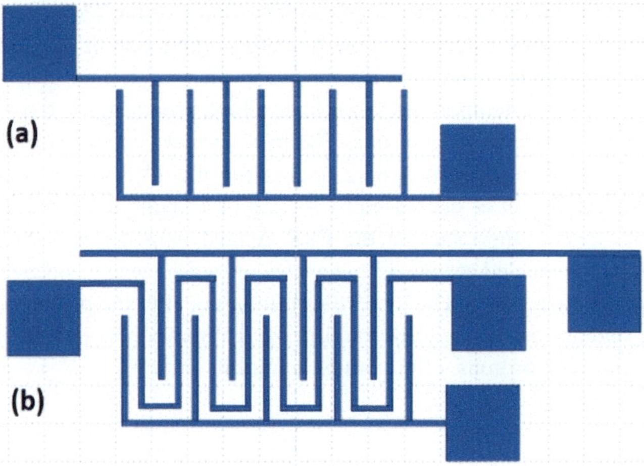

Fig. 16.21 Sketch showing common metallization test structures that can be used for TVS measurements. (**a**) Two terminal metal comb. (**b**) Metal comb with intertwined serpentine structure (four terminals)

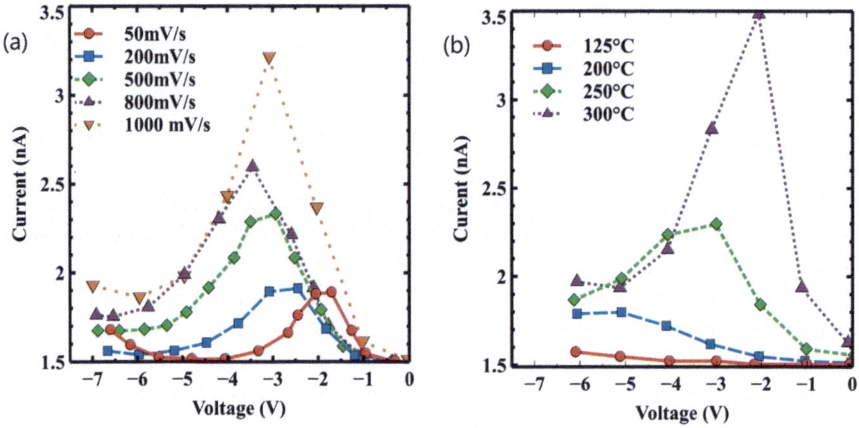

Fig. 16.22 TVS plot of mobile ion contamination measured on a metal comb. (**a**) TVS plot showing increasing TVS signal with increasing ramp rate shown as volts per second at 300 °C. (**b**) TVS plot showing increasing TVS signal with increasing temperature

ion in this reference is assumed to be copper, requiring a higher temperature relative to other mobile ions.

TVS is extremely useful for the characterization of copper metallization, enabling the optimization of barrier metallization and other techniques to prevent copper drift and diffusion, improving the reliability of the metallization, and increasing the probability of meeting the chip's lifetime requirements. Figure 16.23 shows the

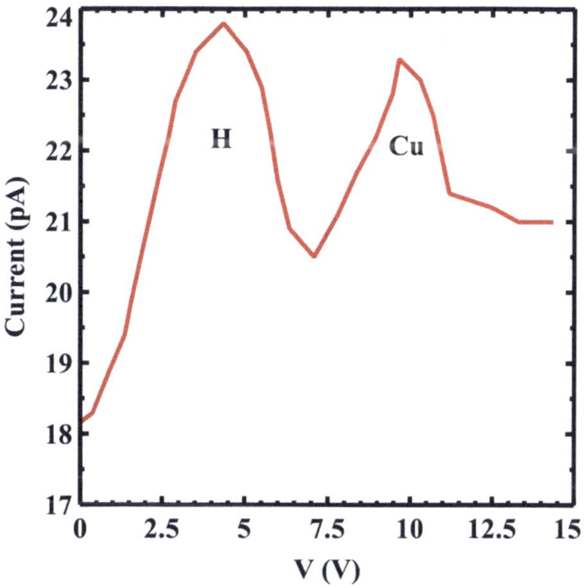

Fig. 16.23 TVS plot showing the simultaneous detection of hydrogen and copper in a low-k dielectric

simultaneous TVS detection of both copper and hydrogen protons (Ciofi et al., 2006a).

TVS can be useful during technology development as an experimental tool for process optimization experimentation. For example, Fig. 16.24 compares different copper damascene barrier metals by TVS (Wojick et al., 2011). Different barrier metals and deposition methods are compared, generating a metric to compare copper metallization. The ruthenium physical vapor deposition group has the worst performance for copper migration, generating a prominent TVS peak. The tantalum nitride physical vapor deposition performs best with a small TVS signal. An aluminum control sample is used as a reference, detecting only a small amount of sodium and no copper. Figure 16.24b shows a current versus time plot for a time-dependent dielectric breakdown (TDDB) test measured at 5 MV/cm at 350 °C for the same barrier metals. The TDDB results correlate well with the TVS results, showing the same experimental group dependence and demonstrating the value of TVS reliability prediction (Wojick et al., 2011).

16.10.6 Thermally Stimulated Ionic Current Measurements

Thermally stimulated ionic current measurements (TSIC) are a technique that is closely related to TVS. However, instead of ramping the voltage, the temperature is ramped. A constant voltage is applied to a MOS, and the temperature is ramped

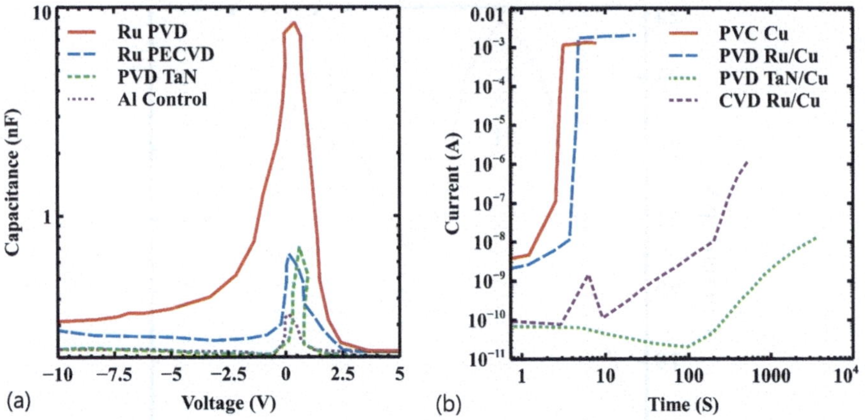

Fig. 16.24 TVS results of a copper barrier experiment comparing different barrier metal materials surrounding a copper damascene line. (**a**) different processes are compared, including ruthenium physical vapor deposition (Ru PVD), ruthenium plasma enhanced chemical vapor deposition (Ru-C PECVD), ruthenium plasma enhanced atomic layer deposition (Ru-C PEALD), and Physical vapor deposition of tantalum nitride (TaN PVD). A control group without copper (Al) is shown with a small mobile ion signature. (**b**) Current versus time plot for a time dependent dielectric breakdown (TDDB) test showing a good correlation of dielectric lifetime with the TVS results

while measuring the current. This measurement setup for TSIC is identical to the one for TVS shown in Fig. 16.15. TVS is generally preferred over TSIC because the temperature ramps for the TSIC measurement tend to be more time-consuming than TVS measurements at a constant temperature. In addition, it is often challenging to keep probe tips on the bondpads during a TSIC measurement as the probe tips expand as the temperature increases.

Figure 16.25 shows an example of a TSIC measurement for samples intentionally contaminated with sodium and potassium (Nauta & Hillen, 1978). The shape of the TSIC peak varies in the plots because of a contamination concentration difference throughout the dielectric. If the capacitor is prestressed at a constant voltage at a high temperature and then ramped down to room temperature under bias, the mobile ions can be driven to one interface, giving a clearer signal during the measurement. The TSIC current peaks at a specific temperature for each contaminant, with the potassium peak at a higher temperature than the sodium peak due to the larger atomic radius. The width of the peak indicates the spread in activation energies for the mobile ion contaminant. The activation energy can be extracted directly from the TSIC curve, determining the activation energy at each temperature of the temperature ramp (Mori et al., 1994).

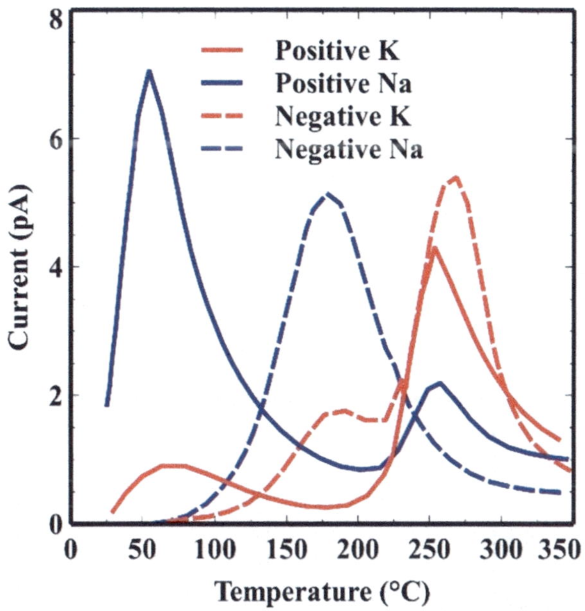

Fig. 16.25 TSIC curves for intentionally contaminated MOS capacitors measured at a positive and negative bias at 2.5 V for a 150 nm SiO_2 dielectric

16.10.7 Mobile Ion Detection Using Capacitance-Voltage Techniques

Simple MOS capacitor measurements are a valuable technique for measuring mobile ion contamination using the capacitance-voltage characteristics of an MOS capacitor by detecting shifts in the high to low capacitance transition (flatband) region of the C-V curve. The shift in the position of the capacitance along the voltage axis indicates the presence of mobile contamination. Gate metallization may block mobile ions introduced after the gate is formed from reaching the gate dielectric, preventing the detection of mobile ions.

A specific measurement algorithm shown in Table 16.6 is used to detect mobile ions. The electrical measurements typically need to be measured at room temperature because the semiconductor thermal generation can overwhelm the carrier concentration, hiding the depletion and inversion regions, and eliminating the ability to detect the threshold voltages in the measurements (See Fig. 16.17). This effect requires that the capacitor be measured at room temperature, with the bias across the capacitor maintained as the temperature decreases from the high-stress temperature to hold the mobile ions at either the semiconductor or gate interfaces, enabling detection. The capacitor is stressed at a high temperature under alternating constant field biases to drive the mobile ions to one interface.

Table 16.6 C-V measurement methods to detect mobile ions

Measure the initial C-V curve at room temperature.
Ramp wafer to high temperature and voltage stress for several minutes.
Ramp temperature back down to room temperature while holding voltage stress.
Measure the C-V curve at room temperature.
Ramp wafer to high temperature and voltage stress with opposite polarity for several minutes.
Ramp temperature back down to room temperature while holding voltage stress.
Measure the C-V curve at room temperature.

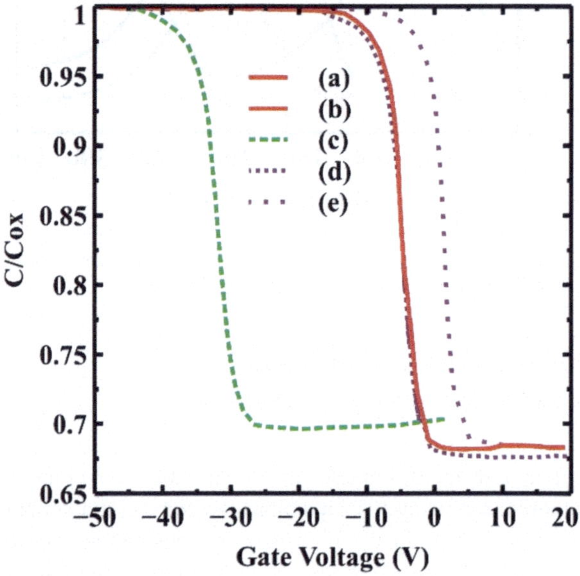

Fig. 16.26 Example C-V measurement for the detection of mobile ions for a 200 nm SiO_2 dielectric. (**a**) Initial curve before stress measurement. (**b**) After 5 min. at −10 V, 150 °C. (**c**) After 5 min. at +10 V, 150 °C. (**d**) After 5 min. at 150 °C with the terminals shorted (no bias). (**e**) A predicted theoretical curve is shown on the right side of the plot

Figure 16.26 shows an example of mobile ion detection using the C-V technique for a MOS capacitor (Snow et al., 1965). First, an initial curve (a) is generated, followed by a negative −10 V gate stress, shifting the curve (b) a negligible amount. Next, a positive +10 V gate stress is performed, generating a significant C-V curve shift (c) to the left. The large asymmetry in the positive and negative shifts indicates that the mobile ions were initially concentrated near the gate dielectric- semiconductor interface. Finally, an unbiased bake with the capacitor electrodes shorted diffuses the mobile concentration to a distribution close to its initial distribution (d). The mobile ion concentration can be directly determined from the C-V plot using the simple calculation $Q = C\ \Delta V/q$, where C is the capacitance measured midway on the y-axis, ΔV is the shift in gate voltage on the x-axis, and q is the electronic charge (1.602×10^{-19} C).

16.11 Mobile Ion Degradation of Nonvolatile Memories

For several reasons, nonvolatile memories (NVM), including 3D NAND, NOR, Flash, etc., are highly susceptible to mobile ions. First, very few electrons are stored as data in an NVM cell, as low as a few electrons per cell up to a few hundred electrons per cell (Prall, 2007). A similar number of mobile ions is sufficient to destroy the data stored in the cell. Second, multilevel cells are commonly used, dividing the small electron count into fractions for each data state (Aritome, 2016). Third, the charge is stored semi-permanently in an NVM cell where it can attract mobile ions even in an unbiased state, unlike other types of CMOS chips where there is no attractive force for mobile ions in an unpowered state. Mobile ions thermally diffusing through the dielectrics in an NVM CMOS chip are attracted to the charge storing the data state, and they can be captured by the charge trapping layer or floating gate. The trapped positive mobile ions near the floating gate compensate and neutralize electrons stored in the NVM cell, shifting the cell's threshold voltage and causing the data state to be corrupted. Additionally, NVM chips often use NVM cells to generate a reference current for data sensing. Mobile ion driven Vt shifts of the reference cells can cause all the data stored in the chip to be corrupted (Sim et al., 2005). Metadata containing data for many critical chip operations may also be stored in the NVM array.

The accidental charge stored on NVM cells due to plasma damage, as described in Chap. 8, can create the conditions for mobile ions to be attracted during wafer processing, affecting the NVM cell's behavior after fabrication. In addition, thermal processing steps can thermally erase the cells, leaving the NVM chip with the cells in a neutral, uncharged Vt state with the mobile ions embedded in the array, where they can quickly degrade the chip later while it is in operation.

Mobile ion contamination in nonvolatile memories has a unique and specific characteristic, making identification of mobile ion contamination straightforward (Liou et al., 2003). The interaction of the positive mobile ions with the negative electrons stored in the trapping layer or floating gate causes a recognizable behavior in the cell with a program, erase, and bake sequence applied to the cell.

Figure 16.27 shows typical sequences used in an NVM cell to detect mobile ions. During processing, the NVM cell often undergoes plasma damage unintentionally programming the cell, raising the V_t, by storing electrons in the trapping or floating gate layer, as shown in Fig. 16.27a. During subsequent thermal processing after the plasma damage step, thermal energy drives mobile ions through the dielectrics. As a result, positive mobile ions near the NVM cell are attracted to the negative charge in the cell and migrate onto the dielectrics surrounding the floating gate or trapping layer, as shown in Fig. 16.27b. The wafer may complete processing with both

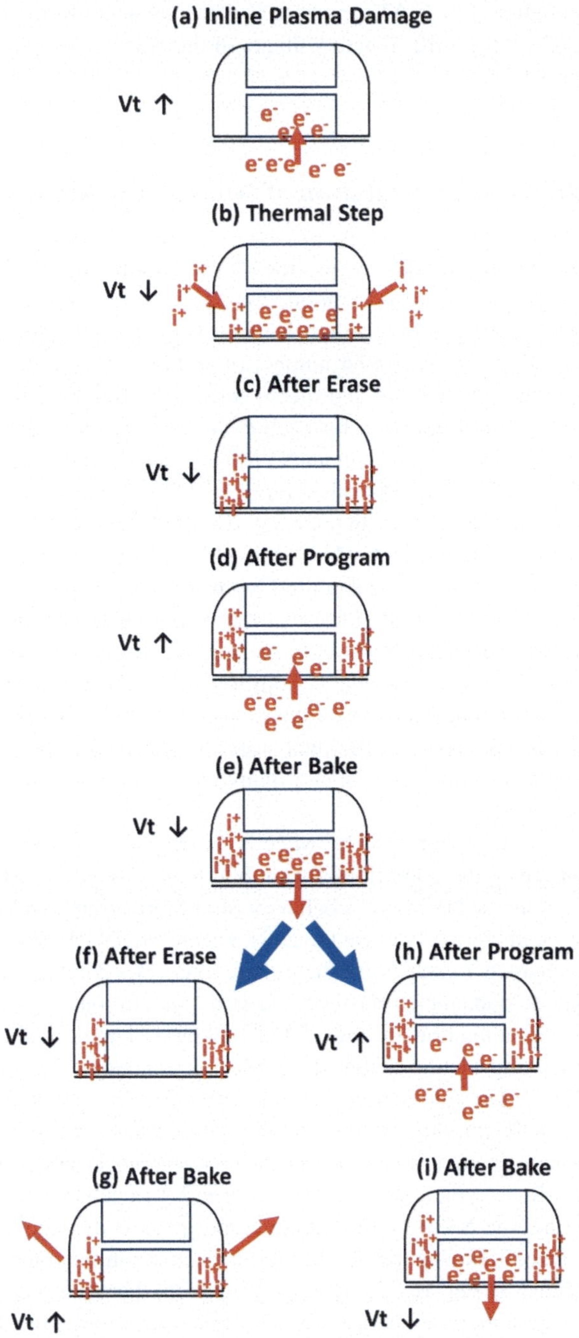

Fig. 16.27 Sketch of a NVM cell showing the interaction of electrons on the floating gate interacting with mobile ions. The UV steps (ultraviolet light exposure) shown in the sketch are typically performed with an electrical erase to remove electrons from the cell

electrons and mobile ions trapped in the cell, as shown in Fig. 16.27b. Typically, the first electrical operation on the cell erases the cell, removing electrons to put the cell in a low V_t state to achieve a known cell condition. Optionally, exposure to UV light can achieve the same result. The erase or UV step removes the electrons from the trapping or floating gate layer, leaving the mobile ions. As a result, the mobile ions are no longer attracted to the negative electrons and are more mobile (Fig. 16.27c). Next, in the sequence shown in Fig. 16.27d, the cells are programmed and baked (Fig. 16.27e).

Finally, in the sequence, some cells are programmed (Fig. 16.27h), and some cells are erased (Fig. 16.27f) in the same array in close proximity. Both the programmed and erased cells are then baked (Fig. 16.27g–i). The mobile ions are driven away from the erased cell by diffusion as they are no longer held by the negative electrons, resulting in an increased threshold voltage. The programmed cells lose electrons or mobile ions (or both), resulting in a decreased V_t. This unique behavior through the sequence shown is typically considered proof of mobile ion contamination.

Back-end-of-line processing temperatures (typically 400–500 °C) can provide enough thermal energy to cause mobile ion drift and diffusion. These temperatures may partially or entirely thermally erase the cells, thermally ejecting electrons from the region where they are trapped. Higher temperatures (~1000 °C) can be expected to completely erase the cells, removing all electrons from the trapping region.

The cell bake conditions used in this example were 24 h at 250 °C (Liou et al., 2003). The bake conditions need to be determined empirically for a given technology. Chip package thermal constraints may limit the maximum temperature.

The sequence of the threshold voltage increases and decreases through this testing procedure provides strong evidence of a mobile ion problem in an NVM process. The critical piece of evidence is a reduced threshold voltage when the NVM cell is baked in the programmed state and an increased threshold voltage when the NVM cell is baked in the erased state, indicating mobile ion drift toward the cell in the programmed state and mobile ion diffusion away from the cell in the erased state. Figure 16.28 shows an example of the test sequence shown in Fig. 16.27, demonstrating positive results of the NVM mobile ion test sequence where both charge gain and loss are observed for the final bake (Liou et al., 2003).

An additional signature of mobile ion contamination in NVM is that the data pattern or topology in an array can control the motion of mobile ions. For example, a data pattern in an array consisting of stripes of programmed and erased cells can shift the mobile ion distribution. The program stripes attract mobile ions, while erased stripes do not attract mobile ions, concentrating the mobile ions in the programmed stripes of the array. Figure 16.29 shows an x-y map of an intentionally contaminated NVM array generated by time-of-flight secondary ion mass spectroscopy (TOF-SIMS). The mobile sodium ions are concentrated in the striped regions containing programmed cells with electrons on the floating gate that attract the ions. The erased stripes of cells have a minimal number of electrons or net positive charge that does not attract mobile ions. As a result, the sodium atoms have migrated from the erased regions, increasing the mobile ion concentration in the programmed regions.

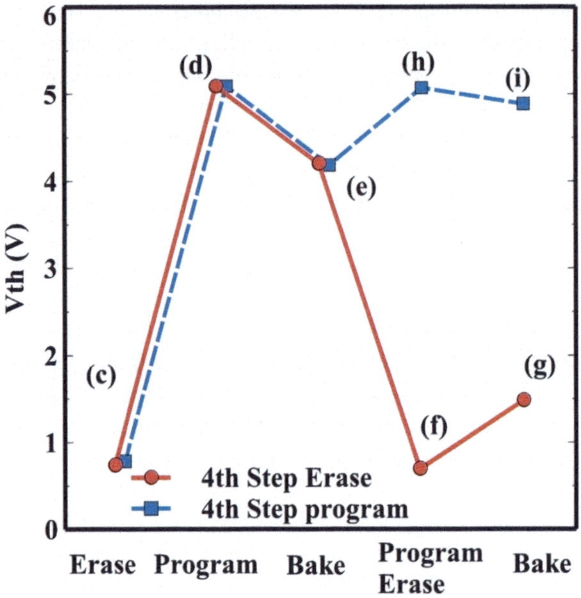

Fig. 16.28 Plot of NVM memory cell threshold voltage through a mobile ion detection sequence consisting of the following steps: Erase (electrical or UV), Program, Bake (250 °C 24 h), Program or Erase, Bake (250 °C 24 h). The letters show the sequence from Fig. 16.27

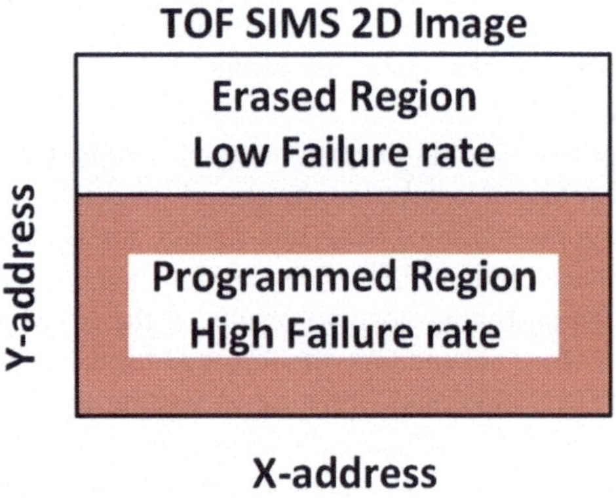

Fig. 16.29 Time of flight secondary ion mass spectroscopy (TOF-SIMS) X-Y image of a contaminated flash array programmed with a stripe pattern of programmed and erased cells baked at 250 °C showing a high concentration of sodium in the programmed region and a low concentration of sodium in the erased region (Kuge et al., 2016)

16.12 Die Seal Techniques to Prevent Mobile Ion Contamination

A CMOS chip has structural features designed to prevent mobile ion contamination from the outside world. The surface of the chip is covered with a mobile ion barrier passivation layer to keep ions, water, etc., out of the chip. Holes are etched into the passivation over the bond pads to allow electrical connection to the pads from the

Fig. 16.30 Scanning electron microscope (SEM) image of scribe line damage between two CMOS chips

outside world. A significant overlap of the bond pads (usually several microns) by the passivation limits mobile ion penetration around the pads.

The passivation at the edge of the chip is often damaged, where individual chips are singulated from the wafer for packaging. Lasers, saws, etc., are commonly used to singulate or dice the chips mechanically, creating structural damage in the scribe line adjacent to the chip edge. Figure 16.30 shows an SEM image of laser damage in a scribe line between chips (Marks et al., 2020). This kind of damage creates a significant amount of damage, generating a path for mobile ion ingress.

A continuous die-seal structure is added to the perimeter edge of the chip, creating a ring of protection surrounding the chip (Chen & Ker, 2005). A die seal typically attempts to create a solid wall of metal at the chip edge from the silicon surface to the final passivation, taking advantage of the fact that most metals provide a mobile ion barrier. Figure 16.31 shows an example sketch of a CMOS die-seal. As shown in the figure, continuous metal lines for all metal layers are routed around the chip perimeter. In addition, the contacts and vias are patterned as a continuous trench between the metal layers instead of the usual single square contact pattern (Hastings, 2005).

In this example, the die seal is connected to the substrate through a p+ diffusion. The die seal can be connected to ground, preventing the attraction of mobile ions toward the die seal. Alternatively, the die seal can be connected to an n + diffusion with the die seal biased at the positive Vcc, repelling mobile ions away from the chip.

16.13 Analytical Chemical Detection of Mobile Ions

Analytical chemical analysis (ACA) of mobile ion contamination problems is often used to detect mobile ion contamination in CMOS wafers for several reasons. ACA can be performed on simple test wafers without the more complicated process necessary to create electrical test structures, allowing the testing of individual tools and materials and isolating problem solving to a few process steps or tools. A vast spectrum of potential materials characterization techniques is available for detecting

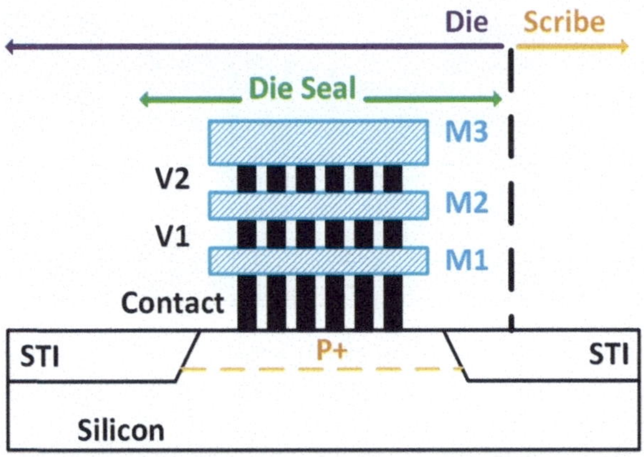

Fig. 16.31 Cross-section sketch of the edge of a CMOS chip showing the structure of a die seal to prevent migration of mobile ions and moisture into the chip

Table 16.7 Comparison of the sensitivity of common material analysis techniques in silicon

Technique	Maximum volume sensitivity (atoms/cm^3)	Maximum areal sensitivity (atoms/cm^2)
AES	5E20	
XPS	5E19	
Dynamic SIMS	1E13	
ICP-MS	1E11	
LEIS		1E10
TXRF		5E8
Static SIMS		1E7
VPD-ICPMS		1E6

mobile ions (Brundle et al., 1992). Some of the more common ones are shown in Table 16.7. The figure compares the measurement capability of the techniques in silicon (van der Heide, 2014, pg. 11). A description of the most common techniques used for mobile ion detection follows.

16.13.1 Secondary Ion Mass Spectrometry

However, a couple of methods are commonly used for many applications in CMOS chips, along with the detection of mobile ions. The various flavors of secondary ion mass spectrometry (SIMS) are one of the most commonly used techniques (van der Heide, 2014). It is a destructive technique that drills a hole in the wafer with an ion beam and analyzes the atoms ejected from the wafer. The exact atom and isotope can be determined and isolated, giving an accurate concentration. The beam or spot size can be relatively small in the range of 1–100 μm, allowing for good spatial

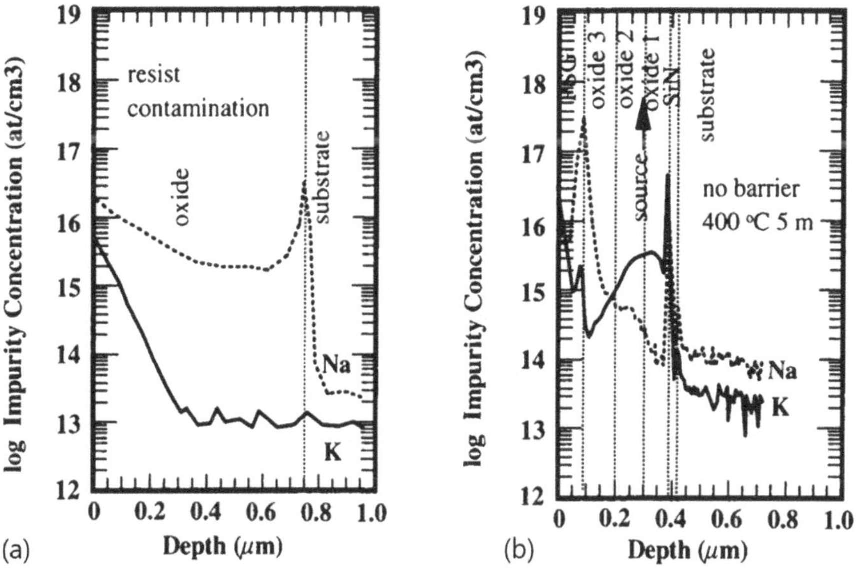

Fig. 16.32 SIMS plots showing mobile ion contamination in a wafer. (**a**) Sodium and potassium contamination in a resist layer on silicon. (**b**) Sodium and potassium contamination in several dielectric films deposited on the wafer after a 5 min. 400 °C anneal. The films were phosphorous silicate glass (PSG) on top, with three layers of oxide on top of a silicon nitride barrier. Samples were contaminated by resist processing and chemical mechanical polishing

resolution across the wafer, enabling an x-y plot across the wafer such as that shown in Fig. 16.29. In addition, SIMS can provide a depth profile into the wafer by analyzing the atoms ejected from the wafer as a function of the sputtering time. This advantage allows the determination of which films in a wafer are contaminated and the level of contamination versus the depth into the wafer. SIMS can provide an accuracy of sodium detection of about 5×10^{12} atoms/cm^2, which is comparable to the electrical sensitivity shown in Fig. 16.13 (van der Heide, 2014).

As an example of SIMS analysis, Fig. 16.32 shows SIMS mobile ion measurements in a CMOS wafer (Gambino et al., 1999). The concentration is plotted as a function of depth into the wafer. Figure 16.32a shows a dielectric contaminated by a resist layer on a wafer after the resist layer was removed, indicating that the resist layer is heavily contaminated. Figure 16.32b shows the mobile concentration in five layers of films deposited on a wafer with a 400 °C 5 min heat step measured from the front side of the wafer. The sodium and potassium contamination is distributed throughout the five films, but the left-most phosphorous silicate glass (PSG) film partially getters the mobile ions, trapping them into the film. The right-most silicon nitride film functions as a mobile ion barrier, blocking the penetration of mobile ions. The substrate's low contamination levels on the right side of the plot are likely the mobile ions' forward knock-on during the SIMS sputtering process. An additional SIMS measurement from the wafer backside is needed to determine if any mobile ions reached the silicon substrate (van der Heide, 2014).

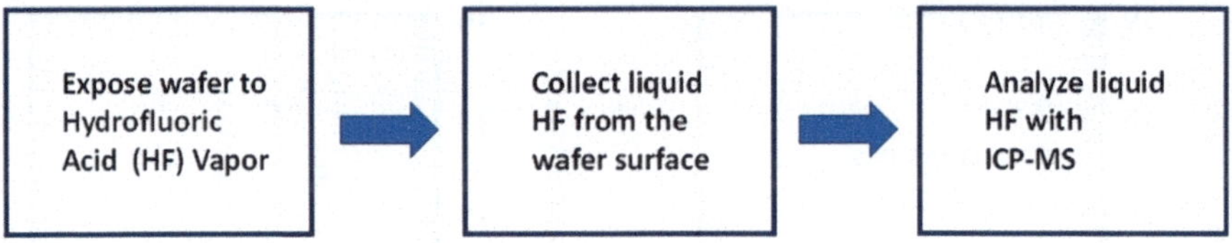

Fig. 16.33 Sketch showing the use of vapor phase decomposition to remove materials from the surface of a wafer for analysis

16.13.2 Vapor Phase Decomposition

Vapor phase decomposition (VPD) and inductively coupled mass spectroscopy (ICP-MS) used together are powerful techniques used to detect low-level contamination in the materials on a CMOS wafer. It is used widely in the semiconductor industry for routine monitoring of fabrication lines, qualification of process equipment, etc. It has parts per trillion sensitivity and more than nine orders of magnitude capability. It can detect almost everything on the periodic table (Boukhayma, 2018).

Vapor phase decomposition (VPD) is a technique used to remove materials from the surface of a wafer for analysis, as shown in Fig. 16.33 (Yonezawa et al., 2018). One method is to expose the wafer to a vapor such as hydrogen fluoride (HF) that etches the materials, such as native oxide, from the surface of the wafer into the vapor, followed by a transfer into an analysis system. A second method scans a chemical droplet across the wafer that etches the surface material. The droplet is transferred into the analysis system.

Inductively coupled plasma mass spectroscopy (ICP-MS) exposes the aerosolized sample to an argon plasma. The plasma ionizes all of the elements in the sample. The ions are analyzed by a mass spectrometer system that separates the ions by mass using a spectrometer. As a result, the concentration of each contaminant ion can be determined, yielding a sensitive measurement capable of detecting very low levels. Table 16.8 shows an example of the areal detection limits for an ICP-MS tool by element. Detection of sodium and potassium can be as low as $10^6/cm^2$ (Yonezawa et al., 2018).

16.14 Mobile Ion Mitigation

16.14.1 Historical Contamination

In the early days of MOS fabrication, after the discovery of mobile ion problems, many techniques were used to mitigate mobile ion contamination, in addition to eliminating the root cause of the contamination by removing mobile ions from the offending consumable or tool. Table 16.9 lists some common historical techniques to mitigate mobile ions by gettering the mobile ions or preventing ingress to the

Table 16.8 Example of ICP-MS detection limits (ions/cm^2)

Li	Na	Mg	Al	K	Ca	Ti	Cr	Mn	Fe
1.8E+05	1.1E+06	2.0E+06	3.0E+06	8.9E+05	3.3E+06	3.3E+06	1.6E+06	2.0E+05	3.0E+06
Ni	Co	Cu	Zn	Ge	Ba	Hf	Ta	W	Pb
7.9E+05	2.2E+06	8.1E+05	2.8E+06	2.7E+06	2.1E+06	1.9E+05	1.9E+05	1.9E+05	1.9E+05

Table 16.9 Common historical techniques to mitigate mobile ion contamination

Technique	Reference
Use of chlorine-containing species in furnace gas ambient for fabrication and tube cleaning	(Barbottin & Vapaille, 1986)
Double-walled furnace tubes are used to prevent mobile ion ingress.	(Barbottin & Vapaille, 1986)
Use of phosphorous silicate glasses and boron phosphorous silicate glasses to getter mobile ions	(Shideler, 1996)
Use of passivation layers to prevent mobile ion ingress into the chip	(Gambino et al., 1999)
Wet cleaning techniques to remove contamination from the wafer surface	(Anderson et al., 1995; Reinhardt & Kern, 2018)
Use of metal barriers to prevent ion migration	(Gupta, 2009)
Use of dielectric mobile ion barriers	(Gambino et al., 1999)

electrically active parts of the wafer. Some techniques, like the use of mobile ion barriers and passivation, wet cleans, etc., are still in use today. However, some techniques, like the use of chlorine, have been discontinued in advanced processes as they are no longer necessary and may have significant downsides.

16.14.2 Modern Mitigation

Mobile ion mitigation is an essential function of a semiconductor organization that is ingrained in all aspects of the operation, including incoming materials and equipment, monitoring and operating procedures, testing, qualification, etc. All consumables are screened and tested for contamination by the supplier and the semiconductor manufacturer using industry-wide standards such as those published by Semiconductor Equipment and Materials International (SEMI) (Reinhardt & Kern, 2018). These contamination control procedures are highly successful, resulting in the total elimination of the problem except for the very rare failure of these systems.

16.14.3 Mobile Ion Barriers

Hydrogen barriers are discussed in Chap. 14. Any barrier capable of preventing the ingress of single protons likely blocks the much bigger atoms such as copper, sodium, or potassium (see Chap. 14). The most common barrier is silicon nitride, which is often used for the final passivation in a CMOS chip, capable of blocking

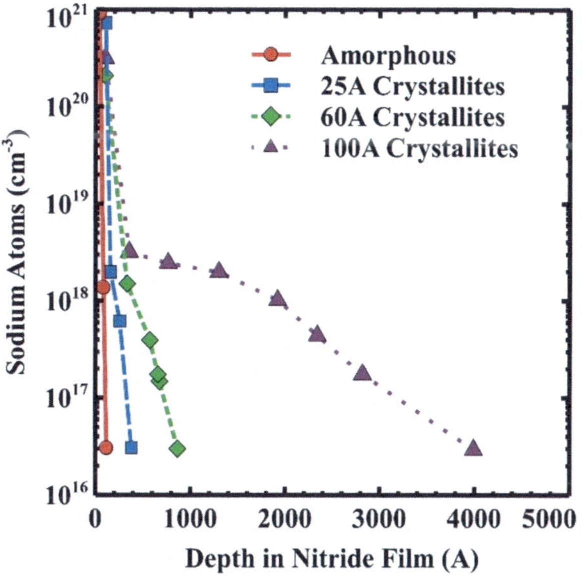

Fig. 16.34 Sodium diffusion through different silicon nitride barrier layers with different crystallinity shows increasing sodium penetration for more crystalline films. The sodium penetration into the silicon nitride film is measured by radioactive tracer profiling

mobile ions, as shown in Fig. 16.32. However, as Chap. 14 describes, silicon nitride may also cause detrimental effects by blocking or trapping hydrogen or water-related molecules. In addition, the crystallinity of the film affects the barrier properties, as shown in Fig. 16.34, where the amorphous film is the most effective mobile ion barrier, and the most crystalline film is the least effective barrier. This result indicates that mobile ion diffusion along grain boundaries dominates mobile ion migration.

Doped glasses such as boron phosphorus-doped silicate glass (BPSG) and phosphorous silicate glass (PSG) can getter mobile ions, trapping them inside the film (Liou et al., 2001). Figure 16.35 shows a SIMS plot of the sodium concentration versus depth for various films in a flash memory process. The sodium is trapped in the PSG film and mobile ion diffusion into the films is limited. Also, the sodium is blocked in the nitride layer, although the exact location is not clear.

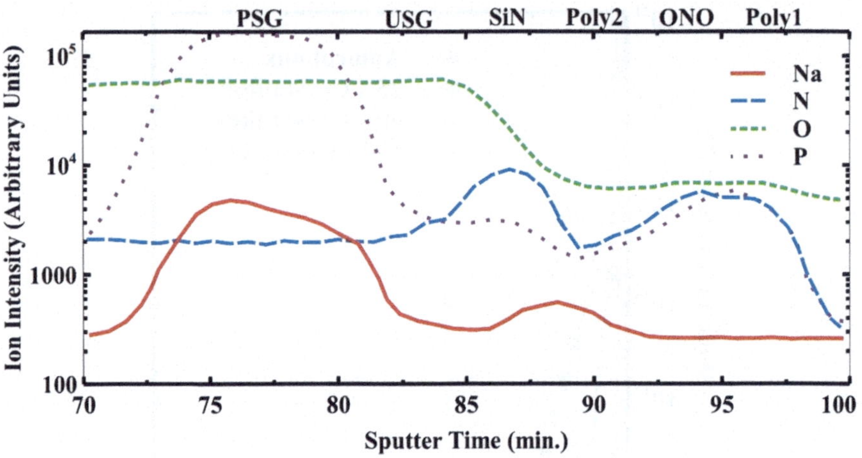

Fig. 16.35 SIMS plot showing sodium concentration in various films in a non-volatile memory process as a function of SIMS sputtering time. The films are phosphorous silicate glass (PSG), undoped silicate glass (USG), silicon nitride (SiN), polysilicon (poly2), oxide nitride oxide (ONO), and polysilicon (poly1)

References

Anderson, L., et al. (1995). A case study in a 100X reduction in sodium ions in a 0.8um BiCMOS process using triangular voltage sweep. *International Integrated Reliability Workshop*. https://doi.org/10.1109/IRWS.1995.493574

Aritome, S. (2016). Nand flash memory technologies. *IEEE Press*. https://doi.org/10.1002/9781119132639.ch3

Barbottin, G., & Vapaille, A. (1986). *Instabilities in silicon devices silicon passivation and related instabilities* (Vol. 1). North Holland.

Boukhayma, A. (2018). *Ultra low noise CMOS image sensors*. Springer. https://doi.org/10.1007/978-3-319-68774-2

Brundle, R. C., Evans, C. A. J., & Wilson, S. (1992). *Encyclopedia of materials characterization*. Butterworth-Heinemann.

Chen, C. H., & Ker, M. D. (2005). Investigation on seal-ring rules for IC product reliability in 0.25-um CMOS technology. *Microelectronic Reliability, 45*, 1311–1316. https://doi.org/10.1016/j.microrel.2005.07.012

Chen, F. F. (2012). Experimental confirmation of electron fluence driven, Cu catalyzed interface breakdown model for low-k TDDB. In *IEEE international reliability physics symposiyum*. https://doi.org/10.1109/IRPS.2012.6241802

Chonko, M. A., et al. (1994). Measurement of mobile ion contamination in multilayer metallization by triangular voltage sweep. *Journal of the Electrochemical Society, 141*(7), 1862–1866. https://doi.org/10.1149/1.2055018

Ciofi, I., et al. (2006a). Detection of copper and water in low-k dielectrics by triangular voltage sweep measurements. In *Materials Research Society Symposium - Proceedings*. https://doi.org/10.1557/PROC-0914-F02-02

Ciofi, I., et al. (2006b). Water and copper contamination in SiOC:H Damascene: Novel Characterization Methodology based on Triangular Voltage Sweep Measurements. Int. Interconnect Technology Conference, https://doi.org/10.1109/IITC.2006.1648682.

Coffa, S., et al. (1987). Diffusion and solubility of gold implanted in silicon. In *European solid state device research conference*.

Gambino, J. P., et al. (1999). Diffusion barriers for mobile ions in 256M DRAMS. In *Materials Research Society Symposium - Proceedings*. https://doi.org/10.1557/PROC-564-229

Gupta,T.(2009).*Copperinterconnecttechnology*.Springer.https://doi.org/10.1007/978-1-4419-0076-0

Gutai, L. (1997). Pulsed bias temperature stress—An accurate and fast techniques to determine Mobile ion concentrations in gate and field oxides. *IEEE International Reliability Workshop*. https://doi.org/10.1109/IRWS.1997.660294

Hastings, A. (2005). *The art of analog layout* (2nd ed.). Pearson.

He, M., & Lu, T.-M. (2012). *Metal-dielectric interfaces in gigascale electronics—Thermal and electrical stability*. Springer. https://doi.org/10.1007/978-1-4614-1812-2

Hofstein, S. R. (1966). An investigation of instability and charge motion in metal-silicon oxide - silicon structures. *IEEE Transactions on Electron Devices, 13*(12), 222–237. https://doi.org/10.1109/T-ED.1966.15674

Hofstein, S. R. (1967). Proton and sodium transport in SiO2 films. *IEEE Transactions on Electron Devices, 14*(11), 749–759. https://doi.org/10.1109/T-ED.1967.16102

Huang, C., et al. (2014). The observation of Mobile ion of 40nm node by triangular voltage sweep. In *International symposiyum on the physical and failure analysis of integrated circuits*. https://doi.org/10.1109/IPFA.2014.6898164

Istsumi, M., et al. (2002). *SiO2 in Si microdevices*. Kodansha and Springer.

Kothandaraman, C., & Cohen, S. (2016). Triangular Voltage Sweep (TVS) characterization for Though Silicon Via (TSV) reliability. In *IEEE international reliability physics symposiyum*. https://doi.org/10.1109/IRPS.2016.7574563

Kuge, N., et al. (2016). Investigation into behavior of mobile ions in storage device using ToF-SIMS. In *International symposiyum on semiconductor manufacturing*. https://doi.org/10.1109/ISSM.2016.7934524

Kuhn, M., & Silversmith, D. J. (1971). Ionic contamination and transport of mobile ions in MOS structures. *Journal of the Electrochemical Society, 118*(6), 966–970. https://doi.org/10.1149/1.2408233

Lifshitz, N., & Smolinsky, G. (1989). Detection of water-related charge in electronic dielectrics. *Applied Physics Letters, 55*(4), 408–410. https://doi.org/10.1063/1.101570

Liou, J.-W., et al. (2001). Plasma charging and mobile ions on the data retention of 0.25um flash memory. In *6th international symposiyum on plasma process induced damage*. https://doi.org/10.1109/PPID.2001.929982

Liou, J. J.-W., et al. (2003). Characterization of process-induced mobile ions on the data retention in flash memory. *IEEE Transactions on Electron Devices, 50*(4), 995–1000. https://doi.org/10.1109/PPID.2001.929982

Marks, M. R., et al. (2020). Mechanism study of SiO2 layer formation and separation at the Si Die sidewall during nanosecond laser dicing of ultrathin Si wafers with Cu backside layer. *Applied Physics A, 126*. https://doi.org/10.1007/s00339-020-3322-1

Mori, H., et al. (1994). Analysis of isolation degradation induced by interlayer material in Capacitor Over Bit-Line (COB) DRAM cell. In *International conference on solid state devices and materials*. https://doi.org/10.7567/SSDM.1994.A-12-1

Nauta, P. K., & Hillen, M. W. (1978). Investigation of mobile ions in MOS structures using the TSIC method. *Journal of Applied Physics, 49*(5), 2862–2865. https://doi.org/10.1063/1.325168

Ohring, M. (2015). *Reliability and failure of electronic materials and devices* (2nd ed.). https://doi.org/10.1016/C2009-0-05748-1

Parent, D. W., et al. (2010). High-K dielectric fabrication process to minimize mobile ionic penetration. In *32nd annual international conference of the IEEE EMBS*. https://doi.org/10.1109/IEMBS.2010.5627801

Prall, K. (2007). Scaling non-volatile memory below 30nm. In *IEEE non-volatile semiconductor memory workshop*. https://doi.org/10.1109/NVSMW.2007.4290561

Reinhardt, K. A., & Kern, W. (2018). *Handbook of silicon wafer cleaning technology* (3rd ed.). Elsevier William Andrew Applied Science Publishers. https://doi.org/10.1016/C2016-0-01001-X

Sabin, E., et al. (1997). Using Iddq drift testing to detect hydrogen in MOS devices. In *IEEE international reliability physics symposium*. https://doi.org/10.1109/RELPHY.1997.584237

Schroder, D. K. (2006). *Semiconductor material and device characterization* (3rd ed.). Wiley Interscience. https://doi.org/10.1002/0471749095

Schroen, W., et al. (1965). Accumulation and decay of mobile surface charges on insulating layers and relationship to reliability of silicon devices. In *IEEE annual symposium on the physics of failure in electronics*. https://doi.org/10.1109/IRPS.1965.362327

Shideler, J. A. (1996). Is BPSG an effective sodium barrier? A WLR success story. In *International reliability workshop*. https://doi.org/10.1109/IRWS.1996.583386

Sim, S.-P., et al. (2005). Anomalous charge loss of reference cell in MLC flash memory due to process-induced mobile ion. *ESSDERC*. https://doi.org/10.1109/ESSDER.2005.1546650

Snow, E. H., et al. (1965). Ion transport phenomena in insulating films. *Journal of Applied Physics, 36*(1), 1664–1673. https://doi.org/10.1063/1.1703105

Stauffer, L., et al. (1995). Mobile ion monitoring by triangular voltage sweep. *Solid State Technology, 38*, S3–S8.

van der Heide, P. (2014). *Secondary ion mass spectrometry—An introduction to principles and practices*. Wiley. https://doi.org/10.1002/9781118916780

Williams, C. K., et al. (1995). Low temperature diffusion of alkali earth cations in thin, vitreous SiO2 films. *Journal of the Electrochemical Society, 142*(1), 303–311. https://doi.org/10.1149/1.2043915

Wilson, M., et al. (2006). The present status and recent advancements in Corona-Kelvin non-contact electrical metrology of dielectrics for IC-manufacturing. *ECS Transactions, 3*(3), 3–24. https://doi.org/10.1149/1.2355694

Wojick, H., et al. (2011). Comparison of PVD, PECVD & PEALD Ru(-C) films as cu diffusion barriers by means of bias temperature stress measurements. *Microelectronic Engineering, 88*, 641–645. https://doi.org/10.1016/j.mee.2010.06.034

Yonezawa, S., et al. (2018). Development of extreme microanalysis technology for metallic impurity on a silicon wafer surface. In *International symposiyum on semiconductor manufacturing*. https://doi.org/10.1109/ISSM.2018.8651161

Chapter 17
Fixed Charge

17.1 Introduction

Fixed charge is the term used to describe a charge semi-permanently trapped in dielectrics. This type of charge storage is described as an electret, where the charge is stored in the bulk of the dielectric instead of on the conducting plates in a capacitor.

Fixed charge problems are not common in CMOS technology, but they occur occasionally, and understanding the subject matter is an important topic for an engineer's tool kit. Fixed charge problems tend to occur in thicker oxides, such as shallow trench isolation (STI), due to the larger volume of the dielectric. Thinner dielectrics, such as gate dielectrics, are more immune to fixed charge problems, usually containing a small amount of charge, which typically causes a slight threshold voltage shift. Some of the accumulated fixed oxide charge know-how in the literature is obsolete as it no longer applies to advanced CMOS technology. However, it is difficult to predict what will be important in the future, and this topic is included in this book.

Fixed oxide charge tends to be very stable in CMOS wafers and is unaffected by electrical stress. It tends to be unaffected by thermal stress below 400–500 °C. Fixed oxide charge is not mobile like mobile ions; mobile ion tests can detect no charge motion in the case of fixed charge. Fixed charge does not recover with time after exposure (Grove, 1967). Charge injected into dielectrics by electrical stress tends to recover, unlike fixed oxide charge.

Fixed charge often correlates with the interface state density. The interface state density may dominate the electrical results (see Chap. 5). Optimization of thermal anneals and hydrogen alloys is often required to minimize the interface state density to the point where fixed oxide charge can be studied (Nicollian & Brews, 1982, pg. 778). As an example, Fig. 17.1 shows a plot highlighting the correlation between interface state density and fixed charge density, indicating common mechanisms (Hori, 1997, pg. 157).

K. Prall, *CMOS Plasma and Process Damage*,
https://doi.org/10.1007/978-3-031-89029-1_17

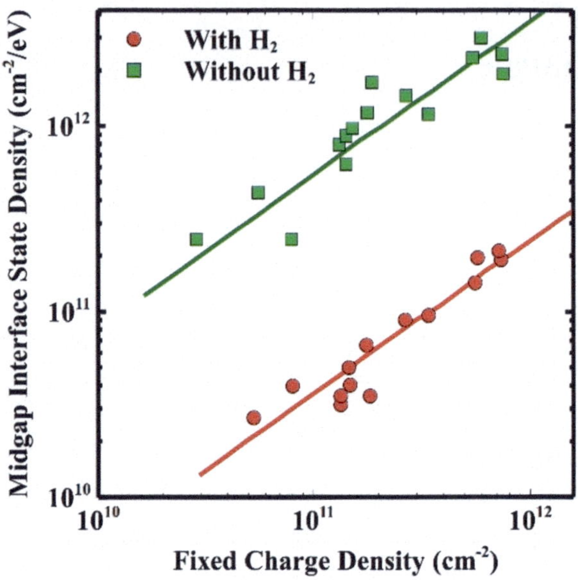

Fig. 17.1 Midgap interface state density plotted as a function of fixed charge density for silicon dioxide with and without hydrogen annealing (alloying)

17.1.1 Electrical Symptoms of Fixed Oxide Charge

Fixed charge occasionally causes electrical problems in CMOS chips, generally through threshold voltage shifts, affecting first isolation devices, as described in Chap. 5. Large amounts of fixed charge can shift active transistors, affecting circuits through threshold voltage shifts. Figure 17.2 shows a sketch of a capacitance-voltage plot showing a shift to the left with a positive fixed charge (Q_{total}) on an n-type wafer (Boogaard, 2001. pg. 99). Fixed charge can be detected on short-loop capacitor wafers, full-flow CMOS wafers, or test wafers with a dielectric on the surface. In addition, fixed charge can be detected with the inline measurement techniques shown in Chap. 10.

The fixed charge can be positive or negative depending on the materials used and the dielectric processing conditions. Optimized processes create a minimal amount of fixed charge with a negligible impact on the CMOS chip.

Specific processes and materials or combinations of process steps typically cause fixed charge problems. For example, the final thermal steps in a CMOS process flow can cause problems as the charge is created and cannot be removed at lower temperatures.

The origin of fixed oxide charge is attributed to multiple mechanisms, including excess silicon in the film, impurities, oxygen vacancies, dipoles, etc. (Barbottin & Vapaille, 1989, pg. 241–242).

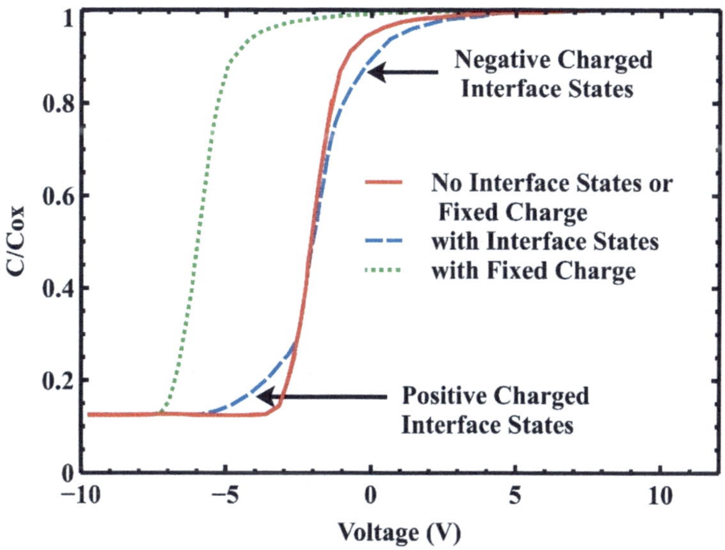

Fig. 17.2 Sketch of a high-frequency capacitance-voltage (C-V) plot showing the shift due to fixed oxide charge. The curve shifts to the left for a positive fixed charge for an n-type wafer. The curve is normalized to accumulation capacitance. The effect of positively and negatively charged interface states is shown

The shallow trench isolation (STI) in a CMOS chip commonly fails first when a fixed charge problem occurs, shorting all devices together and resulting in excessive leakage at the chip level. The isolation leakage can occur at the field edge, as described in Chap. 5, or the entire field region can be inverted without an edge signature. At the product level, isolation failure may cause standby current problems.

Failing isolation can cause anomalous behavior of device structures in the chip as the parametric measurements detect the isolation leakage instead of the expected behavior. Diode breakdown and leakage can increase or decrease depending on the polarity of the fixed charge, as the STI leakage dominates the junction properties. Contact string measurements may show an unexpected reduction in contact resistance as current flows through the isolation, reducing the resistance over the normal current path through the contact string. These are just a few examples of abnormal parametric results caused by isolation inversion and failure. Many other possibilities exist.

Typically, the threshold voltages of the NMOS and PMOS isolation devices shift in opposite directions depending on the polarity of the fixed charge. A positive fixed charge increases the absolute value of the PMOS field Vt and decreases the NMOS field Vt, while a negative fixed charge increases the NMOS field Vt and decreases the absolute value of the PMOS field Vt.

An important signature of fixed charge isolation failure is that field devices covered with a polysilicon or metal gate often do not shift and fail, while ungated field devices and field devices covered with metal (not gate metal) shift and fail. The explanation is that the gate protects the field device from the electric field generated by the fixed charge above the gate or blocks the diffusion of the atoms responsible for the fixed oxide charge so that they do not reach the substrate. This symptom also indicates that the source of the fixed charge originates from the process steps after the gate metal is formed. If all isolation devices fail (gate metal, no gate, interconnected metal-gated), it indicates that the source of the fixed oxide charge originates from process steps before the gate metallization is formed.

Shifts in the electrical behavior of the active transistors often are not observed while the isolation fails. This result is because the gate metal also blocks the electrical field or blocks diffusion in the same manner as the isolation devices. The thin dielectrics used for the active transistors are much less susceptible to fixed charge than the thick isolation dielectrics, as shown in Chap. 16 for mobile ions. Fixed charge has an identical dielectric thickness sensitivity as that of mobile ions described in Chap. 16.

17.2 Wafer Crystal Orientation

The wafer crystal orientation is a significant factor in fixed charge generation (Deal, 1967). The fixed charge as a function of orientation for common silicon crystal orientations follows in decreasing amount of fixed charge as the crystal orientation changes from <111> to <110> to <100>. The <111> orientation has the highest amount of fixed charge. The <110> orientation has an intermediate amount of fixed charge. The lowest amount of fixed charge occurs for the <100> crystal orientation. The ratios of fixed oxide charge for <111>, <110>, and <100> are approximately 3:2:1.

CMOS wafers universally use starting material wafers with a <100> orientation because of the reduced amount of fixed charge, among other reasons such as mobility. For example, Fig. 17.3 shows the fixed charge generated by the oxidation of silicon for two different substrate orientations (Deal, 1967). Wafers are oxidized in two different conditions (920 °C and 1200 °C). For the <111> crystallization orientation, the fixed oxide charge increases substantially for thicker oxides. However, for a <100> crystal orientation, the fixed oxide charge is constant as a function of oxide thickness.

The fixed charge in the dielectric as a function of crystal orientation correlates with the interface state density, as shown in Fig. 17.4. Both defects are related to the bond density difference between crystal planes in the silicon lattice (Hori, 1997, pg. 158). The plot shows the advantages of using <100> wafers in CMOS technology to minimize the fixed oxide charge. Interface states are covered in Chap. 5.

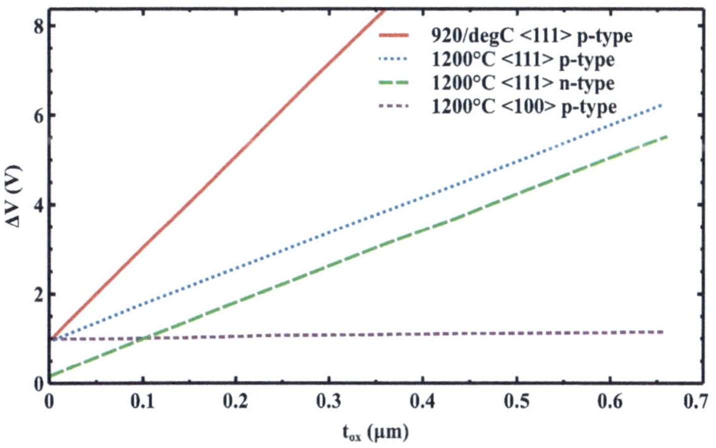

Fig. 17.3 Plot showing the shift in threshold voltage (ΔV) of a MOS capacitor as a function of the oxide thickness (t_{ox}) for <111> and <100> wafer crystal orientations for different oxidation conditions

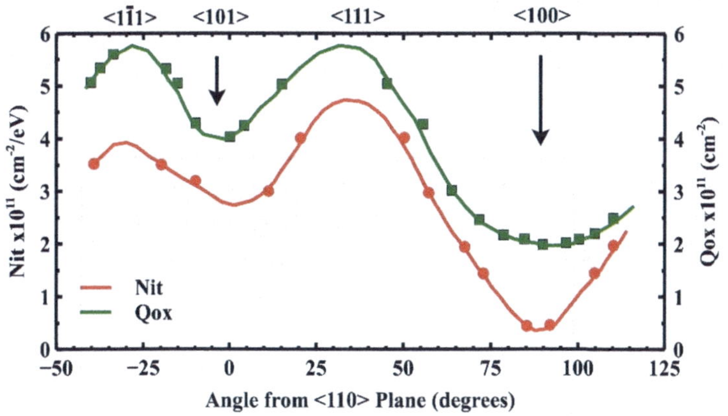

Fig. 17.4 Plot showing fixed charge (Q_{ox}) and interface state density (N_{it}) as a function of crystal orientation

17.3 Thermal Silicon Dioxide Fixed Charge

Historically, silicon dioxide has been widely used in wafer processing, and some of the first fixed charge problems were discovered in SiO_2. Thermally grown silicon dioxide often contains fixed charge that is usually positive. The oxidation conditions for the growth of silicon dioxide on a wafer play a substantial role in the generation of fixed charge. There is a significant difference in the fixed charge as a function of the oxidation temperature, as shown in Fig. 17.3, where 1200 °C growth conditions have less fixed oxide charge than 920 °C. The fixed charge as a function of temperature is shown in Fig. 17.5 for thermal oxidation (Grove, 1967, pg. 343). The fixed charge for dry oxidation is a strong function of temperature. The fixed charge for wet oxidation is a weaker function of temperature. Lower temperatures tend to be worse for fixed oxide charge. Also, there is a weak dependence on p- and n-type wafers, with p-type wafers having slightly more fixed oxide charge.

Post-oxidation annealing can affect the fixed oxide charge concentration. Annealing in a nitrogen or noble gas ambient can reduce the fixed oxide charge to lower levels, as shown in Fig. 17.6. The fixed oxide charge concentration is reversible. Wet or dry oxidation, followed by an anneal has a low fixed oxide charge. Subsequent oxidation again increases the fixed oxide charge (Deal, 1967). This effect can be repeated with cycles of oxidation and annealing.

The final thermal steps in a CMOS process flow determine the fixed oxide for the completed wafers. Therefore, optimizing the final steps to minimize the fixed oxide charge is essential to achieve optimum device performance. In addition, anneals often follow dielectric deposition and growth steps to improve the film quality, and the anneals often change the fixed charge density of the films. For example, Fig. 17.6 shows the fixed charge density for a 35 nm thermal oxide grown and annealed at 975 °C in N_2. Initially, the fixed charge density begins to decrease as the oxide is annealed, reaching a minimum at around 20 minutes. For times beyond 20 minutes, the fixed oxide charge increases. The increased charge with increased anneal time is attributed to the nitridation of the dielectric and silicon surface in the N_2 ambient.

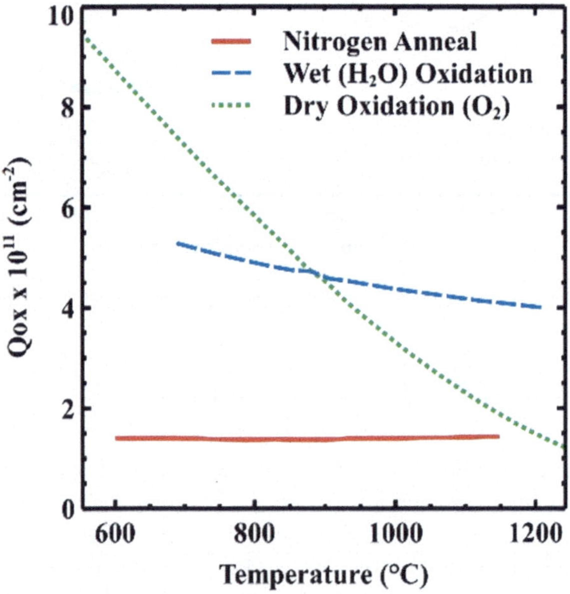

Fig. 17.5 Plot showing fixed oxide charge density as a function of temperature for dry (O_2 only) and steam oxidation (H_2O) for a silicon dioxide dielectric grown on a <111> substrate. The fixed charge is reduced with a post-oxidation nitrogen anneal

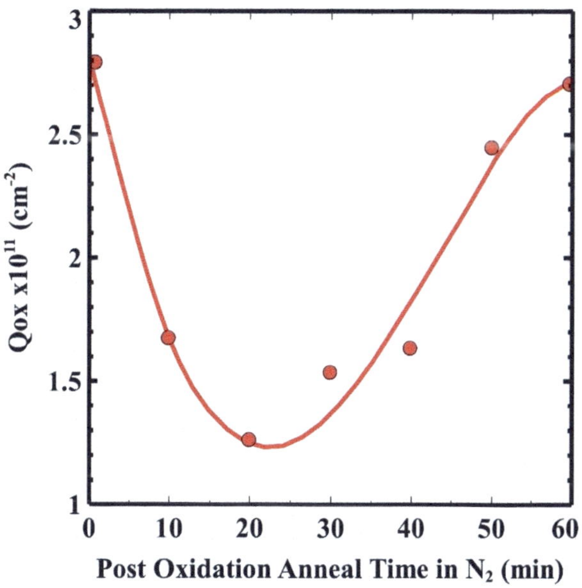

Fig. 17.6 Fixed oxide charge as a function of the post-oxidation anneal time in nitrogen for a 35 nm thermal oxide grown at 975 °C and annealed at 975 °C. The zero time point is an unannealed control

Anneals with noble gases such as argon do not show an increased charge with increased time (Vyas et al., 1982).

Historically, excursions during wafer manufacturing where the wafers are accidentally exposed to oxygen or room air during high-temperature processing can cause isolation failure by shifting the threshold voltage to the point where the isolation leakage becomes catastrophic. Leaks in processing tools can occasionally cause fixed charge problems.

Fixed oxide charge in thermally oxidized films is a function of the stress in the film, as shown in Fig. 17.7, where higher temperature growth conditions yield a film with increased stress and a reduced fixed oxide charge (Barbottin & Vapaille, 1989, pg. 242).

17.4 Fixed Charge in Deposited Dielectrics

17.4.1 Introduction

In addition to thermally grown silicon dioxide, silicon dioxide is commonly deposited using several different techniques. There are multiple advantages to using deposited oxides. Low-temperature processes have a reduced impact on the CMOS transistors. Low-temperature processes are used for intermetal dielectrics where

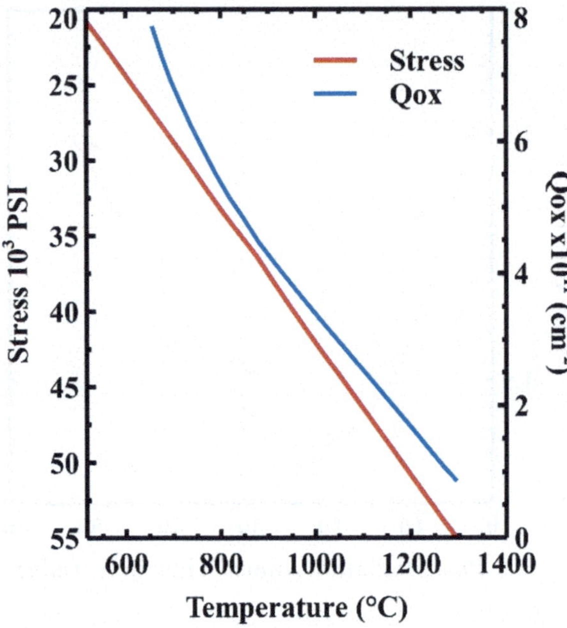

Fig. 17.7 Plot showing the relationship between stress and fixed oxide charge density as a function of the film stress and growth temperature

Table 17.1 Common dielectric films with reported fixed oxide charge problems

Film	Reference
Spin-on-glass (silicate and siloxane) (SOG)	Pramanik et al. (1989)
Plasma-enhanced tetraethyl orthosilicate glass (PETEOS)	Boogaard (2001)
Boron phosphorus tetraethyl orthosilicate glass (BPTEOS)	Mori et al. (1994)
Ozone-tetraethyl orthosilicate glass	Butler et al. (1990)

high temperatures cannot be used without damaging the metal on the wafer. Higher deposition rates can be obtained with deposited dielectrics such as plasma-enhanced chemical vapor deposition (PEVCD), resulting in low costs. Deposited dielectrics can have high step coverage properties, allowing gap fill between features.

Deposited dielectrics often differ substantially from thermal oxide with different bonding, composition, contamination, etc., which can change the fixed oxide charge in the films. Table 17.1 shows some common silicon dioxide-related processes reported to have fixed oxide charge problems (Enigmatic, 2023). Fixed charge problems in deposited dielectrics occur under specific process conditions and do not always occur.

Fixed charge incorporation into dielectrics is often accidental. Sometimes, fixed charge is introduced intentionally into a film to influence the electrical properties of CMOS devices. For example, fixed charge optimization is used to optimize the electrical performance of through-silicon vias (TSVs), as described in Sec. 17.8. It is crucial to determine if the fixed charge is stable and nonvariant through time, temperature, and electric field to eliminate reliability risks.

17.4.2 Fixed Charge Caused by Carbon Incorporation

Some deposited oxides use organic precursors containing carbon. Examples of an organic precursor include tetraethyl orthosilicate glass (TEOS) using the precursor tetraethoxysilane ($Si(OC_2H_5)_4$), which is commonly deposited using plasma-enhanced chemical vapor deposition (PECVD). Its uses include dielectrics between metal layers, passivation, etc. Another example is spin-on-glass processes (SOG). SOG is a common deposition method used for gap fill to planarize wafers over surface structures such as shallow trench isolation, gate metal patterns, and metallization. Many different types of dielectric materials can be deposited by SOG processes, including low-dielectric constant films (Baklanov et al., 2007, pg. 33). SOG chemicals commonly contain carbon-based or organic precursors. Non-organic films are carbon-free but can also have fixed charge problems.

Carbon can be a source of fixed charge in CMOS wafers. Carbon can diffuse through silicon (Chap. 15) and silicon dioxide, particularly at higher temperatures. Carbon has been shown to pile up at the silicon interface, where it can generate a fixed charge (Mori et al., 1994). For example, Fig. 17.8 shows a secondary ion mass spectroscopy (SIMS) profile of carbon diffusing from a deposited PECVD TEOS

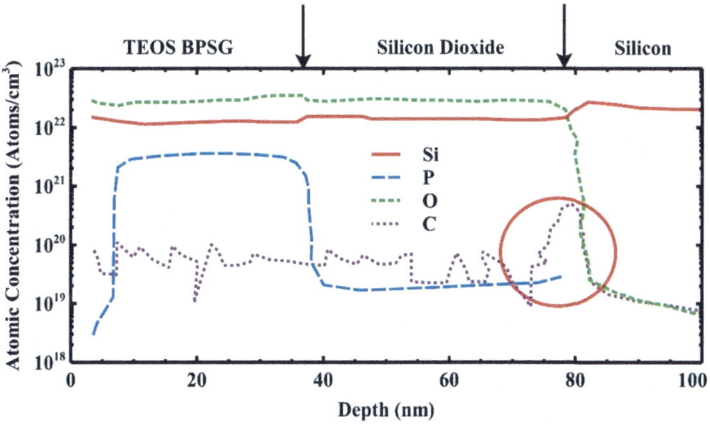

Fig. 17.8 Secondary ion mass spectroscopy (SIMS) of carbon migration from a PECVD TEOS Boron phosphorous silicate glass (BPSG) film into a silicon dioxide film after an anneal. Carbon pileup at the SiO_2-Si interface is shown after a 1000 °C anneal (highlighted by the red circle)

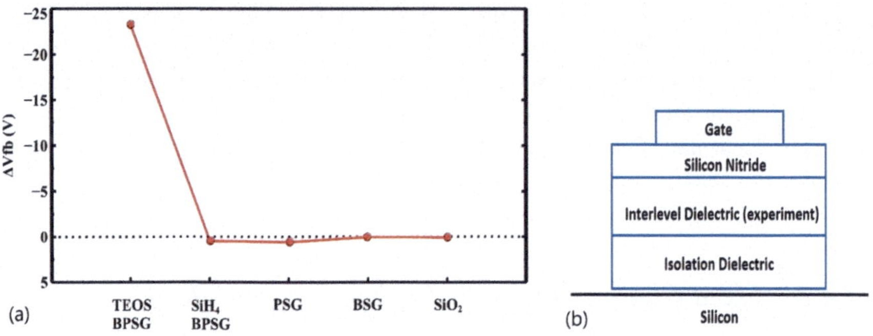

Fig. 17.9 (**a**) Plot showing the flatband voltage shift measured on a MOS capacitor for various deposited dielectrics on top of a silicon dioxide layer under a silicon nitride layer comparing TEOS BPSG, silane BPSG, phosphorous doped silicate glass (PSG), boron doped silicate glass (BSG), and silane-based silicon dioxide. (**b**) Cross-section of the structure used in this experiment

BPSG film into an underlying silicon dioxide layer. The carbon piles up at the silicon dioxide silicon interface after a 1000 °C anneal, forming a peak of heavy concentration (circled in red). The carbon concentration at the silicon surface in the annealed sample shifted the flatband voltage by around 35 volts relative to that of the unannealed sample. This level of fixed charge would be catastrophic to a CMOS chip, causing a total loss of isolation.

As an example of fixed charge problems in films deposited with organic precursors, Fig. 17.9 compares the fixed oxide charge V_t shift for organic and inorganic films (Mori et al., 1994). The TEOS BPSG-based organic film has a sizeable flatband voltage shift of over 20 volts. All the other films deposited in the experiment used inorganic silane (SiH_4) as the precursor and had a negligible flatband voltage shift, indicating a minimal fixed charge.

17.4.3 Interaction with Hydrogen and Diffusion Barriers

Fixed charge problems such as the one shown in Fig. 17.8 are heavily influenced by the presence of a diffusion barrier like silicon nitride which can block the diffusion of carbon, hydrogen, etc. The presence of a diffusion barrier can trap the carbon in the wafer, increasing the concentration of carbon at the surface and, subsequently, the fixed charge, as shown in Fig. 17.10b (Mori et al., 1994). The presence of the barrier prevents carbon from outgassing from the wafer during thermal steps, trapping it in the wafer. Figure 17.10a shows the nitride layer's impact on the isolation failure. The addition of the nitride layer causes the isolation layer to electrically fail at a larger dimension than when the nitride layer is absent. The threshold voltage of the isolation device drops toward zero voltage with decreasing dimensions.

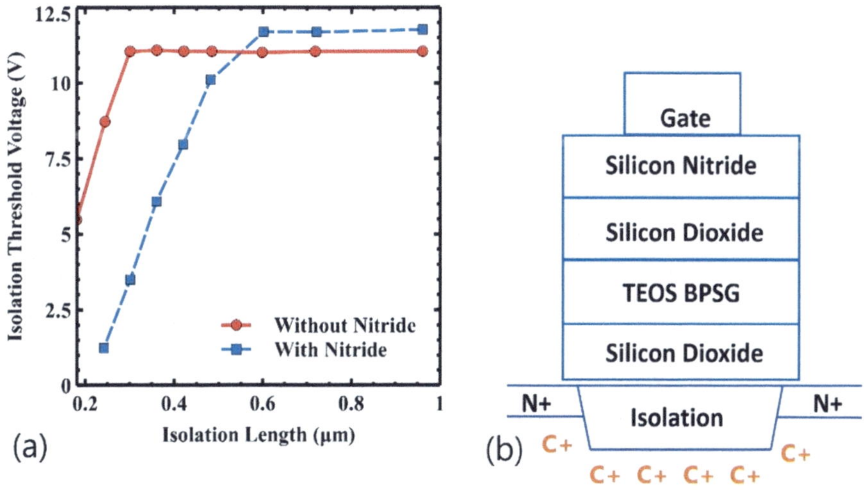

Fig. 17.10 Sketch and plot showing the interaction of carbon diffusion with a silicon nitride layer. (**a**) Plot of threshold voltage plotted against isolation length with and without a silicon nitride capping layer. (**b**) Cross-section sketch showing the presence of silicon nitride layer prevents the carbon from out diffusing from the wafer allowing carbon to be trapped at the dielectric silicon interface

In addition to trapping carbon in the wafer, silicon nitride may emit hydrogen into the underlying films or trap water and hydrogen, as described in Chap. 14. This may impact the electrical isolation performance.

17.4.4 Fixed Charge in Spin-On Dielectrics

Fixed charge has also been reported in spin-on dielectrics (SOD) with similar results as the previous sections show. Both organic and inorganic (SOD) dielectrics have been reported to have fixed oxide charge problems (Hsu et al., 1993; Pramanik et al., 1989). Interactions between SOD films and TEOS-deposited dielectrics and interactions with nitride passivation and hydrogen have been reported.

17.4.5 Fixed Charge in Silane-Based Dielectrics

Fixed charge problems in silane-based dielectrics have also been reported. This data indicates that the problem is not only caused by the presence of carbon (Butler et al., 1990).

17.5 Fixed Charge in Nitrided Dielectrics

Nitridization of a dielectric is used to harden dielectrics for applications such as nonvolatile memories where improved reliability can be obtained (Hori, 1997, pg. 209). It is also used as a gate dielectric for older CMOS processes, where it is used to block boron diffusion and increase the dielectric capacitance in addition to reliability improvements (Wolf, 2002, pg. 112). The presence of nitrogen in a dielectric can generate fixed charge. Nitridation is performed by exposing a silicon dioxide layer to a nitrogen-containing gas such as NO, NH_3, or N_2O in a thermal or plasma system. The nitrogen diffuses through the SiO_2 layer to the interface, where it bonds in the dielectric.

Figure 17.11 shows an example of fixed charge generation during a silicon dioxide gate dielectric nitridation. Figure 17.11a shows the results of nitridation processes in a rapid thermal processing system (RTP) and a furnace system. The fixed charge initially increases, reaches a maximum, and then decreases through time. The initial increase is attributed to the fixed charge generated by nitrogen, and the reduction is attributed to an annealing effect when the dielectric is saturated with nitrogen (Hori, 1997, pg. 238).

A post-nitridation annealing step is commonly used to reduce the fixed charge and interface state density of a nitrided SiO_2 layer (Hori, 1997). Figure 17.11b shows an example of the fixed charge behavior with such an anneal for lightly nitrided films (NOS and NO) and heavily nitrided films. The fixed charge density with annealing is reduced in the lightly nitrided films and increased in the heavily nitrided films. This effect sets an upper limit on nitridation for a viable process.

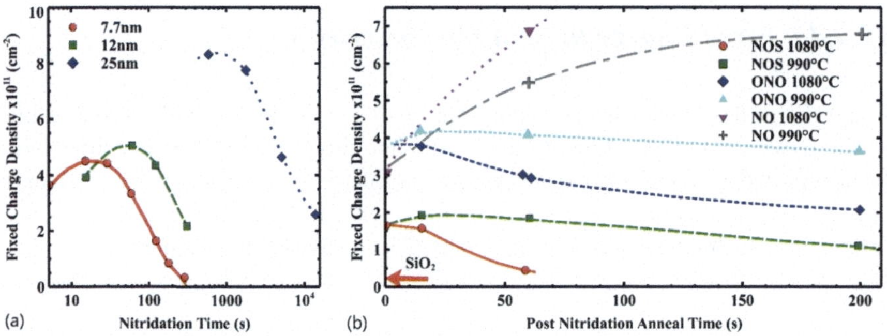

Fig. 17.11 Plot showing fixed charge as a function of the nitridation process and post-nitridation annealing. (**a**) Fixed charge density for a SiO_2 dielectric as a function of nitridation time for a rapid thermal process (RTP) and a furnace process (25 nm). (**b**) Fixed charge density as a function of post-nitridation anneal time for a lightly nitrided film (NO and ONO) and heavily nitrided films (NOS). The fixed charge for an unnitrided SiO_2 film is shown as a reference on the y-axis (red arrow)

17.6 Fixed Charge Polarity

Fixed oxide charge can be positive or negative depending on the atomic composition of the dielectric. The polarity is strongly dependent on the deposition method and processing conditions. Both positively and negatively charged films can be deposited with variations in the recipe conditions. Some dielectrics tend to have a specific fixed charge polarity. For example, thermal silicon dioxide usually has a positive fixed oxide charge, and aluminum oxide usually has a negative fixed oxide charge. Recipes can be optimized to favor one polarity, or different dielectrics can be combined to obtain a specific polarity. As an example, Fig. 17.12 compares the fixed oxide charge as a function of the dielectric thickness (Boogaard, 2001, pg. 123). Thermal silicon dioxide has a positive fixed oxide charge. A PECVD silane-based silicon dioxide, followed by a 900 °C anneal has near-zero fixed charge below 10 nm and a positive fixed charge for thicker films. A 7 nm thermal oxide, followed by PECVD silane oxide has a slightly negative fixed charge that is a slight function of thickness. A PECVD silane oxide has a negative fixed oxide charge reaching a level of -1.0×10^{12} cm^{-2}. The film with the most negative fixed oxide charge is obtained by oxidizing silicon with plasma oxidation, exposing the wafer to an Ar-N_2O plasma without silane, followed by PECVD with silane. The plasma oxidation and silane PECVD have a high propensity to create negative fixed charge. The results of this experiment indicate that there are two competing mechanisms for a silane PECVD process. The mechanisms are oxidation of the silicon substrate and dielectric deposition. Increased fixed charge at the lower thicknesses indicates that plasma oxidation of the silicon surface is the dominant mechanism and is likely due to an excess of negatively charged oxygen atoms near the silicon surface.

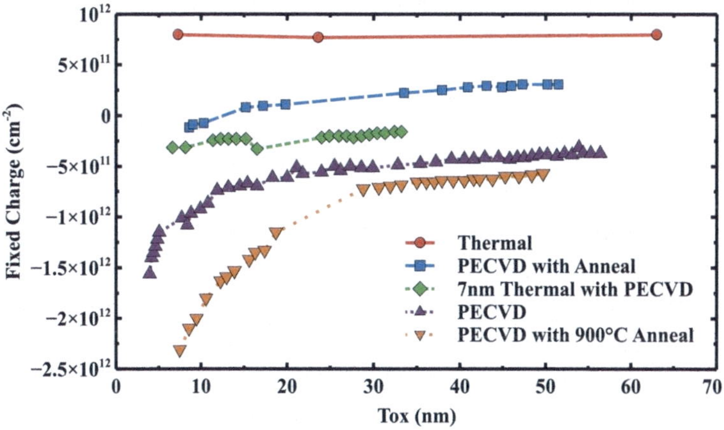

Fig. 17.12 Plot showing the fixed oxide charge density as a function of oxide thickness comparing several different SiO_2 deposition methods including thermal silicon dioxide, PECVD oxide, plasma oxide followed by PECVD oxide, thermal oxide followed by PECVD oxide, PECVD oxide followed by a post-oxidation anneal

17.7 Negative Fixed Charge in Alumina

Alumina (Al_2O_3) films are well known to have a significant amount of negative charge. Alumia is an interesting example of a film that can be used to engineer the fixed charge polarity and concentration in addition to its barrier properties described in Chap. 14 (Kotipalli et al., 2013). For the intentional design of a fixed charge layer, a single or multiple-layer dielectric can be selected for its fixed charge properties. For example, Fig. 17.13 compares the capacitance-voltage plots of several different dielectrics, highlighting the differences in fixed charge. The curves shifted to the left (Si_3N_4 and SiO_2) contained a positive fixed charge, and the alumina film plots shifted to the right and contained a negative fixed charge.

The alumina deposition and post-deposition thermal conditions can be optimized to target a specific charge density (Batra et al., 2015). Figure 17.14a shows the results of varying the deposition temperature for an alumina thermal atomic layer deposition (ALD) process. The chapter's authors state that the fixed oxide charge is caused by the excess negatively charged oxygen atoms. Annealing the films, as shown in Fig. 17.14b, indicates that thermal annealing neutralizes the fixed oxide charge.

The preclean before alumina is deposited can significantly impact the fixed charge density, as shown in Table 17.2, shifting the threshold voltage by more than 1 volt. The cleans are a hydrofluoric acid last clean (HF last), HF clean with hot deionized water with hydrochloric acid (HCl), and a Radio Corporation of America (RCA) clean, followed by hot deionized water with hydrochloric acid (HCl) clean (Tofflinger, 2014, pg. 94). The initial fixed charge $Q_{ox,eff}$ is measured on the first C-V

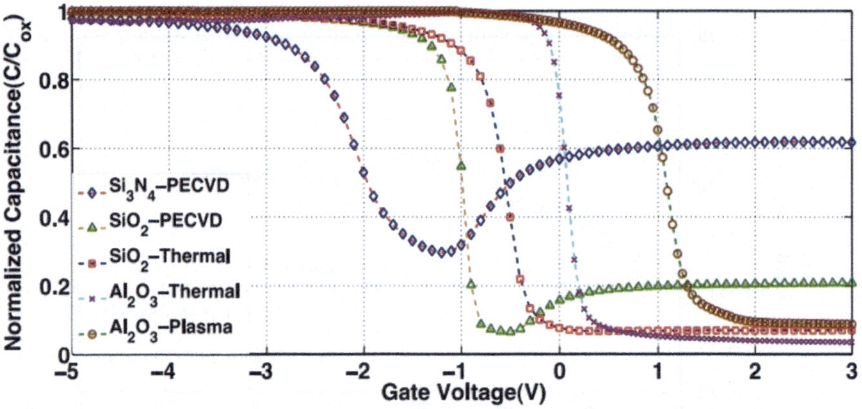

Fig. 17.13 Normalized capacitance-voltage (C-V) plot comparing the fixed charge of several dielectrics, including silicon dioxide, silicon nitride, and alumina. The experiment is performed on p-type wafers

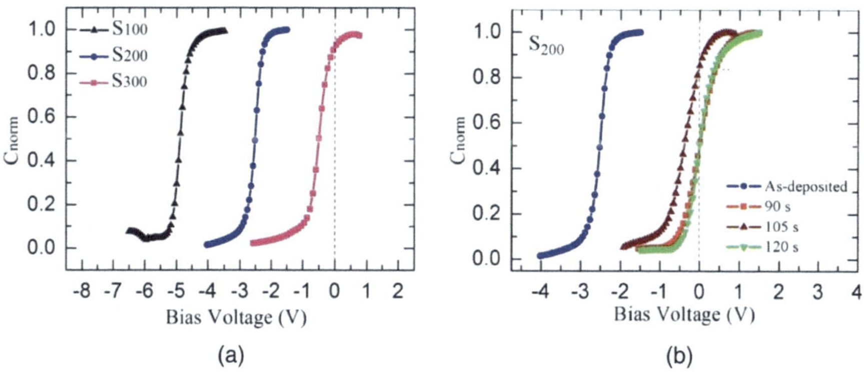

Fig. 17.14 Normalized capacitance-voltage plot for thermal atomic layer deposition (ALD) (**a**) alumina films deposited at different temperatures (100 °C, 200 °C, and 300 °C) and (**b**) after a 400 °C post anneal for the 200 °C recipe. The experiment is performed on n-type wafers

Table 17.2 Fixed charge impact of the silicon preclean before the deposition of an alumina film, followed by a silicon nitride film with and without annealing

c-Si surface	State	Initial $Q_{ox,eff}$ (10^{12} cm^{-2})	Stable ($Q_{ox,eff}$) (10^{12} cm^{-2})
HF last	As-deposited	−5.5 ± 1.0	−0.7…-1.9
	After anneal (425 °C 15 min air)	−4.5 ± 1.0	−4.5 ± 1.5
HF + hot DI water 80:HCL	As-deposited	+0.6…-2.0	−1.9
	After anneal (425 °C 15 min air)	−3.1…-4.0	−0.6…-4.0
	After firing (800 °C 3 s)	−5.0 ± 1.0	−2.0
RCA + HF+ hot DI water 80:HCL	After firing (800 °C 3 s)	−5.0 ± 0.5	−1.9

sweep, and the stable $Q_{ox,eff}$ is measured after multiple sweeps. Note that the C-V plots do not become stable until after several sweeps.

17.8 Fixed Oxide Charge in Through-Silicon Vias

The recent development of through-silicon vias (TSVs) to enable chip stacking and wafer backside contacts provides an interesting study of fixed charge. TSVs have several advantages for CMOS chips. Multiple chips can be stacked into a small volume package. Signal transmission times can be reduced by taking a shorter path through the wafer instead of across the surface of the chip. Backside power distribution (BPD) is a recent technology that routes power and ground on the back of a wafer in addition to the metal routing on the front of the wafer. BPD can be expected

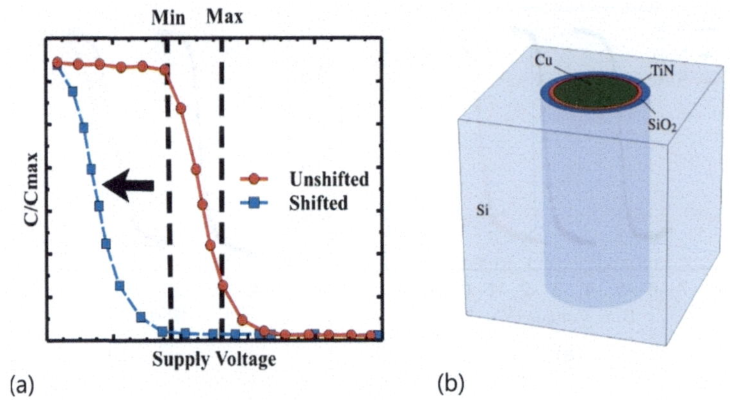

Fig. 17.15 (**a**) sketch of a MOS capacitor capacitance-voltage (C-V) plot showing the variation in the TSV capacitance over the supply voltage range (min to max). The variation is reduced by shifting the C-V curve so that the variation is minimized. (**b**) sketch of the cylindrical MOS capacitor formed by a TSV passing through a silicon wafer. The minimum and maximum supply voltage show the variation in capacitance as the supply voltage varies with the TSV operating without fixed charge engineering

to have similar plasma damage problems as frontside metallization and TSV technology.

TSVs connect the metallization on the front of the wafer through the wafer to the back of the wafer. A copper cylinder is formed in a hole etched in the wafer. The copper cylinder must be surrounded by a dielectric to separate it electrically from the silicon substrate. A barrier metal is placed between the dielectric and copper for adhesion and copper migration prevention. The metal cylinder surrounded by a dielectric, as shown in Fig. 17.15b, forms the gate of a MOS capacitor, and the silicon wafer functions as the semiconductor (Rovitto, 2016, pg. 11). This cylindrical structure behaves identically as a planar MOS capacitor.

TSVs penetrate the wafer well below the ordinarily active surface CMOS devices into possibly lightly doped regions doped during the crystal growth of the silicon substrate. The capacitor's flatband and threshold voltages are controlled by the carrier concentration induced by the added dopants during the crystal or epitaxial growth. CMOS wafers are usually lightly doped and p-type. The natural C-V behavior of this device is sketched in Fig. 17.15a (Katti et al., 2010). The MOS capacitor naturally transitions between the accumulation, depletion, and inversion operating regions over the voltage ranges used in a CMOS chip. The TSV signal voltages range from ground to the supply voltage (V_{dd}), as shown in the figure. Operating the TSV in this voltage range with the capacitance changing is problematic as the TSV capacitance varies with voltage, frequency, and temperature (Zhang, 2014).

The solution to this problem is to shift the C-V curve left or right so that the CMOS operating voltages only operate in the accumulation or inversion regions.

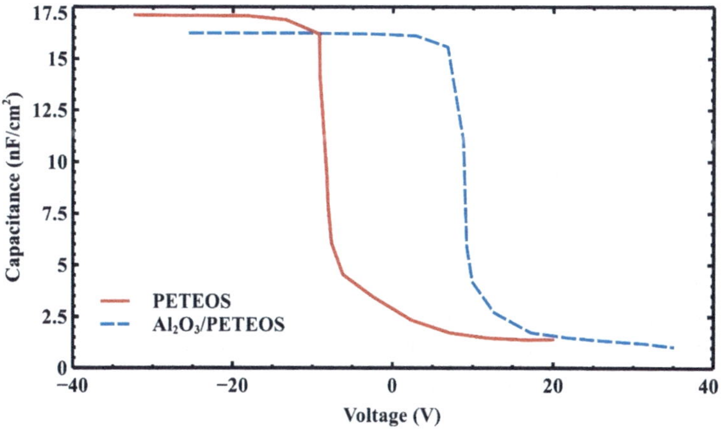

Fig. 17.16 Capacitance-voltage plot showing fixed charge optimization to enable TSV operation in the accumulation with the addition of an aluminum oxide layer

Typically, the higher capacitance accumulation region is chosen because it provides the most stable capacitance. The typical methods of adjusting the flatband and threshold voltages in CMOS devices are challenging to implement in a TSV. Ion implantation is problematic, as it is difficult to implant ions in the TSV sidewalls with an angled implant due to the height and diameter of the TSV. Additionally, the ability to use thermal anneals required to activate the implants is severely limited by the presence of metal on the wafers when the TSV is formed.

Optimization of fixed oxide charge is commonly used to set the flatband voltage of the TSV MOS capacitor (Zhang, 2014, pg. 108). It is a practical, low-cost solution to the problem. For example, Fig. 17.16 shows a capacitance-voltage plot of a TSV where the dielectric between the TSV Cu metal and the silicon substrate has been optimized to set the flatband and threshold voltages. The plot compares a PECVD TEOS silicon dioxide dielectric with a two-layer dielectric consisting of a PECVD TEOS silicon dielectric and an aluminum oxide (Al_2O_3) dielectric deposited with atomic layer deposition. Adding the Al_2O_3 layer used the propensity of alumina to contain a negative fixed oxide charge to shift the C-V curve (see Sec. 17.7) to the point where the TSV MOS capacitor is always in accumulation.

17.9 Fixed Charge Mitigation Techniques

Numerous known factors affect fixed charge problems. A partial list of factors is shown in Table 17.3, along with a reference. Chip manufacturers typically avoid process conditions that create fixed charge to avoid fixed oxide charge problems.

Table 17.3 Common factors affecting fixed oxide charge

Factor	Reference
Avoid plasma damage	This book, (Barbottin & Vapaille, 1989, pg. 228)
Avoid wafer cooling down in oxygen. Cooldown in an inert ambient. Rapid cooldown is better.	Nicollian and Brews, (1982, pg. 787)
Dry oxidation is better than wet oxidation.	Nicollian and Brews, (1982, pg. 787)
For dry oxygen and nitrogen ambient, a lower oxygen percentage is better.	Nicollian and Brews, (1982, pg. 791)
Use high temperature for oxidation or lower temperature with post-oxidation anneal.	Nicollian and Brews (1982, pg. 788–790)
Avoid nitrogen for extended high-temperature anneals (>1 h at 1000+ °C)	Nicollian and Brews (1982, pg. 790)
Optimal precleans can reduce fixed charge.	Barbottin and Vapaille (1989, pg. 225)
Avoid annealing in oxygen.	Barbottin and Vapaille (1989, pg. 225)
Use high-purity chemicals and gases.	Barbottin and Vapaille (1989, pg. 228)
Boron incorporation in dielectrics can make the problem worse.	Barbottin and Vapaille (1989, pg. 228)
Nitridization of dielectrics can make the problem worse	Hori (1997, pg. 239)

References

Baklanov, M., Green, M., & Maex, K. (2007). Dielectric films for advanced microelectronics. *Wiley*. https://doi.org/10.1002/9780470017944

Barbottin, G., & Vapaille, A. (1989). *Instabilities in silicon devices silicon passivation and related instabilities* (Vol. 2). Elsevier Science Publishers B. V.

Batra, N., et al. (2015). Influence of deposition temperature of thermal ALD deposited Al2O3 films on silicon surface passivation. *AIP Advances, 6*, 067113. https://doi.org/10.1063/1.4922267

Boogaard, A. (2001). Plasma-enhanced chemical vapor deposition of silicon dioxide-optimizing dielectric films through plasma characterization [Disertation University Twente].

Butler, J., et al. (1990). Electrical characterization of doped and undoped PECVD TEOS oxides. In *International VLSI multilevel interconnection conference*. https://doi.org/10.1109/VMIC.1990.127907

Deal, B. E. (1967). Characteristics of the surface-state charge (Qss) of thermally oxidized silicon. *Journal of Electrochemical Society, 114*, 266–274. https://doi.org/10.1149/1.2426565

Enigmatic, C. (2023). Fixed charge in deposited oxides. http://www.enigmatic-consulting.com/semiconductor_processing/selected_shorts/Charge_in_dep_oxides.html.

Grove, A. S. (1967). Physics and technology of semiconductor devices.

Hori, T. (1997). *Gate dielectrics and MOS ULSIs principles, technologies, and applications*. Springer. https://doi.org/10.1007/978-3-642-60856-8

Hsu, S.-L., et al. (1993). Field inversion generated in the CMOS double-metal process due to PETEOS and SOG interactions. *IEEE TRED, 40*(1), 49–53. https://doi.org/10.1109/16.249423

Katti, G., et al. (2010). Through-silicon-via capacitance reduction technique to benefit 3-D IC performance. *IEEE EDL, 31*(6), 549–551. https://doi.org/10.1109/LED.2010.2046712

Kotipalli, R., et al. (2013). Passivation effects of atomic-layer-deposited aluminum oxide. *EPJ Photovoltaics, 4*, 45107. https://doi.org/10.1051/epjpv/2013023

Mori, H., et al. (1994). Analysis of isolation degradation induced by interlayer material in capacitor over bit-line (COB) DRAM cell. In *International conference on solid state devices and materials*. https://doi.org/10.7567/SSDM.1994.A-12-1

Nicollian, E. H., & Brews, J. R. (1982). *MOS (metal oxide semiconductor) physics and technology*. Wiley Interscience.

Pramanik, D., et al. (1989). Field inversion in CMOS double metal circuits due to carbon based SOGS. In *International VLSI multilevel interconnection conference*. https://doi.org/10.1109/VMIC.1989.78037

Rovitto, M. (2016). *Electromigration reliability issue in interconnects for three-dimensional integration technologies* [Disertation, Technische Universitat Wien]. https://www.iue.tuwien.ac.at/phd/rovitto/node18.html.

Tofflinger, J. A. (2014). *Interface investigations of passivating oxides and functional materials on crystalline silicon* [Disertation, Technischen Universität Berlin].

Vyas, H. P., et al. (1982). Fixed charge density (Qss) at the Si-SiO Interface for thin oxides. *Journal of the Electrochemical Society, 129*, 1757–1760.

Wolf, S. (2002). Silicon processing for the VLSI era - *Vol.* 4 *Deep-Submicron Process Technology*. Lattice Press.

Zhang, L. (2014). *Through-silicon-via (TSV) design, fabrication and characterization for 3D IC applications* [Disertation, Nanyang Technological University, Singapore].

Index

A
Abnormal dielectric leakage, 153
Accelerated degradations, 205–208
Acceptor neutralization, 332
Acceptor passivation, 321, 332–334, 336
Access transistor, 175, 180, 381
Across power domains, 23–25, 33
Activation energy, 320–324, 339, 359, 377, 393, 395, 396, 418
Active area (AA), 10, 12, 59, 69–71, 119, 293, 294, 300, 302–304
Alignment marks, 236
Alkali metals, 355, 390–393, 396, 414
Alloy, 44, 70, 71, 111, 113, 137, 218, 260, 261, 273, 277, 316, 319, 326, 327, 329, 330, 343, 344, 357, 398, 435
Alumina, 342–343, 448–449, 451
Aluminum (Al), 122, 294, 330, 333, 342, 350, 356, 376, 415, 417, 429
Aluminum oxide (Al_2O_3), 342, 343, 447, 448, 451
Ammonia (NH_3), 190, 191, 345, 446
Amorphous carbon, 189
Amorphous-crystal boundary, 215
Amorphous film, 344, 349, 431
Amorphous region, 214, 215
Analytical chemical analysis (ACA), 425–428
Angled ion implant, 156
Angular distributions, 250
Anion, 134
Annealing, 111–115, 132, 153, 213–215, 264, 331, 333, 336, 337, 339, 345, 374, 379, 436, 440, 446, 448, 449, 452
Antenna, 6, 9, 29, 44, 59, 109, 184, 206, 276, 283, 299, 313
Antenna ratio (AR), 10–12, 15–17, 21, 25, 26, 38, 59, 68, 71, 78, 87, 90, 92, 109, 110, 119–121, 184–188, 196, 200–201, 207, 223, 224, 248, 262, 263, 286–288, 294, 295, 299–303, 305
Anti-reflective coating, 170, 172, 242
ARC, 59, 220, 274
Arcing, 2, 6, 59, 61, 73, 239, 290
Argon (Ar), 66, 67, 89, 91, 111, 115, 147, 211, 246, 256, 257, 336, 337, 428, 441, 447
Argon annealing, 336
Argon plasma, 66, 89, 91, 147, 257, 428
Argon sputter, 115
Assembly process, 225, 261, 384, 395
Atomic hydrogen, 319, 325, 326, 328, 393
Atomic layer deposition (ALD), 135, 343, 418, 448, 449, 451
Atomic level defects, 5, 60, 131–140, 143, 149, 150, 165–168, 174, 177, 179, 339

B
Back-end-of-line (BEOL), 64, 66, 311, 415, 423
Back of the wafer, 74, 165, 230, 231, 237, 262, 450
Backside, 73, 165–167, 170, 172, 175, 197, 226, 230, 236–239, 261, 263, 277, 336, 358, 360–362, 382, 384, 426, 449
Backside films, 167
Backside imagers, 165, 166

K. Prall, *CMOS Plasma and Process Damage*,
https://doi.org/10.1007/978-3-031-89029-1

Backside power distribution, 261, 449
Back sputtering, 115
Bandgap, 34, 35, 66, 87, 98–100, 150, 265, 323, 335–337, 340, 341, 360, 364–367, 371–373, 378, 382
Bandgap energy barrier, 364
Band offsets, 98, 100
Band-to-band tunneling, 35, 73, 75
Barrier height, 277, 364
Beam current, 208, 210, 211
Bias temperature instability, 323, 338, 351
Bimodal distribution, 46
Bipolar chuck, 230, 231
Bipolar electrostatic chuck, 230
Bitline, 160, 161, 182, 258, 381
Blocking dielectric, 144, 155
Bond pad, 4, 14, 31, 102, 208, 221, 222, 225, 246, 263, 273, 283–286, 305, 312, 313, 424, 425
Bond pad etch, 246
Bond pad rules, 305
Border traps, 133
Boron doped silicate glass (BSG), 444
Boron-hydrogen effects, 121
Boron neutralization, 332, 342
Boron phosphorous silicate glass (BPSG), 242, 431, 443, 444
BOX thicknesses, 198
Breakdown event, 61, 62
Breakdown reduction, 189
Broken bonds, 60, 111, 114, 173, 195
Brownian motion, 83
Built-in self-testing (BIST), 88
Bulk defects, 169, 179, 180
Buried oxide (BOX), 36, 38, 197–200
Burn-in, 43, 52, 88, 114–116, 367, 374, 391, 397
Burn-in failure rate, 52, 114, 116
Burn-in failures, 52, 115

C

Calcium (Ca), 392, 429
Capacitance-voltage (C-V), 57, 60, 63–65, 89–91, 111, 218, 219, 250, 265, 266, 278, 280, 330, 331, 336, 400, 402, 407, 409, 412, 415, 419–420, 436, 437, 448–451
Capacitance-voltage (C-V) curve, 65, 266, 330, 401, 402, 419, 450, 451
Capacitance-voltage (CV) plot, 64, 65, 89, 91, 280, 330, 331, 402, 420, 436, 437, 448–451
Capacitative coupled antenna effects, 17–18
Capacitively coupled plasma (CCP), 91
Capture cross-section, 364, 366, 369
Carbon (C), 70, 189, 242, 322, 332, 337, 443–445
Carbon incorporation, 443–444
Carrier generation, 14, 74, 359
Carrier lifetime, 79, 87, 120–121, 133, 218, 219, 243, 275, 277, 315, 355, 359, 369–372, 374, 379, 380, 391
Catalytic reaction, 328, 329
Cation, 134
Cavity etching, 167
Cell plate, 183, 258, 381
Cell to cell leakage, 175
Cerium (Ce), 372
Channel edge plasma damage, 71
Channel hot carrier (CHE), 116, 118
Channel initiated secondary electron (CHISEL), 116, 118
Channel length, 57, 71, 72, 79, 80, 97, 98, 117
Charge-coupled devices (CCDs), 5, 58, 113, 114, 117, 132, 164–174, 267, 268, 384
Charge gain, 58, 117, 132, 149, 154, 423
Charge to breakdown (QBD), 100, 119, 120, 206, 207, 216, 217, 240, 241, 245, 246
Chemical mechanical processing (CMP), 216, 285
Chemical vapor deposition (CVD), 70, 343, 377, 396, 418
Cheung, K.P., 1, 6, 19, 67, 109, 111, 145, 206, 226, 227, 233, 238, 242, 244, 251, 253, 264, 300, 313
Chip operating margin, 85
Chlorine (Cl), 125, 430
Closure, 157–158
Cluster formation, 358
CMOS imagers, 3, 5, 58, 84, 113, 143, 164–174, 212, 221, 278, 306, 349, 355, 358, 359, 369, 384–386
Cobalt (Co), 286, 349, 350, 355–357, 429
Cobalt silicide, 350
Common CMOS process defects, 6
Commonly damaged structures, 3–5, 59
Conductance, 66, 67, 120, 311
Conduction band, 335, 359, 366, 370, 372
Constant current stress, 24, 55, 56, 68, 100, 112, 119, 313
Constant voltage stress, 62, 112, 119, 313
Contact etch, 6, 36, 40, 48, 81, 89–91, 258–260

Contact leakage, 73
Contact metallization, 242, 358
Contact resistance degradation, 58, 89
Contact rules, 301, 303–305
Contacts, 3, 6, 9, 20, 36, 38–41, 48, 49, 58, 61, 71–73, 76, 81, 89–91, 167, 182, 183, 194, 213, 214, 242, 244, 246, 256–260, 284, 286, 289–291, 294, 296, 299, 301–305, 311, 355, 357, 358, 360, 425, 437, 449
Contact string, 437
Contaminants, 103, 133, 223, 247, 322, 336–337, 355, 356, 358–370, 372–381, 383, 392, 393, 418, 428
Contaminated furnace, 337
Continuity tests, 310
Copper (Cu), 199, 261, 286, 342, 355–358, 361, 362, 365, 368, 369, 372–377, 379–381, 384, 392, 395, 396, 403, 404, 416–418, 429, 430, 450, 451
Copper-boron pairs, 373
Copper metallization, 357, 375, 376, 404, 416, 417
Copper migration, 357, 375, 376, 392, 404, 417, 450
Copper silicide, 373
Corner radius, 63
Corona, 275, 278, 279, 390
Corona discharge, 390
Correlated double sampling (CDS), 166, 168
Corrosion, 123, 125, 126
Crystalline film, 349, 431
Crystallinity, 342, 349, 395, 431
Cumulative damage, 45–50
Current stress, 2, 3, 24, 50, 55, 63, 68, 69, 100, 112, 119, 151, 211, 237, 313
Cycled, 151, 182
Cycling life, 157, 158
Cycling window shift, 58, 117, 132, 149, 157–158
Czochralski wafer fabrication, 337

D

Damage mechanisms, 2, 56, 184, 212, 246, 247, 264
Damage recovery, 111–114, 267
Dangling bond, 64, 327, 332
Dark current, 58, 113, 114, 117, 132, 168–173, 267, 278, 279, 349, 384–386
Dark current noise, 113
Dark current recovery, 114, 173, 267
Data retention, 58, 117, 132, 149, 151–154, 175, 178, 181, 397
Datasheet, 88, 122, 151, 159, 160, 181, 182, 382
Data storage, 151, 152, 183
Deep-level transient spectroscopy (DLTS), 337
Defective pixels, 169, 385, 386
Degraded mismatch, 82
Deionized (DI) water, 1, 205, 208, 217–218, 273, 358, 448, 449
Depassivation, 321, 322, 326–329, 332–335, 338
Design rules, 12, 15, 20, 23, 24, 29, 44, 69–71, 82, 119, 123, 188, 192, 220, 248, 257, 263, 274, 283, 284, 286–288, 292, 294, 295, 299–306, 316
Deuterium, 111, 323, 343, 344, 350–351
Deuterium alloy, 343, 344, 350
Diagnostic analysis, 309, 315–316
Diagnostic stress, 44, 312, 314, 316
Dielectric breakdown, 16, 57, 60–63, 73, 99, 119, 125, 184, 187–189, 196, 206, 215, 216, 222, 231, 235, 259, 313, 315, 359, 375, 391, 396, 397, 403, 417
Dielectric breakdown-induced epitaxy (DBIE), 61
Dielectric conduction, 67, 265
Dielectric leakage, 15–17, 73, 77, 79, 91–94, 98, 118, 139, 150, 153, 189–191, 210, 256, 258, 263, 265, 277, 295, 311, 312, 358, 359, 397, 402, 404, 414
Dielectric thickness, 18, 24, 63, 64, 67, 75, 77, 90, 92–96, 120, 138, 139, 150, 151, 184, 186–189, 216, 242, 256, 261, 265, 277, 314, 395, 399, 408, 438, 447
Die seal, 424–425
Differential antenna effects, 14
Differential pair amplifier, 82
Diffusion coefficient, 325, 367
Diffusion length, 343, 359, 362, 364, 369–372, 374, 375, 380, 381
Diffusivity, 342, 364, 366–368, 370, 373
Dimers, 132
Diode capacitance, 288, 325, 326
Diode leakage, 30, 34–36, 57, 73, 75–77, 103–104, 113, 132, 174, 175, 265, 311, 360, 361
Diode operating characteristics, 34
Diode protection, 30–33, 36–40, 75, 77, 148, 196, 197, 291, 301, 302

Dipole ring magnet (DRM), 235
Direct antenna effects, 10–11
Direct tunneling, 62, 63, 67, 73, 92, 93, 147, 210, 364
Disassociation, 320, 322, 339, 346, 380
Disassociation energy, 346
Dislocations, 6, 73, 103, 213–215, 323, 341, 358, 373
DI spin rinse, 218, 219
Disturb cycles, 159
Dog-bone layout, 97, 293
Donor passivation, 321, 334–335
Donut devices, 98, 284, 293–294
Dopant-related defects, 114
Dose rates, 101–102, 323, 351–352
Double trap, 136, 139
Drain to source leakage, 73, 75, 293
DRAM datasheet, 88, 181, 182, 382
DRAM leakage mechanisms, 175
DRAM memories, 174–183, 381
DRAM refresh, 175, 345, 359, 369, 381–384
DRAM retention time, 132, 176–180
Dry cleans, 257
Dry etch, 71, 189
Dry oxidation, 440, 452
Dual inline memory module (DIMM), 178
Dummy gate, 69, 254, 283, 284
Dummy wafer, 240
Dynamic random access memories (DRAM), 3, 5, 22, 40, 57, 58, 85, 88, 117, 132, 143, 174–183, 240, 241, 258, 295, 306, 345, 355, 357–359, 369, 381–384

E
Elastic trap assisted tunneling, 138
Electrical stress, 55, 56, 103, 125, 230, 259, 389, 391, 435
Electric field enhancement, 63, 96
Electromigration lifetime, 123, 126, 127
Electron beam, 69, 70, 223, 224
Electron capture, 360
Electronegativity, 218, 392
Electron emission, 360
Electron-hole pairs, 34, 35, 166, 264, 329, 360, 362
Electron-hole recombination, 360
Electron spin resonance method (ESR), 329
Electron trapping, 66–68, 110, 111
Electrostatic chuck, 230–232
Electrostatic discharge (ESD), 2, 4, 6, 56–60, 102, 217, 220, 223, 225, 263, 305, 306
Electrostatic discharge (ESD) degradation, 56, 102
Electrostatic voltmeter, 274
Elevated temperatures, 34, 44, 152, 177, 205, 316, 360, 390, 397, 400, 401, 409
Emission probability, 364
End of range defects (ERD), 6, 103, 214
Endpoint, 233, 234, 247, 248
Endurance, 58, 117, 132, 149, 157–158
Energy dispersive x-ray (EDX), 357
Energy level, 66, 86, 87, 136, 143, 150, 335, 340, 341, 364–367, 371, 372
Enhanced field, 63, 96
Enhanced low dose rate sensitivity (ELDRS), 101, 102, 323, 351
Epitaxial growth, 6, 214, 450
Equivalent circuit, 13, 19, 186, 226, 237, 242
Erase disturb, 149, 159–161
Erase slowdown, 157
Erase speed, 155, 156
Erratic behavior, 155, 158, 159, 194
Erratic program and erase, 132, 149, 158–159
Erratic programming, 58, 117, 132, 149, 158–159
Etch byproducts, 125, 247
Etch damage, 6, 73, 81, 94, 194, 195, 254–256, 279
Etch rate, 189, 249, 253, 292
Etch variation, 248
EUV light, 158
Excursions, 43, 49–51, 312, 441
Exposed silicon, 113, 236, 237
Extreme ultraviolet light (EUV), 158, 266

F
Faceting, 244
Fast programming, 156
Fast states, 133
Ferroelectric memories, 342
Ferroelectrics, 342
Field edge, 96–98, 213, 293, 437
Field edge leakage, 95, 284, 293–294
Field isolation failure, 97
Field threshold voltage, 345
Filamentary conducting path, 139
Film density, 347–349, 395
Film deposition, 31, 32, 48, 208, 225, 241–244, 345
Film stress, 442
Finfet floating gate, 49
Finfet parameter distributions, 45
Finfets, 3, 46, 48–50, 63, 66, 69, 70, 85, 96, 98, 143, 145, 164, 193–196, 215,

228, 229, 240, 241, 254, 255, 292, 304, 355
Finger capacitors, 125
Finger spacing, 69
Fixed charge, 6, 75, 89, 95, 96, 133, 136, 264, 359, 435–452
Floating body, 38, 197–200
Floating gate, 4, 48, 49, 144, 146, 147, 150–152, 154–156, 161, 163, 164, 228, 229, 232, 421–423
Floating substrate, 234
Flood gun, 211–212, 220
Fluorine (Fl), 126
1/f noise, 83, 84, 339
Focus ring, 238
Fowler-Nordheim (F-N) tunneling, 62, 63, 67, 73, 92, 93, 147, 210, 313
Free electron concentration, 336
Front-opening universal pod (FOUP), 350
Frontside imagers, 165, 166
Furnace anneal, 217

G

Gain (g_m) degradation, 64–67, 121
Gap-fill, 244, 245, 442, 443
Gate all around transistors (GAA), 195
Gate density, 194, 303
Gate design rules, 69
Gate dielectric leakage, 15–17, 73, 91–94, 210, 256, 258, 263, 359
Gate etch, 31, 81, 86, 88, 155, 156, 167, 256, 279, 280
Gate-induced diode leakage (GIDL), 35, 36, 57, 73, 75–78, 174, 175
Gate-last, 254
Gate length, 71, 98, 293
Gate metallization, 71, 80, 209, 215, 246, 357, 358, 401, 419, 438
Gate oxide thickness, 75, 90, 91
Generation lifetime, 372
Getter, 427, 430
Giant random telegraph signal (GRTS), 132, 139–140, 163–164
Gold (Au), 337, 355, 356, 368, 392
Grounded substrate, 234

H

Hafnia, 134, 135, 355
Hafnium contamination, 366
Halogens, 256, 381
Handprint, 390
Hard breakdown (HBD), 61–63, 119
Healing steps, 111
Hf, 323, 356, 429
High cycle count, 157
High-density plasma deposition (HDP), 184, 187, 244
High-energy, 1, 2, 12, 34, 35, 44, 61, 67, 87, 89, 116, 120, 145, 148, 171–173, 189, 208, 223, 241, 242, 244, 245, 247, 249, 250, 256, 263–267, 284, 288
High energy breakdown, 61
High energy ions, 89
High-energy light damage, 12, 171–173, 244, 284
High-energy photons, 35, 171, 189, 223, 242, 246, 247, 264–266
High-k, 3, 62, 75, 76, 81, 86, 98–100, 133–136, 184, 206, 207, 254, 265, 357, 358, 401, 404
High-k MIM capacitors, 184
High noise floor, 166
High-power, 2
High-temperature, 44, 61, 112, 113, 152, 177, 205, 215, 219, 245, 265, 311, 316, 325, 341, 346, 347, 367, 382, 405, 408, 409, 418–420, 441, 442, 452
High-temperature anneal, 112, 113, 452
High-temperature retention, 152
Hole capture, 360
Hole emission, 360
Hole injection, 116, 326
Hole trapping, 68, 99
Hot-carrier, 116–118, 120–121, 242, 313–315, 351
Hot-carrier lifetime, 79, 120–121, 243, 315
Hot electron damage, 88
Hot-electron induced punch-through (HEIP), 81, 117, 118
Hot ion implantation, 215
H-Si bond, 346
Hydrochloric acid (HCl), 448
Hydrofluoric acid, 448
Hydrogen, 6, 44, 70, 109, 180, 218, 288, 316, 319, 356, 392, 435
Hydrogen alloy, 70, 71, 113, 218, 260, 261, 319, 329, 330, 435
Hydrogenation, 325, 326, 332–334, 336, 350
Hydrogen barriers, 341–345, 349, 430
Hydrogen-based acceptor, 332, 334, 336, 340
Hydrogen-based donor, 340
Hydrogen bonding, 111
Hydrogen-carbon complexes, 337

Hydrogen-containing film, 326
Hydrogen diffusion, 320, 324–326, 332, 347, 348, 352
Hydrogen drift, 325, 326, 393
Hydrogen effusion, 346
Hydrogen emission, 345–349
Hydrogen-emitting film, 343
Hydrogen fluoride, 356, 428
Hydrogen interaction, 117, 335–337, 349–350
Hydrogen passivation, 319, 322, 323, 326, 328, 329, 332, 334, 337
Hydrogen-related noise, 339
Hydrogen-silicon reaction, 329
Hysteresis, 329–331, 358, 359

I

Identical layout, 82, 83
Imager pixel, 84, 165, 167, 169, 171, 278, 355, 385
Impact angle, 219, 250
Implantation, 6, 51, 59, 103, 155, 167, 194, 208–216, 219–221, 223, 273, 290, 334, 358, 385, 451
Implantation damage, 59
Indirect antenna effects, 11–12
Inductively coupled plasma (ICP), 67, 80, 85, 86, 91, 428
Inductively coupled plasma mass spectroscopy (ICP-MS), 426, 428
Inelastic trap assisted tunneling, 138
Initial trapping rate, 109
Inline measurements, 275–277, 284, 436
Inline pad, 285, 286
In situ strip, 126
Integrity checks, 310
Interconnect metallization, 189, 295
Interface defects, 132, 180, 330
Interface state passivation, 320, 326, 329
Interface states, 6, 9, 55–57, 62–67, 70, 75, 76, 79, 81, 85–87, 89, 95, 102, 111, 112, 121, 135, 136, 138, 170, 197, 215, 218, 245, 314, 319–321, 326–330, 338, 339, 342, 345, 435, 437–439, 446
Intermetal dielectric degradation, 123–125, 359
Intermetal dielectrics, 62, 109, 117, 123–125, 243, 295, 311, 312, 359, 376, 377, 395, 396, 404, 415, 441
International Symposium on Plasma and Process-Induced Damage, 1
Interstitial, 6, 115, 133, 134, 195, 214, 257, 320, 322, 325, 340, 367, 368, 373, 378–380
Interstitial diffusion, 325, 367, 368
Interstitial diffusivity, 373
Intrinsic distribution, 153, 169, 176
Ion beam, 208–211, 223–224, 250, 426
Ion bombardment, 189
Ion energy, 103
Ionic bonds, 100, 134
Ion implantation, 51, 59, 103, 155, 208–212, 214–216, 219, 220, 223, 273, 334, 358, 385, 451
Iron (Fe), 337, 355–357, 361, 365, 368, 378–381, 384, 429
Iron-boron pair, 378–380
Isolation dielectrics, 36, 62, 96, 266, 399, 400, 438
Isolation edge leakage, 95–98
Isolation failure, 97, 98, 399–400, 437, 441, 444

J

Joint electron device engineering council (JEDEC), 315
Junction isolation, 311
Junction leakage, 35, 58, 73, 75, 132, 133, 167, 179, 359, 373, 382, 391, 397
Junction spiking, 6, 103

K

Kinetic ion damage, 103
Knock-on, 6, 195, 212, 215, 223, 246, 427
Kurtosis, 43, 46, 155, 169

L

Latent defects, 45, 88, 156
Latent reliability, 59, 259
Latent reliability problems, 1, 4, 259
Lateral surface diode, 198
Lattice damage, 195, 212
Layout methodology, 82
Lead zirconate titanate (PZT), 342
Leakage, 5, 12, 30, 45, 57, 113, 132, 149, 209, 277, 284, 302, 311, 351, 358, 391, 437
Lifetime degradation, 109, 120–121, 359, 372, 374
Lift pins, 230, 231

Light damage test structures, 288
Lightly doped drain (LDD), 12, 77–81, 120
Lightning-like marks, 59
Lightning strike, 60
Lithium (Li), 355, 392, 394, 395, 429
Lithography, 158, 171, 172, 239, 274, 285
Local heating, 59
Local leakage, 59
Lorentz force, 235
Low dielectric constant (low-k), 4, 189–194, 312, 376, 443
Low-k dielectrics, 3, 4, 6, 123, 126, 127, 143, 189–194, 242, 295, 359, 377, 395, 396, 404, 417
Low light detection, 164
Low-pressure chemical vapor deposition (LPCVD), 245, 246, 377, 396
Low-temperature diffusion, 367
Low-temperature retention, 153

M

Magnesium (Mg), 392, 429
Magnetically enhanced reactive ion etch (MERIE), 155, 156
Magnetic field, 173, 174, 218, 235–236
Marangoni drying, 216
Mask edge defect (MED), 6, 214
Masking material, 208, 218, 219, 221, 247, 248, 290, 358
Masking pattern, 220
Mask material, 219
Matched transistors, 31, 82
Matching issues, 82
Melting point, 61, 132
Memory array, 3, 24, 39, 40, 82, 131, 138, 139, 149, 152, 153, 159, 160, 177, 295
Memory wordlines, 39–41
Metal antenna, 17, 18, 21, 22, 64, 65, 76–78, 82, 85–87, 121, 184–187, 199, 210, 221, 228, 284, 287, 291, 300, 301, 303, 304, 306
Metal comb, 284, 292, 415, 416
Metal contamination, 73, 103, 173, 337, 355–359, 362–364, 368, 372, 381–386
Metal design rules, 123, 257, 284, 303–305
Metal etches, 31, 32, 65, 75, 83, 85, 87, 90, 91, 94, 95, 222, 225, 240, 246, 252
Metal etching, 83, 167
Metal filaments, 103
Metal fingers, 11, 252
Metal-insulator-metal (MIM) capacitors, 3, 4, 17–19, 38, 39, 57, 117, 122, 143, 175, 183–188, 284, 294–295, 305, 306, 311
Metal-insulator-metal (MIM) rules, 305
Metal jumper, 303
Metallization, 3, 4, 9, 12, 17, 24, 36, 50, 56, 71, 80, 103, 117, 122–127, 165, 184, 189, 205, 209, 214, 215, 242, 244, 246, 284, 286, 294–296, 303–306, 311, 312, 327, 342, 355–358, 373, 375, 376, 392, 401, 404, 415–417, 419, 438, 443, 450
Metallization resistance, 123
Metal oxide semiconductor (MOS) capacitor, 57, 58, 64, 89, 91, 117, 211, 249, 261, 275, 276, 331, 360, 362, 402, 406, 410, 412, 415, 419, 420, 439, 444, 450, 451
Metal pitch, 192, 299
Metal plate pixel, 169, 171
Metal precipitates, 357, 359
Metastable defects, 325
Metrology marks, 236
Midgap interface state density, 436
Minimum refresh specification, 382
Minority, 133, 175, 218, 219, 275, 277, 355, 359, 362, 369–374, 379, 380, 391, 392
Minority carrier lifetime, 133, 218, 219, 275, 277, 355, 359, 369–372, 374, 379, 380, 391
Mismatch, 2, 57, 82, 83, 284
Mismatch structures, 284
Misplaced atom, 132
Mobile ion barriers, 344, 349, 424, 425, 427, 430, 431
Mobile ion contamination, 389–432
Mobility, 79, 83, 86, 116, 120, 133, 259, 328, 358, 359, 373, 389, 393–395, 410, 412, 415, 438
Mobility reduction, 358, 359
Molecular dynamic simulation, 256
Molecular hydrogen, 319, 325, 326, 345
Molybdenum (Mo), 170, 356
Multimodal noise, 84
Multi-state VRT, 177

N

NAND memory, 149, 153
NAND read disturb, 162–163
Nanocrystalline, 135, 137

Native oxide, 89, 91, 356, 428
Negative bias instability, 117, 118, 121–122, 188, 314, 315, 338
Negative bias temperature instability (NBTI), 117, 118, 121–122, 188, 314, 315, 338, 346
Negative fixed oxide charge, 447, 451
Negative hydroxyl (OH-) molecules, 415
Neutral hydrogen atom, 329
N-H bonds, 344, 346, 348
Nickel (Ni), 337, 349, 355–358, 361, 384, 429
Nickel silicide, 349
Nitride composite spacer, 71
Nitrided dielectrics, 446
Nitride liner, 98–99
Nitride passivation, 156, 406, 445
Nitride spacer, 173, 174, 214
Nitride spacer etch, 173, 174
Nitride storage layer, 153
Nitrogen-rich, 346, 347
Nitrogen to silicon ratio, 347
Noise, 2, 22, 24, 39–340, 358, 359, 379, 385, 410
Noise metrics, 86, 167, 339
Noise power, 86, 87, 339
Nonnormal distributions, 45
Nonuniformity, 43, 147, 219, 227, 228, 235, 237, 280
Nonuniform plasma, 227, 228, 236
Nonvolatile memories (NVMs), 3, 4, 22, 40, 50, 57, 58, 85, 117, 125, 126, 132, 138, 140, 143–164, 166, 170, 194, 229, 231, 266, 267, 306, 397, 420–424, 446
NOR architecture, 160
NOR flash, 155, 156
NOR Read Disturbance, 161
NVM memory (NVM), 157, 158, 424
NVM mobile ion test, 423

O

Offset voltage, 82
Off-state leakage, 75, 98
Off-state stress (OSS), 116, 118
Operating window, 157
Ostwald ripening, 214
Outgassing, 219, 346, 444
Over-erase, 132, 149, 154–156, 161
Overetch, 40, 79, 248–249, 254, 257
Over programming, 58, 132, 149, 154–156, 160, 161
Oxide–nitride–oxide (ONO), 170, 278, 432, 446
Oxygen, 35, 41, 191, 192, 256, 257, 322, 330, 335, 336, 345, 373, 378, 436, 441, 447, 448, 452
Oxygen clusters, 335, 336, 373
Oxygen-enhanced diffusion, 345
Oxygen plasma, 35, 191, 192
Oxygen vacancies, 330, 335, 436

P

Pad sizes, 82, 225, 283, 305
Palladium (Pd), 358
Parasitic capacitance, 12, 13, 19
Parasitic charge trapping, 58, 117, 149, 164
Parasitic leakage region, 96
Partial recovery, 112, 113, 266
Passing wordline, 160, 181–183
Passivation, 155–157, 165, 208, 225, 283, 305, 319–323, 326–329, 332–337, 342, 349, 405, 424, 425, 430, 443, 445
Passivation layer, 155, 156, 165, 424, 430
Pass transistors, 82
Pb center, 327, 329
Peak range defect (PRD), 214
Peak substrate current, 116
PECVD, 156, 262, 343, 377, 396, 418, 443, 447, 451
Periodic table, 355, 356, 364, 390, 393, 428
Phosphorous-hydrogen complex, 325
Photoconduction, 34, 35, 242, 265
Photoconductivity, 265
Photodiode, 165, 168–171, 358
Photoinduced diode leakage, 35
Photon damage, 171, 173, 246, 247, 264, 266
Photoresist, 208, 212, 221
Physical damage, 1, 2, 9, 57, 59, 61, 77, 253
Physical vapor deposition (PVD), 208, 244, 245, 343, 376, 418
Pinned photodiode, 169
Pixel noise, 84
Planar nonvolatile memories, 143, 164
Plasma ashing, 82, 206, 207, 221–223
Plasma dielectric deposition, 19, 184, 241–244, 265
Plasma-enhanced chemical vapor deposition (PEVCD), 70, 343, 377, 396, 418, 442, 443
Plasma-enhanced intermetal dielectric deposition, 109
Plasma-enhanced tetraethyl orthosilicate glass (PE-TEOS), 243, 412–414, 442
Plasma etching, 71, 208, 219, 246, 337
Plasma extinction, 233
Plasma hydrogenation, 332, 334

Plasma ignition, 233
Plasma nitridation, 240, 241
Plasma nonuniformity, 147
Plasma processing induced damage (PPID), 1
Plasma start conditions, 231
Plasma treatments, 89, 126, 127, 190–192, 326, 335
PMOS chips, 389
Polymers, 6, 189, 247
Polysilicon, 61, 104, 111, 113, 144, 155, 158, 221, 242, 245, 246, 294, 330, 342, 432, 437
Polysilicon etch, 103, 104, 113
Porosity, 189, 190, 342, 376
Porous films, 189
Positive bias instability (PBTI), 117, 118, 122
Positive fixed charge, 436, 437, 447, 448
Positive fixed oxide charge, 338, 447
Post-oxidation anneal, 440, 441, 447, 452
Post-package repair, 178
Potassium (K), 355, 390, 392–394, 411, 412, 418, 427–428, 430
Power lift, 277
Pre-clean, 132, 448, 449, 452
Preventative maintenance, 43, 51
Probe tip, 274, 286, 310, 409, 418
Product-level performance, 109
Product yield, 115, 118
Program disturb, 58, 117, 149, 159–161
Program-erase window, 157, 158
Program-erase window shift, 157
Program speed increase, 157
Protected wordline, 40
Proton, 320, 323, 325, 328–331, 340, 392, 405, 414, 415, 417, 430
Proton memory, 323, 330
Proximity correction, 283
Pseudo-ground, 125
Punch-through, 58, 73, 74, 81, 117, 118, 311
Punch-through resistance, 81, 117

Q

Quantox™, 275, 276
Quantum barrier, 161
Quasistatic capacitance-voltage, 111

R

Radiation damage, 2, 6, 56–58, 67, 101, 223, 293, 339, 340, 351–352
Radiation hardness, 56, 58, 101, 102, 117
Radiation hardness degradation, 101
Radioactive tracer profiling, 431
Radio Corporation of America (RCA) clean, 448, 449
Radio frequency (RF), 115, 226, 227
Radius of curvature, 63–64, 96, 123
Random thermal signal (RTS) noise, 83, 84, 86, 88, 132, 166, 339, 379
Rapid thermal processing (RTP), 34, 215, 358, 446
Reactive ion etch (RIE), 80, 155, 156, 235
Read disturb, 58, 117, 149, 159–163
Read disturbance, 161
Recessed gate, 240, 241
Recipe power, 157
Recipe transients, 233–234
Recombination lifetime, 371, 372
Recovery, 44, 111–114, 173, 267, 314, 316, 332, 333, 390, 405
Recrystallization, 214, 215
Redundant cells, 88, 176
Refractive index (RI), 189, 344, 347, 348
Refresh degradation, 175, 384
Refresh failures, 177, 182, 359, 382
Refresh frequency, 183
Refresh time, 175–179, 381, 384
Reliability, 1, 18, 38, 44, 55, 109, 132, 144, 242, 284, 305, 309, 319, 367, 391, 443
Reliability hazards, 2, 64, 67, 109, 118, 120, 125, 295
Remote antenna effects, 26–27
Remote chamber, 171
Remote hydrogen plasma, 326
Remote plasma, 170, 319
Replacement gate, 254, 255
Residuals, 6, 112, 113, 125, 126, 247
Resist, 6, 9, 35, 48, 87, 120, 158, 170–172, 190, 191, 216, 218–223, 248, 250, 259, 260, 274, 275, 277, 284, 290–291, 319, 357, 358, 427
Resistance-capacitance (RC), 190
Resist ashing, 48, 170, 190, 191
Resist overlap, 220, 221
Resist pattern, 219–221, 274
Resist strip, 35, 87, 171, 221, 222, 259, 260, 280
Retention specifications, 151, 177
Revealing stress, 23, 51, 109, 312–315
Reverse antenna effects, 25–26
Rhenium (Re), 358
Ring oscillators, 58, 88–89, 284, 288, 311
Row hammer, 58, 117, 132, 181–183, 381
Ruthenium (Ru), 376, 417

S

Sacrifical transistor, 29
Salicide, 71, 73, 89, 103, 214, 260, 261, 357–359
Saturation region, 116, 311
Scaling, 5, 84, 98, 120, 123, 131, 138, 143, 164, 165, 194, 195
Scanner, 113, 171, 239
Schockley-Read-Hall (SRH) defects, 87, 212, 359–362, 365–367, 369, 372, 375, 378, 382–385
Schottky diode, 325, 326
Scribe line, 225, 236, 425
Scribe line structures, 236
Secondary emission, 209, 210, 267
Secondary ion mass spectrometry (SIMS), 344, 350, 357, 405, 423, 424, 426–427, 431, 432, 443
Secondary x-rays, 246
Self-interstitial, 133
Semiconductor Equipment and Materials International (SEMI), 430
Sense amplifiers, 82, 292
Serpentine, 284, 295, 296, 410, 415
Shockley-Read-Hall (SRH) centers, 87, 88, 212, 358–362, 365–367, 369, 372, 375, 378, 382–385
Shorted junctions, 358
Short loop, 276, 277, 279–280, 406, 436
Shorts tests, 310
SiC quartzware, 337
Signatures of process damage, 43–52, 55–104
Silane (SiH_4), 345, 347, 443–447
Silicide block dielectric, 71
Silicides, 61, 71, 259, 349, 350, 355, 373
Silicon-dielectric interface, 132, 167, 169, 170
Silicon dioxide, 64, 75, 98–100, 119, 132–135, 184, 197, 218, 265, 266, 295, 324, 327, 329, 338, 350, 390, 393–395, 407, 408, 436, 439–444, 446–448, 451
Silicon embrittlement, 323, 341
Silicon nitride, 98, 99, 157, 242, 324, 326, 342–349, 426, 427, 430–432, 444, 445, 448, 449
Silicon on insulator (SOI), 3, 5, 13, 14, 36–38, 143, 164, 196–201, 329, 331, 401
Silicon oxide nitride oxide silicon (SONOS), 4, 144, 148, 151, 153, 158, 266
Silicon oxynitrides, 242
Silicon-rich, 346, 347
Silicon surface mobility, 328
Silsesquioxane (SSQ), 189
Silver (Ag), 92
Single atom contaminant, 358
Single electron, 148, 149, 183, 392
Single trap, 139
Single wafer cleans, 216
Slow states, 133
Snow-melting salt, 391
Sodium (Na), 392, 394, 429
Soft breakdown, 61–63, 119, 359
SOI Antenna Ratios, 800–801
SOI diode, 36, 198
Solid-phase epitaxial regrowth (SPEG), 213, 214
Solid solubility, 364, 368–370
Source-drain, 12–17, 26, 27, 49, 57, 58, 73, 75, 77–81, 85, 155, 179, 198, 253, 255, 287–288, 300, 302–303, 305
Source drain antenna effects, 12–17
Source-drain design rules, 12, 15, 302–303, 305
Source-drain edge, 81
Source-drain overlap, 77, 79, 85, 179
Source-drain region, 12, 75, 77–79, 81, 155, 198, 254, 305
Source-drain resistance, 57, 79, 80
Source follower transistor, 165
Spacer etch, 173, 174, 254–256
Special contacts, 36, 38
Special finfet structures, 145
Spike anneal, 215
Spin chuck, 217, 218
Spin dryers, 216
Spin-on dielectrics (SOD), 445
Spin-on-glass (SOG), 70, 224, 442, 443
Spreading resistance, 334, 335
Sputter clean, 261
Sputter deposition, 208, 244–246, 384
Sputtering, 115, 167, 219, 223, 244–246, 267, 427, 432
Stable chamber pressure, 233
Stacked gate etch, 155
Stacking faults, 213, 215, 358, 359
Standby current, 405, 437
Statistical distributions, 45, 46
Statistical signature of process damage, 45
Steel, 356, 378
Stepper job, 239
STI edge, 73, 74, 95–98, 103, 213, 214, 293, 294, 360, 361
STI edge leakage, 74, 96, 98, 293, 294
Stoichiometric, 347

Stress-induced leakage current (SILC), 57, 58, 62, 63, 73, 77, 95, 115, 117, 119, 125, 136–139, 149–153, 175, 351
Stress migration, 123, 126–127, 295, 311, 312, 315
Stressor, 6, 71, 214, 242, 260
Strontium bismuth tantalate (SBT), 342
Structural damage, 56, 60, 61, 73, 212–215, 223, 425
Substitutional diffusion, 367, 368
Substitutional impurity, 133
Substrate, 6, 13, 14, 17–22, 24, 30, 32, 36, 38, 39, 61, 73, 75–77, 81, 86, 89, 90, 103, 113, 116, 120, 124, 125, 134, 146, 179, 182, 197, 198, 209, 212–215, 234, 236–239, 247, 253, 257, 258, 275, 284, 296, 316, 324, 330, 335, 348, 357, 360, 425, 427, 438, 440, 447, 450, 451
Substrate bias, 24, 116
Substrate contacts, 38, 39, 76, 258
Substrate current, 81, 86, 116, 120
Subthreshold conduction, 57, 73–75, 174, 224, 345
Subthreshold conduction degradation, 57, 74–75
Subthreshold plot, 75, 78, 345
Subthreshold slope, 57, 74, 75, 77, 101, 311, 358, 359
Subthreshold slope degradation, 57, 75, 101
Surface charge, 253, 259–260, 273, 274
Surface etch damage, 256
Surface generation, 168, 359, 362
Surface implant, 169
Surface photovoltage (SPV), 275, 277, 278, 381
Surface recombination, 359, 362, 373, 374
Symptoms of plasma damage, 43, 44, 316

T

Tantalum nitride (TaN), 376, 417
Temporal read noise (TRN), 166
Tendrils, 60, 220
Test structure, 3, 4, 31, 44–50, 52, 59, 115, 119, 125, 126, 225, 228, 229, 232, 252, 283–296, 309, 316, 405, 406, 415, 425
Tetraethoxysilane, 443
Tetraethylorthosilicate glass (TEOS), 90, 242, 343, 412–414, 442–445, 451
Thermal annealing, 111–113, 153, 158, 213, 333, 336, 342, 347, 374, 435, 448, 451
Thermal damage, 103
Thermal desorption spectroscopy (TDS), 346
Thermal donors, 322, 335–336
Thermally stimulated ionic current (TSIC), 406, 417–418
Thermal noise, 83
Thermionic emission (TE), 152, 153
3D NAND, 40, 41, 151, 153, 162, 164, 241, 258, 420
3D NAND wordline, 40
Threshold hysteresis, 330
Threshold voltage, 14, 32, 39, 49–51, 57, 67–72, 75, 76, 89, 95–98, 101, 110, 133, 143–148, 152–157, 159, 161, 162, 164, 197, 198, 221–224, 228, 231, 237, 238, 240, 261, 265, 311, 313, 329, 330, 338, 339, 345, 351, 358, 389, 391, 396–402, 405–409, 415, 419–421, 423, 435–437, 439, 441, 444, 445, 448, 450, 451
Threshold voltage shift, 39, 50, 51, 57, 68–72, 95, 101, 145, 147, 198, 222–224, 237, 240, 265, 313, 339, 351, 391, 396–401, 406, 435, 436
Through silicon vias (TSVs), 208, 261–264, 284, 294, 295, 301, 306, 357, 358, 372, 401, 443, 449–451
Time constant, 85–88, 177
Time-dependent chip instability, 389
Time-dependent dielectric breakdown (TDDB), 117–119, 189, 196, 315, 403, 417
Time of flight secondary ion mass spectrometry (TOF-SIMS), 357, 423, 424
Time zero, 4, 38, 39, 50, 52, 55, 57, 59–64, 101, 109, 112, 115, 122, 125, 178, 188, 216, 259, 309, 314
Time-zero dielectric breakdown (TZDB), 216
Titanium (Ti), 342, 349, 355–357, 368, 369, 429
Titanium silicide, 350
Tokyo electron (TEL), 235
Topology effects, 240–241
Tracer element, 343
Transfer gate (TG), 165, 173
Transient decay, 326
Transient nonuniformity, 228
Transistor channel length, 71
Transistor clamp, 33–34
Transistor leakage, 71–81, 97, 212
Transistor mobility, 133, 328
Transition metal oxides, 100, 134, 357
Transition metals, 322, 336, 337, 355, 392

Transit time, 395
Trapped holes, 98
Trapped moisture, 345
Trapping, 4, 35, 45, 57, 58, 63, 64, 66–71, 75, 85, 86, 94, 95, 99, 100, 109–112, 114, 118, 138, 144–155, 161, 164, 170, 177, 197, 260, 278, 328, 364, 366, 421, 431, 444, 445
Trapping curve, 67, 68
Trapping layer, 144–147, 150–152, 154, 155, 161, 164, 421
Trap to trap tunneling, 95, 152
Triangular voltage sweep (TVS), 375–377, 393, 406, 410–418
Triboelectric charging, 216–217
Triple wells, 21–25, 32, 33, 284, 300, 302
Tritium barrier experiments, 342
Tungsten, 61, 126, 286, 355, 357, 384–386
Tungsten vias, 126
Tunnel dielectric defects, 163
Tunnel oxide leakage, 138, 151
Two-state VRT, 177

U
Uncycled, 151
Undoped silicate glass (USG), 242, 243, 432
Uniform gate stress, 117
Unipolar chuck, 231
Unipolar electrostatic chuck, 230, 232
Unpassivated, 156–158, 328
Unselected cells, 159–161
Unstable chamber pressure, 233
UV laser, 114, 171, 172
UV light, 34, 44, 113, 148, 158, 170–172, 266, 267, 316, 391, 423
UV scanner, 113

V
Vacancy, 6, 330, 335, 378, 436
Vacuum annealing, 345
Valence band, 359, 370–372
Vapor phase decomposition (VPD), 356, 428
Variable retention time (VRT), 58, 117, 132, 176–180, 381
Via, 3, 31, 47, 65, 115, 153, 208, 284, 301, 312, 357, 401, 443
Via chain, 123, 284, 296
Via etch, 153, 154, 237, 238, 247, 254, 256–258, 261–263
Via rules, 258, 284, 304, 305
Via void, 127
Vibrating Kelvin probe, 275
Victim rows, 183
Virtual antenna effects, 18–19
Virtual fork, 252
Voltage to breakdown (VBD), 119

W
Wafer bevel, 236–239
Wafer curing, 223
Wafer haze, 357, 359
Wafer orientation, 438–439
Wafer scrubbing, 216
Wafer thinning, 165, 384
Wafer to wafer bonding, 165
Water-based oxidation, 345
Well coupling, 19–24, 33, 302
Well damage, 23, 33
Well design rules, 302
Wet chemical, 1, 216, 360
Wet etch, 65, 71, 260
Wet metal etch, 83
Wet oxidation, 440, 452
White pixel, 117, 132, 173, 174
Wordline diodes, 40
Write-erase cycles, 155

X
Xenon (Xe), 246
X-ray, 2, 12, 56, 101, 102, 145, 246, 267, 356, 357, 361
X-ray fluorescence (TXRF), 356, 361, 426

Y
Yield, 1, 2, 4, 43, 52, 55, 59, 68, 82, 85, 114–116, 118, 123, 152, 155, 215–217, 220, 221, 231, 237, 254, 260, 265, 286, 355, 415, 441

Z
Zirconia, 355

The manufacturer's authorised representative in the EU is Springer Nature Customer Service Centre GmbH, Europaplatz 3, 69115 Heidelberg, Germany. If you have any concerns regarding our products, please contact ProductSafety@springernature.com

Printed and bound by CPI Group (UK) Ltd, Croydon, CR0 4YY
01/10/2026
02225929-0014